Current Progress in Biochemistry

Current Progress in Biochemistry

Edited by Artie Weissberg

MURPHY & MOORE

www.murphy-moorepublishing.com

Published by Murphy & Moore Publishing,
1 Rockefeller Plaza,
New York City, NY 10020, USA
www.murphy-moorepublishing.com

Current Progress in Biochemistry
Edited by Artie Weissberg

International Standard Book Number: 978-1-63987-140-7 (Hardback)

Cataloging-in-Publication Data

Current progress in biochemistry / edited by Artie Weissberg.
    p. cm.
Includes bibliographical references and index.
ISBN 978-1-63987-140-7
1. Biochemistry. 2. Biology. 3. Chemistry. 4. Biochemical engineering.
I. Weissberg, Artie.
QH345 .C87 2022
572--dc23

TABLE OF CONTENTS

Permissions

List of Contributors

Index

PREFACE

The science concerned with the study of chemical processes related to or occurring in living organisms is known as biochemistry. It is concerned with the understanding of how biological molecules contribute to the processes occurring within and between cells. This helps to further the scientific understanding of the structure, function, and organization of tissues and organs. Enzymology, metabolism, and structural biology are the chief disciplines within biochemistry. The applications of biochemistry are diverse, extending from medicine and nutrition to agriculture. This book contains some path-breaking studies in the field of biochemistry. It outlines the principles, techniques, and applications of this field in detail. This book is a vital tool for all researching or studying this discipline as it gives incredible insights into emerging trends and concepts.

This book unites the global concepts and researches in an organized manner for a comprehensive understanding of the subject. It is a ripe text for all researchers, students, scientists or anyone else who is interested in acquiring a better knowledge of this dynamic field.

I extend my sincere thanks to the contributors for such eloquent research chapters. Finally, I thank my family for being a source of support and help.

Editor

Demonstration of a Multiplex Milk Allergen ELISA using Oligonucleotide-tethered Principal Component Proteins

Robert S. Matson*

Robert S. Matson, QuantiScientifics LLC, 1920 E. Katella Ave. Suite S, Orange, CA 92867, USA

***Corresponding author:** Robert S. Matson, QuantiScientifics LLC, 1920 E. Katella Ave. Suite S, Orange, CA 92867, USA,*
E-mail: rsmatson@quantiscientifics.com

Abstract

The A-Squared (A²®) Microarray System (QuantiScientifics, Orange, CA, USA) is a micro plate-based, microarray platform for quantitative multiplexed immunoassays. Arrays of capture oligonucleotides are covalently attached to the bottom of the plate well. Complementary strands are in turn linked to proteins such as antibodies or antigens. The resulting Oligo-Antibody (-Antigen) conjugates are pooled and applied to the wells. Through the process of hybridization, capture oligo-oligo protein hybrid pairs undergo self-assembly to form an "Oligonucleotide-Tethered" antibody or antigen array in each well.

In this study, we demonstrate the ability to tether known "Allergenic" proteins found in cow's milk to create a multiplex immunoassay useful in the diagnosis of Serological (IgE-Mediated) cow's milk allergy. Purified milk proteins whose epitopes are known to be associated with the allergy, as well as, crude milk allergen extracts were conjugated to oligonucleotides to create the panel. The panel included F2 crude milk protein extract, as well as, principal component proteins Beta-Lacto Globulin, alpha-lactalbumin, IgG, Bovine Serum Albumin, Lactoferrin, and 3 caseins. Patient serums were screened for the presence of specific IgE (sIgE) providing a profile and subsequent identification of principal component allergen.

Keywords: Milk; F2; Allergen; Multiplex; Microarray; Micro plate; ELISA; Fluorescence; Immunoassay; Oligonucleotide; Tethered; Conjugation; Self-Assembly; Hybridization; Principal Component; Specific Ige; Crude Extract; A² Microarray System

Introduction

Microarrays, as a multiplex immunoassay platform, have been used in the assessment of allergen specific-IgE (sIgE) for over the past 15 years. Wiltshire, et al. [1] prepared extracts of cat hair, dust mites and peanut, and then arrayed these onto an activated glass slide (microarray substrate). Sera with known levels of sIgE based upon Auto Cap (Pharmacia) values were applied. Detection of the sIgE on the microarrays was accomplished using ImmunoRCA (Rolling Circle Amplification) and the results compared to the CAP values. The microarray method was found to have a higher positive predictive value, higher specificity and in most cases greater sensitivity. Likewise, Kim, et al. [2] dispensed

prepared extracts of various allergens onto a nitrocellulose slide (FAST Slide, Schleicher & Schull). The presence of sIgE In sera For Allergens (Dermatophagoides pteronyssinus (Dp), Egg White, Soybean, Milk, and Wheat) was determined by immune fluorescence laser scanning of the microarray. The microarray results correlated well with those determined using UniCap (Pharmacia & Upjohn Diagnostics AB).

Bacarese-Hamilton, [3] prepared allergen microarrays from extracts onto activated glass slides and evaluated sensitivity and specificity using the Tyramide Signal Amplification (TSA) catalyzed by HRP (Molecular Probes). A lower limit of detection of the microarray bound sIgE < 1 fg was achieved from 100 µl sera corresponding to about 10 fg/ mL equivalent to ~ 4.1 X 10⁻⁶ IU/ mL (1 IU/ mL IgE = 2.44 ng/ mL) or a cut-off value at 4.1 k IU/ L.

Lebrun, [4] adapted the standard ELISA for quantitative determinations on a nitrocellulose substrate (Zeta-Grip, Miragene) using a flat-bed scanner for colorimetric detection of the developed microarray. Sensitivities for sIgE were achieved that fell below the World Health Organization (WHO) cut-off value of 0.35 k IU/ L.

Hiller, et al. [5] first immobilized purified or recombinant allergenic proteins (rather than crude extracts) onto amine reactive glass slides (VBC-Genomics, Vienna) to measure allergen sIgE. This approach subsequently led to the co-development with Phadia of the ISAC (Immuno Solid-phase Allergen Chip) microarray ushering in Component-Resolved Diagnostic (CRD) for allergen testing. The ImmunoCAP ISAC chip products are commercially available from Thermo Fisher Scientific for the assessment of sIgE.

Ott, et al. [6] used purified allergenic proteins from milk and eggs spotted onto a microarray slide in an effort to assess the clinical predictive value of component resolved diagnosis. The study encompassed 130 children with allergy to milk or eggs and included correlation of the microarray to that of UniCap (Phadia). The authors concluded, " that microarray-based IgE quantification was accurate in predicting clinical reactivity to allergenic proteins in our study population".

Martinez-Aranguren, et al. [7] evaluated the ImmunoCAP ISAC 112 chip kit comprising 5 slides with each slide containing 4 sub-arrays of 112 protein components covering 51 allergens. The validation study was based upon manufacturer's calibrators, as well as, sera obtained from 19 patients and included inter-assay, intra-assay and inter-lab analyses. The mean Coefficient of Variation (CV) for calibrators was 9.42%. The Intra-Class Coefficient Correlation (ICC) was used as a measure of reproducibility and repeatability. For inter-assay (reproducibility), determinations for IgE in 94/112 (83.9% of allergens) had very good reproducibility (ICC > 0.90), as was found for the reproducibility (intra-assay) of this set. The inter-laboratory agreement was very good for 73/112 allergens (ICC > 0.90) and good for 22 (ICC = 0.71-0.90). While the microarray performed well, there were several instances noted in which the expected values were not achieved leading the authors to conclude that, "due to the low accuracy obtained in some (reported 7-8 components) of the studied allergens, the application of this semi-quantitative technique for diagnosis in clinical situations where results may have a major impact on the therapy prescribed may not be advisable".

However, Moreira, et al. [8] evaluated sera of patients from Brazil having grass pollen allergy using ImmunoCAP ISAC microarrays. In their study, an ISAC chip containing 103 components was used that included 9 allergens for timothy grass pollen and 1 allergen for Bermuda grass. In this case, 77 of 78 patients tested positive for timothy grass pollen. Thus, these researchers suggest that the ISAC allergen microarray could be useful in diagnosis and immunotherapy monitoring for patients suffering from Brazilian grass pollens.

Studies on the development of microarrays in general and specifically for allergen testing have been largely based upon surface activated glass or nitrocellulose coated glass slide substrates. Micro plate-based microarray platforms are now commercially available and being adopted for multiplex immunoassays including allergy testing. The A^2 Microarray System (QuantiScientifics) is a multiplex quantitative immunoassay platform based upon a 96-well polypropylene plate to which proteins maybe tethered via an oligonucleotide bridge to create antibody or antigen microarrays. For example, Robbins, et al. [9] created an A^2 plate cytokine antibody array in order to measure the release of cytokines resulting from the immuno-stimulatory effects of lipid-delivered siRNAs in primary CD34+progenitor–derived hematopoietic cells.

Here, we examine the utility of the A^2 multiplex assay platform in the development of component-resolved diagnostics for allergens based upon oligonucleotide tethering of allergenic proteins. In these studies, we use both purified milk protein allergens and crude extracts as a model [10].

Materials and Methods

The following reagents were obtained from QuantiScientifics

(Orange, CA) and used according to the manufacture's kit instructions:

A2 Capture Oligo Plate (A21002)

Enzyme Linked Assay Kit (A21003) containing Hybridization Buffer (A21003-HYB), Biotinylated Reference Oligo (A21003-REF), streptavidin alkaline phosphatase, AP Enzyme Substrate (A21003-SUB), wash buffer; Conjugation Kit: oligo-antibody coupling (A21004) containing spin-columns, solid-phase (Butyl-Sepharose), initiator (iminothiolane), binding & elution buffers.

Binding Buffer: sodium phosphate, 20mM containing sodium sulfate, 1M, pH 7.5.

Elution Buffer: sodium phosphate, 20mM, pH 7.5.

Activated oligonucleotides

4-(N-Maleimido methyl) cyclohexane-1-carboxylic acid N-hydroxy succinimidyl-terminated oligonucleotides (A21005-QS1, -QS3, -QS5, -QS9, -QS13 & -QS17). Oligonucleotides are complementary to the microarray capture oligos covalently attached to the A2 Capture Oligo Plate. Length, 30mer; Melt Temperature, Tm ~ 64°C.

Wash buffer: Tris Buffered Saline with Tween 20, pH 8.0 (TBST: Sigma-Aldrich, T9039), 0.05M Tris, 0.138M NaCl, 0.0027M KCl, Tween 20, 0.05%.

A2 microarray reader (A21001): Optical Cube A installed, Ex 545 nm, Em 650 nm and Optical Cube B installed, Ex 480 nm, Em 535 nm

Proteins

Cow's milk derived proteins were obtained in purified form from Sigma-Aldrich (see Table 1). Streptavidin-alkaline phosphates conjugate (Thermo Scientific, Pierce # 21324); Goat anti-human IgE alkaline phosphatase conjugates (Sigma-Aldrich, A3525).

Serum Samples: Hycor Biomedical (Garden Grove, CA) provided milk allergenic positive patient sera and spiked control sera for evaluation.

Preparation of oligonucleotide-protein conjugates

2OD$_{260nm}$ units of an SMCC (4-(N-maleimidomethyl) cyclohexane-1-carboxylic acid N-hydroxysuccinimidyl) activated oligonucleotide are reacted with 100 μg of purified protein. The process involves adsorption (hydrophobic interaction) of the protein onto a solid-phase held within a spin-column. Next, initiator is added to modify the protein with reactive sulfhydryl (-SH) groups. Finally, the SMCC-oligonucleotide is added which covalently couples to the thiolate protein to form the oligo-protein conjugate. The resulting conjugate still bound to the solid-support is rinsed free of reactants; and subsequently eluted in purified form from the column ready for use.

Preparation of oligo- F2 (1) β-Lacto globulin conjugate from purified protein

Protein modification: 100 μL Butyl-Sepharose 4 Fast Flow

Table 1: Description of Principal Protein Components.

Comp #	Component	Molecular Weight	Source	SMCC-Oligo
F2	F2 crude extract	NA	Hycor	QS17
F2-1	β-lactoglobulin	18,300	L3908	QS1
F2-2	α-lactalbumin	14,200	L6010	QS3
F2-3	IgG	150,000	I5506	QS5
F2-4	Bovine Serum Albumin (BSA)	66,300	A9085	QS9
F2-5	Lactoferrin	80,000	L9507	QS1
F2-6	αS-casein	25,000	C6780	QS3
F2-7	β-casein	24,000	L6905	QS5
F2-8	r-casein	10,000	L0406	QS9
	Goat anti-IgE	150,000	Immuno Reagents GtxHu-002-D	QS13

List of known allergenic proteins associated with milk allergy used in this study.
Comp# (Milk Allergen Principal Component nomenclature for proteins used in this study) Molecular Weight (Daltons), Source (Sigma-Aldrich catalog number unless otherwise stated), SMCC-oligo assignments QS1,3,5,9,13,17 refer to unique sequences of SMCC terminated oligo nucleotides for covalent coupling to specified component proteins.

(GE Healthcare Bio-Sciences AB, Uppsala, Sweden) was placed in a fritted spin column to achieve a compressed bed volume of 0.1 mL. The column was centrifuged at 2000 rpm for 30 seconds to remove liquid.

The resin was re-suspended in 600 μL of deionized water using a pipette; and the column washed by centrifugation. The centrifugation-rinse process was repeated, followed by 2 rinses in binding buffer.

Following the equilibration in binding buffer, the bottom of the column was carefully sealed with a cap.

600 μL of binding buffer was used to re-suspend the resin. Then, 50 μL = 100 μg of β-Lacto globulin (Sigma-Aldrich, L3908, MW 18,300 Daltons) from stock solution of 2 mg/ mL in TBST buffer was added. The resin was then mixed by Pipetting up & down and column recapped.

Following end-to-end tumbling of the column for 10 minutes, the caps were removed and the column centrifuged to remove the solution.

700 μL of freshly prepared protein activator solution (Iminothiolane Hydrochloride, Sigma-Aldrich, 16256) was added to the column bed; and the resin re-suspended. The column was capped and then mixed by tumbling for 1 hour.

The caps were then removed and the column centrifuged and the bed re-suspended, then rinsed 6 times with binding buffer as previously described.

Finally, 600 μL of fresh binding buffer was added, the column bottom capped and the resin re-suspended.

Coupling of the oligonucleotide to thiol-protein

100 μL of SMCC-oligo (2 ODs) prepared in binding buffer was added and the resin re-suspended and column capped. The spin-column was mixed by tumbling for 24 hours at room temperature.

Following removal of the solution, the column was rinsed 5 times in binding buffer in order to remove reactants.

Elution of the oligo-protein conjugate

The column was then placed in a clean 1.5-mL Eppendorf micro-centrifuge tube for collection of the conjugate.

100 μL of elution buffer was added to the column with gentle mixing to re-suspend the resin. The column was incubated for 3 minutes and then centrifuged. The eluted fraction was collected as the source of the conjugate.

This process was repeated 4 more times with the collection of 100 μL fractions in separate tubes for a total of 5 fractions (500 μL).

The eluted fractions were combined and the solution stored at 4°C prior to use.

Preparation of the oligo-F2 (2); -F2 (3); -F2 (4); -F2 (5); -F2 (6); -F2 (7) and -F2 (8) conjugates

Additional milk proteins, F2 crude extract, as well as, the goat anti-Human IgE were conjugated to SMCC-oligonucleotides as described in the previous section detailing the preparation of oligo-F2 (1) conjugate. The assignments regarding choice of oligo used in coupling are provided in Table 1. The selected Placement

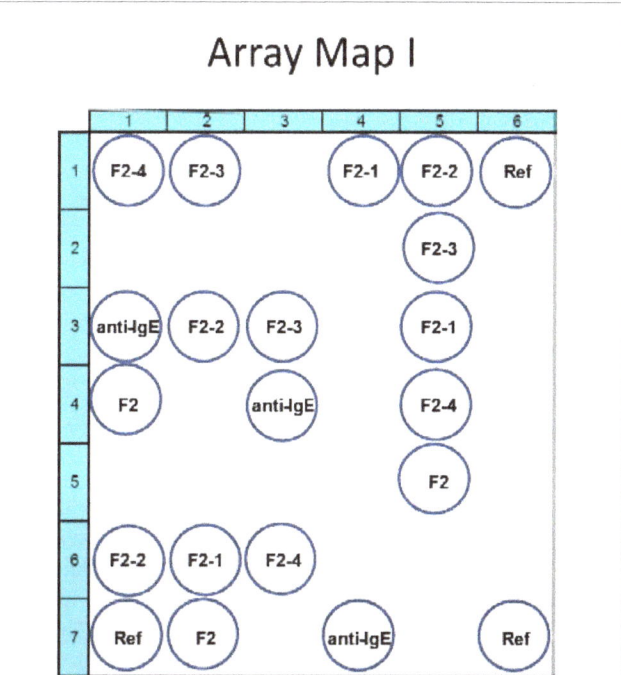

Figure 1: Array Format I
Array I well map for placement of microarray spots in each well corresponding to the various hybridized oligo-proteins listed in Table 1.

Table 2: Estimation of the Specific IgE Content.

F2#	Protein	S1	S2	S3	S4	S5	S6	S7	S8	S9	S10	S11	S12	S13
1	β-lactoglobulin		0.428		0.062		2.348			0.698			2.458	0.326
2	α- lactalbumin		1.174				1.373			0.258			0.951	0.064
3	IgG	0.206					1.123							0.244
4	BSA						3.164			0.127		0.364		0.459
5	Lactoferrin	0.525					0.082							
6	αS-casein		0.611		0.688					0.725			0.627	0.442
7	β-casein		0.098											0.911
8	r-casein		0.154		0.397		0.454							1.133

Component specific IgE (sIgE) in serum samples (S1 to S13) determined using the A² Multiplex Immunoassay.
Mean Fluorescent Intensity (MFI) signal values obtained from the microarray spots were converted to International Units (IU) based upon the IgE standard curve (MFI vs. IU/ mL) provided in Figure 3. Values ≤ 0.35 IU/mL are reported for indication only as a Minimal Detectable Dose (MDD).

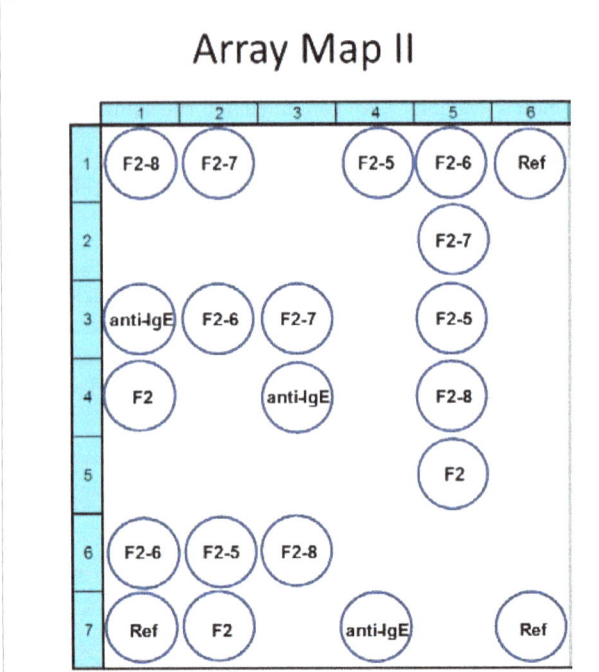

Figure 2: Array Format II
Array II well map for placement of microarray spots in each well corresponding to the various hybridized oligo-proteins listed in Table 1.

within the array well are for format I (Figure 1) or array format II (Figure 2).

Preparation of the microplate microarray with oligonucleotide-tethered milk allergens

The oligo-protein conjugates were pooled in hybridization buffer at 2 μg/mL of each for application to the A2 capture oligo plate. Following a 5 minute soak and rinse of the plate wells with 3 x 200 μL TBST, 55 μL of the oligo-protein hybridization cocktail was delivered to each well using a pipette. The plate was incubated at ambient temperature with shaking for 1 hour. The wells were rinsed 3-times in TBST in preparation of the immunoassay.

Development of the multiplex allergen immunoassay

Serum samples (1-13) containing sIgE, negative control serum without IgE, and reference control sera containing varying amounts of IgE (0.35 IU/mL to 100 IU/mL) as standards (1 IU = 2.44 ng), as well as, the biotinylated reference oligo (fiducial) were delivered to wells of the microplate at 55 μL per well. The plate was incubated for 1 hour with shaking. Following TBST buffer rinse, a secondary antibody-alkaline phosphates conjugate (goat anti-human IgE) was applied in buffer at 55 μL per well. Reference fiducials of the biotinylated oligo were detected by the addition of streptavidin alkaline phosphatase to the mix. The plate was incubated for 1 hour with shaking, and then extensively rinsed in buffer. Next, 55 μL of AP Enzyme Substrate was added to each well. The plate was covered with foil and incubated for 30 minutes without shaking. The solution was carefully removed to semi-dry but not rinsed. 200 μL of fresh buffer was added to the developed plate and the plate scanned using the A2 Reader. Note: the proprietary enzyme substrate forms a fluorescent precipitate (similar to the tyramide signal amplification process) which is detected at 535 nm.

Analysis

An ELISA standard curve (Figure 3, 4PL2 plot, AssayFit, IVDtools, Inc. Nijmegen, The Netherlands) was generated from serial dilution of the reference control sera. Fitted Curve:Response (MFI, mean fluorescent intensity) vs. Dose (Human IgE, IU/ mL). The IgE content in serum samples, negative serum and buffer was calculated from the curve. The A² Microarray System reader reports in pg/ mL units. Conversion to International Units (IU) was made based upon: 1 IU/ mL IgE = 2.44 ng/ mL = 2440 pg/ mL.

Results

The total IgE as measured from signal on arrayed spots (Figure 4) containing oligo-gt anti IgE was comparable across both well array formats (r = 0.9764), Figure 5.

Likewise, signal (MFI) corresponding to sIgE was obtained for all samples. Negative control serum (MFI = 337 ± 67) and buffer blank (MFI= 259 ± 53) values were used to establish a cut-

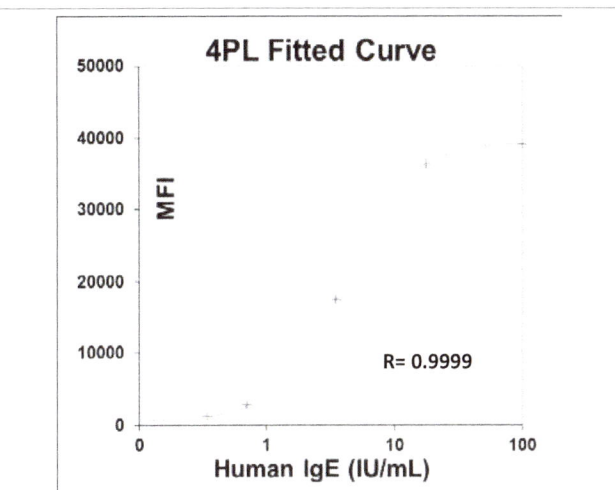

Figure 3: IgE ELISA Standard Curve
dose-response 4PL plot for estimating the IgE content in the serum samples.
4PL (4 Parameter Logistic nonlinear regression model).MFI (mean fluorescence intensity). Range: 0.35-100 IU/mL. MDD (minimal detectable dose) ≤ 0.35 IU/mL. The A2 MicroArray System reports in pg/ mL. Conversion to IU/mL is based upon 1 IU/mL IgE = 2440 pg/mL.

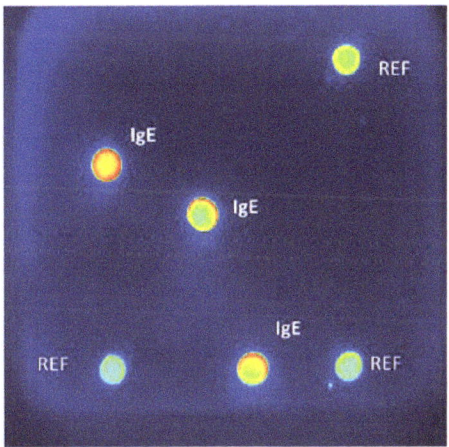

Figure 4: screen capture of an A2 Plate well image from the A2 Micro-Array System.
REF (reference: biotinylated oligo). IgE (oligo-goat anti IgE capture antibody). Signal developed using streptavidin-alkaline phosphatase conjugate (REF signal) + goat anti-Hu IgE alkaline phosphatase (IgE signal) with a proprietary fluorescent substrate that forms a precipitant detectable at 535 nm.

off baseline for non-specific binding at MFI = 538 corresponding to the negative control, MFI+3SD. In this study, the lowest IgE control used was 0.35 IU/mL yielding a signal, MFI = 1249 or about 2.3-fold above the cut-off value. Using the baseline cut-off, the identity and level of the principal "allergenic" component present in each sample could be more easily recognized in a composite profile (Figure 6).

Array I wells detected the presence or absence of sIgE in sera for immunological response to cow's milk proteins, principal components: F2-1 (β-lactoglobulin), F2-2 (α-lactalbumin), immunoglobulin G (IgG), and Bovine Serum Albumin (BSA). Spots containing immobilized goat anti-human IgE were used to estimate total IgE content in the sample. Very high levels (> 200,000 pg/mL or 200 ng/mL ~ 82 IU/mL) of total IgE were found in samples #3, #7, #9 and #11. These particular samples exhibited very low or non-detectable levels of sIgE (Figure 7). In comparison, samples #6 and #12, while exhibiting low total IgE levels were found to contain mostly sIgE. Sample #6 contained sIgE toward F2-1, F2-2, F2-3 and F2-4; while sample #12 was prominent for F2-1(Figure 8, β-lactoglobulin) and F2-2.

From Array II, sample #1 was predominate for F2-5 (Figure 9, lactoferrin), while sample #13 contained modest amounts of sIgE for F2-7 (β-casein) and F2-8 (r-casein). Samples #2, #6, #9, #12 and #13 had detectable levels of F2-6 (αS-casein) above background. Coupling of oligonucleotide (QS17) to a crude milk protein extract was achieved providing the signal profiling sIgE's in serum samples but with a lowered response (Figure 10).

A summary of the estimated content of specific IgE in the patient serum samples in provided in Table 2. Values ≤ 0.35 IU/

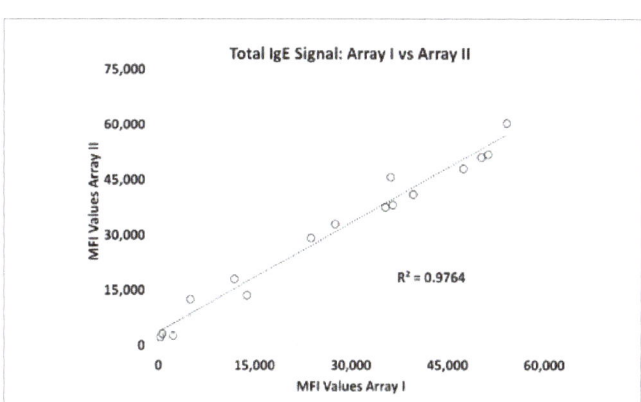

Figure 5: Correlation for Total IgE Signal between Arrays
Correlation between array I and array II signal intensity obtained from microarray spots at various concentrations of IgE standard input.

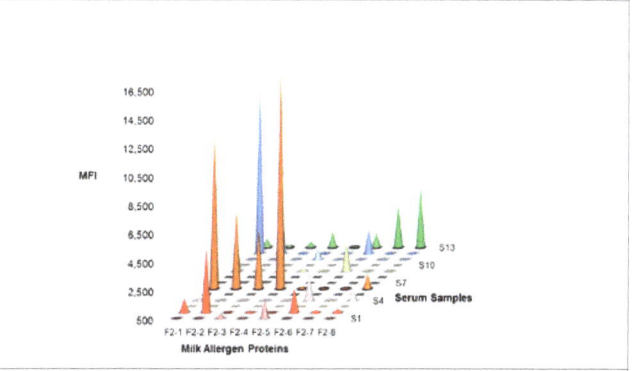

Figure 6: Milk Allergen Principal Components Profiles in Patient Serum
Identification of major allergenic milk proteins associated with the IgE levels found in serum samples.

mL are the result of extrapolation below the lowest standard and should be regarded for indication only.

Discussion

The A^2 Microarray technology is an open platform based upon a pre-printed and fixed pattern (microarray) of capture oligonucleotides in a 96-well micro plate, the A^2 Plate. Multiplex ELISAs once designed by the end-user are constructed using the complementary oligo linked to the selected proteins. A mixture of different oligo-proteins may then be hybridized to the microarray to create a protein array. For example, the tethering of oligo-antibody to form an antibody array is the primary application for this technology platform. Since, hybridization is a thermodynamically driven; self-assembly process the number of oligo/oligo-protein hybrids formed on the surface is limited and relatively constant. This results in improved spot uniformity within the array and reduces the variability commonly found with direct printing of proteins. Moreover, the oligo-protein conjugates are reagents that can be stored separately and combined as needed to create different plate assays upon demand.

Allergens represent a complex mixture of proteins and other biomolecules. It is most often difficult to resolve those proteins that are associated with a particular "allergenic" response.

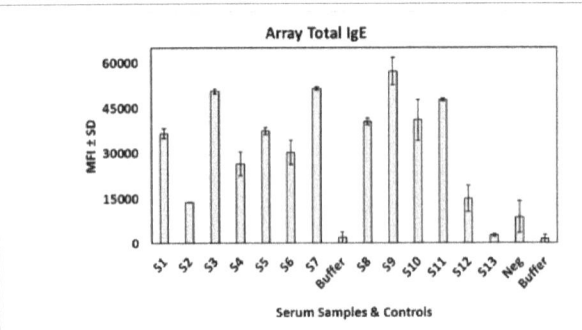

Figure 7: Total IgE in Samples
Total IgE profile determined from QS13 anti-human IgE signal. Mean Fluorescence Intensity (MFI) from array I & array II spots. Standard Deviation (SD) for n =2 determinations.

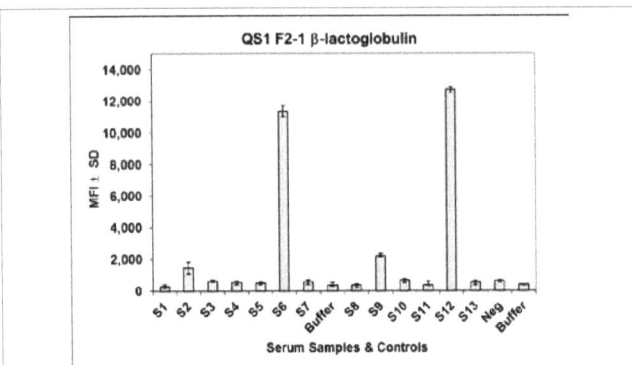

Figure 8: Array Profile for F2-1
sIgE serum profiles determined from QS1oligo tetheredβ-lactoglobulin. Mean Fluorescence Intensity (MFI) from array I spots. Standard Deviation (SD) for n =2 determinations.

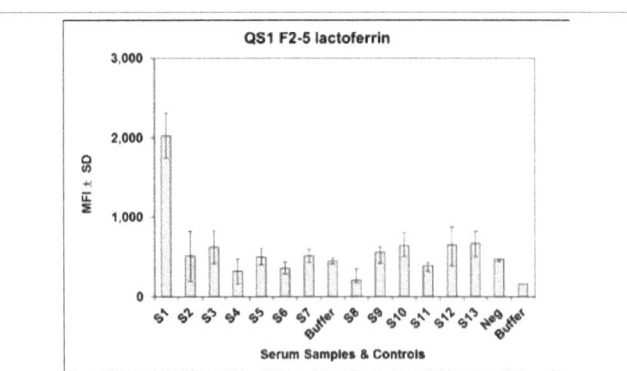

Figure 9: Array Profile for F2-5
sIgE serum profiles determined from QS1 oligo tethered lactoferrin. Mean Fluorescence Intensity (MFI) from array I spots. Standard Deviation (SD) for n = 2 determinations.

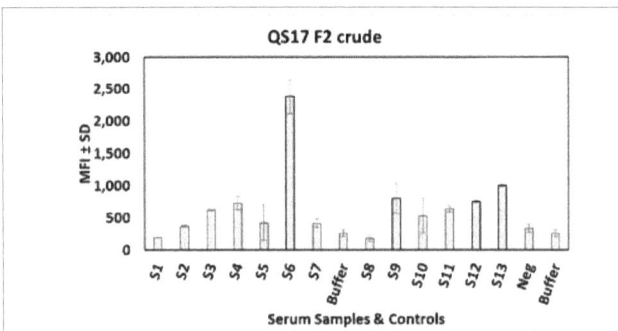

Figure 10: Array Profile for F2 Crude Extract
sIgE serum profiles determined from QS17 oligo tethered F2 crude protein extract.
Mean Fluorescence Intensity (MFI) from array II spots. Standard Deviation (SD) for n = 2 determinations.

Because of such complexity, as well as, the vast number of allergens present in our environment it remains an arduous task in diagnosis. Thus, the well- known skin patch test based upon the use of allergen extracts is still prevalent even though a migration and adoption of the immunoassay has been underway for several decades. As previously discussed in the introduction, the development of immunoassays has progressed from e.g. the use of the Radio Allegro Sorbent Test (**RAST**), to performing a sIgE ELISA using slides or micro plates of printed extracts, to the use of automated diagnostic platforms such as ImmunoCAP or the ImmunoCAP ISAC for component resolved diagnostic platforms that were developed by Phadia (now part of Thermo Fisher Scientific).

In this study, we examined the potential use of the A^2 multiplex ELISA in the development of a component resolved methodology for detection of food-based allergens based upon determination of sIgE. Milk allergy was selected as a model because the allergy is prevalent, and the component proteins are readily available in purified form from commercial sources.

The oligo-protein conjugates were successfully prepared, assembled onto the microarray and the IgE ELISA demonstrated. The A^2 multiplex ELISA was used to quantify both total IgE and

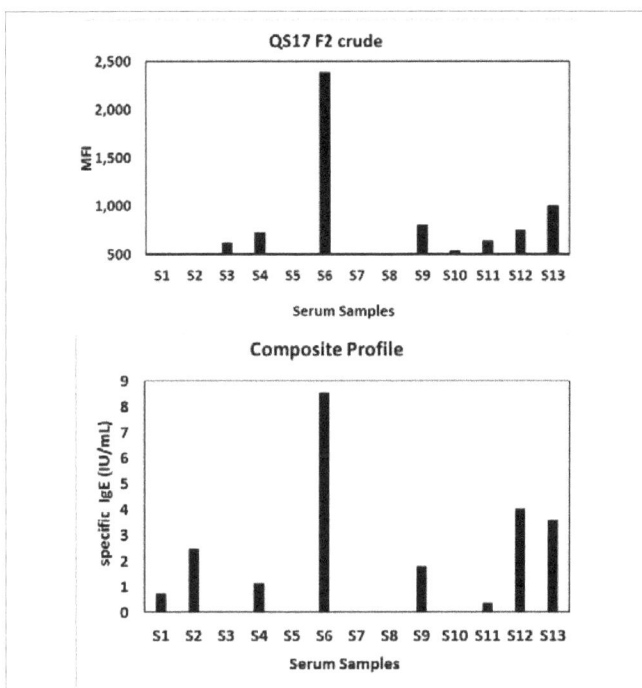

Figure 11: Comparison of sIgE profiles
comparison of sIgE profiles for serum samples based upon reconstruction of a composite profile from calculated component IgE content (Table 2) to that obtained using a milk allergen crude extract (Figure10).

sIgE in serum samples over the range of 0.35 -100 IU/mL In addition to evaluating conjugates of component proteins, the study included a conjugate prepared from the crude milk allergen extract. A profile (Figure 10) of milk allergen content (sIgE) across the serum samples was obtained, albeit at reduced signal intensity relative to individual component proteins. Moreover, it was possible to partially reconstruct the profile (Figure 11) by summation of the sIgE content (IU/ mL) of the major component proteins found in each sample listed in Table 2. This provides a further validation that the observed IgE levels detected using the oligo-tethered crude extract reflect the relative contribution of the individual component protein IgE response within the sample.

In summary, this model study demonstrates that oligonucleotide-tethering of proteins of broad molecular weight range (10,000 to 150,000 Daltons) and diverse physical-chemical properties are possible using the A^2 technology. Moreover, the microarray bound oligo-allergens were found to be accessible to interaction with serum sIgE allowing the development of a multiplex immunoassay for milk allergens. The A2 multiplex ELISA permitted screening for sIgE using an oligo-tethered milk allergen extract, as well as, quantitation of component specific IgEand total IgE content of the sample within the same well.

Acknowledgement

Thanks to Dr. Jang B. Rampal for assistance with experiments; and Hycor Biomedical for providing milk allergen standards and serum samples for this study.

References

1. Wiltshire S, O'Malley S, Lambert J, Kukanskis K, Edgar D, Kingsmore SF, et al. Detection of multiple allergen-specific IgEs on microarrays by immunoassay with rolling circle amplification. Clin Chem. 2000;46(12):1990-3.

2. Kim TE, Park SW, Cho NY, Choi SY, Yong TS, Nahm BH, et al. Quantitative measurement of serum allergen-specific IgE on protein chip. Exp Mol Med. 2002;34(2):152-8.

3. Bacarese-Hamilton T, Mezzasoma L, Ingham C, Ardizzoni A, Rossi R, Bistoni F, et al. Detection of allergen-specific IgE on microarrays by use of signal amplification techniques. Clin Chem. 2002;48(8):1367-70.

4. Lebrun SJ, Petchpud WN, Hui A, McLaughlin CS. Development of a sensitive, colorometric microarray assay for allergen-responsive human IgE. J Immunol Methods. 2005;300(1-2):24-31.

5. Hiller R, Laffer S, Harwanegg C, Huber M, Schmidt WM, Twardosz A, et al. Microarrayed allergen molecules: diagnostic gatekeepers for allergy treatment, FASEB J. 2002;16(3):414-6. Doi. 10.1096/fj.01-0711fje.

6. Ott H, Baron JM, Heise R, Ocklenburg C, Stanzel S, Merk HF, et al. Clinical usefulness of microarray-based IgE detection in children with suspected food allergy. Allergy. 2008;63(11):1521-8. Doi: 10.1111/j.1398-9995.2008.01748.x.

7. Martínez-Aranguren R, Lizaso MT, Goikoetxea MJ, García BE, Cabrera-Freitag P, Trellez O, et al. Is the Determination of Specific IgE against components using ISAC 112 a reproducible technique? PLoS One. 2014;9(2):e88394. Doi: 10.1371/journal.pone.0088394.

8. Moreira PF, Gangl K, Vieira Fde A, Ynoue LH, Linhart B, Flicker S, et al. Allergen microarray Indicates Pooideae sensitization in Brazilian grass pollen allergic patients. PLoS one. 2015;10(6):e0128402. Doi: 10.1371/journal.pone.0128402.

9. Robbins MA, Li M, Leung I, Li H, Boyer DV, Song Y, et al., Stable expression of shRNAs in human CD34+ progenitor cells can avoid induction of interferon responses to siRNAs in vitro. Nature biotech. 2006;24(5):566-571. Doi:10.1038/nbt1206.

10. Matson RS. Multiplex immunoassay for allergens using oligonucleotide tethering. 15th Int'l Congress of Immunology; 2013; Milan, Italy;2769-P; doi: 10.3252/pso.eu.15ici.2013.

2

Lysozymes, Proteinase K, Bacteriophage E Lysis Proteins, and some Chemical Compounds for Microbial Ghosts Preparation: a Review and Food for Thought

Amro Abd Al Fattah Amara*

The head of the Protein Research Department, Genetic Engineering and Biotechnology Research Institute, City for Scientific Research and Technological Applications, Universities and Research Centre District, New Borg El-Arab, Egypt

Corresponding author: Amro Abd Al Fattah Amara, The head of the Protein Research Department, Genetic Engineering and Biotechnology Research Institute, City for Scientific Research and Technological Applications, Universities and Research Centre District, New Borg El-Arab, Egypt, E-mail: amroamara@web.de

Abstract

Microbial Ghosts (MGs) is a new term that describes evacuated and dead microbes. Apparently, MGs will be able (soon) to substitute another term the "bacterial ghosts (BGs)". A new protocol for preparing MGs was introduced using the critical concentration or amounts of some chemical compounds and enzymes. In principle any compound such as SDS and NaOH or enzyme such as lysozyme and Proteinase K that could induce a pore(s) in the microbial cell wall could be used. "Evacology" might be a name for a new science that deals with living cells and viruses' evacuation. In addition, the bio-critical concentration of H_2O_2 enables Virus Ghosts (VGs) preparation. The bacteriophage E lysis gene based protocol is restricted only to the gram-negative bacteria. The Sponge-like Protocol (SL) has opened the window to nearly all microbes and all biological cells and viruses to be prepared as ghosts. In this shift point this review aims to cover the most important information about such a topic. SL protocol is based on determining the critical concentration of compounds that can kill, make pore(s), evacuate the cells, but did not deform or affect the cell wall or their antigens (under such concentration). Lysozyme has been used in the original protocol to complement any deficiency result in survive of any of the E. coli cells. Lysozymes and Proteinas K can stand-alone or can be combined with the other possible chemical compounds. The SL protocol for ghosts preparation is simple, inexpensive, in house, reliable, safe and cause pores starting from outside the cells to their inside. The future will show rising interest with such simple protocol, which could allow us to prepare our vaccine and drug delivery different ghosts' related formula in kitchen. In this, review most of the experiences gained from practicing experiments in ghosts' preparation and some idea about such subject were summarized and discussed.

Keywords: Bacterial Ghosts; Microbial Ghosts; Biological Ghosts; Sponge-Like protocol; Bacteriophage E lysis gene

Abbreviations

BGs: Bacterial Ghosts; MGs: Microbial Ghosts; MGC: Minimum Growth Concentration; MIC: Minimum Inhibition Concentration; SL: Sponge-Like; VCG: *Vibrio cholerae* ghost

Introduction

MGs in principle are empty microbial cells but with correct 3D structure. The cytoplasmic constituents of the MGs were come out, because of a pore was made due to the activity of the expression of the E lysis gene or by using the critical concentration of some chemical compounds and enzymes. The main idea is to remove the cytoplasm and its constituents without damaging the microbes' 3D structure or deforming their cells surface antigen. After being evacuated, the microbes are empty and dead. They are safe while they are unable to replicate. For that, they are the best candidates that can be used as vaccines. On the other hand, because they have each an empty internal space, they can be used in the drug delivery applications. Many microbial species were prepared as ghosts using both of the E lysis gene and the SL (Sponge-Like) protocol. Nevertheless, the bacteriophage E lysis gene is restricted only to the gram-negative bacteria. Alternatively, the SL protocol enables the preparation of MGs from both of the gram-negative and positive bacteria, yeasts, and even viruses. Amara(2015) has been suggested that such protocol (SL) and their modification using both of chemical compound at critical concentrations and enzymes at critical activity will emerge a new science might be carrying the name "Evacology"[1]. This review is a collection of the tools, idea, success, improvement, advantage, disadvantage, future, applications, and many facts and idea about the BGs, MGs, and biological cell ghosts. Such protocol might be able to establish a new science, which might be given the name "Evacology".

The first evacuation process

It is important to track the first root of evacuating a biological structure. The old civilizations somehow knew how to evacuate a complete mammalian organism or the human body (corpse). They know also that water is an essential factor for any living or enzymatic process. For that, they remove water correctly without damaging the cells. Then they prevent water or moisture from their ability to re-enter to the dried organisms. They know also about many fine structures which were existing around them such as, fungi particularly species which are visible to the nacked eye. Therefore, they had added some potent antimicrobial/water prevention agents such as resins. Additionally, they know how to use yeast for producing beer, wine, and bread. They know that boiled, smoked; salted etc food could survive (for different periods). In addition, they gained knowledge, experiences, and practice to avoid the bad effect of such spoiled food. In general, they should have the knowledge to control the microbial deterioration to design such perfect protocol for mummification.

Egyptian mummy

The ancient Egyptian might be the first to evacuate the corpse carefully from their organs. Then the evacuated corpse was subjected to careful treatment to remove its water content(s). Some reports have been described early that they have first treated the corpse with wines. Ethanol as one of the major constituent of the wine is a dehydrating agent. Other, bio-natural and chemical compounds were used for extending the protection and to avoid any physical, chemical or microbial degradation. Resins also were used. The Egyptian have used to wrap, embalmed corpses [2]. Then the corpse coated with cotton or any suitable Bandages and isolated by painting it with resins and putting it in well-isolated container. The process has given the name Mummification, which identified as: the process for embalming or artificially preserving lifeless bodies [1].

The microscope and the first empty observed microbe

Perhaps the first microbe, which was seen as a ghost was in the first sample, which was shown, using, lenses by the first microscopic examination to a water sample by Antonie van Leeuwenhoek in 1670s. Many microbiologists especially those, who were using the light microscope on a regular basis to show microbial specimens, should saw dead, evacuated, or partially damaged MGs. Even during the use of simple stain (crystal violet) one should remark that some microbes do not stain at all and some are deeply stained and some are only their wall edge were stained. Those, which are not stained (or get faint stain), are probably still life and resist the staining process that is clear in case of yeast when stained by crystal violet either when dead or viable. Such simple staining (using dilute crystal violet) could distinguish between the live and the dead microbes. Some researchers prefer to use special dye, like Trypan blue.

Ghost different names

Living cells after being evacuated can be given the name ghosts. For that, the term MGs should not be given to the dead microbes or cells but only to the dead and evacuated ones. There are different definitions and terms were used with "ghost" to give special meaning in the English language and literature. For examples, RBCGs ghosts for the RBCs cytoplasm free cells; BGs for bacterial ghosts and MGs for microbial ghosts; Phage ghosts were given for the first time to the genomic free bacteriophage and might be apparently the first name to be given for empty

virus. Virus ghosts, recently introduced after evacuating the Newcastle virus "Newcastle virus ghosts" from its genomic RNA [3]. For small parasite or a stage of them such as the trophozoit, it can be suggested to put the name followed by ghosts such as "trophozoit ghosts".

Did the mish (Mesh) cheese have been used for immunization?

Like any creature, each microbe has a life cycle, which end with death. The microbes' cell walls after their death could stay longer and resist decaying [4-7]. However, due to the process of the natural decay, environmental effect, enzymatic activity of other microbes, or any other expected mechanism for a pore or pores formation, natural loss of the cytoplasm or in better words, natural microbial ghosting will be happened. As being parts of the nature, microbes after being dead they are affected by their surrounding ecological environment. For that, cell ghosts and microbial ghosts are produced daily in our bodies (inside, or outside). They are produced in the lung, in the stomach, in the surface of our skin, in our aged food and so on. Therefore, dead and ghost microbes are natural phenomena.

In addition, they play different roles in immunization. Natural Ghosting phenomenon in our bodies plays important roles in immunization. Aged foods used by old civilization might have been used for the aim of immunization. The ancient Egyptian invented some type of cheese which named old cheese or Mish (Mesh) which combine between being so old so if any harm microbe is existed as a contaminate it is either being attenuated or dead. Such old cheese or Mish/Mesh cheese might be invented not as a type of cheese only but to protect the Egyptian from many diseases.

BGs and the mummy

When something (Physical, chemical or biological) cause microbes or cells to loss their cytoplasm contents, this will let their membrane or cell walls to stay longer (Figure 1 and 2). Ghost cells contain less or no cytoplasm constituents (protein and carbohydrates, salts etc.). The existence of the cytoplasm in the dead cells either enhance their degradation by the activity of the internal enzymes or by supporting the growth of other microbes which can secrete enzymes could degrade the cell wall or the membranes of the dead ones to be able to utilize the nutrients in such cytoplasm. Being dead, they become as food for other microbes. Additionally, as the mummy MGs have no water content. Some microbial cells which contain less protein resist the degradation process. For example, yeast contains only three percent of its cell wall as protein. For that Baker yeast can be preserved at room temperature. However, most of its cytoplasm content are protein in their nature. Being empty cells MGs show great resistance to the degradation process that was proved by their long shelf life.

Pore formation by the φX174 bacteriophage E lysis protein

The mechanism of causing pore(s) in the bacterial cells infected by bacteriophage become clear by the power of the

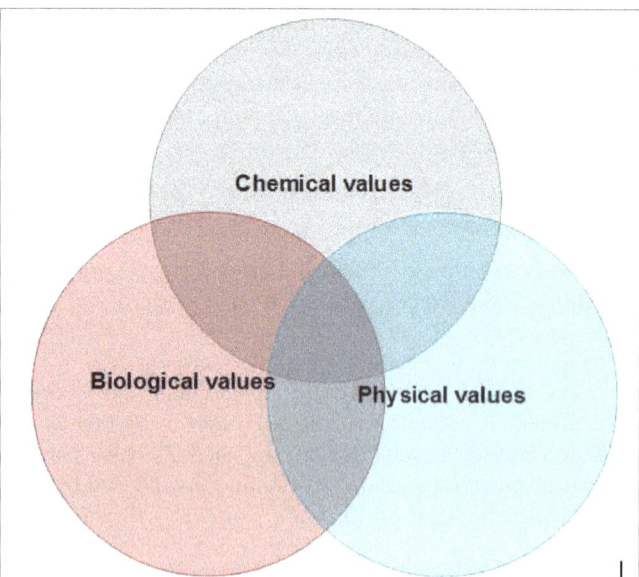

Figure 1: Chemical, biological and physical values interact during the MGs preparation.

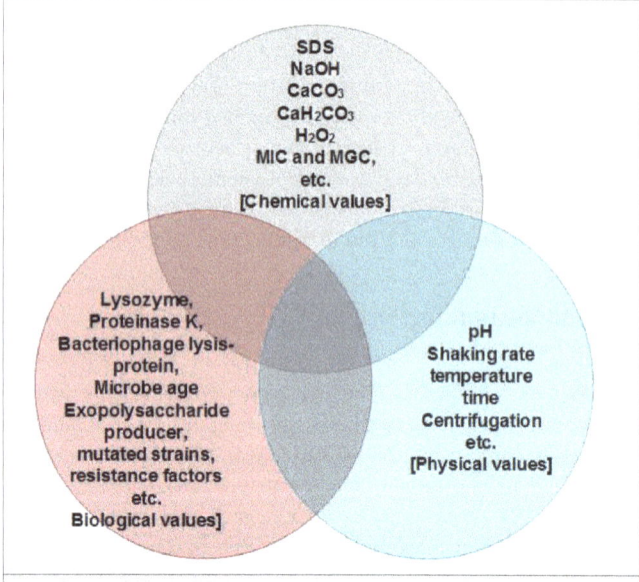

Figure 2: The most important values involved in MGs preparation.

electron microscope [8-12]. After the pore formation the bacterial cells wall is opened and the bacterial cells lose their cytoplasm content [13-23].

The φX174 E lysis protein cause E specific transmembrane tunnel structure built through the cell envelope complex [24-27]. It form a fusion through the inner and outer cell membranes, forming a specific transmembrane tunnel structure. Through such tunnel or pore, the bacterial cytoplasmic content is passing out. Genetic engineering and molecular biology tools enables better control for the bacteriophage E lysis gene based protocol.

The produced lysis enzyme enable pores in the bacterial cell wall. There is still a need for degrading the residue of the DNA, plasmid, lysozymes, proteinase K and so on, which are still contaminating the opened bacterial cells.

E lysis gene, its activity is restricted only to the gram-negative bacteria. The gene *E* was cloned and expressed in different gram-negative hosts. Such expression has been controlled by a heat sensitive promotor, which allows the expression of the *E* lysis gene. Using the heat sensitive promotor and regulator were used for better control. Nevertheless, its main weak point that it is restricted only to the gram-negative bacteria [4, 28].

Gene *E* codes for 91 amino acids

Gene *E*: The amino acids content of the E lysis protein sequence of the Enterobacteria phage φX174 and its nucleotides content were well identified (as below). One can obtain uncut phage DNA from the market and clone the *E* lysis gene. Alternatively, one can isolate the phage and made DNA isolation then clone the *E* lysis gene or cut it with suitable enzyme and made gene library.

E **lysis gene nucleotides sequence:** atggtacgctggactttgtgggataccctcgctttcctgctcctgttgagtttattgctgccgtcattgcttattatgttcatcccgtcaacattcaaacggcctgtctcatcatggaaggcgctgaatttacggaaaacattattaatggcgtcgagcgtccagttaaagccgctgaattgttcgcgcttaccttgcgtgtacgcgcaggaaacactgacgttcttgctgacgcagaagaaaacgtgcgtcaaaaattgcgtgcagaaggagtga

Amino acids sequence: MVRWTLWDTLAFLLLLSLLLPSL-LIMFIPSTFKRPVSSWKALNLRKTLLMASSVQLKPLNCSRLPCVYAQETLTFLLTQKKTCVKNCVQKE

One pore is enough!: A pore formation in the microbes' cell walls will lead to the removal of the cytoplasm which come out due to the cell wall imbalanced pressure force, osmosis differences, mechanical pressure, etc. The external medium can diffuse through the lysis tunnel filling the inner cell space of the still rigid BGs [27]. Apparently, and after observing many of the electron microscope images only one pore is usually existed. That might be due to the force happened as the result of the existence of the first pore. When first pore is being opened; it imbalance the internal pressure lead to getting rid of the cytoplasm contents and the rest of the E lysis protein so there is no chance to form another pore [29].

Foreign surface antigen and drug delivery: Ghost cells can be performed after the expression of foreign antigen, loading drugs, DNA, plasmid etc to the cells. However, in case of using MGs in the drug delivery, after such loading to the drug, the existed pore must be closed or in better word must be sealed [30]. Recently, another tactic have been introduced where *Saccharomyces cerevisiae* cell ghosts were loaded with dissolved ghosts gossypol acetic acid, then the dissolved ghosts gossypol acetic was allowed to crystallize inside the yeast cells. That will give the chance to load drugs without sealing the cells.

SL and the idea of the critical chemical concentrations

SL protocol introduce the idea of using the critical concentration of chemical compounds and recently enzymes for the ghosts preparation. The used steps were selected preciously to do in sequence steps enable full evacuation for the treated microbes or cells from their cytoplasmic contents. The future might show a more perfect chemical compounds or a strong modification in the protocol but the Acknowledge should be given to the original six used chemical compounds. They are, NaOH, SDS, $CaCO_3$, H_2O_2, NaCl and Ethanol. Recently $NaHCO_3$ was used instead of $CaCO_3$ to produce yeast ghosts. H_2O_2 has stood alone to degrade Newcastle RNA using bio-critical concentration. Enzymes, which can affect on the microbial cell wall such as lysozyme and protinase k are under optimization to give the equal results of the SL protocol [data not shown] but chemical compounds are more cost affective and omit the risk of incorrect immunization due to the use of enzymes which are protein in their nature [31].

The main idea of the SL protocol: The SL protocol main idea is simple and applicable; it depend on determining the (MIC) Minimum Inhibition Concentration and the (MGC) Minimum Growth Concentration of the used compounds or enzymes. The minimum killing effect in case of using MIC of the used compounds or enzymes they should cause minimum effect on the dead cells. In case of MGC the cells still alive. However, by using another MIC/MGC for the another compound(s) plus the physical effect of the repeated centrifugation steps, that all will lead to empty dead cells but with correct 3D structure and correct surface antigens which enables correct immunization upon the use of the experimental animals. SL protocol gives the chance to prepare ghosts from gram-positive and gram-negative bacteria, yeast, and virus and so on. Such a concept could be pass the microbes to any other biological cell forms. Many other forms from the biological system will join the ghost family after their preparation by this protocol (critical chemical concentration) soon.

Using critical chemical concentration for microbial killing is a natural phenomenon: *P. aeruginosa* is a dominant microbe in the hospitals while it is – in one word - hydrocarbon biodegradable microbes. H_2O_2 could turn any of the *P. aeruginosa* to mucoid strain hence to alginate producer. *P. aeruginosa* is able to be mutated and to produce a huge amount of exopolysaccharide, mainly the alginate if exposed to the H_2O_2. That also could be happened in the patients' lungs. As a defense mechanism against the *P. aeruginosa* infection the lung produces H_2O_2. How could the lung cells adjust the amount of the H_2O_2 to kill the *P. aeruginosa* but not to kill their own cells? In general, the lung cell must produce (somehow) H_2O_2 in critical concentration to kill such microbes and not to damage or to kill its own cells [32]. For one or another purpose, some of the *P. aeruginosa* cells were not killed by the killing dosage of the H_2O_2, but were exposed instead to less amounts of the H_2O_2 by one or another mechanism. Less amount of H_2O_2 will induce mucoid mutant. Such mutants are able to produce a huge amount of exopolysacharid. From such natural phenomena one can understand some facts about our "biological system" and how it could use the concept of "the critical chemical concentration" intelligently. That includes:

1. Our biological system knows how to produce chemical compounds in critical concentrations.

2. Concentrations less than the MIC could cause mutation, where such concentrations could keep the pathogens affected but alive.

3. Compounds produced by our biological system such as the oxidants and free radicals particularly those which produce to control pathogens should be given more concerns.

4. Misused of antioxidants could deteriorate our endogenous oxidant defense mechanism. For that, antioxidants should be taken wisely.

5. The biological system still proves that it is designed in perfect and intelligent way, so mechanisms used by such system should be given priority.

6. Lysozymes, proteases, DNases and other natural enzymes which are part of our defense system should be given more concerns to find more intelligent mechanism to control such pathogens.

7. Potant compound could be used upon dilution to do minimum side effect.

El-Baky and Amara(2014) combine between such phenomena and the idea of the SL protocol to *in vitro* degrade the Newcastle virus RNA to turn it to ghosts, which might be optimized to be *in vivo* protocol for controlling some pathogens without harming our biological system [3].

MGs omit the most important virulence factor: When foreign microbe enters to our body; our immune system starts to react with it and the battle will base on their number, type and our immune system quality. However, some microbes have extra virulence factors, which could be collectively stronger than our immune system. So, even, we are strong enough, the microbe can be fatal and exceed the speed of our immune system. Such conditions cause death, or severe illness or disorder. For that, scientists prepare killed microbes or attenuated ones to be sure that the immune system will manage the situation and control the invasion. Such in / less-active microbe(s) give our immune system the chance and the time to react with it. However, there are many reports prove that attenuated microbes can be turned to virulence ones in some conditions.

For that scientists spent time to reduce the virulence factors of the pathogenic strains or to find alternative solutions. Such alternative solutions might be summarized as follow:

1. Repeated cultivation, in media did not maintain the microbe's virulence factors.

2. Cell aging.

3. Viruses can be grown in an unspecific host, which induce safer mutants.

4. The use of recombinant strains enable expressing antigen(s) of some pathogenic microbes on their cell wall surfaces.

5. Genetically modified pathogens with less or completely deactivated virulence factors can be used.

6. Safe strains could induce antibodies which can protect against some virulence pathogens.

7. Close species are used which are able to induce immunization against certain targeted pathogens.

8. Using MGs (Figure 3).

9. Totally killed microbes but with with a suitable of effective surface antigens (Some killing process affect severely on the microbes surface antigen).

After that, our immune system produces the suitable antibodies. And, become ready for any pathogenic invader.

MGs are truly dead cells and they could not replicate. For that, if they are pathogens they will lose the chance to win the battle against our immune system (by increasing their numbers). For that, they are perfect candidates for activating safely our immune system. Old civilizations have aware by the knowledge and the tools of the vaccination. They know how to vaccine against the smallpox, the disease which kill millions of peoples. They use unsuccessful virus legion of smallpox to vaccine-uninfected patients. The original name of the vaccination is variolation or inoculation. Even such reports prove that the main idea was transferred to the new world, only after doing some improvement in a method from the Middle East and Africa, which was known as ventilation. We still -until nowadays- use the same old technique. Such simple tools saved the life of unknown number of peoples. It is an African practice is transferred to America nearly in 1706-1721 by a Sudanese slave [33]. However, it might be a shift in the concept when cowpox was used instead of the smallpox to give the full immunization against the later.

Some concerns about our immune system: There are some points that should be considered for those who are involved in the battle of the pathogens control. Some of such important concerns can be summarized as follows:

Figure 3: The unique criteria for M.

1. Our immune system is active in certain ages and less active in another. For that, we should be vaccinated in the correct time.

2. However, in case of using dead cells like BGs the equation is different. The vaccine types and the rout of administration is an important issue. Weak immune system, should give a correct time, formula (dead/attenuated microbes), site of administration and so on. MGs even an empty dead cells but it still have the correct 3D structure, so the immune system react with it with nearly the same power when it attack living cells. But one should consider the correct used number of cells. For that better and safer immunization were reported in case of using the MGs particularly for immunization.

3. One could take the maximum dosage from MGs safely while they are dead cells.

4. The genetic engineering and the molecular biology tools enable introducing or expressing the protein of the fatal microbes in the surface of recombinant safe microbe(s). After that the microbes turned to ghosts and become dead and safe.

5. BGs can be given safely in newly root of administration such nasal or on the surface of the wounded skin.

6. As a perfect biological package targeted by different immunological cells particularly the macrophage, they become the best choice for the gene therapy. BGS have increased our understanding to some biological factors such as our understanding to the resistance where intake cells (but not dead) might acquire new genetic material made them survive. Alternatively, before they loses their content under certain condition they might hybrid with other MGs.

7. Exopolysaccharide microbes could give false MIC where some microbes are survived due to biofilm formation. For that, microbes, which can be affected or mutated by the used chemical compounds, are recommended to be killed by NaOH using its MIC concentration first H_2O_2, SDS can induce exopolysacchaide production. As the other empty similar containers, MGs have more respective size and stable envelope and can be used in the field of the drug delivery.

8. BGs can be used in the diagnostic Kit as a reference antigen, where it should give positive reaction with serum containing the proper antibodies [34].

9. BGs will be in its better form after removing their cytoplasmic contents including the DNA, RNA, and the protein and so on.

The elements of the evacuation process

The protocol can use either the introduced chemical compounds or enzymes in critical amounts or the combination between both. The original SL protocol and its reduced and modified forms were succeeding to prepare BGs from gram-negative,

Gram-positive, yeast and viruses till nowadays. In fact, the concept of the protocol enables preparing ghosts from any microbe or even from any biological cells. That because it based on some chemical compounds or enzymes were selected based on their ability to kill the microbes and induce pores in their cell walls, degrade/or remove their DNA, RNA and the protein. The compounds were used in concentrations that enable induction of minimal effect on the cell wall and the protocol itself use physical parameters to get rid of the cytoplasm such as the shaking and the cell pressing using centrifugation. For that, it is given the name SL protocol. The lysozyme and proteins K which are able to lysis most types of different microbes will give another chances for improving a commercial protocol for ghosts preparation. In fact, most of the molecular biologist have used such enzymes to prepare the DNA from various microbes. Only, they are in need to be used in critical activities enables ghosts preparation rather than the lysis of the microbes.

The first used chemical compounds*:* The selected compounds, which were used to prepare BGs using SL protocol, are NaOH, SDS, $CaCO_3$, and H_2O_2 and both of NaCl and Ethanol are included. Lysozyme and proteinase K upon their use in a critical amount also give the same results. Some physical factors are involved (Centrifugation, shaking and temperature). Biological parameters are playing the central role in the success of the protocol such as, the type of the used cell or the microbe and their age. The ghosts' preparation condition(s) are effective factor(s). Such as cells density during the preparation. The cells quality should be monitored either by light or electron microscopes. The ability of the prepared ghosts' cells to induce correct immunization should be investigated. The sequence of the treatment is also an important issue. For example, exopolysaccharide-producing microbes must be treated with NaOH firstly. After preparing MGs, the cells must be investigated for the existence of any viable colonies. In addition, the viable cells should be deactivated, converted again to ghosts or the overall batches should be sterilized (in case of fatal microbes). It is might be interesting to highlight that the vaccine and the immunological technologies are fine technology and in some cases grams from correctly prepared MGs is in need to satisfy the demand.

Why NaOH and Why SDS? : NaOH and SDS are two well-known compounds used for plasmid isolation from gram-negative bacteria. The protocol is given the name "alkaline lysis protocol for plasmid isolation". The lysis buffer made from ten percentage NaOH (not autoclave) and SDS (need to be autoclaved) as stock solutions (400 µl of the ten percentage SDS on 3400 µl water and then add 80 µl of ten percent NaOH). Both of NaOH and SDS are used to prepare the lysis buffer by mixing them with water. SDS introduces pores in the bacterial cells. For that, it is added to most of the toothpaste, used in lysis buffer for the plasmid isolation and for some of the DNA isolation protocol. If one use SDS in its MIC that might give minimum effect and introduce pores in the bacterial cells, such concept was one of the successful keys to isolate the DNA from any gram-negative microbes without a need for buying expensive kits. Increasing the temperature and the exposure time were additional option in case

of gram-positive bacteria [35]. NaOH prove to be effective on the cell wall and for that, it is better to use its MGC. However, it is a potent bacterial killer. For that, it is recommended to use it at the first to deactivate strains able to be mutate and to resist the ghost preparation steps.

Why MIC and MGC?*:* MIC is a critical point where the used chemical, drug, enzymes or the antimicrobial agents are able to kill the microbe under the investigation condition with minimum side effect on their cells. In the serial dilution experiment, the tube after the MIC, which shows the first growth and which, given the name MGC such concentration should still effective on the microbe somehow. Using the concept that chemical compounds and enzymes were used in their MIC in a time or a combination between the MIC and MGC of other compounds this will support the concept of generating the minimum effect on the cell wall of the microbial strains. In another word, the used combinations collaborate to do relaxed job by complementing each other.

Experimental Design-Plackett-Burman, Box-Behnken and the excel solver for optimization: Experimental design has been used to optimize many of the biological processes. It has been also used to optimize the production of the BGs. Experimental design needs two levels for each variable, one is low and the other is high. And, it is preferable to use from four to ten variables. In the original protocol twelve experiments are conducted. Each experiment contains either the high level or the low level of each of the used microbe. All of the used experiments must be different and followed the Plackett-Burman design. Therefore, concerning the used chemical compounds concentration; the MIC is the higher one +1 and the MGC is the lowest one -1 [36].

$CaCO_3$ or $NaHCO_3$: $CaCl_2$ is used in the competent cells preparation. CaCl2 is able to facilitate the movement of the plasmid from outside the cells toward its inside. It can do the same with the SDS and other compounds.

Using another stronger "Ca" based compound will give better results with some microbes. Some could have dual activities as a membrane transfer and as an alkali such as $CaCO_3$ and $NaHCO_3$. $CaCO_3$ was selected because it has another unique property that it is poorly water dissolved even after its autoclavation and it can be used as a suspension.

$NaHCO_3$ was used with the first Eukaryotic prepared as ghosts using the idea of the critical chemical concentration [37,38].

Ethanol*:* Ethanol can be used to precipitate the DNA and protein if used as 90 percentage. If it is used as 70-percent concentration that enables both of precipitation and salts elimination, (The 30 percent water content enables that). However, if it is used in concentration less than 70 percentage it could eliminate the salt, DNA, RNA and the soluble protein as well. In the SL protocol, 60 percent of the ethanol was used to evacuate the bacterial cells content from their DNA, RNA and the salts.

Lysozyme: Lysozyme is an enzyme found in some tissues and secretions and is considered as a part of the defense mechanism against pathogens while it is able to lyse some microbes. The mechanism of its activity is mainly by its targeting to the bacterial exopolysaccharide existed in the cell wall causing osmotic shock or lysis. The most famous source is the hen egg, saliva, milk and bacteriophage T4. Its substrate is consisting of alternate residues of (1-4)-linked β-N-acetylemuramic acid (MurNAc) and β-N-acetylelucoseamine (GlcNAc). Lysozyme hydrolysis the bond between C-1 of β-N-acetylemuramic acid and C-4 of GlcNAc. Chitin (β-1-4-linked GlcNAc) is also a substrate [31].

Proteinase K: One of the endopeptidases able to digest keratin and it is a broad-spectrum serine proteinase. It is activated by calcium. It is used to remove protein contaminating nucleic acid. The enzyme's activity is stimulated by denaturants such as SDS. It is always used side by side with lysozymes. Proteinase K can be used to improve the MGs preparation [39].

Centrifugation: What can be happened if the pore in the microbial cell wall is so small and the bacterial cells still able to maintain their content (even the microbe have such a small pore). For solving such a problem, centrifugation will be the best process for pressing the microbial cells.

After introducing single pore (small or big) or more than single pore the microbe cells might still be able to maintain. Their cytoplasmic content due to the preplasmic membrane, the centrifugation will be able to press the cells to get rid of their contents. One can imagine the cells coming down to the test tube bottom due to the centrifugation force. After being settled and aggregated. The cells pressed like the sponge and the cells continue to get rid of their contents. Therefore, the protocol has been given the name SL protocol. So one should use a suitable centrifugation speed. One should not exceed 4000 rpm/min speed during the ghosts preparation. Speed from 2000 to 3500 rpm/min will give better cell quality.

Washing: Washing is an important step that because after the cultivation and during the centrifugation step; the microbe surface and biomass trap debris and fatty acids. Such cells contaminating elements should be eliminated. In most cases, such cytoplasmic constituent or the rest of the growth condition could neutralize the effect of the used chemical compounds or enzymes. Also, washing enable getting rid from the elaborated cytoplasmic contents. So several washing steps using saline solution could improve the MGs quality. In fact after each centrifugation step samples from each supernatant should be investigated spectrophotometerically or by using gel electrophoresis to monitor the remove of the DNA and the protein. After the success of the protocol, its steps can be reduced and optimized.

The BGs Quality: The above step is concerned with monitoring the remove of cytoplasm which is not an indication about that the microbial cell walls are safe and did not deteriorate or damage. One question is immerged, how to evaluate the quality of the MGs cell walls during the ghosts preparation steps? The quality of the prepared MGs is an important issue. Light microscope should give correct judgment. However, electron microscope will give sharp evaluation for the bacterial 3D structure. To evaluate the quality of the BGs one should count randomly in a certain area ten bacteria and count the number of the cells which have correct 3D structure and use them as a percentage. Hemocytometer

can be used for more precise results. Both of the scanning and transmission electron microscopes can be used to evaluate the quality of the MGs.

Electron microscope is absolutely proof that the prepared BGs or MGs are excellent, very good, good, or damaged cells. In case of using the electron microscopes, simply one can spread one drop of distilled water on the surface of the microscope slide. After that the diluted MGs have been added, spread gently and left for air dry. After being well dried one should rewash the microbial smear to remove any of the salt crystals (to get better electron microscope result) [40]. The microbial smear then dried again. In case of transmission microscope the standard protocol is used.

Stocks and Microbial Cells Preparation

Stocks

10 percentage NaOH: Ten percentage of NaOH is sterile by itself and did not need to be autoclaved.

10 percentage SDS: Ten percentage of SDS need to be autoclaved.

30 percentage H_2O_2: Thirty percentage of H_2O_2 can be purchased from any local pharmacy and did not need to be autoclaved. But one should be sure that it does not contain any other ingredients.

10 percentage of NaHCO3: Ten percentage of $NaHCO_3$ need to be autoclaved. $NaHCO_3$ has been used in the preparation of the yeast ghosts instead of $CaCO_3$.

In case of $CaCO_3$, it did not cause any apparent toxicity to the bacterial cells in high concentrations, so it is used either 1.05 μg/mL or 0.35 μg/mL.

The First Prepared Bacterial Strain Using The SL Protocol

Microbial cells preparation

1. Prepare a pure and an identified microbial strain, simply by doing streak method for spreading the cells in the proper medium or on the selective medium plate. You should be care that the used medium does not mutate your strain.

2. Pick up single colony and re-spreads it on the surface of suitable medium and remark its phenotype.

3. Test the bacterial cells using light microscope for being sure of its type and purity. Both of simple stain and gram stain should be used.

4. Pick up a single pure colony either by sterile needle or by tooth pick and inoculate it to 25 ml flask containing 10 ml of the proper broth medium. And incubate for overnight. Fresh cells will give better MIC and MGC result.

Determination of the MIC and MGC for NaOH, SDS, and H_2O_2, enzymes etc.

1. Prepare several test tubes containing 5 ml or 4.5 ml of the proper medium aiming to conduct the serial dilution experiment.

2. Add 0.5 ml from the above solutions each in the first test tube of a set of tubes (7 tubes) and transfer 0.5 ml from one to second tube as in the standard protocol of the serial dilution. Remove 20 µl of each tube and add 20 µl of about 10^8 of the overnight cell culture in each tube (for each of the above chemical compound related experiment). Any change in the dilution should be considered. Incubate the tubes at 37°C for overnight (or at the growth temperature, which is recommended for the microbe under investigation). Calculate the MIC and the MGC for each of the used chemical compounds.

3. In case of $CaCO_3$ it was used either as 1.05 µg/mL or 0.35 µg/mL was used.

MGs Evacuation Protocol

Microbial aging

The culture of the used microbe was cultivated in one liter flask contains 500 mL NB or any other suitable medium under static condition at 37°C for 72 hr. Some microbes need adding blood to support their growth. Long time cultivation was used aiming to give old culture and thick cell wall. That could be better for ghost preparation. Further more old culture reduced the virulence of the used microbe (in most cases).

The five used strategies for MGs preparation

After collecting, the aged cells by centrifugation and washing them with saline several times; upon the type of the used microbe one can follow one of the below strategies.

Strategy No. 1 (The original SL protocol with twelve Plackett-Burman experiments protocol) (Figure 4).

The original used method is based on using 5x concentration strategy. NaOH, SDS and $CaCO_3$ were used in one step. Then the washed cells treated with H_2O_2 in the second step. After enough washing and centrifugation steps, 60 percent ethanol is used in the third step. Each of the twelve used experiments is following the Plackett-Burmen design as in the original protocol [41].

Strategy No. 2 (The Plackett-Burman reduced protocol) (Figure 5) same as in the strategy no-1., exactly but only the best two experiments were selected. Such reduced protocol was designed to be used only for the similar strains such as *E. coli* BL21 and JM109. However, it has been used for other microbes and proved to be effective [42].

Strategy No. 3 (The 2x strategy) (Figure 6), one can use 2x concentration of the used chemical compounds each in separate step instead of 5 x. For example; add one ml of the suspended bacterial cells to one ml of 2x of the NaOH (NaOH 10 percentage is sterile by itself but if diluted one should avoide any contamination). The 2x stock (prepared according to the results of the MIC and the MGC as recommended, in Plackett-Burman reduced protocol) are prepared and added to the bacteria. That depends on the experiment one or two in the reduced protocol. After finishing the treatment, centrifuging the cells and washing, the treatment with the 2x of each compound of MIC/MGC is started. That

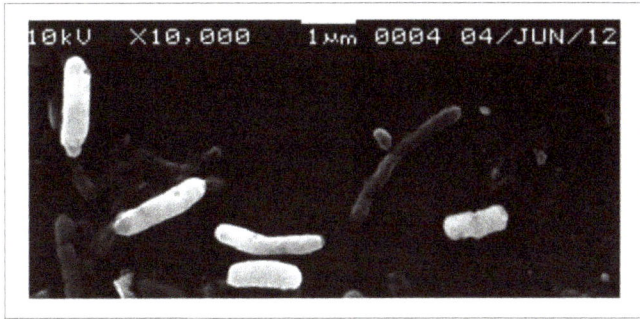

Figure 4: *E. coli* Ghosts.

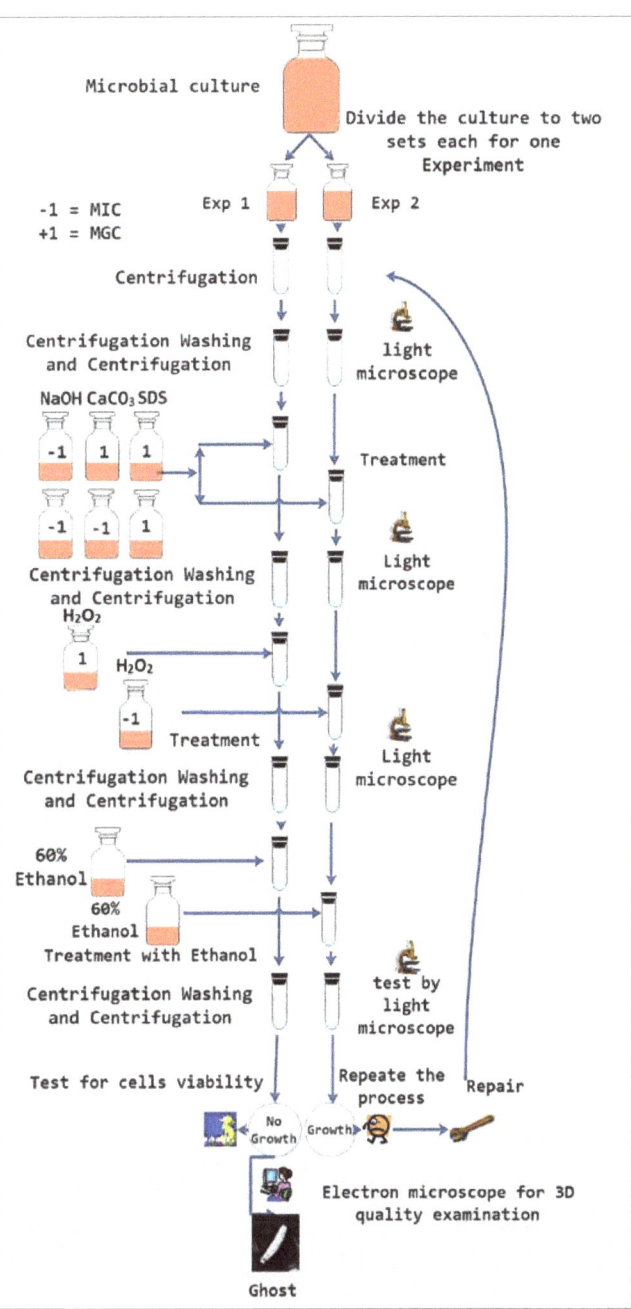

Figure 5: SL reduced protocol reduced-5x strategy (Strategy No 2).

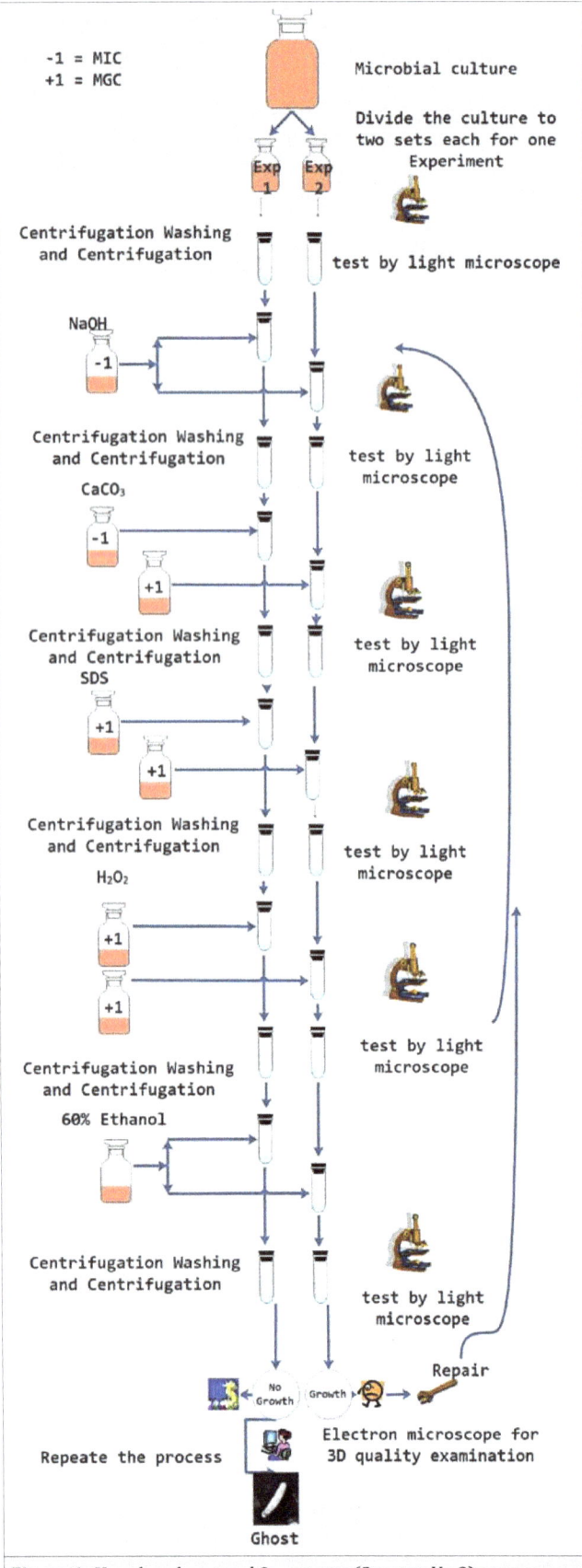

-1 = MIC
+1 = MGC

Microbial culture

Divide the culture to
two sets each for one
Experiment

Exp 1   Exp 2

test by light microscope

Centrifugation Washing
and Centrifugation

NaOH

-1

Centrifugation Washing
and Centrifugation

test by light
microscope

CaCO₃

-1

+1

Centrifugation Washing
and Centrifugation
SDS

test by light
microscope

+1

+1

Centrifugation Washing
and Centrifugation

test by light
microscope

H₂O₂

+1

+1

test by light
microscope

Centrifugation Washing
and Centrifugation

60% Ethanol

Centrifugation Washing
and Centrifugation

test by light
microscope

No Growth   Growth

Repair

Repeat the process

Electron microscope for
3D quality examination

Ghost

Figure 6: SL reduced protocol 2x strategy (Strategy No 3).

enables us to use NaOH alone with the expolysaccharide producing microbe to kill it and to take from the cells the chance to produce the exopolysaccharide.

Strategy No. 4, (The special protocol for mutated strains such as exopolysaccharide producing strain (for example *P. aeruginosa*). After working with an expolysaccharide producing strains, or eukaryotic and viruses strains. It becomes clear that H_2O_2 might induce resistance. In such case, NaOH must be used in the first treatment to kill the microbial cells and to prevent it from producing the exopolysaccharide. And in the same case the MIC must be test correctly while exopolysaccharide will give wrong result. In case of obtaining wrong MIC one can use the MIC which which is calculated the *E. coli* or use lesser concentration while exopolysaccharide will give high MIC value [43].

Strategy No. 5 Using enzymes after determining their MIC and MGC side by side with the used chemical compounds particularly with some gram-positive and halophilic strains [under optimization].

General Advices for a Better BGs Preparation

The SL protocol is a simple, inexpensive and in-house protocol. However, one aim of this review is to give the reader the most tested tactic, problems, solutions and image during his experiment design. This review should be read in detailed steps to enable the reader to implement or design his own protocol for a particular microbial strain. The protocol can be used to prepare ghosts from nearly all the bacterial strains, yeasts, and viruses. One can use any of the strategies explained in this protocol and included in this review. It is important to determine the MIC of the used chemical compounds and enzymes to start the protocol. One should also know some important information about the used microbes either gram-negative or positive, exopolysaccharide producer, the strains ability to be mutagenized and so on.

The following advices will give the reader some image during his experiment design. This review should be read carefully to gain all the idea within.

1. You should do all your experiment under aseptic condition and well-planned microbiological experiment in a microbiological lab containing all the needed facility. For example any contamination with a spore former microbe such as *Bacillus sp* will give wrong MIC and MGC.

2. For fatal pathogenic microbes, special care should be taken. One should refer to the technical advices concerning the handling of the microbe(s) under investigation.

3. You should know suitable information about the microbial strain, which you are going to evacuate its cytoplasm (gram-positive or negative, spore former, exopolysaccharide producer, yeast, fungi, trophozoit and so on).

4. You should use pure strain. Better to recheck the strain purity.

5. You should know its morphology on plats and under the light microscope.

6. Some microbes are sensitive to the chemical compounds used in the protocol such as *P. aeruginosa*, which is sensitive to the H_2O_2 and can be turned to be mucoid (exopolysaccharide producer). For that, it is recommended to use NaOH at the first step.

7. You can change the protocol by using each compound in separate steps (2x strategy). For that, you can simply prepare 2x stock from both of the MIC and the MGC you are going to use.

8. In some special cases where MIC and MGC could not be calculated correctly one can use either the *E. coli* MIC and MGC or that of another related strains.

9. In the serial dilution step, you should use glass tube to get correct MIC and MGC result. Plastic tubes might give wrong judgment.

10. You should avoid aggregation or clamping, so suitable volume should be used to allow correct contact between the microbial cells and the used chemical compounds.

11. Prepare double (or more) concentration of the MIC and the MGC. Therefore, you can reach the correct concentration after adding the solution which contains the microbe under investigation. In such case it is recommended that one might use the 5x strategy as in the original protocol. However, for the beginner it is recommended to apply NaOH, SDS, H_2O_2 and $CaCO_3$ using 2x protocol separately.

12. Better to age the microbe to get stronger cell wall and in some cases to reduce the microbe's virulence factors. Alternatively, some chemical compounds can induce cell rigidity.

13. One should monitor the release of the DNA and the protein, spectrophotometrically and by using gel electrophoresis.

14. Use gentle centrifugation not to exceed 3500 rpm.

15. Wash several times after each step to get rid of the residue of the chemical compounds by using saline (0.5-0.9 percentage) as well as to remove more DNA and protein (etc.).

16. Investigate for the exopolysaccharide producing strain and start the protocol with NaOH to kill them first.

17. You can keep the test tube of the serial dilution for more than one day and you can reevaluate the concentration you use. That because some microbes grow slowly.

18. Calculate the concentration correctly. Any wrong calculation may damage the cells. Light microscope is a suitable tool to monitor the quality of the cells during the adjustment of the MIC and the MGC effect.

19. You can also change the exposing time of the microbe to the used chemical compounds or enzymes. However, regular control for the cell quality by using light microscope should be followed to prevent cell-lysis.

20. You can extend the microbial exposure to H_2O_2 for overnight.

21. Do not forget that NaOH in the original protocol was used in both experiments (1 and 2) as -1, which mean the MGC. For a certain microbe, you can increase the concentration of the NaOH until the MIC but under the control of the light microscope. And in such case one can reduce the exposure time. It is recommended to conduct both of the experiment number ones and two in the SL reduced protocol. For the design of experiment one and two refer to Amara et al. (2013) [42].

22. After completing the MGs preparation steps, you should examine the cell viability. In case of the existence of viable cells in one or both of the conducted experiments, you should repeat the ghost preparation steps but you still be able to use your incomplete prepared MGs again. There is no need to prepare new cells if the cells are in good quality (investigate the cell quality using the light microscope).

23. In complicated microbes, experimental design should be used. Plackett-Burman, Box Behnken and Excel solver can be used in sequence to get the best-expected optimization. In fact, experimental design could map the critical points involved in the cell ghosts preparation and could be able to optimize perfectly any complicated process. For that MIC for the used compounds in the SL protocol might be used in strains differentiation. Similar strains such as in case of *E. coli* Bl21 and JM109 prove that they have different MIC and MGC.

Optimizing the gap between MIC and the MGC that can be achieved using Plackett-Burman, Box-Behnken and the Excel solver.

24. Spectrophotometer and electrophoresis can be used to monitor the evacuation of the cytoplasm content.

25. The best MGs are ones which are dead, empty from their cytoplasmic content, have correct 3D structure and are able to induce the immune system upon their treatment.

26. Polyacrylamide gel electrophoresis can be used to show the differences between the viable cells and the ghosts. In original protocol viable cells, which existed after running this protocol (did not turn to dead ghosts and still viable), were subjected to lysis by inducing the lysozyme gene carried on pLysS plasmid.

MGs Definition

After the introducing of the critical chemical concentration and activity method for preparing the MGs, many facts were changed and wide range of microbial and mammalian cells as well as other biological containers could be evacuated using such protocol. Therefore, new definition of the MGs can be introduced as follows:

"MGs are empty and dead microbial cells or viruses (envelopes) devoid of cytoplasmic contents or any internal fluidized or any ge-

Figure 7: The most important applications for MGs .

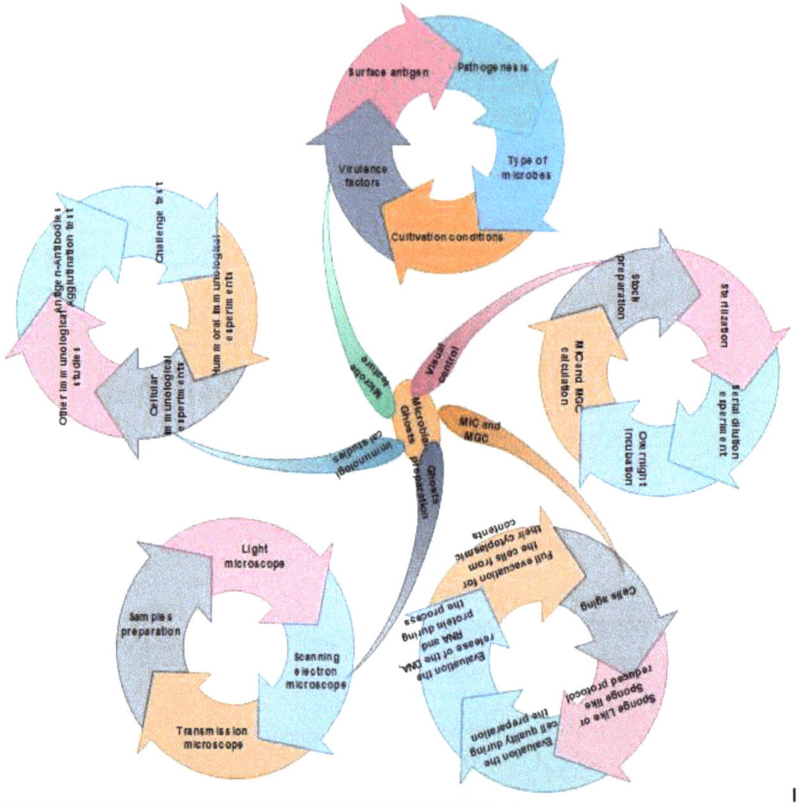

Figure 8: Five steps for produce high quality MGs : Microbe feature; Visual control; MIC and MGC; Ghosts preparation; immunological studies.

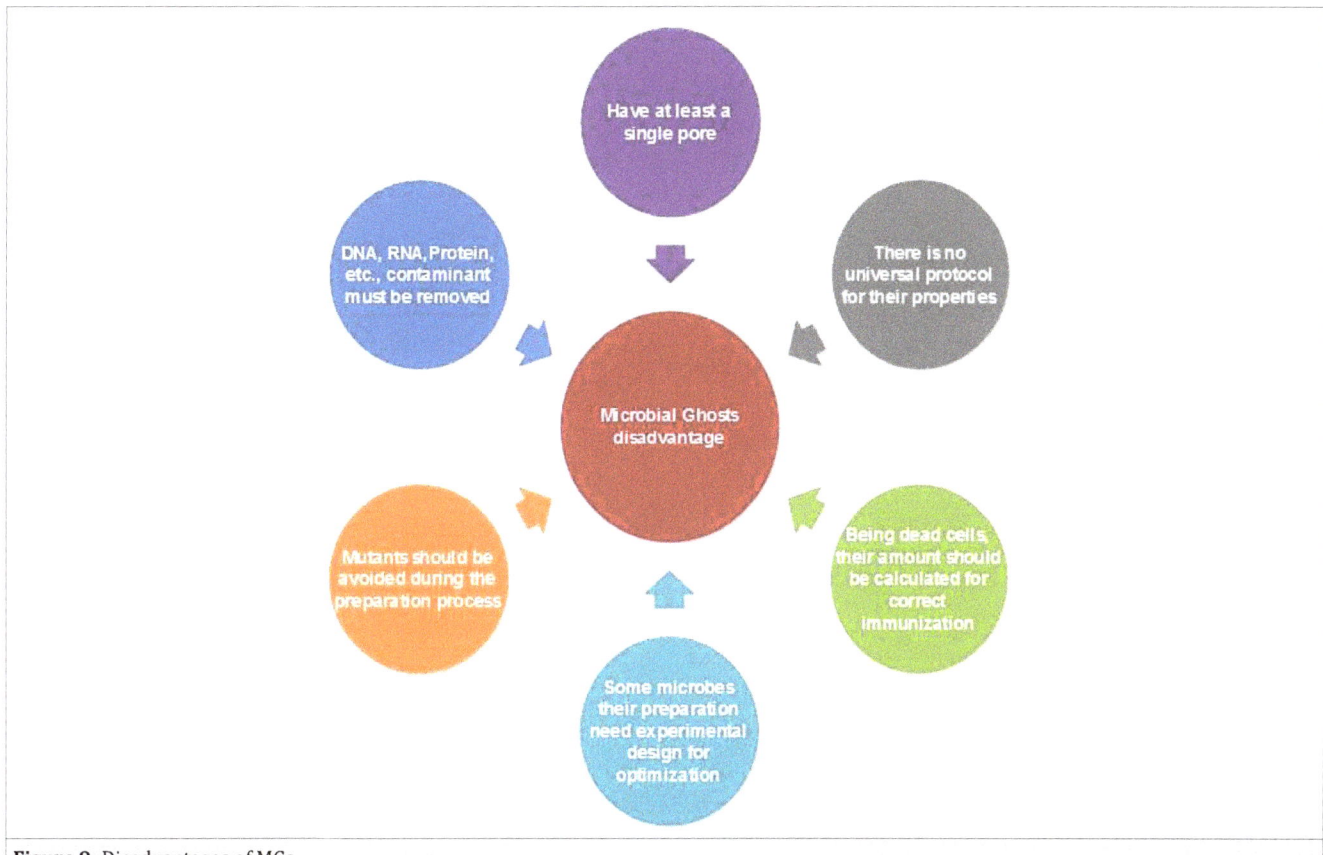

Figure 9: Disadvantages of MGs.

netic element. That were caused by one or more than one pore hap- *pened in their cell walls. Or direct remove of the genetic elements. They have correct 3D structure, morphology and native surface antigens structure able to induce the immune system of the delivered host to produce specific antibodies that could react correctly with the mother viable cell or viruses. Additionally as being empty cells they can be used as drug delivery system for various drugs, genes and antigen or surface antigen expressed protein from another potent pathogenic bacteria. The critical chemical concentration and enzymes activity methods have extended the spectrum of the BGs from the gram-negative bacteria to all types of bacterial strains including gram-positive, Archaea as well as Eukaryotic such as yeast ghosts. Viruses were achieved and parasites also are expected to join the Ghosts Family. The future of the MGs is bright"*

One should observe that the previously introduced E lysis protocol was restricted only to the gram-negative bacteria. However, the critical chemical concentration and the critical enzymes activities enables ghosts preparation from any microbes.

The advantage of MGs (Figure 7,8)

After introducing the SL protocol most microbes become target to be prepared as ghosts. Microbial or Biological Ghosts have some advantages which could be concluded in the following points:

1. After purification, MGs could be subjected to long-term storage at ambient room temperature as lyophilized products.

2. Genetically engineered or surface expressed antigens can be targeted to the inner or outer membrane.

3. Before being prepared as MGs, they were alive active cells, so they can be manipulated through the molecular and genetic engineering tools such as loading surface protein antigen of a potent microbial pathogenic strain. That can be happened on the surface of a suitable recombinant strain such as *E. coli*. Such ghosts could be used as vaccine against such protein or such strain. As carriers, there is no limitation in the size of foreign antigens that can be inserted in the membrane and any spaces in the cells.

4. Being dead cells make them safe during their handling, transfer, transport, processing and so on.

5. Being dead cells enable many new root of administration including oral, inhalation, superficial, ocular, wound, skin ulcer, beside the classical ones such as subcutaneous, aerosol, intradermal, intramuscular, intravenous, intraperitoneal, intragastric, rectal and intravaginal and so on.

6. MGs can be fractionated further to smaller fragment for certain applications, particularly if foreign surface antigen was introduced on the surface of recombinant strain (for example *E. coli*), hence such antigen can be re-purified and reused for research or immunization purposes. BGs can be produced inexpensively in large quantities.

7. They are dead cells, so they cannot replicate or do any metabolic activity.

8. They proved to be recognized by the immune system in manner similar to the living cells, so they can be used for application aim to target certain cells particularly the macrophage. Drugs of different types, compounds and proteins can be loaded to the internal lumen or periplasmic space of the BGs.

9. High potential to target different cell types; such as dendritic cells, macrophages, tumor cells, endothelial cells and epithelial cells.

10. MGs can be an alternative to the adjuvant, so if they are used as a drug delivery they can do more than one function.

11. Initial aerosol vaccinations showed that this route of administration can establish protective immunity and are able to target the mucosal immune system.

12. Ghosts introduced to the experimental animals are able to induce specific humoral and cellular immune responses against microbial components.

13. Large microbial ghost cells can be used for immobilizing small one, in certain industrial applications.

14. Multiple antigenic determinants can be presented simultaneously.

15. They can be suspended after being prepared as ghost in solvent such as ethanol which may contain an active ingredient such as gossypol acetic acid and then the ethanol is evaporated and the gossypol acetic acid is crystallized inside the cells. Water which did not dissolve the gossypol acetic acid can then be used to wash the outer cells crystal of the gossypol acetic acid. This strategy can be used to preparer some types of drug delivery products [38].

16. Multiple antigens of the native BG envelope and recombinant protein or DNA can be combined in a single type of MG.

17. No cytotoxic and genotoxic impacts on the viability and metabolic activity of cells recognizing BGs.

18. Simple and high-dose manufacturing process of MGs can be easily and quickly performed either in disposable fermenters, small laboratory steel fermenters or in large-scale fermenters.

19. They can be used in small or large quantities.

20. They have broad spectrum of possible safe applications.

21. They can induce several potent immune regulatory cytokines.

22. Their quality can be validated, using both the light and electron microscope for validating their 3D structure. Regular electrophoresis and spectrophotometer (260-280 nm) methods can be used to validate their content (as contaminant) of DNA and protein. But the most important point is to prove that they are still able to induce correct immunization. The research on eukaryotic cells and viruses has been started using the critical chemical concentration method and proved some advantages over the old used method [3,37,38].

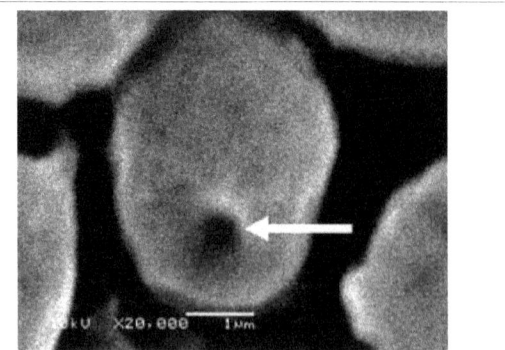

Figure 10: *Saccharomyces cerevisiae* Ghosts.

23. After adding the critical activity of the enzymes more relaxed protocol are expected to cover a wide range of microbes.

Disadvantage of the BGs (Figure 9)

By screening the literature published so far about the BGs, apparently, there is no mentioned disadvantage; however, MGs still have some disadvantages such as:

1. They have one open pore (at least), for that; they need to be sealed in case of their use as drug delivery system.

2. Even, their preparation become clear and straightforward, but aggregation should be avoided to avoid any survival cells.

3. Exopolysaccharide producing strains can give some wrong MIC result.

4. H_2O_2 could induce mutation if not calculated correctly. For that, some microbes such as *P. aeruginosa*, which mutated upon their exposure to H_2O_2 can produce mucoid strains. Such microbes should be treated with NaOH firstly.

5. Being dead cells, they should be manually counted preferably by using the hemocytometer.

6. Being dead, they should be given with higher dosage to equalize the effect of the active cells

7. Being empty cells, they can return some of the last solution, they suspended on, so they should be carefully washed and dried.

8. They should be gently subjected to the drying process to maintain the feature of their 3D structure.

9. After being empty cells, they can lose their 3D structure not by deforming the cell wall but by shrinking or getting rid of the empty space [43].

10. In case of preparing MGs using enzymes, which are protein in nature, enzymes should be removed after finishing the process. Any contamination with protein will induce the host immune system upon treatment.

Bacterial Cells Prepared As Ghosts Or Their Surface Protein Expressed In Ghosts

By revising the literatures there are many microbial cells that were prepared as ghosts as well as yeasts, viruses and cells.

E gene mediated lysis was achieved in a variety of gram-negative bacteria while SL protocol has extended the spectra to the eukaryotic and viruses. That includes *Acholeplasma laidlawii [E lysis][44]* ; *Actinobacillus pleuropneumoniae [E lysis] [45] [46] [47]*; *Aeromonas hydrophila [E lysis] [48]*; *Bacillus megaterium [E lysis][49]*; *Chlamydia trachomatis [E lysis] [50] [51]*; *Clostridium botulinum [E lysis](Ellison, Mattern, et al. 1971)*; *E. coli (EHEC) O157:H7 [E lysis] (Mayr, Haller, et al. 2005) [52]* ; *E. coli (EPEC) E2348 /69 [E lysis] [53]*; *E. coli BL21 (DE3) pLysS (Promega) [SL]* (Figure 3) *[54, 55]*; *E. coli JM109 [SL][36]*; *E. coli K12 [E lysis][56]* ; *E. coli Nissle 1917 [E lysis][57]*; *E. coli NK9373 [E lysis][58]*; *E. coli NM522 [E lysis][30, 59]*; *E. coli O157:H7 [E lysis][52]*; *E. coli O26:B6 [E lysis][60]*; Enteropathogenic *E. coli (EPEC) E2348/69 [E lysis] [53]*; *Edwardsiella tarda [E lysis] [61]*; *Flavobacterium columnare [E lysis] [62]*; *Haemophilus influenzae [E lysis][63]*; *Haemophilus parasuis serovar five reference strain Nagasaki [E lysis][64]*; *Helicobacter pylori [E lysis] [65]*; *Mannheimia haemolytica [E lysis] [66] [67] [68]*; *Mycobacterium phlei [E lysis] [69] [70]*; *Neisseria sicca [E lysis] [71]*; *Pasteurella multocida [E lysis] [72] [72]*; *Pseudomonas aeruginosa [E lysis and SL] [73]*; *Salmonella* Enteritidis *(and Yersinia enterocolitica) [E lysis][26, 74]*; *Salmonella gallinarum [E lysis][75]*; *Salmonella typhi Ty21a [E lysis][76]*; *Salmonella typhimurium [E lysis][60, 77]*; *Salmonella typhimurium ATCC 14028 [SL][34]*; *Salmonella typhimurium C5 [E lysis][60]*; *Streptococcus fecalis [E lysis][78]*; *Trichosurus vulpecula (pest) [E lysis] [59]*; *Vibrio cholerae [E lysis][79]*; *V. cholerae O1 or O139 [E lysis] [80]*; *Vibrio anguillarum [E lysis][81]*; *Yersinia enterocolitica [E lysis][82] [83]*; *Yeast [84]*; *Saccharomyces cerevisiae [SL] [37, 38]* (Figure 9) and Viruses such as bacteriophage *[85] and Newcastle virus [3].*

Ghosts Future

The protocols introduced in this review show that there are big opportunities to prepare different types of microbial and viruses ghosts. The most interesting one is that which based on using the critical concentration of some chemical compounds or the critical activity of some enzymes. Additionally, the future will show extending the ghosts preparation to nearly all types of cells. For that such protocols as well as the different types of ghosts will flourish the science of the immunology and pharmacology. The problem of any vaccine will be solved and new intelligent drug delivery systems will be introduced. There will be no need to go through the risk of using alive or attenuated vaccines. MGs will make a break through in many scientific fields and will be extended to the other forms of cells. This will enable many industrial, biotechnological, pharmaceutical, medicinal etc applications. In addition, it will be a great chance to study dead microbes and cells but still have their correct 3D structures.

Conflict Of Interest Statement

The author declares that there is no conflict of interest in this review article.

References

1. Amara AA. Kostenlos viral ghosts, bacterial ghosts microbial ghosts and more: Schuling Verlag - Germany, 2013.

2. Britannica E. Encyclopedia Britannica. INC. Yayınları, C 2000;9.

3. El-Baky NA, Amara AA. Newcastle disease virus (LaSota strain) as a model for virus Ghosts preparation using H_2O_2 bio-critical concentration. International Science and Investigation journal 2014;3:38.

4. JACOB F, FUERST CR. The mechanism of lysis by phage studied with defective lysogenic bacteria. J Gen Microbiol. 1958;18(2):518-26.

5. Maaloe O, Ingraham GL, Maaloe O, Neidhardt FC. Growth of the bacterial cell. Sinauer Association, Inc., Sunderland, MA. There is no corresponding record for this reference 1983.

6. PETHICA BA." Bacterial Lysis. Discussion Meeting in) J Gen Microbiol. 1958;18(2):473-80.

7. SALTON MR. The nature of the cell walls of some Gram-positive and Gram-negative bacteria. Biochim Biophys Acta. 1952;9(3):334-5.

8. Bachrach U, Friedmann A. Practical procedures for the purification of bacterial viruses. Appl Microbiol. 1971;22(4):706-15.

9. Dluzewski AR, Rangachari K, Wilson RJ, Gratzer WB. Properties of red cell ghost preparations susceptible to invasion by malaria parasites. Parasitology. 1983;87 (Pt 3):429-38.

10. Weidenbach H, Massmann J. [Electron microscopic study on the "ghost bodies" in experimental arteriosclerotic lesions of the vascular wall (author's transl)]. Exp Pathol (Jena). 1975;10(5-6):251-7.

11. Yamaguchi T, Tamura G, Arima K. Substructure of the cytoplasmic membrane of *Bacillus megaterium*. I. Method for the fractionation of "Ghosts". J Bacteriol. 1967;93(1):483-9.

12. Zitzer A, Palmer M, Weller U, Wassenaar T, Biermann C, Tranum-Jensen J, et al. Mode of primary binding to target membranes and pore formation induced by *Vibrio cholerae* cytolysin (hemolysin). Eur J Biochem. 1997;247(1):209-16.

13. Cornett JB. Spackle and immunity functions of bacteriophage T4. J Virol. 1974;13(2):312-21.

14. Duckworth DH. Biological activity of bacteriophage ghosts and "take-over" of host functions by bacteriophage. Bacteriol Rev. 1970;34(3):344-63.

15. Fukuma I, Kaji A. Effect of bacteriophage ghost infection on protein synthesis in *Escherichia coli*. J Virol. 1972;10(4):713-20.

16. Gershanovich VN, Avdeeva AV, Goldfarb DM. [Release of enzymes of the fluence of even t-phage ghosts]. Biokhimiia. 1963;28:700-8.

17. Goldfarb DM, Avdeeva AV, Borisova NB. Phage ghost-induced spheroplasts of *E. coli* 'B' as a system for phage reproductions. Nature. 1962;195:1202-3.

18. Okaichi K, Seki T, Ohnishi T, Nozu K. Effects of infection with UV- or X-ray-irradiated T2 phage ghosts on RNA synthesis in *Escherichia coli*. J Radiat Res. 1977;18(3):247-50.

19. Ou CT, Matsumoto I, Rozhin J, Tchen TT. Enzyme assay in cultures of *Escherichia coli* by a continuous flow method based on lysis from without by a phage ghost. Anal Biochem. 1978;88(2):357-66.

20. Snustad DP, Tigges MA, Parson KA, Bursch CJ, Caron FM, Koerner JF, et al. Identification and preliminary characterization of a mutant defective in the bacteriophage T4-induced unfolding of the *Escherichia coli* nucleoid. J Virol. 1976;17(2):622-41.

21. Swift RL, Wiberg JS. Bacteriophage T4 inhibits colicin E2-induced degradation of *Escherichia coli* deoxyribonucleic acid. II. Inhibition by T4 ghosts and by T4 in the absence of protein synthesis. J Virol. 1973;11(3):386-98.

22. Takeishi K, Kaji A. Protein synthesis in bacteriophage ghost-infected cells. J Virol. 1976;18(1):103-10.

23. Winkler HH, Duckworth DH. Metabolism of T4 bacteriophage ghost-infected cells: effect of bacteriophage and ghosts on the uptake of carbohydrates in *Escherichia coli* B. J Bacteriol. 1971;107(1):259-67.

24. Eko FO, Witte A, Huter V, Kuen B, Fürst-Ladani S, Haslberger A, et al. New strategies for combination vaccines based on the extended recombinant bacterial ghost system. Vaccine. 1999;17(13-14):1643-9.

25. Haidinger W, Szostak MP, Jechlinger W, Lubitz W. Online monitoring of *Escherichia coli* ghost production. Appl Environ Microbiol. 2003;69(1):468-74.

26. Szostak MP, Mader H, Truppe M, Kamal M, Eko FO, Huter V, et al. Bacterial ghosts as multifunctional vaccine particles. Behring Inst Mitt. 1997;(98):191-6.

27. Witte A, Wanner G, Sulzner M, Lubitz W. Dynamics of PhiX174 protein E-mediated lysis of *Escherichia coli*. Arch Microbiol. 1992;157(4):381-8.

28. Carlton RM. Phage therapy: past history and future prospects. Arch Immunol Ther Exp (Warsz). 1999;47(5):267-74.

29. Lubitz W, Witte a, Eko FO, Kamal M, Jechlinger W, Brand E, et al. Extended recombinant bacterial ghost system. J Biotechnol. 1999;73(2-3):261-73.

30. Kudela P, Paukner S, Mayr UB, Cholujova D, Kohl G, Schwarczova Z, et al. Effective gene transfer to melanoma cells using bacterial ghosts. Cancer Lett. 2008;262(1):54-63. Doi: 10.1016/j.canlet.2007.11.031.

31. Smith CAWEJ. Biological Molecules "Molecular and cell biochemistry": Chapman and Hall-London, 1991.

32. Amara AA. Opportunistic pathogens and their biofilm Food for thought. Science against microbial pathogens: communicating current research and technological advances. Microbiology Book Series 2011;3:813.

33. Kenneth K. The life and times of cotton Book .Mther, harper and Raw, New York;1984:339.

34. Amro AA, Neama AJ, Hussein A, Hashish EA, Sheweita SA. Evaluation the Surface Antigen of the *Salmonella typhimurium* ATCC 14028 Ghosts prepared by "SLRP". Scientific World Journal. 2014;2014:840863. Doi: 10.1155/2014/840863.

35. Awang R, Ahmad S, Ghazali R. Properties of sodium soap derived from palm-based dihydroxystearic acid. Journal of Oil Palm Research 2001;13:33.

36. Amara AA, Salem-Bekhit MM, Alanazi FK Plackett–Burman randomization method for Bacterial Ghosts preparation form *E. coli* JM109. Saudi Pharm J. 2014;22(3):273-9. Doi: 10.1016/j.jsps.2013.06.002.

37. Amara AA. *Saccharomyces cerevisiae* Ghosts Using the Sponge-Like Re-Reduced Protocol SOJ Biochem 2015:1.

38. Amara AA. Bacterial and Yeast Ghosts: *E. coli* and *Saccharomyces cerevisiae* preparation as drug delivery model ISIJ Biochemistry 2015;4:11.

39. Morihara k, Tsuzuki H. Specificity of proteinase K from *Tritirachium album* Limber for synthetic peptidase. Agric. Biol. Chem. 2010;347:233.

40. Amara AA, Serour EA. Wool quality improvement using thermophilic crude proteolytic microbial enzymes. American-Eurasian Journal of Agricultural & Environmental Sciences 2008;3:554.

41. Amara AA, Salem-Bekhit MM, Alanazi FK. Sponge-like: a new protocol for preparing bacterial ghosts. Scientific World Journal. 2013;2013:545741. Doi: 10.1155/2013/545741.

42. Amara AA, Salem-Bekhit MM, Alanazi FK. Preparation of bacterial ghosts for *E. coli* JM109 using sponge-like reduced protocol. Asian J Biol Sci 2013;6(8):363-69. DOI: 10.3923/ajbs.2013.363.369

43. Amara AA. Kostenlos viral ghosts, bacterial ghosts microbial ghosts and more: Schuling Verlag - Germany, 2015.

44. Brunner H, Dörner I, Schiefer HG, Krauss H, Wellensiek HJ. Lysis of *Acholeplasma laidlawii* by antibodies and complement. Infect Immun. 1976;13(6):1671-7.

45. Felnerova D, Kudela P, Bizik J, Haslberger A, Hensel A, Saalmuller A, et al. T cell-specific immune response induced by bacterial ghosts. Med Sci Monit. 2004;10(10):BR362-70.

46. Katinger A, Lubitz W, Szostak MP, Stadler M, Klein R, Indra A, et al. Pigs aerogenously immunized with genetically inactivated (ghosts) or irradiated *Actinobacillus pleuropneumoniae* are protected against a homologous aerosol challenge despite differing in pulmonary cellular and antibody responses. J Biotechnol. 1999;73(2-3):251-60.

47. Huter V, Hensel A, Brand E, Lubitz W. Improved protection against lung colonization by *Actinobacillus pleuropneumoniae* ghosts: characterization of a genetically inactivated vaccine. J Biotechnol. 2000;83(1-2):161-72.

48. Chu W, Zhuang X, Lu C. [Generation of Aeromonas hydrophila ghosts and their evaluation as oral vaccine candidates in Carassius auratus gibelio]. Wei Sheng Wu Xue Bao. 2008;48(2):202-6.

49. Yamaguchi T, Tamura G, Arima K. Substructure of the Cytoplasmic Membrane of *Bacillus megaterium* I. Method for the Fractionation of "Ghosts". J Bacteriol. 1967;93(1):483-9.

50. Macmillan L, Ifere GO, He Q, Igietseme JU, Kellar KL, Okenu DM, et al. A recombinant multivalent combination vaccine protects against Chlamydia and genital herpes. FEMS Immunol Med Microbiol. 2007;49(1):46-55.

51. Eko FO, Lubitz W, McMillan L, Ramey K, Moore TT, Ananaba GA, et al. Recombinant *Vibrio cholerae* ghosts as a delivery vehicle for vaccinating against Chlamydia trachomatis. Vaccine. 2003;21(15):1694-703.

52. Mayr UB, Haller C, Haidinger W, Atrasheuskaya A, Bukin E, Lubitz W, et al. Bacterial ghosts as an oral vaccine: a single dose of *Escherichia coli* O157:H7 bacterial ghosts protects mice against lethal challenge. Infect Immun. 2005;73(8):4810-7.

53. Liu J, Wang WD, Liu YJ, Liu S, Zhou B, Zhu LW, et al. Mice vaccinated with enteropathogenic *Escherichia coli* ghosts show significant protection against lethal challenges. Lett Appl Microbiol. 2012;54(3):255-62. Doi: 10.1111/j.1472-765X.2011.03202.x.

54. Amara AA, Salem-Bekhit MM, Alanazi FK. Sponge-like: a new protocol for preparing bacterial ghosts. ScientificWorldJournal. 2013;2013:545741. Doi: 10.1155/2013/545741.

55. Tuntufye HN, Ons E, Pham AD, Luyten T, Van Gerven N, Bleyen N, et al. *Escherichia coli* ghosts or live *E. coli* expressing the ferri-siderophore receptors FepA, FhuE, IroN and IutA do not protect broiler chickens against avian pathogenic E. coli (APEC). Vet Microbiol. 2012;159(3-4):470-8. Doi: 10.1016/j.vetmic.2012.04.037.

56. Henning U, Höhn B, Sonntag I. Cell envelope and shape of *Escherichia coli* K12. The ghost membrane. Eur J Biochem. 1973;39(1):27-36.

57. Stein E, Inic-Kanada A, Belij S, Montanaro J, Bintner N, Schlacher S, et al. In Vitro and In Vivo Uptake Study of *Escherichia coli* Nissle 1917 Bacterial Ghosts: Cell-Based Delivery System to Target Ocular Surface Diseases *Escherichia coli* Nissle 1917 Bacterial Ghosts. Invest Ophthalmol Vis Sci. 2013;54(9):6326-33. Doi: 10.1167/iovs.13-12044.

58. Abtin A, Kudela P, Mayr UB, Koller VJ, Mildner M, Tschachler E, *Escherichia coli* ghosts promote innate immune responses in human keratinocytes. Biochem Biophys Res Commun. 2010;400(1):78-82. Doi: 10.1016/j.bbrc.2010.08.013.

59. Walcher P, Cui X, Arrow JA, Scobie S, Molinia FC, Cowan PE, et al. Bacterial ghosts as a delivery system for zona pellucida-2 fertility control vaccines for brushtail possums (*Trichosurus vulpecula*). Vaccine. 2008;26(52):6832-8. Doi: 10.1016/j.vaccine.2008.09.088.

60. Mader HJ, Szostak MP, Hensel A, Lubitz W, Haslberger AG. Endotoxicity does not limit the use of bacterial ghosts as candidate vaccines. Vaccine. 1997;15(2):195-202.

61. Kwon SR, Kang YJ, Lee DJ, Lee EH, Nam YK, Kim SK, et al. Generation of *Vibrio anguillarum* ghost by coexpression of PhiX 174 lysis *E* gene and staphylococcal nuclease A gene. Mol Biotechnol. 2009;42(2):154-9. Doi: 10.1007/s12033-009-9147-y.

62. Zhu W, Yang G, Zhang Y, Yuan J, An L. Generation of biotechnology-derived *Flavobacterium columnare* ghosts by PhiX174 gene E-mediated inactivation and the potential as vaccine candidates against infection in grass carp. J Biomed Biotechnol. 2012;2012:760730. Doi: 10.1155/2012/760730.

63. Riedmann EM, Kyd JM, Smith AM, Gomez-Gallego S, Jalava K, Cripps AW, et al. Construction of recombinant S-layer proteins (rSbsA) and their expression in bacterial ghosts–a delivery system for the nontypeable *Haemophilus influenzae* antigen Omp26. FEMS Immunol Med Microbiol. 2003;37(2-3):185-92.

64. Hu M, Zhang Y, Xie F, Li G, Li J, Si W, et al. Protection of piglets by a *Haemophilus parasuis* ghost vaccine against homologous challenge. Clin Vaccine Immunol. 2013;20(6):795-802. Doi: 10.1128/CVI.00676-12.

65. Panthel K, Jechlinger W, Matis A, Rohde M, Szostak M, Lubitz W, et al. Generation of Helicobacter pylori ghosts by PhiX protein E-mediated inactivation and their evaluation as vaccine candidates. Infect Immun. 2003;71(1):109-16.

66. Paukner S, Kohl G, Lubitz W. Bacterial ghosts as novel advanced drug delivery systems: antiproliferative activity of loaded doxorubicin in human Caco-2 cells. J Control Release. 2004;94(1):63-74.

67. Ebensen T, Paukner S, Link C, Kudela P, de Domenico C, Lubitz W, Bacterial ghosts are an efficient delivery system for DNA vaccines. J Immunol. 2004;172(11):6858-65.

68. Kudela P, Paukner S, Mayr UB, Cholujova D, Kohl G, Schwarczova Z, et al. Effective gene transfer to melanoma cells using bacterial ghosts. Cancer Lett. 2008;262(1):54-63. Doi: 10.1016/j.canlet.2007.11.031.

69. Lee SK, Kim YS. Current Concepts and Occurrence of Epithelial Odontogenic Tumors: II. Calcifying Epithelial Odontogenic Tumor Versus Ghost Cell Odontogenic Tumors Derived from Calcifying Odontogenic Cyst. Korean J Pathol. 2014;48(3):175-87. Doi: 10.4132/KoreanJPathol.2014.48.3.175.

70. Rastogi N, Labrousse V. Extracellular and intracellular activities of clarithromycin used alone and in association with ethambutol and rifampin against *Mycobacterium avium* complex. Antimicrob Agents Chemother. 1991;35(3):462-70.

71. Dajani AS, Law DJ, Bollinger RO, Ecklund PS. Ultrastructural and biochemical alterations effected by viridin B, a bacterocin of alpha-hemolytic streptococci. Infect Immun. 1976;14(3):776-82.

72. Marchart J, Dropmann G, Lechleitner S, Schlapp T, Wanner G, Szostak MP, et al. Pasteurella multocida-and *Pasteurella haemolytica*-ghosts: new vaccine candidates. Vaccine. 2003;21(25-26):3988-97.

73. Doig P, Franklin AL, Irvin RT. The binding of *Pseudomonas aeruginosa* outer membrane ghosts to human buccal epithelial cells. Can J Microbiol. 1986;32(2):160-6.

74. Meyer-Bahlburg A, Brinkhoff J, Krenn V, Trebesius K, Heesemann J, Huppertz HI, et al. Infection of synovial fibroblasts in culture by *Yersinia enterocolitica* and *Salmonella enterica* serovar Enteritidis: ultrastructural investigation with respect to the pathogenesis of reactive arthritis. Infect Immun. 2001;69(12):7915-21.

75. Jawale CV, Chaudhari AA, Lee JH. Generation of a safety enhanced *Salmonella Gallinarum* ghost using antibiotic resistance free plasmid and its potential as an effective inactivated vaccine candidate against fowl typhoid. Vaccine. 2014;32(9):1093-9. Doi: 10.1016/j.vaccine.2013.12.053.

76. Wen J, Yang Y, Zhao G, Tong S, Yu H, Jin X, et al. Salmonella typhi Ty21a bacterial ghost vector augments HIV-1 gp140 DNA vaccine-induced peripheral and mucosal antibody responses via TLR4 pathway. Vaccine. 2012;30(39):5733-9. Doi: 10.1016/j.vaccine.2012.07.008.

77. Szostak MP, Hensel A, Eko FO, Klein R, Auer T, Mader H, et al. Bacterial ghosts: non-living candidate vaccines. J Biotechnol. 1996;44(1-3):161-70.

78. Abrams A, Nielsen L, Thaemert J. Rapidly synthesized ribonucleic acid in membrane ghosts from *Streptococcus fecalis* protoplasts. Biochim Biophys Acta. 1964;80:325-37.

79. Eko FO, Hensel A, Bunka S, Lubitz W. Immunogenicity of *Vibrio cholerae* ghosts following intraperitoneal immunization of mice. Vaccine. 1994;12(14):1330-4.

80. Eko FO, Schukovskaya T, Lotzmanova EY, Firstova VV, Emalyanova NV, Klueva SN, Evaluation of the protective efficacy of *Vibrio cholerae* ghost (VCG) candidate vaccines in rabbits. Vaccine. 2003;21(25-26):3663-74.

81. Kwon SR, Kang YJ, Lee DJ, Lee EH, Nam YK, Kim SK, et al. Generation of *Vibrio anguillarum* ghost by coexpression of PhiX 174 lysis E gene and staphylococcal nuclease A gene. Mol Biotechnol. 2009;42(2):154-9. Doi: 10.1007/s12033-009-9147-y.

82. Huppertz HI, Heesemann J. Experimental Yersinia infection of human synovial cells: persistence of live bacteria and generation of bacterial antigen deposits including "ghosts," nucleic acid-free bacterial rods. Infect Immun. 1996;64(4):1484-7.

83. Meyer-Bahlburg A, Brinkhoff J, Krenn V, Trebesius K, Heesemann J, Huppertz HI, et al. Infection of synovial fibroblasts in culture by *Yersinia enterocolitica* and *Salmonella enterica* serovar Enteritidis: ultrastructural investigation with respect to the pathogenesis of reactive arthritis. Infect Immun. 2001;69(12):7915-21.

84. Alvarez P, Sampedro M, Molina M, Nombela C. A new system for the release of heterologous proteins from yeast based on mutant strains deficient in cell integrity. J Biotechnol. 1994;38(1):81-8.

85. Konopa G, Taylor K. Isolation of coliphage lambda ghosts able to adsorb onto bacterial cells. Biochim Biophys Acta. 1975;399(2):460-7.

3

Moroccan Formulation of Oils for the Care of Hair: Chemical Composition and Antibacterial Activity

Tarik Ainane[1]*, Said Gharby[2], Mohammed Talbi[3], Abdelmjid Abourriche[4], Ahmed Bennamara[4], Naoual Oukkache[5], Hassan Lamdini[6] and Mohamed Elkouali[3]

[1]Superior School of Technology - Khenifra (EST-Khenifra), University of Moulay Ismail, PB 170, 54000 Khenifra, Morocco
[2]Laboratory of Chemistry of Plants, Organic Synthesis and Bioorganic, Faculty of Science, University Mohammed V-Agdal, Rabat, Morocco
[3]Laboratory of Analytical Chemistry and Physical Chemistry of Materials, Faculty of Sciences Ben Msik, University Hassan II, BP 7955 Casablanca 20660, Morocco
[4]Biomolecules and organic synthesis laboratory, Faculty of Sciences Ben Msik, University Hassan II, BP 7955 Casablanca 20660, Morocco
[5]Laboratory of Venoms and Toxins, Pasteur Institute of Morocco, 1 Place Louis Pasteur, Casablanca 20360, Morocco
[6]Department of Infectious Diseases, IbnRochd Hospital University Center, Casablanca 20270Morocco

*Corresponding author: Tarik Ainane, Superior School of Technology - Khenifra (EST-Khenifra), University of Moulay Ismail, PB 170, 54000 Khenifra, Morocco, E-mail: ainane@gmail.com

Abstract

The objective of this work is studying the chemical composition and the antibacterial activity of a formulation used in the Moroccan tradition for hair care composed of two vegetable oils (argan oil and olive oil) and three essential oils (*Thymus vulgaris, Nigella sativa,* and *Allium sativum*). At first, we analyzed the physicochemical parameters for both vegetable oils, such as: acidity, peroxide index, saponification index, iodine index, absorption coefficients in UV at 232 nm and 270 nm, and humidity, we also determined their fatty acids compositions, sterols compositions and tocopherols contents. On the other hand, the essential oils components were identified by GC/MS, the results of these analyzes showed that the major constituents of the essential oils of *Thymus vulgaris* and *Nigella sativa* were monoterpene hydrocarbons and phenolic monoterpenes, and the major constituents of the essential oil of *Allium sativum* were diallyl sulfides and methyl allyl sulfides. Finally, the antibacterial activity of the vegetable oils, essential oils and formulation were determined against strains bacteria, using a well diffusion agar method, where the results of this antibacterial test indicated the effectiveness of the formulation prepared during this study for cosmetics and pharmaceutical preparations.

Keywords: Formulation; Vegetable oils; Essential oils; Chemical compositions; Antibacterial activities

Introduction

Since antiquity, the use of natural products presents considerable interest for humans, but nowadays with the industrial spirit, scientific advance is being used in the development of innovative natural products in medicine, pharmaceutics, and cosmetics industries with thorough and targeted studies [1-5]. Several public and private sectors, industries and laboratories focus their research towards the development of natural products (Bio-products) because it is provide interesting turnover and millions of dollars in revenue [6-7]. Also, most consumers are choosing natural products to avoid any danger source of synthetic chemicals products that could be harmful to their health. Besides, taking into account the growing environmental issues, bio-products help to protect the environment by the ease of degradation [8-9].

On the other hand, the industrial development of natural resources in Morocco was begun for ten years, while several special industries were installed to produce Bio-products and exploit these resources [10], because Morocco is undoubtedly both a well-known name and a significant producer in the world of oils (essential oils and vegetal oils). This is the result of several main factors: The geography and climate, which is governed by the Mediterranean Sea, the Atlantic Ocean, the desert in the south and its three main mountain ranges. Morocco hosts a complete range of Mediterranean climates and soils that favour an extremely rich biodiversity, including an impressive variety of aromatic plants (both Mediterranean classics and endemic species) [8,11].

Consequently, many species of plants produce seeds containing fats which are used as a food reserve for the developing seedling and they are quite often present in sufficient quantities to make their extraction, in the form of oil, worthwhile. Vegetable oils are produced from nuts, seeds, grains and beans. They are sometimes referred as fixed oils because they are not as volatile (easily evaporated) as essential oils [12,13]. Vegetable oils particularly the argan oil and olive oil have a wide range of uses, and whilst many of these involve processes that are too technical for small scale ventures, there are still many ways in which we can employ them as a cosmetic or pharmaceutics products [14]. Also, the essential oils are the subject of intensive scientific research and attract attention of cosmetic and pharmaceutical industries due to their potential as active pharmacological compounds or natural preservatives. Enormous diversity of this group of natural compounds and wide spectrum of biological properties make them attractive for many industries. Regardless from sensory properties of essential oils, antimicrobial and antifungal activities are the goal of research [15,16].

This work is a chemical and biological study of a natural formulation from vegetable oils (argan oil and olive oil) and essential oils (*Thymus vulgaris, Nigella sativa*, and *Allium sativum*) of Moroccan tradition for hair care. This formulation has been used for several centuries in the rural areas, the Sahara and the Atlas mountains.

Materials and Methods

Vegetable oils

Argan oil was obtained from the cooperative of GIE TARGANINE. Olive oil was acquired from a supermarket (Casablanca -Morocco).

Analytical determination

The chemical and physical parameters (acidity, peroxide index, saponification index, and iodine index, absorption coefficients in UV at 232 and 270 nm, and humidity) were analysed, in triplicate following the analytical methods described in Regulations EC 2568/91 [17].

Fatty acid composition was determined on their corresponding methyl esters by gas chromatography on a CPWa x 52CB column (30 m × 0.25 mm i.d.) using He (flow rate 1 ml/min) as a carrier gas. Oven, injector, and detector temperature were set at 170, 200, and 230°C respectively. Injected quantity was 1 μl for each analysis.

Sterol composition was determined after trimethylsilylation of the crude sterol fraction using a Varian 3800 instrument equipped with a VF-1 ms column (30 m × 0.25 mm i.d.) and using helium (flow rate 1.6 ml/min) as carrier gas. Column temperature was isothermal at 270°C; injector and detector temperature was 300°C. Injected quantity was 1 μl for each analysis [18].

On the basis of the AOCS Official method Ce 8-89, tocopherols content was determined by HPLC using Shimadzu instruments equipped with a C18-Varian column (25 cm 94 mm). Detection was performed using a fluorescence detector (excitation wavelength 290 nm, detection wavelength 330 nm). Eluent used was a 99:1 isooctane/ isopropanol (V/V) mixture, flow rate 1.2 ml/min [19].

Essential oils

Aerial parts of the *Thymus vulgaris,* seeds of *Nigella sativa,* and fruits of *Allium sativum* were purchased from the local market in Casablanca (Morocco) and identified at Department of Biology, Faculty of sciences Ben M'sik University of Hassan

II - Casablanca. Voucher specimen of the plants were dried and deposited at the herbarium of laboratory. Origins of species are displayed in (Table 1).

Dried biomasses were submitted to steam distillation in a Clevenger-type apparatus for 4 h. The essential oils obtained were separated from water and dried over anhydrous Na_2SO_4 then stored at 4°C until use.

The qualitative analysis of essential oils is done by gas chromatography coupled to mass spectrometry (GC/ MS: Hewlett Packard 5971A). Determining the relative proportions of various molecules obtained by gas chromatography coupled with flame ionization (GC/FID: Hewlett Packard 5890A). Analysis by GC/MS and GC/FID are made under identical conditions. GC/ MS were performed on a DB-5 column (5% phenyl methyl siloxane) whose dimensions are: length: 30 m; diameter: 250 μm; film thickness 0.32 microns. The applied temperature program was 40°C for 5 min, 40 to 20 °C at 3°C/ min then held at 200°C for 5 min. The carrier gas was helium (pressure: 49.9 kPa, flows: 1ml/ min). The source of the mass spectrometer to a temperature of 230°C and the mass range is scanned from 50 to 350 amu [20].

Preparation of formulation

In a graduated flask of 100 ml, was added 1 ml of each essential oil. Afterwards, the flask completed until gauge by the mixture of argan and olive oils (1:1), and finally stored in a refrigerator at 4°C before analysis or use.

Antibacterial activities

The method used is the well diffusion agar described by T. Ainane and A. Abourriche, et al. [21]. This method can quickly observe effects of a substance by bacterial growth. Screening for antibacterial activity of the products was determined by agar well diffusion method. The oils and formulation were dissolved in DMSO (Dimethyl sulfoxide) 5%. Ten microliter of crude extract (2 mg/ ml) was loaded onto well (diameter 6 mm). Fresh colonies of *Streptococcus faecalis, Escherichia coli, Staphylococcus aureus* and *Pseudomonas aeruginosa* on supplemented MH (Mueller Hinton) agar, were inoculated in supplemented MH broth and incubated overnight under aerobic condition. The bacterial suspensions were adjusted to McFarland standard at 0.5 and spreaded onto supplemented MH agar plates. The seeded plates and incubated at 37°C for 24 h under aerobic condition. The diameters of the inhibition zones were measured and the mean was recorded. Experiments were done in triplicate. Bacterial culture with 1% DMSO was used as negative control. In addition, tetracyclin and streptomycin used a positive control.

Table 1: Origins of the medicinal plants used in formulation for the care of hair.

Plants	Region	Extracted part
Thymus vulgaris	Oujda (Eastern Morocco)	Aerial parts
Nigella sativa	Beni Mellal (Atlas median)	Seeds
Allium sativum	Meknes (Atlas median)	Fruits

Results and Discussion

After a survey carried out on natural products used in the Moroccan tradition, particularly cosmetics, we selected a formulation for the care of hair known in the population of the Atlas, where it's composed by the vegetable oils and the essential oils. This formulation is a composition of two vegetable oils: argan oil and olive oil, also made-up of three essential oils of *Thymus vulgaris*, *Nigella sativa* and *Allium sativum*.

Analysis of Vegetable Oils

The first part of this work was devoted to the chemical composition of this formulation. So we started with the arganoil and olive oil. (Table 2) shows the physicochemical parameters of two oils used such as: acidity, peroxide index, saponification index, iodine index, absorption coefficients in UV at 232 and 270 nm, and humidity. Acidity is an important factor in assessing the quality of an oil and is widely used both as a test of classification of olive and argan oils, and also a factor that informs the oil alteration by hydrolysis. The acidity of argan oil and olive oil are respectively 0.30% and 0.62%. The peroxide value, saponification value and iodine value are parameters depend by the physicochemical properties and stability of fatty acids. The peroxide value of argan oil and olive oil are respectively: 1.1 meq/ kg and 2.1 meq/ kg, saponification values are respectively: 189.9 mgKOH/ g and 194.5 mgKOH/ g, and the iodine value are respectively: 98.3 and 87.7. Other parameters studied are the absorption coefficients in UV at 232 and 270 nm, because the conjugated diene have a strong absorption band in the ultraviolet near 232 nm and the triene have a triple band at 270 nm. Determining the absorbance in the vicinity of two wavelengths allows for the detection and evaluation of primary and secondary oxidation products. The results obtained for argan oil are: K_{232} = 1.19 and K_{270} = 0.20 and for olive oil are: K_{232} = 1.71 and K_{270} = 0.16. Humidity is the moisture and volatile matters is the weight loss experienced by the product after heating to 103°C ± 2°C in the operating conditions, the results obtained of argan and olive oils are respectively: 0.06% and 0.04%. According to the parameters recommended in the literature and the authorized values (norm) [17], both vegetables oils used are fresh and they having a good quality.

From another side, vegetable oils are essentially defined by their major composition of fatty acids and their minor composition of sterols and tocopherols. The study of the fatty acid composition of argan oil and olive show that oleic acid (46.9 % and 74.6 %) and linoleic acid (33.3 % and 10.7 %) are the majority fatty acids followed by palmitic acid and stearic acid. The analysis of total sterol gives very interesting results, 169 mg/ 100g for argan oil and 207 mg/ 100g for olive oil, also tocopherols have an minor composition for both oils, where the total tocopherol for argan oil is 738 mg/ kg in which a predominant amount of γ-tocopherol, and the total tocopherol for olive oil is 182 mg / kg in which a predominant amount of α-Tocopherol. All results of analyzes of fatty acids, total sterols and tocopherols are displayed in (Table 3).

Analysis of essential oils

After the distillation of essential oils, the determination of yield of each oil was calculated and then the values found were: 1.12 ± 0.21% for *Thymus vulgaris*, 0.83 ± 0.03% for *Nigellasativa*, and 0.52 ± 0.17% for *Allium sativum*. Therefore, qualitative analysis of essential oils of *Thymus vulgaris*, *Nigella sativa*, and *Allium sativum* made by gas chromatography coupled to mass spectrometry GC/ MS are shown respectively in (Table 4, Table 5 and Table 6).

Analysis results of essential oil of *Thymus vulgaris* give identification of 90.8% of these constituents. The majors compounds identified are: thymol (47.4%) and p-cymen (17.0%), and other compounds also detected with interesting percentages: β-caryophyllene (3.5%), carvacrol (3.2%), linalool (2.4%), α-thujene (2.2%), γ-terpinen (2.1%), terpinen-4-ol (1.9%), cadinene (1.8%), camphene (1.8%), β-myrcene (1.4%), borneol (1.3%), and α-pinen (1.2%).

The second analysis of essential oil of *Nigella sativa* shows the presence of p-cymen (60.5%) as the major compound of the oil, also, the analysis confirms existence the other compounds with remarkable percentages such as: α- thujene (6.9%), thymoquinone (3%), carvacrol (2.4%) and β-pinene (2.4%), and other compounds with low yields. All of the identified compounds of this essential oil a yield of the order of 87.5%.

Finally, analysis of essential oil of *Allium sativum* shows that all detected compounds are types of diallyl sulfide and methyl allylsulphides. The indentified total compounds present 74%. The major compounds are: diallyl disulfide (18.8%), methyl allyl trisulfide (16.3%) and diallyl trisulfide (15.9%), also other

Table 2: Physicochemical parameters of argan oil and olive oil used in this study.

Parametersa	Argan oil	Authorized values for argan oil [17]	Olive oil	Authorized values for olive oil [17]
Acidity (g/100 g)	0.30 ± 0.01	<0.8	0.62 ± 0.01	< 0.8
Peroxide value (meq/ kg)	1.1 ± 0.1	<15	2.1 ± 0.5	< 20
Saponification value (mgKOH/ g)	189.9 ± 0.2	189–199.1	194.5 ± 0.1	184–196
Iodine value	98.3 ± 0.5	91–110	87.7 ± 1.0	75–94
K_{232}	1.19 ± 0.06	-	1.71 ± 0.01	< 2.5
K_{270}	0.20 ± 0.04	<0.35	0.16 ± 0.01	< 0.22
Humidity (%)	0.06 ± 0.01	<0.1	0.04 ± 0.01	< 0.2

Table 3: Fatty acid, sterol, and tocopherol composition of argan oil and olive oil.

	Composition	Argan oil	Olive oil
Fatty Acids (%)	Palmitic acid	12.1 ± 1.5	9.2 ± 1.5
	Stearic acid	6.2 ± 1.0	2.9 ± 0.5
	Oleic acid	46.9 ± 1.5	74.6 ± 2.5
	Linoleic acid	33.3 ± 1.5	10.7 ± 1.5
	Linolenic acid	0.08 ± 0.10	0.9 ± 0.1
Sterols (% total sterols)	Total sterols (mg/ 100 g)	169 ± 10	207 ± 10
Tocopherols (mg/ kg)	α-Tocopherol	49.5 ± 6.0	167 ± 15
	β-Tocopherol	1.5 ± 0.6	10.5 ± 2.5
	γ-Tocopherol	651.4 ± 2.0	2.3 ± 0.3
	δ-Tocopherol	57.3 ± 6.0	20.1 ± 6
	Total	738 ± 26	182 ± 30

Table 4: Percentages of chemical compositions of the essential oil *Thymus vulgaris.*

Comp.	Structure	Percentage	Comp.	Structure	Percentage
1		α-humulene 0.2%	11		β-caryophyllene 3.5%
2		α-Pinen 1.2%	12		γ-Terpinen 2.1%
3		Sabinen 0.7%	13		Terpinen-4-ol 1.9%
4		β-Pinen 0.4%	14		α-thujene 2.2%
5		1,8-Cineol 0.9%	15		β-myrcene 1.4%
6		α-Terpinen 0.8%	16		Carvacrol 3.2%

Comp.	Structure	Name	Comp.	Structure	Name
7		p-Cymen 17.0%	17		Borneol 1.3%
8		Cadinene 1.8%	18		Camphene 1.8%
9		Germacrene 0.4%	19		Linalool 2.4%
10		Thymol 47.4%	20		Thymol methyl ether 0.2%

Table 5: Percentages of chemical compositions of the essential oil *Nigella sativa*.

Comp.	Structure	Percentage	Comp.	Structure	Percentage
1		α-Thujen 6.9%	9		1,8-Cineol 0.1%
2		α-Pinen 1.7%	10		γ-Terpinen 3.5%
3		Sabinen 0.9%	11		Terpinen-4-ol 2.1%
4		β-Pinen 2.4%	12		p-Cymen-8-ol 0.2%

5		Myrcen 0.1%	13		Thymoquinon 3.0%
6		α-Terpinen 1.0%	14		Carvacrol 2.4%
7		p-Cymen 60.5%	15		Longifolen 0.9%
8		Limonen 1.4%	16		Thymohydro-quinon 0.4%

Table 6: Percentages of chemical compositions of the essential oil *Allium sativum.*

Comp.	Structure	Percentage	Comp.	Structure	Percentage
1		Dimethyl disulfide 2.3%	8		Diallyltrisulfide 15.9%
2		1,2-dithiacyclo-pentane 0.4%	9		Diallyltetrasulfide 0.8%
3		Diallyl sulfide 1.3%	10		3-Vinyl-[4H]-1,2-dithiin 7.3%
4		Methyl allyl disulfide 4.8%	11		2-vinyl thiophene 0.2%
5		Dimethyl trisulfide 1.9%	12		3-methylthio-propanal 0.2%
6		Diallyl disulfide 18.8%	13		Allyl 2,3-epoxypropyl-sulfide 0.1%
7		Methyl allyltrisulfide 16.3%	14		2-Vinyl-[4H]-1,3-dithiin 3.7%

Table 7: Antibacterial activity of the vegetable oils, the essential oils, the formulation, the tetracycline and streptomycin.

Product	S. faecalis	E. coli	S. aureus	P. aeruginosa
Argan oil	+	+	+	+
Olive oil	-	+	+	-
T. vulgaris	+	++	++	-
N. sativa	+	++	++	+
A. sativum	+	+++	+	++
Formulation	+	++	++	+
Tetracycline	++	++	+++	++
Streptomycin	+++	++	+++	-

Key: -: no inhibition, +: less than 10mm diameter inhibition, + + inhibition diameter between 10 and 15mm, + + + greater than 15mm diameter inhibition.

compounds were detected with an important yields such as: 3-vinyl-[4H]-1,2-dithiin (7.3%), methyl allyl disulfide (4.8%), 2-vinyl-[4H]-1,3-dithiin (3.7%) and dimethyl disulfide (2.3%).

Antibacterial activities

The vegetable oils and essential oils are the main products of the preparation of the formulation for the care of the hair. They are evaluated for antimicrobial activity against four strains *Streptococcus faecalis*, *Escherichia coli*, *Staphylococcus aureus* and *Pseudomonas aeruginosa* (Table 7) gives the results obtained during the antibacterial tests of these four strains by well diffusion agar method with a concentration of 2 mg/ ml for all products.

All oils showed important activity against the four strains, except olive oil which doesn't present an activity against *Streptococcus faecalis*, and *Pseudomonas aeruginosa*, also, no activity for essential oil of *Thymus vulgaris* against *Pseudomonas aeruginosa*. Altogether, formulation prepared during this work showed remarkable activity against four strains. Finally, the positive results of this antibacterial test are achieved by the constituents of the chemical composition of all the oils and formulation, particularly fatty acids, thymol, p-cymen and sulphides; they exhibit important activities according to the literature [20,22-24].

Conclusion

This work was devoted under investigation of the chemical composition and the antibacterial activity of a formulation of the Moroccan tradition, for the care of the hair. This formulation was prepared of two vegetable oils: argan oil and olive oil, and three essential oils: *Thymus vulgaris*, *Nigella sativa*, and *Allium sativum*. The physicochemical analysis of vegetable oils shows that two oils had good qualities according to the international recommendations. Also, their chemical compositions give a majority composition of fatty acids and a minority composition of sterols and tocopherols. The chemical analyses of the components of essential oils shows that the three oils had interesting compounds bioactifs, from where the essential oil analysis of *Thymus vulgaris* gives a majority composition of thymol and p-cymen, the essential oil analysis of *Nigella sativa*

gives a majority composition of p-cymen, and essential oil analysis of *Allium sativum* gives a majority composition of diallyl disulfide, methyl allyl trisulfide and diallyl trisulfide.

Finally, the antibacterial test of the formulation gives an important activity against *Streptococcus faecalis*, *Escherichia coli*, *Staphylococcus aureus* and *Pseudomonas aeruginosa*. The appearance of this activity is caused by bioactive compounds present in components of vegetable oils and essential oils.

References

1. Khan IA, Abourashed EA. Leung's encyclopedia of common natural ingredients: used in food, drugs and cosmetics. John Wiley & Sons; 2011.

2. Achilladelis B, Antonakis N. The dynamics of technological innovation: the case of the pharmaceutical industry. Research Policy. 2001;30(4):535-588. DOI:10.1016/S0048-7333(00)00093-7.

3. Eshun K, He Q. Aloe Vera: a valuable ingredient for the food, pharmaceutical and cosmetic industries—a review. Crit Rev Food Sci Nutr. 2004;44(2):91-6. DOI: 10.1080/10408690490424694.

4. Lubbe A, Verpoorte R. Cultivation of medicinal and aromatic plants for specialty industrial materials. Industrial Crops and Products. 2011;34(1):785-801. DOI:10.1016/j.indcrop.2011.01.019.

5. Ainane T, Elkouali M, Ainane A, Talbi, M. Moroccan traditional fragrance based essential oils: Preparation, composition and chemical identification. 2014;6(6):84-89.

6. Burssens S, Ingelbrecht I, Van Montagu M, De Oliveira D, Pertry I. Green biotechnology applications for industrial development: opportunities and challenges for cooperation between the EU and the Mercosur. Mercosur European Union dialogue. 2013;80-97.

7. Bekatorou A, Plessas S, Mantzourani I. Biotechnological Exploitation of Brewery Solid Wastes for Recovery or Production of Value-Added Products. Advances in Food Biotechnology: John Wiley & Sons; 2015.

8. Daughton CG, Ternes TA. Pharmaceuticals and personal care products in the environment: agents of subtle change? Environ Health Perspect. 1999;107;6:907-38. DOI: 10.1016/j.watres.2009.12.032.

9. Kumar S. Exploratory analysis of global cosmetic industry: major players, technology and market trends. Technovation. 2005;25(11):1263-1272. DOI:10.1016/j.technovation.2004.07.003.

10. Faysse N, Errahj M, Imache A, Kemmoun H, Labbaci T. Paving the way for social learning when governance is weak: Supporting

dialogue between stakeholders to face a groundwater crisis in Morocco. Society & Natural Resources. 2014;27(3):249-264. DOI: 10.1080/08941920.2013.847998.

11. Jarlan L, Driouech F, Tourre Y, Duchemin B, Bouyssié M, Abaoui J, et al. Spatio-temporal variability of vegetation cover over Morocco (1982–2008): linkages with large scale climate and predictability. International Journal of Climatology. 2014:34(4):1245-1261. DOI:10.1002/joc.3762.

12. Nasir, M. Taxonomic perspective of plant species yielding vegetable oils used in cosmetics and skin care products. African journal of biotechnology. 2005;4(1):36-44.

13. Ryan E, Galvin K, O'Connor TP, Maguire AR, O'Brien NM. squalene, tocopherol content and fatty acid profile of selected seeds, grains, and legumes. Plant Foods Hum Nutr. 2007;62(3):85-91. DOI: 10.1007/s11130-007-0046-8.

14. Charrouf Z, Guillaume D. Argan oil: Occurrence, composition and impact on human health. European Journal of Lipid Science and Technology. 2008;110(7):632-636. DOI: 10.1002/ejlt.200700220.

15. Bassolé IH, Juliani HR. Essential oils in combination and their antimicrobial properties. Molecules. 2012;17(4):3989-4006. Doi: 10.3390/molecules17043989.

16. Salem M.Z, Zidan YE, Mansour MM, El Hadidi NM, Elgat WAA. Antifungal activities of two essential oils used in the treatment of three commercial woods deteriorated by five common mold fungi. International Biodeterioration & Biodegradation. 2016;106:88-96. DOI:10.1016/j.ibiod.2015.10.010.

17. Commission Regulation (EEC) 2568/91 on the characteristics of olive oil and olive-residue oil and on the relevant methods of analysis. Official Journal of the European Communities. 1991;248:1–82.

18. Gharby S, Harhar H, Guillaume D, Haddad A, Matthäus B, Charrouf Z, et al. Oxidative stability of edible argan oil: A two-year study. LWT-Food Science and Technology. 2011;44(1):1-8. DOI:10.1016/j.lwt.2010.07.003.

19. American oil chemist's society. Determination of tocopherols and tocotrienols in vegetable oils and fats by HPLC. Uniform Methods Committee. AOCS Official Method Ce 8e89. 1993; Champaign, II: AOCS.

20. Ainane T, Askaoui Z, Elkouali M, Talbi M, Lahsasni S, Warad I, et al. Chemical composition and antibacterial activity of essential oil of Nigella sativa seeds from Beni Mellal (Morocco): What is the most important part, Essential Oil or the rest of seeds?. Journal of Materials and Environmental Science. 2014;5(6):2017-2020.

21. Ainane T, Abourriche A. Brown Seaweed Bifurcaria bifurcata: Bioguided Fractionation of Extracts by Antibacterial Activity and Cytotoxicity Test. Biosciences Biotechnology Research Asia. 2014;11(3):1081-1085. DOI: 10.13005/bbra/1492.

22. Zheng CJ, Yoo JS, Lee TG, Cho HY, Kim YH, Kim WG, et al. Fatty acid synthesis is a target for antibacterial activity of unsaturated fatty acids. FEBS Lett. 2005;579(23):5157-62. doi:10.1016/j.febslet.2005.08.028.

23. Gavaric N, Mozina SS, Kladar N, Bozin B. Chemical Profile, Antioxidant and Antibacterial Activity of Thyme and Oregano Essential Oils, Thymol and Carvacrol and Their Possible Synergism. Journal of Essential Oil Bearing Plants. 2015;18(4):1013-1021. DOI: 10.1080/0972060X.2014.971069.

24. Casella S, Leonardi M, Melai B, Fratini F, Pistelli L. The role of diallyl sulfides and dipropyl sulfides in the in vitro antimicrobial activity of the essential oil of garlic, Allium sativum L, and leek, Allium porrum L. Phytother Res. 2013;27(3):380-3. Doi: 10.1002/ptr.4725.

4

Understanding Urea Assimilation and its Effect on Lipid Production and Fatty Acid Composition of *Scenedesmus* Sp

Saumya Dhup,[1*] **Dheeban C. Kannan**[2] **and Vibha Dhawan**[2]

[1]*Centre for Bioresources and Biotechnology, TERI University, 10, Institutional Area, Vasant Kunj, New Delhi 110070, India*
[2]*Biotechnology and Management of Bioresources Division, the Energy and Resources Institute (TERI), India Habitat Centre, Lodhi Road, New Delhi 110003, India*

Corresponding author: *Saumya Dhup, Centre for Bio resources and Biotechnology, TERI University, 10, Institutional Area, Vasant Kunj, New Delhi 110070, India, E-mail: saumya.dhup.18@gmail.com*

Abstract

Much advancement has been made in reducing the cost of large-scale production and harvesting of bio fuels. Nutrients contribute to one of the major cost components in algal production; hence, it is essential to understand the importance of minimizing costs at the nutrient level. In the current study, evaluation of urea as a low-cost and efficient source of nitrogen was investigated. The effect of its uptake mechanisms on growth, change in lipid productivity and fatty acid composition of *Scenedesmus* sp. was explored. Total Disappearance Rate (TDR) and total uptake rate were used to study the efficiency of urea as a nitrogen source. It was found that the nutrient uptake efficiency of urea was higher than that of nitrate. In addition, urea showed an increase in biomass productivity and lipid productivity by 26% and 45%, respectively, when compared to control. Results also demonstrated degradation of urea into ammonia upon uptake by algal cells, which is then diffused out of the cells into the medium. Ammonia present in the medium is converted into ammonium ion with simultaneous degradation into nitrate. Eventually, both ammonia and nitrate are absorbed by cells in the final growth phase. These results suggest that urea can be effectively used as an alternative nitrogen source.

Keywords: Scenedesmus sp; Urea uptake mechanism; Nutrient uptake efficiency; Lipid productivity; Fatty acid composition

Introduction

Microalgae are known for their remarkable potential as a feedstock for biofuels and proteins due to their environment-friendly conversion and enhanced yields. However, the challenges faced while using microalgae at every step of production add to the cost of the end F product. Thus, realizing the importance of overcoming these challenges, constant efforts are being made to develop efficient methods for reducing the cost. High lipid productivity of dominant, fast-growing algae is a major prerequisite for the commercial production of microalgal biodiesel.

Biomass generation by photosynthetic microalgae varies depending on a number of environmental factors, including nitrogen concentration, temperature and light intensity. Nitrogen is one of the essential nutrients critical for the cultivation of algae due to its role in growth and regulation of metabolism [1]. It is well documented that there are three different sources of nitrogen such as nitrate, ammonia and urea, which are absorbed by microalgae [2,3]. Urea is known to contain approximately 46.7% nitrogen content, that is, 0.467 Kg of urea–N and hence it is considered as a low-cost and efficient form of nitrogen when compared to other sources [4,5]. Under typical water conditions, urea gives rise to nutrients such as N, P and C (100%) and thus is supplied as a chemical fertilizer to the culture medium for algal uptake [4]. Conversely, nitrate (sodium or potassium salt) provides only 16% of nitrogen, 100% of which is readily available for the culture due to its efficient water-soluble property [5]. The above-mentioned increase in the availability of nitrogen source through urea as compared to other sources of nitrogen is observed due to the presence of two nitrogen atoms per mole of urea. Urea also proves to be an efficient alternative source to other nitrate forms as its cost of production and price per kg of N available is much lower (price per kg of sodium nitrate is 3.1 times that of urea; Table 1).

The mechanism of uptake of each nitrogen source varies, eventually changing its pathway for metabolism. Previous studies suggest that the nitrate and ammonium sources have the simplest uptake mechanisms, as they are taken up as nitrate and ammonium ion, respectively, from the medium under sufficient sunlight during photosynthesis; however, much less is known about urea uptake [6]. In addition, urea, as a nitrogen

Table 1: Comparison of cost per kg of urea and sodium nitrate with respect to availability of nitrogen (N)[23,24].

	Cost (Rs/ kg)	N available (%)	Cost of N available (Rs/ kg)
Urea	5–9	46.7	11.56–19.26
Sodium nitrate	31–37	16.5	187.8–224.2

source, cannot be directly assimilated, thus involving a complex mechanism [7]. Suggested that carbon dioxide and ammonia released upon urea degradation are eventually absorbed by microalgae [7]. Conversion of urea, its removal and uptake from the culture have seldom been reported.

$$NH_2-CO-NH_2 + H_2O \qquad CO_2 + 2NH_3$$
$$NH_3 + H_2O \qquad NH_4^+ + OH^-$$

Both ammonia-based chemical fertilizers and urea accumulate ammonia in water in an unionized form. Therefore, in order to study the effect of an alternative source of nitrogen such as urea on the algal cell growth and understand the mechanism of conversion of urea into other nitrogen sources and their uptake, experiments were performed in a semi-continuous culture system of *Scenedesmus* sp. The effect of urea on growth, lipid productivity and fatty acid composition were studied to evaluate its efficacy as a major nitrogen source.

Materials and Methods

Microalgal strain and culture conditions

The freshwater microalgae *Scenedesmus* sp. (TX3) was procured from Hauz Khas lake park, Hauz Khas, India. The culture was maintained in Bold's basal medium (BBM) which consisted of the following components (g/l): KH_2PO_4 (0.175), $CaCl_2.2H_2O$ (0.025), $MgSO_4$ (0.075), K_2HPO_4 (0.075), NaCl (0.025), H_3BO_3 (0.115), $NaNO_3$ (0.25), Na_2EDTA (0.001), KOH (6.2), $FeSO_4$ (4.98) and conc. H_2SO_4 (1 ml) [8]. The pH of the medium was adjusted to 6.8. The medium was sterilized at 121°C for 15 min. The axenic cultures were maintained in 250-ml glass jars as stock and incubated at 25 ± 2°C.

Experimental design

Experiments were designed with four biological replicates in two similar batches – experimental and control in a semi-continuous mode. The factors differentiating the batches were based on alternate sources of nitrogen such as urea in experimental batch and nitrate in control batch. Duplicates for each biological replicate were taken for measurement at every step. Sodium nitrate has already been reported to be a typical nitrogen source for BBM; therefore, it was considered as control [8]. A pre-incubation step was carried out in which sodium nitrate was replaced with urea for acclimatization of the strain in medium with nutrient urea. This culture was then used as an inoculum and added into a 1-L Erlenmeyer flask containing a working volume of 800 ml. An initial inoculum of the strain was added at a proportion of 10% (v/v) into the solution. As shown in Figure 1, the experiment was conducted for three passages. The results of the third passage were recorded for analysis. The inoculum used for the third phase was 50% (v/v) of the previous passage. Therefore, the culture for the final phase was inoculated at a proportion of 50% (v/v). The culture flasks were maintained under continuous agitation at 160 rpm under 16:8 h light–dark photoperiod conditions. Removal of different sources of nitrogen from the medium including ammonium, nitrate and urea was evaluated. The obtained data were analysed statistically using one-way analysis of variance (ANOVA).

Analytical method

Growth Measurements: Culture growth was observed by measuring the optical density at 680 nm using a Shimadzu spectrophotometer. Sampling (3 ml) was done every 4 days for growth kinetics and in the late exponential phase for lipid and fatty acid methyl ester (FAME) analyses. The dry cell weight was determined by filtering 3 ml of the cell culture through a 0.22-mm-pore-size glass fibre filter (oleic acid). The filters with the biomass were washed with water and dried in a hot air oven at 100°C. The increase in weight of the dry filter was measured.

Nutrient Analysis: Culture samples (3 ml) were taken at an interval of 4 days until they were harvested for nutrient measurements. The samples collected were filtered through glass fibre filters for analysis. Nutrients were analysed photo metrically using the standard methods developed by the American Public Health Association (APHA). The total disappearance rate (TDR) was calculated by summation of the disappearance rates of all nitrogen forms in a particular medium. The nutrient disappearance rate (NDR, mg-Nutrient/L/day) was calculated as follows:

$$NDR = (C_o - C_t) / t$$

Where C_o and C_t are the nutrient concentrations (mg/ l) at the beginning and end of the experiment, respectively, and t is the time interval (days).

The percentage of stored nitrogen reserves within an algal cell was calculated using a CHN elemental analyser (Thermo Finnigan, San Jose, CA, USA). The nutrient uptake rate (NUR, mg-Nutrient/L/day) refers to the amount of nutrient taken up by unit mass of algae and it was calculated on the basis of dry cell weight as follows:

$$NUR = (D_f \times N_f - D_i \times N_i) / t$$

Where D_f and D_i are the final and initial dry cell weight in mg/l, respectively, and N_i and N_f denote the percentage of the

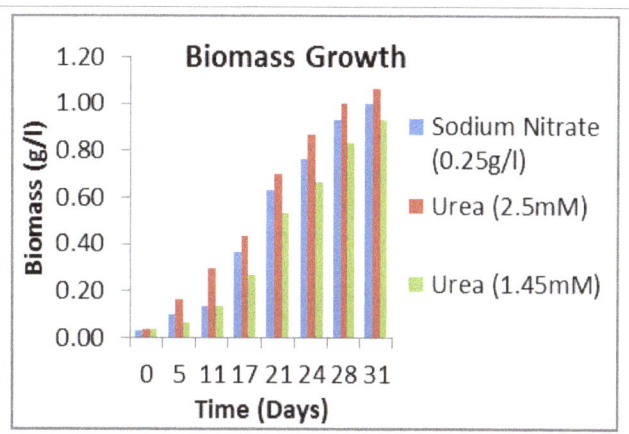

Figure 1: Optimization of two different concentrations of urea, graph represents the biomass growth for sodium nitrate and urea at two different concentrations.

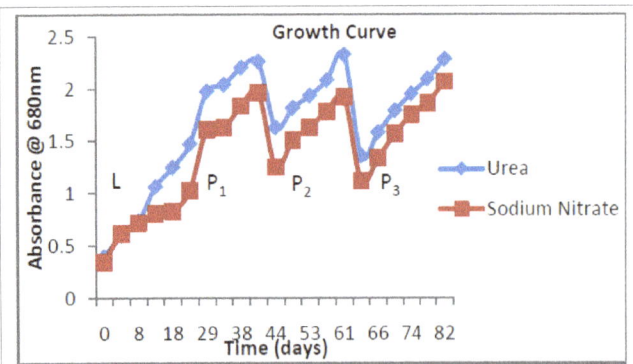

Figure 2: Periodic fluctuation of growth in semi-continuous culture: L represents pre-culture period; P1, P2 and P3 are nutrient recharge and biomass removal intervals.

initial and final concentration of nitrogen, respectively, present in the algal biomass.

Lipid extraction, esterification and fatty acid analysis

The two batches of samples were harvested by continuous centrifugation (8000 rpm for 5 min). The concentrated algal samples were frozen overnight at −80°C and freeze-dried under vacuum. The algal mass was accurately weighed, and lipid extraction was carried out according to the method adopted by [9].

Lipid content (%) and lipid productivity (g/L/day) were calculated as follows:

Lipid content (C_{lipid}) = (wt. of lipid / wt. of sample) × 100

Lipid productivity = (C_{lipid} × DCW) / t

Where, C_{lipid} is the lipid content (%), DCW the dry cell weight (g/l) and t the time interval (days).

Lipid extracts were converted to methyl esters with methanolic HCl and hexane. FAMEs were prepared by adding 1 ml of concentrated HCl along with 5 ml methanol to the methyl esters. The mixture was heated at 80-90°C in a water bath for 30 min. Then, 1 ml of hexane was added to the vial after methylation. The top hexane layer containing the methyl esters was placed into gas chromatography (GC) vials for subsequent GC analysis (Agilent 6890N, Agilent Technologies, Palo Alto, CA, USA) using a DB-5 column (0.2mm ID, 30m, 0.25mm film thickness; Agilent Technologies, Palo Alto, CA, USA) equipped with a flame ionization detector. The temperature programme consisted of an initial temperature of 2°C and followed by an increase of 50°C min^{-1} up to 250°C. The peaks were integrated using the Chemstation and identified by comparing the retention times with the pure standard (Sigma). The system's performance was checked with blanks and standard samples prior to analysis. Concentrations were expressed in mg/ml and then converted to percentage. All the tests were performed in triplicates.

Results and Discussion

Effect of urea on growth kinetics

BBM, best known for culturing green algae, provides an optimal environment for a large variety of microalgal species. It is essential to understand that each algal species has specific culture conditions, and thus its optimization is very crucial. As nutrients are one of the major cost components in algal production, they have to be optimized efficiently for a higher productivity. An optimal nutrient condition is attributed to the combined effect of both concentration and source of nutrient [10]. Conducted a study on *S. bijugata* and reported the significance of selecting a suitable nitrogen source and concentration for rapid growth [10]. The authors also stated that the association between the source of nitrogen and concentration is significant. In this study, an alternative source of nitrogen was considered because nitrogen, being one of the major basic elements, often limits growth and accumulation of lipids. The effect of urea on growth and lipid productivity of *Scenedesmus* sp. was investigated. The study was carried out to evaluate the use of a cost-efficient alternative of nitrogen in the medium. The amount of urea and sodium nitrate added in the medium is indicated in Table 2. The concentration of urea in the medium was optimized previously for growth and lipid productivity. In an earlier study, two different molar concentrations of urea were tested: 1.45 mM (equiatomic to sodium nitrate) and 2.5 mM (equivalent weight basis: 17.6 mM). As the urea concentration (2.5 mM) in the medium was higher than the equiatomic concentration of sodium nitrate, it resulted in higher biomass growth (Figure 1) and lipid productivity (12.4 mg/l/day). The lipid productivity with urea concentration of 1.45 mM and sodium nitrate was 7.3 mg/l/day and 8.8 mg/l/day respectively. It can be inferred that an increased concentration of urea provides an increased amount of nitrogen which leads to higher growth and productivity [10]. Observed that an optimal higher concentration of urea results in better biomass growth [10]. Therefore, the concentration of nitrogen available after conversion of urea is observed to be optimum, thereby enhancing lipid productivity. Urea is also known to enhance growth rate as it acts as a complementary source of organic carbon. Thus, the combined effect of nitrogen and carbon simultaneously leads to an increase in the growth of *Scenedesmus* sp.

The concentration of the nutrient in the medium varies according to the nutrient source used. A change in the nutrient source might lead to an increased concentration in the medium above the tolerance level of the algal cells; such high concentrations are toxic and inhibit cell growth [11]. But an appropriate amount of nitrogen concentration is known to increase the microalgal growth rate [12]. In this study, it was observed that the difference in the available forms of the nitrogen

Table 2: Growth parameters and productivity of *Scenedesmus* sp.

N Source	Urea	Sodium Nitrate
Amount (mM)	2.5	2.9
Biomass productivity (g/l/day)	0.048	0.038
Specific growth rate (day^{-1})	0.027	0.018
Lipid content (%)	8.7 ± 0.4	7.6 ± 0.6
Lipid productivity (mg/l/day)	4.176	2.88

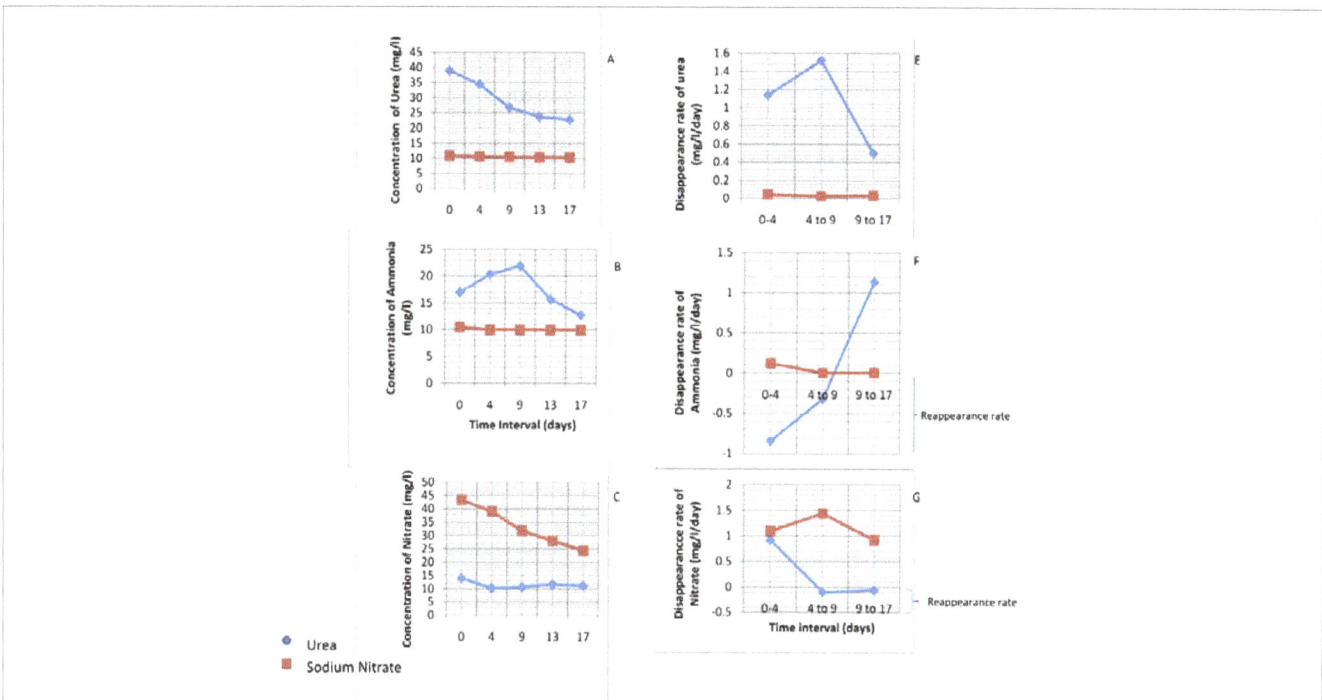

Figure 3: Disappearance of nitrogen forms from the medium containing urea (experimental) versus sodium nitrate (control).
Figures A, B and C show concentrations of urea, ammonium and nitrate at disappearance, respectively
E, F and G show the disappearance rate of urea, ammonium and nitrate, respectively.

for intake leads to a change in the nutrient concentration in the medium and thus causes variations in the assimilation rate. These alterations subsequently affect the growth rate of the algal species.

Growth curves for *Scenedesmus* sp. grown in two nitrogen sources are presented in Figure 2. In the medium containing urea, a higher exponential increase in the growth curve is observed with a specific growth rate of 0.06 day^{-1}. However, in the medium with sodium nitrate, specific growth rate was 0.05 day^{-1}. Therefore, it is evident from Table 2 that the specific growth rate of this species is significantly affected by the nitrogen source used ($P < 0.05$), as the growth rate with urea in the medium is 1.4 times higher than with nitrate. These results suggest that urea is more preferred as a nitrogen source in the medium [13]. Reported a similar growth pattern in the *Chlorella* sp. with urea and sodium nitrate as the two nitrogen sources and their specific growth rates being 0.071 day^{-1} and 0.047 day^{-1}, respectively [10,13]. In their study on *Scenedesmus* sp. also reported that the amount of biomass obtained with urea in the medium was equivalent to that with nitrate [10]. Previous reports on phytoplankton species suggest that urea is degraded into ammonia inside the cells [13]. In this study, ammonia, a product of urea degradation in the medium with urea, is being preferred over nitrate in the batch having sodium nitrate as the nitrogen source [14]. Observed a similar result wherein some microalgal species prefer ammonium ion over nitrate [14,15]. Reported that *Scendesmus* sp. grew faster in the stable phase of growth when the ammonium form of nitrogen was provided in the medium [15]. They stated that in accordance with the specific growth rate, the preference of nitrogen source

is in the order of ammonium < urea < nitrate. Ammonia is considered as a preferred source for intake by the cells because it is directly absorbed into the cell and converted into amino acids which are eventually used for metabolic activities. When nitrate as a nitrogen source is absorbed by the cells, it reaches the cytosol where it gets reduced to nitrite by nitrate reductase. Nitrite produced is transported immediately to the chloroplast where it is further reduced to ammonium ion by nitrite reductase [16]. Therefore, the algal cells prefer a source which is directly assimilated and used for metabolic activities. The results suggest that ammonium ion is a much preferred nitrogen source for *Scenedesmus* sp. Therefore, the production of ammonium ions during urea degradation makes this nutrient more suitable for algal growth than sodium nitrate.

The above-mentioned results show a variation in the growth characteristics based on two different nitrogen sources, and a change in the biomass concentration is also expected. As shown in Table 2, a difference between the biomass productivity of urea and sodium nitrate was observed. The medium with urea showed an increase in biomass productivity by 26% in 17 days with a biomass concentration of 0.832 g/l. Similar results with an enhanced biomass dry weight were reported with urea in *Scenedesmus* sp [17]. They reported that *Scendesmus* sp. performed better in a medium with urea, resulting in a significantly higher biomass concentration and lipid productivity when compared to nitrate. It was also observed that the percentage of nitrogen (calculated using an elemental analyser; Thermo Finnigan, San Jose, CA, USA) inside the cells was higher in the medium with urea (8.179% in medium with urea and 7.219%

Table 3: Nutrient disappearance and uptake efficiency for *Scenedesmus* sp. in medium with urea and sodium nitrate.

	Total Disappearance Rate (mg/l/day)			Total Uptake Rate (mg/l/day)			Total Disappearance Rate	Nutrient Uptake Rate	Nutrient Uptake Efficiency (%)
Days	0–4	4–9	9–17	0–4	4–9	9–17	0–17		
Urea	1.22	1.11	1.58	1.11	1.1	1.54	1.3636	1.3117	96.00%
Sodium Nitrate	1.26	1.47	0.96	1.12	1.45	0.9	1.1845	1.1182	94.00%

with nitrate). This suggests that using urea as the nutrient source leads to a higher nitrogen uptake efficiency, thereby increasing the amount of nitrogen in vivo and also resulting higher biomass concentration.

Uptake mechanism of urea and sodium nitrate

This study focuses on urea as an alternative nitrogen source and its impact on growth, lipid production and fatty acid composition. The impact is observed largely due to the difference in the uptake mechanisms of the nitrogen sources such as urea and sodium nitrate. Studies related to the uptake mechanism of urea by algal species are limited. In 1988, Price and Harrison conducted a study proposing a model of urea uptake and its assimilation by a diatom, *Thalassiosira pseudonana* in nitrate-sufficient and nitrate-starved cells [13]. They have explained the diffusion of urea through an ammonia conversion route using radioactive isotopes. They proposed that upon cellular uptake, urea gets degraded into ammonia which is released. The ammonia reacts with water and is converted into ammonium ion, thereby releasing hydroxide ion in the medium. This mechanism can be further explored in other species in order to assess and confirm the consumption of urea by other phytoplankton communities. In this study, a similar trend was observed using different analytical methods for uptake and assimilation. It has been previously studied that ammonia-based chemical fertilizers such as urea accumulate a unionized form of ammonia in the medium. The unionized form of ammonia is taken up biologically by the algal culture during photosynthesis in sufficient light intensity before it gets oxidized to nitrate, thus maintaining the culture alkalinity [18]. But it was also observed that a small quantity of ammonia which was diffused out from the cells was converted into ammonium ion and nitrate simultaneously [18]. The nitrate and ammonium ion were eventually reabsorbed by the cells. Thus, based on this study, it can be assumed that under light and dark conditions both nitrogen sources, ammonium ion and nitrate, respectively, are simultaneously available in a medium with urea.

The rate of urea disappearance from the medium after 4 days averaged 1.14 mg/l/day. But the maximum disappearance rate of urea from the medium was recorded as 1.52 mg/l/day for an interval of 4-9 days. With the disappearance of urea, there was an exponential increase in the reappearance of ammonium ion in the medium (Figure. 3E and F). This suggests that urea absorbed by the cells was degraded *in vivo* into ammonia. Ammonia was then diffused out into the medium. Degradation of urea inside the cell was carried out in the presence of urease enzyme present in most

algal cells [19]. The ammonia diffused out was eventually taken up as ammonium ion, as during the final interval of cell growth, the rate of disappearance of ammonium ion from the medium was observed to be 1.13 mg/l/day (Figure 3F). While ammonia disappeared during the final phase, simultaneous reappearance of nitrate in the medium was observed at a rate of 0.1 mg/l/day (Figure 3G).

When considering the rate of uptake, the overall disappearance of nitrogen source from the medium was taken into account over the period. The uptake rate in the medium with urea for an interval of 0-4 days was observed to be more than that for a period of 4-9 days (1.11 mg/l/day and 1.10 mg/l/day, respectively). The TDR was also observed to be higher for the interval of 0-4 days (Table 3). In the final phase of growth, the maximum disappearance rate was observed due to the simultaneous uptake of both ammonium ion and nitrate nitrogen from the medium with urea. Thus, an increase in the rate of uptake of the nitrogen source was observed. In the medium with nitrate, the maximum disappearance and uptake rates was observed for an interval from 4 to 9 days. It was also observed that the nutrient uptake efficiency of medium with urea was 96%, while that of medium with sodium nitrate was 94% (Table 3). Thus, the results indicated an enhanced uptake of nutrient from the medium with urea in comparison to the medium with sodium nitrate. It can be concluded that urea is a better nitrogen source as it is being efficiently absorbed and utilized for metabolic functions leading to enhanced growth rate and lipid productivity.

Effect of urea on lipid productivity and fatty acid composition

Biomass productivity is considered as an imperative parameter for determining the effect of a condition. Lipid content is another parameter which is essential for the evaluation of the overall performance of the strain. In the current study, the lipid content and biomass were considered as the twin effects for evaluating the overall lipid productivity under different nutrient sources.

We proceeded to analyse the total lipid content of the strain in both media (experimental and control) by the end of the cycle where minimal nutrients were left in the medium. Harvesting in the late exponential phase leads to a decrease in the nitrogen and phosphorous concentrations but an increase in the lipid content [12]. Table 2 shows the total lipid content and productivity of the two media. It was observed that the total lipid content of the batch with urea (8.7%) was higher than that of sodium nitrate

Table 4: Fatty acid profile of *Scenedesmus* sp. in medium with urea and sodium nitrate.

	Urea (%)	Sodium nitrate (%)
C14	0.36	0.76
C14:1	0.13	0.39
C15	0.41	0.44
C15:1	1.86	2.78
C16	5.40	9.00
C16:1	3.49	17.70
C17	0.73	0.75
C17:1	1.58	2.51
C18	0.00	0.00
C18:1	14.16	8.99
C18:2 TRANS	1.05	5.44
C18:3	61.04	37.84
C18:3 (6)	2.55	4.13
C18:3 (3)	2.79	3.97
C20	2.60	0.96
C22:2	0.58	2.03
C20:2	0.60	0.80
C21:1	0.52	0.71
C24:1	0.16	0.40
SFA	9.50	11.90
MUFA	21.90	33.47
PUFA	68.60	54.23

(7.6). Due to a 14% increase in the lipid content of the strain in the medium with urea, a difference between the lipid productivity of the two media was observed. The lipid productivity in the batch with urea was found to be 1.5 times higher than that with sodium nitrate. It was inferred that the lipid content increases with an increase in biomass productivity. Hence, the lipid productivity of the cells in the batch with urea increased. Similar results were also indicated in previous studies carried out by [20]. Thus, the results listed in Table 2 shows that the lipid content and productivity were significantly dependent on the nitrogen source provided ($P > 0.05$). These results indicate that nitrogen in the medium with urea should be provided in sufficient amount for the cells to grow and store lipids efficiently. More importantly, in the medium with urea, the presence of ammonia makes it possible for the cells to absorb it as a direct nitrogen source for assimilation. Direct assimilation of ammonia facilitates rapid cell metabolism, thereby increasing the lipid productivity.

Fatty acid composition is another key parameter that determines the potential of an algal strain as a biodiesel candidate. In this particular study, fatty acid analysis of the species in the two different media gave interesting insights with a prominent shift in the fatty acid composition. The major saturated, monosaturated and polyunsaturated fatty acids were found to be C16:0, C16:1 and C18:1 and C18:3, respectively. According to the previous

studies conducted on *Scenedesmus* sp., stearic acid was absent in both media [21]. As shown in Table 4, there is a significant visible shift of the fatty acid percentage from C16:0 and C16:1 to C18:1 and C18:3 in the medium with urea; the percentage of the latter increased by 5 and 23, respectively [22] (Table 4). Observed a similar shift from palmitic acid to oleic acid in other feedstock soybean, but stated that such an increase in oleic acid was not due to environment influences [22]. But, herein, the shift to oleic acid with a significant increase in the level of an essential omega-3 fatty acid (alpha-linolenic acid) was observed due to the change in nitrogen source. An increase in oleic acid makes the strain desirable as a potential source for biodiesel, as monounsaturated fatty acids are most suitable when considering cloud point and oxidative stability. In addition, an increased amount of saturated fatty acids poses problems with gelling of fuel in cold weather.

Due to the presence of a high amount of omega-3 fatty acids, this strain is an important source of essential nutrients and also presents health benefits apart from biodiesel production; however, it negatively influences the oxidative stability. Therefore, the percentage of essential fatty acids of the total algal lipids in urea was 94% (significant percentages of α-linolenic acid and oleic acid were 61% and 14%, respectively), whereas in the medium with sodium nitrate, it was 74%. Hence, it can be concluded that with a change in the nitrogen source from nitrate to urea, there is an increase in lipid productivity with a positive shift in the fatty acid composition from non-essential to essential fatty acids.

Conclusion

The growth and lipid productivity of green algae *Scenedesmus* sp. was tested in BBM medium containing a low-cost nitrogen source. Our findings shed light on the use of urea as a nitrogen source, as it proved to be more efficient in terms of growth, lipid productivity and fatty acid composition when compared to sodium nitrate. With regard to the nutrient uptake mechanism and fate of urea in *Scenedesmus* sp., it was observed that the results are in compliance with previous studies conducted on other phytoplankton species, with a reasonably higher nutrient uptake efficiency than nitrate. Although sodium nitrate is a readily available source of nitrate ion, the ammonium ion obtained after urea degradation is a much-preferred source of nitrogen for uptake. In this study, the presence of nitrate as an additional nitrogen source along with ammonium ion was observed. The nitrogen obtained upon degradation of ammonia was also observed to be absorbed by the cells. Hence, urea can be effectively used as an alternative economical substitute of nitrogen for growth and lipid production of *Scenedesmus* sp. at a large scale.

Highlights

- Urea is considered as a low-cost alternative source of nitrogen.

- A higher uptake efficiency of urea (two nitrogen sources including ammonium ion and nitrate) is observed when compared to sodium nitrate.

- A change in the fatty acid composition is observed upon urea uptake.

Acknowledgement

We sincerely thank the Department of Biotechnology, Government of India for funding the research. We also thank the Director General, TERI, for providing the necessary infrastructural services and facilities for this study as well as Late Dr. Prem Dureja, Ms V Devi and Ms Swati Patel for their valuable guidance during the research.

References

1. Xian-Ming Shi, Feng Chen, Jian-Ping Yuan, Hui Chen. Heterotrophic production of lutein by selected *Chlorella* strains. J Appl Phycol. 1997;9(5): 445-50.

2. Ludwig, C. A. The availability of different forms of nitrogen to a green alga. Am J Bot. 1938;25(6):448-58.

3. Prabakaran P, Ravindram AD. Influence of different carbon and nitrogen sources on growth and CO2 fixation of microalgae. Adv Appl Sci Res. 2012;3 (3): 1714-17.

4. Knud-Hansen, Batterson CF TR, CD, McNabb S. Harahat, K. Sumantadinata and H.M. Eidman . Nitrogen input, primary productivity and fish yield in fertilized freshwater ponds in Indonesia. Aquacul. 1991;94: 49-63.

5. Harben PW, Theune C. Nitrogen and nitrates. In: Kogel JE, Trivedi NC, Barker JM. and Krukowski ST. (eds.) Industrial Minerals & Rocks: Commodities, Markets, and Uses, Littleton, CO, Society for Mining, Metallurgy and Exploration, Inc. 2006;671-78.

6. Kim SY, Lee, Hwang SJ. Removal of nitrogen and phosphorous by *Chlorella sorokiniana* cultured heterotrophically in ammonia and nitrate. Intl Biodeterio Biodegrad. 2013;85:511-16.

7. Price NM, Harrison PJ. Uptake of urea C and urea N by the coastal marine diatom *Thalassiosira pseudonana*. *Limnol oceanogr.* 1988;33(4): 528-37.

8. Bischoff HW, Bold HC. Phycological studies IV. Some soil algae from enchanted rock and related algal species. University of Texas Publication. 1963;6318: 1–95.

9. Folch J, Lees M, Sloane Stanley GH. A simple method for the isolation and purification of total lipids from animal tissues. J Biol Chem. 1957;226(1):497-509.

10. Arumugam M, Agarwal A, Arya MC, Ahmed Z. Influence of nitrogen sources on biomass productivity of microalgae *Scenedesmus* sp. Bioresour Technol. 2013;131:246-9. doi: 10.1016/j. biortech.2012.12.159.

11. Xu D, Gao Z, Zhang X, Qi Z, Meng C, Zhuang Z. Evaluation of the potential role of the macroalga *Laminaria japonica* for alleviating coastal eutrophication. Bioresour Technol. 2011;102(21):9912-8. doi: 10.1016/j.biortech.2011.08.035.

12. Dhup S, Dhawan V. Effect of nitrogen concentration on lipid productivity and fatty acid composition of *Monoraphidium* sp. Bioresour Technol. 2014;152:572-5. doi: 10.1016/j.biortech.2013.11.068.

13. El-Sayed, AB, Abel-Maguide AA. Growth response of *Chlorella vulgaris* to acetate carbon and nitrogen forms. Nature Sci. 2011;9(9).53.

14. Kang YH, Park SR, Chung IK. Bio filtration efficiency and biochemical composition of three seaweed species cultivated in a fish-seaweed integrated culture. Algae. 2011;26(1):97-108. DOI: 10.4490/ algae.2011.26.1.097.

15. Xin L, Hong-ying H, Ke G, Jia Y. Growth and nutrient removal properties of a freshwater microalga *Scendesmus* sp. LX1 under different kinds of nitrogen sources. Ecol Eng. 2010;36(4): 379-81.

16. Sakihama Y, Nakamura S, Yamasaki H. Nitric oxide production mediated by nitrate reductase in the green alga *Chlamydomonas reinhartii*; An alternate no production pathway in photosynthetic organisms. Plant Cell Physiol. 2002;43(3):290-7.

17. Ying Shen, Zhijian Pei, Wenqiao Yuan, Enrong Mao. Effect of nitrogen and extraction method on algae lipid yield. Intl J Agricul Biol Eng. 2009;2(1): 51-57. DOI: 10.3965/j.issn.1934-6344.2009.01.051-057

18. Sugiyama M, Kawai A. Microbiological studies on the nitrogen cycle in aquatic environments-VI. Metabolic rate of ammonium nitrogen in a goldfish culturing pond. B Jap Soc Sci Fish. 1979;45(6):785-789. DOI: 10.2331/suisan.45.785.

19. Bekheet IA, Kandil KM, Shaban NZ. Studies on urease extracted from *Ulva lactuca*. Hydrobiol. 1984;116(1): 580-83

20. Ren HY, Liu BF, Ma C, Zhao L, Ren NQ. A new lipid- rich microalga *Scenedesmus* sp. strain R-16 isolated using Nile red staining: effects of carbon and nitrogen sources and initial pH on the biomass and lipid production. Biotechnol Biofuels. 2013;6(1):143. doi: 10.1186/1754-6834-6-143.

21. Choi KJ, Nakhost Z, Krukonis VJ, Karel M. Supercritical fluid extraction and characterization of lipids from algae *Scenedesmus obliquus*, Food Biotechnol. 1987;1(2):263-81.

22. Graef G, LaVallee BJ, Tenopir P, Tat M, Schweiger B, Kinney AJ, et al. A high-oleic-acid and low-palmitic-acid soybean: agronomic performance and evaluation as a feedstock for biofuel. Plant Biotechnol J. 2009;7(5):411-21. doi: 10.1111/j.1467-7652.2009.00408.x.

23. Market price. http://www.needsinfo.com/globalinquirymagazine/ market_price.htm

24. Chemical prices remain steady. Buisness standards. http://www. business-standard.com/article/markets/chemical-prices-remain-steady-111020800160_1.html.

5

Why is Hippocampal CA1 Especially Vulnerable to Ischemia?

Hanbai Liang[1], Shota Kurimoto[1], Kosuke R Shima[2], Hiroki Shimizu[2], Tsuguhito Ota[2], Yoshio Minabe[3] and Tetsumori Yamashima[1,3]*

[1]Department of Restorative Neurosurgery and Psychiatry
[2]Department of Cell Metabolism and Nutrition, Brain/Liver Interface Medicine Research Center
[3]Department of Psychiatry and Neurobiology, Kanazawa University Graduate School of Medical Science.
13-1 Takara-machi, 920-8640, Kanazawa, Ishikawa, Japan

*Corresponding author: Tetsumori Yamashima, Department of Restorative Neurosurgery and Psychiatry, Kanazawa University Graduate School of Medical Science, Kanazawa, Japan, E-mail: yamashima215@gmail.com

Abstract

Since the formulation of 'calpain-cathepsin hypythesis' in 1998, calpain-mediated lysosomal rupture has been accepted to explain the mechanism of neuronal death. As Hsp70.1 contributes to the chaperone function of aged/damaged proteins and membrane stabilization of lysosomes, its depletion can induce neurodegeneration via autophagy failure and lysosomal destabilization. The cleavage assay in vitro previously showed that Hsp70.1 is a substrate of activated μ-calpain, especially after its carbonylation by the lipid peroxidation product - hydroxynonenal (HNE). The hippocampal CA1 neurons are known to be especially vulnerable to the ischemic insult, but the underlying mechanism still remains incompletely elucidated. Here, using various primate brain tissues including thalamus, putamen, medulla oblongata, and CA1, calpain-mediated in-vitro cleavage of Hsp70.1 which eventually leads to the lysosomal rupture was analysed after treatment with or without HNE. In all tissues studied Hsp70.1 cleavage from size 70 kDa to 30 kDa occurred only in the presence of activated μ-calpain, and the increasing HNE treatment caused a stepwise escalation in cleavage. However, in the absence of calpain activation, the protein carbonylation alone, even by the larger HNE doses, failed to cause Hsp70.1 cleavage. Although the pathogenic synergism between calpain activation and Hsp70.1 carbonylation works in concert, μ-calpain is considered to be the principal factor whereas HNE alone could not mediate Hsp70.1 protein cleavage. With the aid of HNE, calpain activation would facilitate lysosomal destabilization by cleaving carobonylated Hsp70.1. Among the 4 brain tissues studied, the CA1 tissue intriguingly showed the minimum Hsp70.1 cleavage. Accordingly, the specific vulnerability of CA1 neurons in the living brain can be explained by their excessive and/or long-standing calpain activation due to the remarkable Ca^{2+} mobilization potential.

Keywords: Neuronal Death; Lysosome; Hsp70.1; Carbonylation; Calpain-Cathepsin Hypothesis

Introduction

Since 1998 when the 'calpain-cathepsin hypothesis' was formulated by Yamashima and his associates [1], it became gradually accepted as a molecular mechanism of necrotic neuronal death [2]. The core of this hypothesis is calpain-mediated lysosomal destabilization/rupture and the resultant release of lysosomal cathepsin enzymes. However, the in-vivo substrate of calpain for inducing lysosomal destabilization had long remained unknown. A molecular chaperone Hsp70.1, a major human Hsp70, also called Hsp72 or HSPA1, is recently known to stabilize lysosomal membrane by recycling damaged proteins and protect cells from oxidative stresses [3,4]. Hsp70.1 is crucial for cell death, because Hsp70.1 gene knockout-mice showed exacerbation infarction size after focal cerebral ischemia [5]. In contrast, enhanced Hsp70 expression in transgenic mice protected the brain and heart from ischemia by an unknown mechanism [6-9].

In both the rodent [10] and primate [11,12]experimental paradigms, Hsp70.1 is susceptible to the oxidative stress-induced modification especially by a lipid peroxidation product hydroxynonenal (HNE). In a rat model of chronic alcohol-induced oxidative stress, Carbone, et al. [13] showed that Hsp72, the inducible variant of Hsp70, treated with 10 and 100 μM HNE caused adduct formation at Cys267 in the ATPase domain of the chaperone by the mass sprectrometrical analysis. Recently, the author's group has suggested that oxidative modification of Hsp70.1 occurs early in the pathogenesis of neuronal death in the postischemic monkey hippocampus. In response to HNE being generated by the oxidative stress, a specific oxidative injury 'carbonylation' occurred at the key site Arg469 of Hsp70.1, which coincides well with the carbonyl increase [11]. Furthermore, analyses of the postischemic hippocampal tissues [14] and the glaucoma-suffered retina [15] in primates showed the same result that Hsp70.1, especially after HNE-mediated carbonylation, is susceptible to cleavage by activated μ-calpain.

Because of its chemical reactivity, HNE can exert pleiotropic effects particularly in cell death. For example, after the ischemia/reperfusion sequence in myocardial infarction, accumulated Reactive Oxygen Species (ROS) promote generation of HNE, which disrupts the actin cytoskeleton, alters Ca^{2+} homeostasis, and triggers cardiomyocyte cell death [16]. HNE induces signaling for apoptosis via both the Fas-mediated extrinsic and the p53-mediated intrinsic pathways [17,18]. So, HNE can trigger the pancreatic β cell apoptosis, induce glucose intolerance and the development of diabetes [19]. As HNE impairs Na^+/Ca^{2+} pumps and glucose and glutamate transporters by modifying membranes, the resultant ionic and energetic disturbances may cause neuronal cell death [20,21]. However, the detailed mechanisms of calpain and HNE synergy as well as contribution of each player for the cell death still remain incompletely elucidated.

The present study aimed at elucidating why the hippocampal CA1 is especially vulnerable to the ischemic insult among the various brain regions by focusing cysteine protease 'calpain' and lysososomal stabilizer 'Hsp70.1'. Here, using four representative brain portions of the non-human primates, the calpain-mediated Hsp70.1 cleavage was compared in-vitro to elucidate (1) [22] how the calpain and HNE synergy affects, (2) [13] whether calpain or HNE alone can induce the Hsp70.1 cleavage, and (3) [17] which is the principal factor. Although the data are artificial, they are helpful for understanding the diverse function of Hsp70.1 protein under the stress condition.

Materials and methods

Using the non-ischemic monkey (*Macaca fuscata*) brain tissues, Western blotting was done to analyse Hsp70.1 cleavage as reported previously [23]. All experimental procedures were performed in strict adherence with the guidelines of the Animal Care and Ethics Committee of Kanazawa University and the NIH Guide for the care and Use of Laboratory Animals. Four young monkeys with a body weight of 6-10kg were bred in air-conditioned cages and allowed free daily access to food and water. Under GOF general anesthesia, the monkeys were sacrificed for the normal brain tissue sampling. We aimed to examine whether activated μ-calpain (calpain-1) can cleave Hsp70.1 being involved in the homogenate tissues of non-ischemic thalamus, putamen, medulla oblongata, and CA1 in vitro, by adding purified μ-calpain (Calbiochem, La Jolla, CA) plus Ca^{2+} mixed with or without HNE (Calbiochem, La Jolla, CA).

For the in-vitro μ-calpain activation, various concentrations (0, 1, 3 mM) of $CaCl_2$ (Wako Pure Chemical, Osaka, Japan) were added with 0.5 units (U) μ-calpain to the homogenate tissue samples (20 or 5 μg) and the recombinant Hsp70.1 protein (200 ng). Calpain buffer comprised of 1 M Tris pH7.2, 100 mM DTT, 10 mM EDTA, and 26 mM EGTA. The μ-calpain activation was stopped by adding 100 μM EDTA and EGTA (Dojindo Laboratories, Kumamoto, Japan).

Subsequently, using various concentrations of HNE (0, 1, 2 mM), it was studied whether oxidative stress-induced Hsp70.1 carbonylation can promote its cleavage by activated μ-calpain. Furthermore, the homogenate brain tissues were incubated for various times (0, 1, 5, 10, 30, 60, 120 m) in 2 mM HNE with activated μ-calpain that was made from 0.5 U μ-calpain after incubation in 3 mM $CaCl_2$. Since the configuration of the Hsp70.1 cleaved bands was sometimes not distinct because of the presence of consecutive bands around 30 kDa (for example, *Figures* 2, 3 and 6), the densitometolytic analysis of the band intensity was not done.

The protein samples from the homogenate brain tissues after given incubation were separated by 15% SDS–PAGE gel (Biocraft, SDG-571) and transferred on the PVDF membrane (ATTO, Tokyo, Japan). Primary antibody was purified mouse anti-human HSP70 that recognizes amino acid 429–640 residue (at a dilution of 1:6,000, BD Transduction Laboratories, 610607, San Jose, California, USA), while the secondary antibody was horseradish peroxidase-conjugated goat anti-mouse IgG (at a dilution of 1:10,000, Santa Cruz Biotechnology, SC-3697, Santa Cruz, USA). The recombinant Hsp70.1 protein (recombinant human Hsp70/ Hsp72, Enzo Life Science, ADI-NSP-555-D) was utilized as a positive control.

Results

First, to determine the optimal amount of purified μ-calpain under 3 mM Ca^{2+} concentration in the absence of HNE, *in-vitro* cleavage of Hsp70.1 was analyzed using 0.2, 0.5 and 1.0 units (U) of μ-calpain to the recombinant Hsp70.1 protein (200 ng) and the thalamus tissue (20 μg). Hsp70.1 cleavage into the 30 kDa fragment occurred by activated μ-calpain alone. Although the calpain-mediated proteolysis was negligible without calpain activation in both the recombinant and the tissue, it increased dose-dependently after calpain treatment (*Figure* 1). There were about 60 kDa protein bands in the recombinant Hsp70.1 cleavage panel, but not in the thalamus tissue (*Figure* 1). This is presumably because 2D structure recombinant protein was prone to calpain cleavage, compared to 3D structure Hsp70.1 protein in the thalamus tissue. Since μ-calpain concentration of 0.5 U appeared to be sufficient in both the recombinant and the tissue for cleaving Hsp70.1 into 30 kDa fragment, 0.5 U μ-calpain was utilized in the following experiments.

Next, the optimal Ca^{2+} concentration necessary for 0.5 U μ-calpain activation was to determined. In all HNE concentrations of 0.5, 1 and 2 mM, 1 mM Ca^{2+} failed to cleave the recombinant Hsp70.1 protein (200 ng)(data not shown), but 3 mM Ca^{2+}

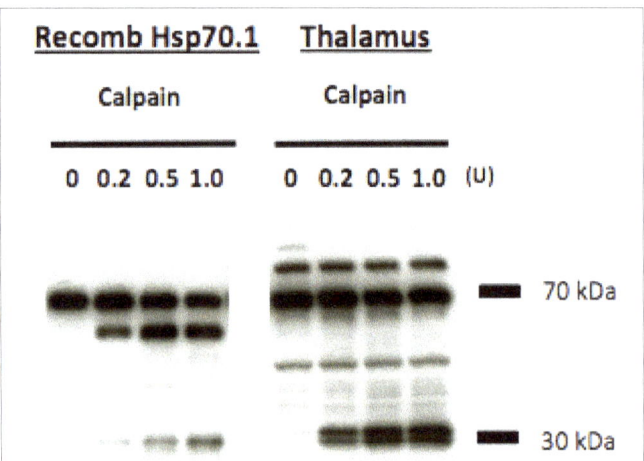

Figure **1:** Using the recombinant Hsp70.1 proteins (200 ng) and the thalamus tissue (20 μg), optimal concentration of purified μ-calpain necessary for Hsp70.1 cleavage (70 kDa to 30 kDa) was estimated to be 0.5 units (U). Activated μ-calpain alone can cleave Hsp70.1 in the absence of HNE.

cleaved the recombinant a little bit at the HNE concentrations of 1 and 2 mM after incubation for 2 hours. In contrast, Hsp70.1 in the thalamus tissue (20 μg) was cleaved sufficiently by 3 mM Ca^{2+} regardless of the HNE concentration. Accordingly, 3 mM Ca^{2+} was utilized in the following experiments. Although the Hsp70.1 cleavage occurred in the absence of HNE, it increased gradually after incubation with 0.5, 1 and 2 mM of HNE (*Figure* 2).

Since 3 mM Ca^{2+} induced sufficient Hsp70.1 cleavage in the thalamus tissues (20 μg), effect of 1 mM Ca^{2+} was studied, but the cleavage was very little (*Figure* 3). So, under 3 mM Ca^{2+} concentration, the effects of HNE concentrations of 0.5, 1 and 2 mM were analysed after incubation for 2 hours. A remarkable Hsp70.1 cleavage into ~30 kDa fragments was seen dose-dependently (*Figure* 3). As the HNE concentration of 2 mM was thought to be sufficient, 2 mM HNE was utilized thereafter.

To clarify whether HNE is indispensable for the Hsp70.1 cleavage,the thalamus tissue (20 μg) was incubated for 2 hours with 0.5, 1 and 2 mM HNE with or without 0.5 U μ-calpain plus 3 mM Ca^{2+}. Even in the absence of HNE, activated μ-calpain could cleave Hsp70.1 into 30 kDa fragment. In contrast, in any concentrations HNE alone could not cleave Hsp70.1 at all (*Figure. 4*). This indicated that HNE is not an indispensable factor for the Hsp70.1 cleavage.

Subsequently, to compare the susceptibility of Hsp70.1 to activated μ-calpain, CA1, thalamus, putamen, and medulla oblongata (in all 4 samples, a smaller amount of 5 μg tissues were utilized for the precise comparison of cleaved band intensities) were incubated with 2 mM HNE for 1~120 min under 3 mM Ca^{2+} concentration plus 0.5 U μ-calpain. First, the most ischemia-vulnerable CA1 (5 μg) was studied with the recombinant Hsp70.1 protein (200 ng) as a positive control. Although the calpain-

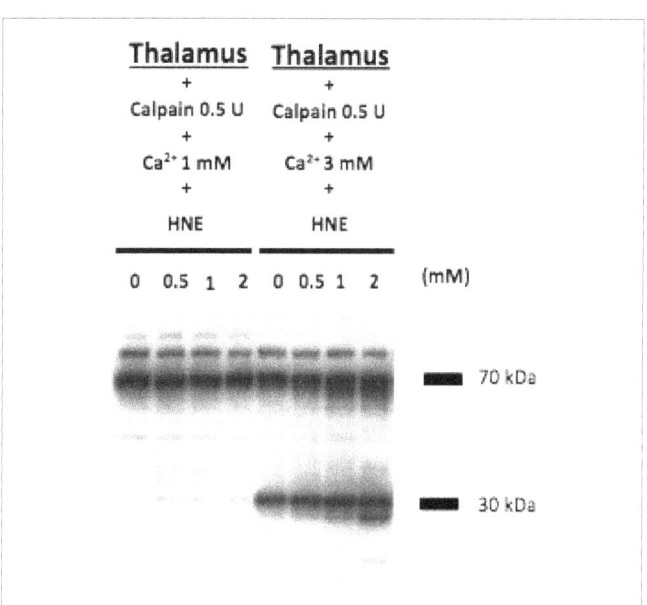

Figure 3: Using the thalamus tissues (20 μg), optimal concentration of hydroxynonenal (HNE) necessary for Hsp70.1 cleavage was estimated to be 2 mM.

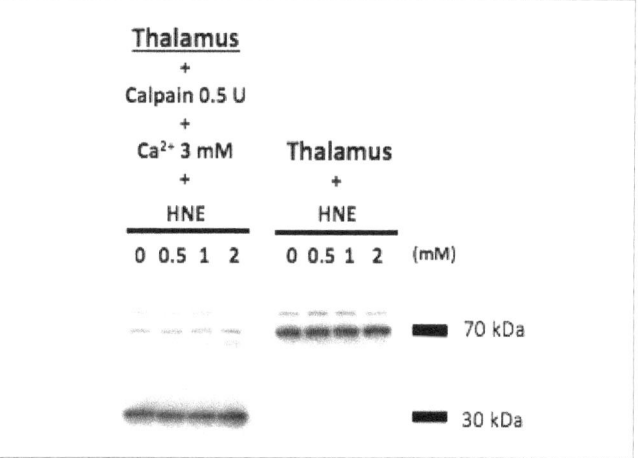

Figure 4: The thalamus tissue (20 μg) showed Hsp70.1 cleavage after incubation with 0.5 U calpain and 3 mM Ca^{2+}, but HNE without calpain and Ca^{2+} showed no cleavage. Accordingly, HNE was thought to be merely an accelerator of the calpain-mediated Hsp70.1 cleavage.

mediated Hsp70.1 cleavage was negligible before HNE treatment in both the recombinant Hsp70.1 protein and CA1, the cleaved band appeared a little bit 5 min after incubation and gradually increased time-dependently (*Figure* 5). Using thalamus (5 μg) (*Figure* 6) and medulla oblongata (5 μg)(*Figure* 7), similar cleavage was shown to occur time-dependently after incubation with 2 mM HNE for 1~120 min under 3 mM Ca^{2+} concentration plus 0.5 U μ-calpain. The same results were obtained using globes pallidus and caudate nucleus (data not shown).

Finally, the cleaved band intensities of the four brain tissues (precisely 5 μg for each) were compared within the same gel electrophoresis after incubation with 2 mM HNE at 10 and 30

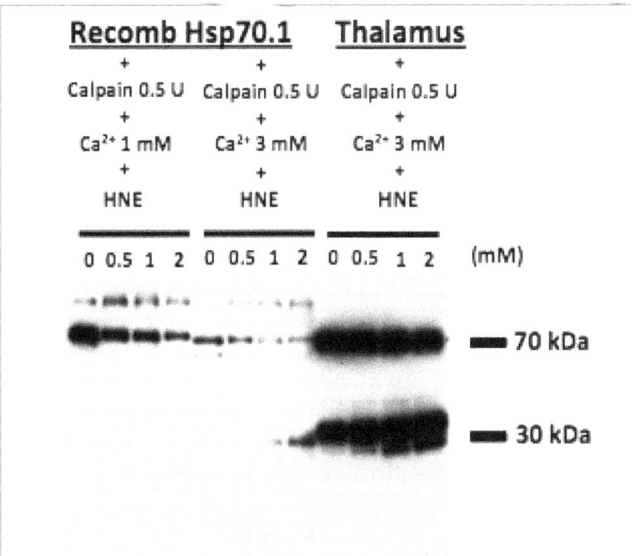

Figure 2: Using the recombinant Hsp70.1 proteins (200 ng) and the thalamus tissue (20 μg), optimal concentration of Ca^{2+} necessary for Hsp70.1 cleavage was estimated to be 3 mM in the present experimental paradigm.

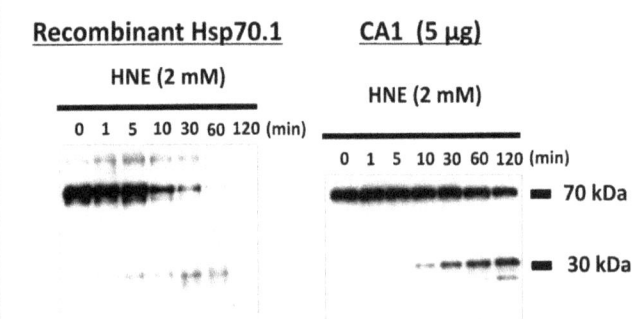

Figure **5**: Calpain-mediated Hsp70.1 cleavage occurs time-dependently in both the CA1 tissue (5 μg) and the recombinant Hsp70.1 protein (200 ng) after incubation with 3 mM Ca^{2+} and 2 mM HNE.

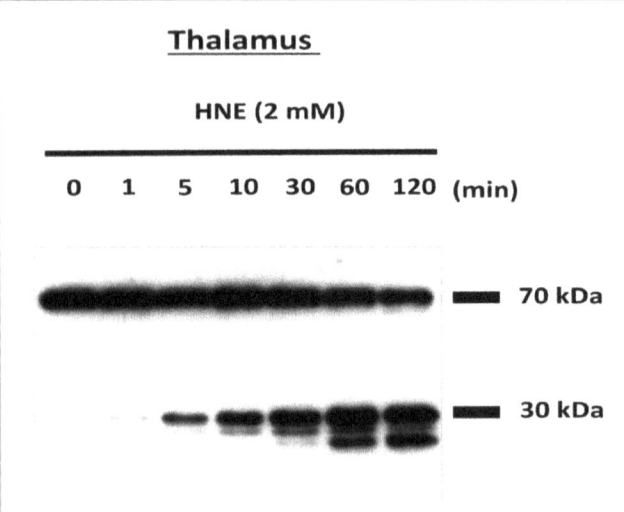

Figure **6**: Calpain-mediated Hsp70.1 cleavage occurs time-dependently in the thalamus tissue (5 μg) after incubation with 3 mM Ca^{2+} and 2 mM HNE.

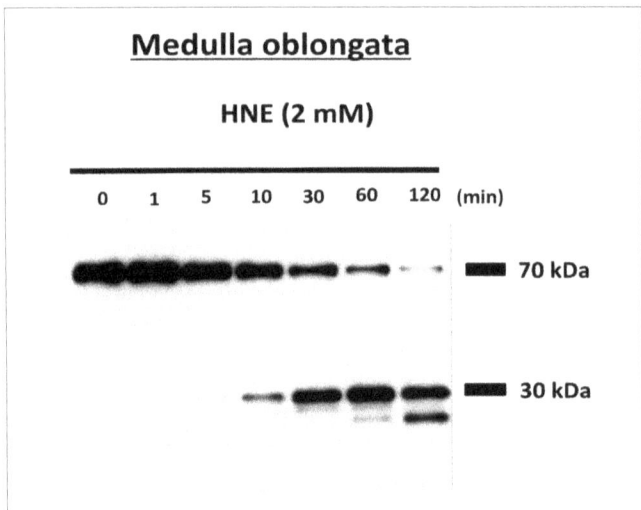

Figure **7**: Calpain-mediated Hsp70.1 cleavage occurs time-dependently in the medulla oblongata tissue (5 μg) after incubation with 3 mM Ca^{2+} and 2 mM HNE.

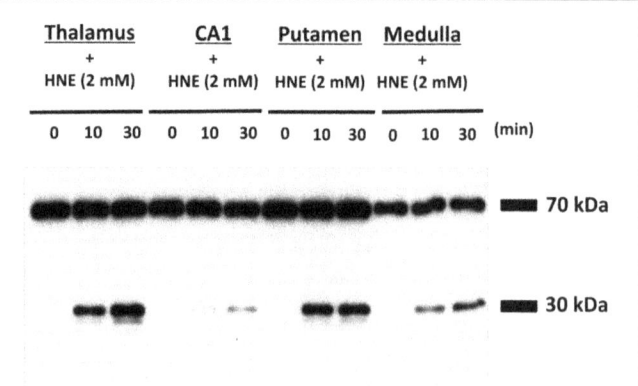

Figure **8**: Comparison of calpain-mediated Hsp70.1 cleavage after incubation with 3 mM Ca^{2+} and 2 mM HNE among 4 tissues; thalamus, CA1, putamen and medulla oblongata (each 5 μg). Interestingly, the CA1 tissue showed the minimum cleavage among them that was confirmed by repeated Western blots.

min under 3 mM Ca^{2+} concentration plus 0.5U μ-calpain. The Hsp70.1 cleavage was maximum in thalamus or putamen, but surprisingly it was minimum in CA1 (Fig. 8). Although the CA1 tissues from the different monkeys were studied repeatedly, the results were the same.

Discussion

Because of the presence of numerous dendrites and very long axon, neurons must maintain considerably large volumes of membrane and cytoplasm, and continually traffic autophagy-related substrates long distances back to the cell body where lysosomes are most active for degradation [24]. Protein quality control, done by a balance between its folding and degradation, is fundamental to the cell homeostasis. Together with optimal co-chaperones, Hsp70.1 recognizes irreversibly aged/damaged proteins and ubiquitinates these proteins, thereby targeting them for degradation via proteasomes. Further, it recognizes proteins containing the marker sequence KFPRQ and sends them for degradation into lysosomes [25]. Hsp70.1 is crucial not only as a molecular chaperone but also as a stabilizer of the limiting membrane. It contributes to lysosomal stabilization by binding to the anionic phospholipid, Bis(Monoacylglycero) Phosphate (BMP), a co-factor essential for sphingomyelin metabolism [4]. Hsp70.1-BMP binding enhances activity of acid sphingomyelinase, which mediates the sphingolipid degradation at the internal membrane in the acidic (pH 4.5) compartment to generate ceramide [26-28]. Ceramide protects the lysosomal limiting membrane from rupturing [4,29,30], presumably because the increased concentration of lysosomal ceramide can facilitate fusion of lysosomes with other intracellular vesicles and membranes, and strengthen limiting membranes [31].

Lipid peroxidation is the oxidative deterioration of polyunsaturated fatty acids containing two or more carbon-carbon (C = C) double bonds. Following lipid peroxidation, HNE and malondialdehyde are the most abundant aldehydes produced, while acrolein is the most reactive [18,32,33]. HNE is a 9-carbon amphiphilic lipid with both water-soluble and lipophilic properties that make it remain associated with the membranes [19]. Thus, HNE-membrane interaction provides a reactivity of HNE with proteins inside and outside the cell [33]. HNE forms adducts with four different side chains in proteins, namely Cys, His, Lys, and Arg. Cys residues displayed by far the highest reactivity, and the order of the molar HNE/amino acid ratio was Cys (0.6) >> His (1×10^{-3}) > Lys (3×10^{-4}) >> Arg (4×10^{-5}) [34]. Accordingly, numerous proteins are modified by HNE, including plasma membrane ion and nutrient transporters; receptors for growth factors and neurotransmitters; mitochondrial electron transport chain proteins; protein chaperones; proteasomal proteins; and cytoskeletal proteins [35,36].

Uncontrolled and/or excessive production of HNE interferes with normal cellular signaling and disrupts ion homeostasis such as Ca^{2+}, impairs Na^+/K^+ ATPase activity, disrupts the microtubule structure, and activates the caspase pathways. A recent review of Perluigi, et al. [32] described the role of lipid peroxidation, particularly of HNE-induced protein modification in neurodegenerative diseases. HNE is a potent modulator of numerous cell processes such as oxidative stress signaling, cell proliferation, transformation, or cell death. Although the effects of HNE have been a focus of the recent research, the detailed mechanism of its effects upon neuronal death had been unknown. Interestingly, however, two-dimensional carbonyl immunoblots of the postischemic monkey hippocampal tissues after immunoprecipitation with anti-Hsp70.1 antibody, showed a remarkable upregulation of carbonylated Hsp70.1. A decrease of its molecular weight from 157.20 to 113.12 indicated oxidative injury of Hsp70.1 [11]. Accordingly, we suggested that 'HNE-induced Hsp70.1 carbonylation' may be a crucial event for elucidating the mechanism of neuronal death [11,12], but whether and how HNE increases the risk of neuronal death have been incompletely understood.

Calpain is Ca^{2+}-regulated cysteine protease, playing an important role in the regulation of cell death [22]. The 'calpain-cathepsin hypothesis' corroborated the role of lysosomal rupture as an executor of programmed neuronal necrosis after transient brain ischemia in the non-human primates [12,14,37-44]. During ischemia, excessive Ca^{2+} mobilization occurs specifically in the CA1 neuron, and μ-calpain is remarkably activated. During reperfusion, oxidation of ω-6 polyunsaturated fatty acids by ROS produces HNE which carbonylates Hsp70.1 at the lysosomal membrane. Then, carbonylated Hsp70.1 is efficiently cleaved by activated μ-calpain, and this leads to the lysosomal membrane destabilization/rupture. Since calpain was found to be activated at the lysosomal membranes [38,41], both calpain activation and Hsp70.1 carbonylation may occur at the same place simultaneously. Consequently, release of hydrolytic enzyme cathepsins from the lysosomal lumen occurs to induce programmed CA1 neuronal necrosis within the CA1 sector [12]. Since the in-vitro Hsp70.1 cleavage was blocked by a calpain inhibitor N-Acetyl-Leu-Leu-Nle-CHO (ALLN) dose-dependently, it is likely that Hsp70.1 can be more efficiently cleaved by activated μ-calpain especially after HNE-induced carbonylation [23,43,44].

Neurons are highly sensitive to ROS, because it contains the highest content of polyunsaturated fatty acids among the body. ROS can attack linoleic and arachidonic acids incorporated into the brain to generate HNE. Toxic properties of HNE have been extensively demonstrated for various neurodegenerative diseases, however, the detailed mechanisms of HNE neurotoxicity was suggested very recently [13,43,44]. As mentioned above, two events of calpain activation and HNE generation contribute to the lysosomal destabilization/rupture. In this study, activated μ-calpain alone could cleave Hsp70.1 whereas HNE alone failed to cleave Hsp70.1. Accordingly, it is conceivable that calpain is the principal factor while HNE is the supportive factor for Hsp70.1 cleavage leading to neuronal death. With the aid of HNE, activated μ-calpain would facilitate lysosomal destabilization by cleaving carobonylated Hsp70.1 sufficiently. In addition, under the same level of Ca^{2+} mobilization, the more the intake of ω = 6 polyunsaturated fatty acids and/or the oxidative stresses, the more the Hsp70.1 cleavage may occur in proportion to the amount of HNE generated.

Hippocampal CA1 is well known to be extremely vulnerable to the ischemic insult. After the transient ischemia, CA1 neurons develop cell death on days 5~7 after ischemia [40]. However, the present data intriguingly showed that calpain-mediated cleavage of carbonylated Hsp70.1 occurred much less in the CA1 tissues, compared to thalamus, putamen, and medulla oblongata. Accordingly, it is suggested that calpain is a principal factor while HNE is a supportive factor for Hsp70.1 cleavage. As calpain is Ca^{2+}-dependent, processes involving the management of intracellular Ca^{2+} can influence the extent and length of calpain activation and mainly determine the cell death fate. Since the CA1 slice showed the greatest Ca^{2+} mobilization during hypoxia-hypoglycemia [37], excessive calpain activation may occur most remarkably there, and this can explain the specific vulnerability of CA1 neurons. Maintenance of Ca^{2+} homeostasis is critical for neuronal viability; however, tight regulation of its intracellular concentrations would be disturbed during ischemia especially in the vulnerable neurons. As the increasing HNE generation can cause a stepwise escalation in Hsp70.1 cleavage, the subjects with decreased cerebral blood flow due to arteriosclerosis and potential calpain activation had better avoid intake of excessive ω = 6 polyunsaturated fatty acids (for example, cheap cooking oils or everything made from it – mayonnaise, margarine, dressing and deep-fried dishes).

In summary, the authors here provided direct evidence by the cleavage assay in vitro that Hsp70.1 protein in the various brain regions is the substrate of cysteine protease μ-calpain, and that calpain-mediated cleavage of oxidized Hsp70.1 causes neurodegeneration in response to the abnormal Ca^{2+} mobilization and HNE accumulation. We speculate that under the stress condition, for example brain ischemia, the cleavage of Hsp70.1 protein similarly occurs anywhere in the brain, but occurs differently in response to the extent of Ca^{2+} mobilization during the stress and HNE accumulation in the corresponding brain regions. The specific vulnerability of CA1 neurons can be explained by their excessive and/or long-standing calpain activation [39,41] due to the remarkable Ca^{2+} mobilization during the stress such as ischemia.

References

1. Yamashima T, Zhao L, Wang XD, Tsukada T, Tonchev AB. Neuroprotective effects of pyridoxal phosphate and pyridoxal against ischemia in monkeys. Nutr Neurosci. 2001;4(5):389-97.

2. Syntichaki P, Xu K, Driscoll M, Tavernarakis N. Specific aspartyl and calpain proteases are required for neurodegeneration in *C. elegans*. Nature. 2002;419: 939–44. doi:10.1038/nature01108.

3. Kirkegaard T, Jäättelä M. Lysosomal involvement in cell death and cancer. Biochim Biophys Acta. 2009;1793(4):746-54. doi: 10.1016/j. bbamcr.2008.09.008.

4. Kirkegaard T, Roth AG, Petersen NH, Mahalka AK, Olsen OD, Moilanen I, et al. Hsp70 stabilizes lysosomes and reverts Niemann-Pick disease-associated lysosomal pathology. Nature. 2010;463:549–53. doi:10.1038/nature08710.

5. Lee SH, Kwon HM, Kim YJ, Lee KM, Kim M, Yoon BW. Effects of Hsp70.1 gene knockout on the mitochondrial apoptotic pathway after focal cerebral ischemia. Stroke. 2004;35(9):2195-9. doi:10.1161/01. STR.0000136150.73891.14

6. Marber MS, Mestril R, Chi SH, Sayen MR, Yellon DM, Dillmann WH. Overexpression of the rat inducible 70-kD heat stress protein in a transgenic mouse increases the resistance of the heart to ischemic injury. J Clin Invest. 1995;95(4):1446–56. doi: 10.1172/JCI117815.

7. Fudaba Y, Tashiro H, Ohdan H, Miyata Y, Shibata S, Shintaku S, et al. Efficacy of HSP72 induction in rat liver by orally administered geranylgeranylacetone. Transpl Int. 2000;13(1):S278-81.

8. Kelly S, Bieneman A, Horsburgh K, Hughes D, Sofroniew MV, McCulloch J, et al. Targeting expression of hsp70i to discrete neuronal populations using the Lmo-1 promoter: assessment of the neuro protective effects of hsp70i in vivo and in vitro. J Cereb Blood Flow Metab. 2001;21(8):972-81. doi: 10.1097/00004647-200108000-00010.

9. Matsumori Y, Hong SM, Aoyama K, Fan Y, Kayama T, Sheldon RA, et al. Hsp70 overexpression sequesters AIF and reduces neonatal hypoxic/ischemic brain injury. J Cereb Blood Flow Metab. 2005;25(7):899-910. doi: 10.1038/sj.jcbfm.9600080.

10. Nakajima E, David LL, Bystrom C, Shearer TR, Azuma M. Calpain-specific proteolysis in primate retina: contribution of calpains in cell death. Invest Ophthalmol Vis Sci. 2006;47(12):5469-75. doi: 10.1167/iovs.06-0567.

11. Oikawa S, Yamada T, Minohata T, Kobayashi H, Furukawa A, Tada-Oikawa S, et al. Proteomic identification of carbonylated proteins in the monkey hippocampus after ischemia-reperfusion. Free Radic Biol Med. 2009;46(11):1472-7. doi: 10.1016/j.freeradbiomed.2009.02.029.

12. Yamashima T, Oikawa S. The role of lysosomal rupture in neuronal death. Prog Neurobiol. 2009;89(4):343-58. doi: 10.1016/j. pneurobio.2009.09.003.

13. Carbone DL, Doorn JA, Kiebler Z, Sampey, BP, Petersen DR. Inhibition of Hsp72-mediated protein refolding by 4-hydroxy-2-nonenal. Chem Res Toxicol. 2004;17(11):1459-67. doi: 10.1021/tx049838g.

14. Zhu H, Yoshimoto T, Yamashima T. Heat Shock protein 70.1 (Hsp70.1) affects neuronal cell fate by regulating lysosomal acid sphingomyelinase. J Biol Chem. 2014;289(40):27432-43. doi: 10.1074/jbc.M114.560334.

15. Koriyama Y, Sugitani K, Ogai K, Kato S. Heat shock. protein 70 induction by valproic acid delays photoreceptor cell death by N-methyl-N-nitrosourea in mice. J Neurochem. 2014;130(5):707–19. doi: 10.1111/jnc.12750.

16. VanWinkle WB, Snuggs M, Miller JC, Buja LM. Cytoskeletal alterations in cultured cardiomyocytes following exposure to the lipid peroxidation product, 4-hydroxynonenal. Cell Motil Cytoskeleton. 1994;28(2):119-34. doi: 10.1002/cm.970280204.

17. Chaudhary P, Sharma R, Sharma A, Vatsyayan R, Yadav S, Singhal SS, et al. Mechanisms of 4-hydroxy-2-nonenal induced pro- and anti-apoptotic signaling. Biochemistry. 2010;49(29):6263-75. doi: 10.1021/bi100517x.

18. Dalleau S, Baradat M, Guéraud F, Huc L. Cell death and diseases related to oxidative stress: 4-hydroxynonenal (HNE) in the balance. Cell Death Differ. 2013;20(12):1615-30. doi: 10.1038/cdd.2013.138.

19. Mattson MP. Roles of the lipid peroxidation product 4-hydroxynonenal in obesity, the metabolic syndrome, and associated vascular and neurodegenerative disorders. Exp Gerontol. 2009;44(10):625-33. doi: 10.1016/j.exger.2009.07.003.

20. Keller JN, Pang Z, Geddes JW, Begley JG, Germeyer A, Waeg G, Mattson MP. Impairment of glucose and glutamate transport and induction of mitochondrial oxidative stress and dysfunction in synaptosomes by amyloid beta-peptide: role of the lipid peroxidation product 4-hydroxynonenal. J Neurochem. 1997;69(1):273-84.

21. Mark RJ, Lovell MA, Markesbery WR, Uchida K, Mattson MP. A role for 4-hydroxynonenal, an aldehydic product of lipid peroxidation, in disruption of ion homeostasis and neuronal death induced by amyloid beta-peptide. J Neurochem. 1997;68(1):255-64.

22. Bevers MB, Neumar RW. Mechanistic role of calpains in postischemic neurodegeneration J Cereb Blood Flow Metab. 2008;28(4):655-73. doi: 10.1038/sj.jcbfm.9600595.

23. Sahara S, YamashimaT. Calpain-mediated Hsp70.1 cleavage in hippocampal CA1 neuronal death. Biochem Biophys Res Commun. 2010;393(4):806-11. doi: 10.1016/j.bbrc.2010.02.087.

24. Lee S, Sato Y, Nixon RA. Lysosomal proteolysis inhibition selectively disrupts axonal transport of degradative organelles and causes an Alzheimer's-like axonal dystrophy. J Neurosci. 2011;31(21):7817-30. doi: 10.1523/JNEUROSCI.6412-10.2011.

25. Malyshev I. Immunity, tumors and aging: The role of Hsp70, Springer Briefs in Biochem and Mol Biol. 2013. DOI: 10.1007/978-94-007-5943-5_2.

26. Linke T, Wilkening G, Lansmann S, Moczall H, Bartelsen O, Weisgerber J, et al. Stimulation of acid sphingomyelinase activity by lysosomal lipids and sphingolipid activator proteins. Biol Chem. 2001;382(2):283-90. doi: 10.1515/BC.2001.035.

27. Linke T, Wilkening G, Sadeghlar F, Mozcall H, Bernardo K, Schuchman E, et al. Interfacial regulation of acid ceramidase activity. Stimulation of ceramide degradation by lysosomal lipids and sphingolipid activator proteins. J Biol Chem. 2001;276(8):5760-8. doi: 10.1074/jbc.M006846200.

28. Kolter T, Sandhoff K. Principles of lysosomal membrane digestion: stimulation of sphingolipid degradation by sphingolipid activator proteins and anionic lysosomal lipids. Annu Rev Cell Dev Biol. 2005;21:81-103. doi: 10.1146/annurev.cellbio.21.122303.120013

29. Petersen NH, Kirkegaard T. HSP70 and lysosomal storage disorders: novel therapeutic opportunities. Biochem Soc Trans. 2010;38(6):1479-83. doi: 10.1042/BST0381479.

30. Petersen NH, Kirkegaard T, Olsen OD, Jäättelä M. Connecting

Hsp70, sphingolipid metabolism and lysosomal stability. Cell Cycle. 2010;9(12):2305-9. Doi:10.4161/cc.9.12.12052.

31. Heinrich M, Wickel M, Winoto-Morbach S, Schneider-Brachert W, Weber T, Brunner J, et al. Ceramide as an activator lipid of cathepsin D. Adv Exp Med Biol. 2000;477:305-15. doi: 10.1007/0-306-46826-3_33.

32. Perluigi M, Coccia R, Butterfield DA. 4-hydroxy-2-nonenal, a reactive product of lipid peroxidation, and neurodegenerative diseases: A toxic combination illuminated by redox proteomics studies. Antioxid Redox Signal. 2012;17(11):1590-609. doi: 10.1089/ars.2011.4406.

33. Schaur RJ, Siems W, Bresgen N, Eckl PM. 4-Hydroxy-nonenal—A bioactive lipid peroxidation product. Biomolecules. 2015;5(4):2247-337. doi: 10.3390/biom5042247.

34. Legards JF, des Rosiers C. 16–18 June 2006. Assay of 4-hydroxynonenal (HNE) adducts with various polyaminoacids (PAA) using gas chromatography-mass spectrometry (GCMS). In Proceedings of the 3rd International Meeting of the HNE-Club, Genova, Italy.

35. Petersen DR, Doorn JA. Reactions of 4-hydroxynonenal with proteins and cellular targets. Free Radic Biol Med. 2004;37(7):937-45. doi: 10.1016/j.freeradbiomed.2004.06.012.

36. Poli G, Biasi F, Leonarduzzi G. 4-hydroxynonenal-protein adducts: A reliable biomarker of lipid oxidation in liver diseases. Mol Aspects Med. 2008;29(1-2):67-71. DOI: 10.1016/j.mam.2007.09.016.

37. Yamashima T, Saido TC, Takita M, Miyazawa A, Yamano J, Miyakawa A, et al. Transient brain ischaemia provokes Ca^{2+}, PIP_2 and calpain responses prior to delayed neuronal death in monkeys. Eur J Neurosci. 1996;8(9):1932-44.

38. Yamashima T, Kohda Y, Tsuchiya K, Ueno T, Yamashita J, Yoshioka T, et al. Inhibition of ischaemic hippocampal neuronal death in primates with cathepsin B inhibitor CA-074: a novel strategy for neuroprotection based on 'calpain–cathepsin hypothesis'. Eur J Neurosci. 1998;10(5):1723-33.

39. Yamashima T, Tonchev AB, Tsukada T, Saido TC, Imajoh-Ohmi S, Momoi T, et al. Sustained calpain activation associated with lysosomal rupture executes necrosis of the postischemic CA1 neurons in primates. Hippocampus. 2003;13(7):791-800. doi: 10.1002/hipo.10127.

40. Yamashima T. Implication of cysteine proteases calpain, cathepsin and caspase in ischemic neuronal death of primates. Prog Neurobiol. 2000;62(3):273–95.

41. Yamashima T. Ca^{2+}-dependent proteases in ischemic neuronal death: a conserved 'calpain–cathepsin cascade' from nematodes to primates. Cell Calcium. 2004;36(3-4):285-93. doi: 10.1016/j.ceca.2004.03.001.

42. Yamashima T. Hsp70.1 and related lysosomal factors for necrotic neuronal death. J Neurochem. 2012;120(4):477-94. doi: 10.1111/j.1471-4159.2011.07596.

43. Yamashima T. Reconsider Alzheimer's disease by the 'calpain-cathepsin hypothesis'--a perspective review. Prog Neurobiol. 2013;105:1-23. doi: 10.1016/j.pneurobio.2013.02.004.

6

Vaccines against Pathogens: A Review and Food For Thought

Amro Abd Al Fattah Amara*

Head of the Protein Research Department and the office of the Scientific Publishing, Genetic Engineering and Biotechnology Research Institute, City for Scientific Research and Technological Applications, Universities and Research Center district, New Borg El-Arab, Egypt

*****Corresponding author:** *Amro Abd Al Fattah Amara, Head of the Protein Research Department, Genetic Engineering and Biotechnology Research Institute, City for Scientific Research and Technological Applications, Universities and Research Center District, New Borg El-Arab Egypt, E-mail: amroamara@web.de*

Abstract

This review is a competition between the formal writing for the constituents of a pragmatic scientific topic, the immunization, and the simplification of describing such constituents without the deviation from the basic facts, knowledge and discussion. The immune system is complicated and not fully understood yet. Understanding either how the components of the immune system are working collectively against the foreigner components (e.g. antigens and pathogens) or the inducers (e.g. adjuvant) will enable establishing better strategies for health protection and disease control. For the immune system, some microbes can be confusable such as the polysaccharide-producing microbes. Other factors, which are able to reduce the immune system efficacy like the age, gender, moral, behaviour, etc should be considered. Particular candidates can train the immune system's components to be more efficient, like vaccines, dead microbes in some type of foods (e.g. aged food), mild infections, etc. All that will be highlighted in the following text. Our immune system is created to protect us from foreign antigens and to clean our bodies from any of them. Some virulence invaders can crack our immune system. Virulence viable microbe, due to its virulence elements and its replication rate is collectively stronger than the immune system. That can be happened if it is attack suddenly and without a previous preparation. In the case of virulence viable microbe our immune system need to be prepared before such microbe attack us. Vaccines for that are important candidates for protecting us. Previously prepared homologous or heterologous antibodies will safe us occasionally until we become ready with our own defenses. This review contains brief notes about the immune system and its different responses to the pathogens. In addition, it contains hints about how is the immune system is work. The factors, which are affecting on our immune system, the vaccine types and the progress of the vaccine technology, will be mentioned. Correct battle with the pathogen using correct elements will reduce the battle cost, time, side effect etc., and will lead to our survive. It is also contains brief information about key points in the vaccine history enabling better understanding for the tactics and tools which have used in ages where most of the existing facilities and instruments nowadays were not available. It was simplified to be readable and understandable to non-specialists, and informative to specialists. It is written to touch daily practices in our life could affect positively or negatively on our immune system, for better understanding and for healthier bodies.

Abbreviations

BCG: Bacillus of Calmette and Guérin; BG: Bacterial Ghosts; CBPP: Contagious Bovine Pleuropneumonia; CDR: Complementarily Determining Region; CFU: Colony Forming Unit; CTL: Cytotoxic T Lymphocytes; FMD: Foot and Mouth Disease; MG: Microbial Ghosts; MGC: Minimum Growth Concentration; MHC: Major Histocompatibility Complex; MIC: Minimum Inhibition Concentration; RBC: Red Blood Cells; RSV: Respiratory Syncytial Virus; UK: United Kingdom

Introduction

Principles of Vaccination

The immune system: The immune system is the system that protects the body from foreign substances and pathogenic organisms by producing the immune response. Like the digestive and the circulating systems, mammalians including us have a system for the protection against foreign components, which is named "the immune system". It has dynamic communicating network of cells, tissues and organs that work together to defence the body against foreigner attacks. It is composed from different lymph nodes (in lymphoid tissues) which are the source of lymph and lymphocyte existed in different parts of the body. The immune system components include the lymph nodes, lymphatic vessels, thymus, spleen, Peyer's patches, bone marrow, appendix, tonsils and adenoids. However, the most importance components of the immune system are the lymphocytes, which are small white blood cells that are the key player in the immune system [1]. For more details, refer to any textbook about the immunology.

The antigen: The antigens are any molecules, or macromolecules, which are foreigner to our body. They were recognized as unwanted components. They can be either molecules or macromolecules normally bigger than 400 Da. That also explain both of the behaviour of low molecular weight (less than 400 Da) poisonous and why our immune system did not protect us from them and in the same time explain the allergy happened after eating some foods (or juices). They can inter to our body in a pure form such as toxins or as a part of lived or unlived microbes such as bacteria and viruses respectively. If the antigens are inter to our body on the surface of the live microbe or non-live but have enough genetic materials to react inside the viable cells as alive (e.g. viruses), in such cases the problem will be exponential. That because deactivating the existed surface antigens is important but not enough. Microbes' killing in the case of those, which are able, to produce toxins is the only solution to stop their replications. Therefore, the treatment will take place in several directions at once. Those directions could be summarized as:

1- Toxin deactivation by using passive immunization with previously prepared antibodies.

2- Using effective antibiotics to kill such microbe.

3- Using vaccines for those not being infected yet to cut the spreading of the pathogen.

4- Isolate the infected individuals from non-infected ones.

5- Stop any physical, biological or environmental factor helping in spreading of such infections.

The active immunity will be based on the intensity of the attack. Different level of attack, the type of the microbes, the route of infection, the individual immune situation and other factors will specify the type of the immune response, which will be triggered. Different factors will lead to trigger different levels of the immune signals and will activate one or both of the humoral and the cellular immune response (or both). For virulence microbes, when we are not immunized against it before, in most cases, the battle will be in the side of the virulence microbe. Virulence microbes were specified as virulence while they have extra components and virulence factors enable them to bypass our immune system defences. Even we are strong enough (or the individual under attack so), our immune system is working efficiently, and everything is prepared for the pathogen/immune system battle but due to the pathogen's virulence factors, it will win the battle against our immune system. For that, a previous preparation for such expected cases are critical and usually made through the vaccinations and through other natural immunization elements such as the mild infections.

Allowing our bodies to resist naturally mild infections of pathogenic microbes will activate our immune system and keep it ready. Even the modern history of the immunization is started by an observation about that the cowpox infection could protecting against the smallpox which is well known by farmers but explained scientifically by a physician [2]. It is also important to mention that our grandfathers were reacting more naturally than us. Nowadays, there are drugs to get rid from our bodies and our immune system signals such as fevers, tiredness etc., just to allow us to work continuously. All such biased reactions will interfere with our immune system quality as well as with our general health conditions. Some microbes could protect against others. And, sensing a repeated mild infection means that some new antigen(s) or virulence factor(s) are existed and are able to activate our immune system again. Accumulating such, responses

to the different mild infections that will build a strong immune response and will reduce most of the side effect of the infections caused by many pathogens (even for unrelated pathogens). The immune system did not protect us against pathogens and foreign antigens only but it is able to identify the foreign tissues carrying non self markers upon transplantation (except an identical twin).

Rough classification for the Antigen

Antigens can be classified roughly based on their degree of virulence and their linkage to the microbes. Not all antigens are virulence. However, some are able to do serious problems. They also either being linked to the microbe surface or they are excreted out free from the microbes. Additionally they are derived or supplied from non-microbial origin. Being on the surface of the microbes, that means efficient activity for both of them and the immune system response. In addition, the more the microbe reproduces the more the antigens are existed. In such cases, the immune system must get rid from all viable microbes. Being free means that, they are more mobile and could reach different parts in the body. Their amount, types and side effect can be lethal. Dead microbes inter-our bodies by different routs could have less virulence even they are carrying toxins because they are missing the most effective virulence factor they have, their replication. That might explain the use of the Egyptian civilization for the aged food and sub-rotten salted fish.

Rough classification for the antigens

1- Free antigens

a. Virulence

b. Non-virulence

2- Linked antigen

c. On the surface of life microbes

d. On the surface of the dead microbes

e. On the surface of virus

f. Combined with other macromolecules such as the lipopolysaccharide

The Immune System Components

Blood serum

Blood serum is the main component, which contain antibodies, and other immunological mobile components. It was used early for preventive or curative aims in both human and veterinary medicine starting from the late 19th Century. Emil von Behring and Emile Roux introduced serum therapy for children suffering from diphtheria in Germany and France in 1894, respectively. Sclavo and Marchoux used serum therapy for anthrax in 1895. Serum from immunized cattle versus foot and mouth disease (FMD) was applied by Friedrich Löffler (1852-1915) in 1897 and used on a large scale in Denmark [3].

Immune cells

Immune cells are comes from the immature stem cells in the bone marrow. They differentiated to different cell types as a response to the different types of the cytokines and other signals. Such differentiation lead to different types of cells such as T cells, B cells and phagocytes. Some type of the immune cells has wide range of different attack process and some are highly specific in their act. The immune cells can contact either by direct physical contact or by releasing chemical messengers [4-7]. One should remark that all of the immunological activities come from cells however; cells can react by themselves directly or by their products (e.g. antibodies and chemical compounds).

B cells

Each B cell is programmed according to the signal it received from the existence of a single epitope to make one specific antibody. B cell is a lymphocyte derived from bone marrow that provides humoral immune response; it recognizes free antigen molecules in solution and matures into many large cells known as plasma cells that secrete immunoglobulin (antibodies) that inactivate the antigens. When the B cell finds its specific antigen, it works as a factory to produce a specific antibody. The plasma cells release their antibodies directly to the bloodstream.

Antibody

The antibodies (or immunoglobulin) are protein molecules produced by B-lymphocytes. The immune responses are generally produced more perfect in response to a live antigen than dead or inactive ones. Surface proteins, are easily recognized by the immune system such as hepatitis B surface antigen. In contrast, surface polysaccharide is less effective antigens. For that, the immune response is less effective with microbes such as *Streptococcus pneumoniae* [8-13]. Antibody is any of a large variety of proteins normally present in the body or produced in response to an antigen, which it neutralizes, thus producing an immune response.

The antibodies are large protein molecules known as immunoglobulin (Ig). Ig has different variants include:

Immunoglobulin G (IgG) can coat the microbes, speed their uptake by other cells; Immunoglobulin M (IgM) is effective in bacteria killing;

Immunoglobulin A (IgA) is concentrate in body fluids, tears, saliva, the secreting of respiratory tract and digestive tract;

Immunoglobulin E (IgE) is able to protect against parasite and responsible for the symptoms of the allergy;

Immunoglobulin D (IgD) is remain attached to the B cells and contributes to the early B-cell response.

The antibody affinity refers to the tendency of an antibody to bind to a specific epitope at the surface of an antigen, i.e., to the strength of the interaction. The avidity is the sum of the epitope specific affinities for a given antigen. It directly relates to its function [14].

There are four major sources of antibody used in human medicine. These are:

1-Homologous pooled human antibody

It is the IgG antibody fraction collected from adult donors. It contains antibodies to many different antigens. It is used mainly for post exposure prophylaxis for hepatitis A and measles and treatment of certain congenital immunoglobulin deficiencies.

2-Homologous human hyper immune globulin

It is high titer of specific antibody. It is a product from the plasma of humans with high levels of particular antibody. However, other antibodies in lesser quantities are existed. Hyper immune globulins are used for post exposure prophylaxis for several diseases, including hepatitis B, rabies, tetanus, and varicella.

3-Heterologous hyper immune serum

Heterologous hyper immune serum is also known as antitoxin. This product is produced in animals, usually horses (equine), and contains antibodies against only one human antigen (based on the used antigen purity).

4- Monoclonal antibody

Nearly all of the antigenic preparation, the used hosts gives rise to a mixture of antibodies. However, in many cases, there is a need for specific pure antibodies from monotype.

In the 1970s, techniques were developed to isolate and "immortalize" (cause to grow indefinitely) single B cells by hybridizing it with myeloma cells. That led to the development of specific cells able to produce monoclonal antibody. It is produced from a single clone of B cells, so these products contain antibody to only one antigen or closely related group of antigens. A producer for direct production of the monoclonal antibodies was introduced where splenic B cells from immunized animal was fused with malignant (immortal) plasma cells, forming a hybridoma. The B cell hybridoma, which is able to secret the desired antibody, then isolated from the other cells by reactivating it with the antigen of interest. The cells then cloned and expended in tissue culture to enable it to reproduce in large quantity and to produce large amount of antibody of a single type, which are specific for single antigens. Those monotype antibodies were given the name "monoclonal antibodies". Monoclonal antibody products have many applications, including the diagnosis of different types of cancer (e.g. colorectal, prostate, ovarian, breast etc.), cancer treatment (B-cell chronic lymphocytic leukemia, non-Hodgkin lymphoma), transplant rejection prevention, and autoimmune diseases treatment (Crohn disease, rheumatoid arthritis), infectious diseases such as Respiratory Syncytial Virus (RSV) infection. It is called palivizumab (Synagis). Palivizumab is a humanized monoclonal antibody specific for RSV [3,15].

The antibody-antigen reaction

One of the basic practices to discover that the body was or is subjected to a microbial infection such as a virus or a certain harmful protein is by making a reaction between the serum of the investigated person, which expected to contain antibodies for such foreigner antigens or the microbe itself. The positive reaction is a clear indication about the existence of antibodies for such pathogen. Positive antibody/ antigen reaction could not tell if that the foreigner components still existed or not but for sure, it proves that, it existed at one time before. However, inspecting the related DNA or RNA or other specific components will prove that such foreigner still existed (live and viable) or not. One should consider the other components which their presence or absence could prove the presence of the invader such as the infection symptoms. However, some virus's infection symptoms could not be detected early. As an example virus C might not be discovered by any symptoms until years are passed, and till a real deterioration to the liver. However, the change in the skin and eye colour to be yellowish, and the darkness urine colour might be positive signals. Only DNA or RNA inspection as well as the protein could prove or disprove the existence of live and viable microbe. The amount of each of DNA and or the RNA particularly will prove or disprove that such pathogen is still active or not and for which extends.

T cells

Dissimilar from the B cells, T cells do not recognize free-floating antigens. T cells surface contain specialized antibody-like receptors. Such receptors are able to recognize fragment of the antigens on the cells surface. T cells play two major essential key roles, they regulate the immune responses and do direct attack to different foreigners. [16] T cells have different forms more than the B cells, such as:

Helper T cells (Th cells): The Th cells are able to coordinate immune response by communicating with other cells. They are able to stimulate the B cells to produce antibody. Some are microbial eradicator and other is able to activate other type of the T cells [17].

Killer T cells (cytotoxic T lymphocytes or (CTLs)): They can perform different actions. Directly attack cells covering certain foreign or abnormal molecules on their surface with granules containing potent chemicals. CTLs can recognize small fragments of viruses coming out from the cell membrane and launch an attack to kill the cell. CTLs only recognize antigen, which carried on the cell surfaces by the body's own Major Histocompatibility Complex molecules (MHC) [16,18]. MHC molecules are proteins recognized by CTLs. In such case, such recognition will distinguish between self and non self. Each of our cells has MHC protein but each person has its own MHC protein. CTLs will destroy any cell has non self MHC surface protein.

Natural killer (NK) [19]

It is a type of the white blood cells or lymphocytes. Like CTLs, NK cells have granules contain potent chemicals. NK cells are not able to recognize MHC molecules and recognize cells having missing or low MHC Class I molecules. They are able to attack different types of molecules.

Phagocytes family member

Phagocytes are large white cells that can swallow and digest microbes and other foreign particles.

Monocytes: Monocytes are phagocytes that circulate in the blood. Upon their migration to the tissue, they become as macrophages specialized to the tissue where they resides. Such tissues include lungs, kidneys, brains and liver. Monocytes produce chemical signals named monokines involved in the immune responses.

Granulocytes [20]: T cells that contains granules of chemicals that can destroy the microbes. In addition, they contain histamine, which contribute in the inflammation and allergy. Granulocytes contain different types of cells including:

1- Neutrophiles have chemical pre-packed to breakdown the ingested microbes.

2- Eosinophils [21] and basophils [22] are able to spray their granulated chemicals onto harmful cells or microbes.

3- Mast cell [23] is a twin of basophil except it is not a blood cell, lining nose and intestinal tract and is responsible for the allergy symptoms.

4- Related structure the Blood platelet is a cell fragment, which also contain granules also. The platelet are responsible for the blood clotting, wound repair and are able to activate some parts of the immune system.

Cytokines [24]: Any of various protein molecules secreted by cells of the immune system that serve to regulate it. It is able to either activate or inactivate certain immune cell types. Interleukin is any of several lymphokines that promote macrophages and killer T cells and B cells and other components of the immune system. Interleukin 2(IL-2) triggers the immune system to produce T cells.

Complement [25-27]: It is one of several blood proteins that work with antibodies during an immune response. The complement system is made up of about 25 proteins that work together to "complement" the action of antibodies in destroying bacteria. Complement also helps to rid the body of antibody-coated antigen (antigen-antibody complexes). Complement proteins which cause blood vessels to become dilated and then leaky, contribute to the redness, warmth, swelling, pain and less of function that characterize an inflammatory response.

Immunity as a term: Different definitions are existed for the "immunity" as a term. Even so, the word sounds firstly for a term about our ability to defence us against pathogens. From the different existed definitions, I suggest this definition for the immunity: "Immunity is the ability of the human and the other creature (included the plants) bodies to allow material indigenous to their bodies ("self") without opposing or prohibiting (tolerate) their presence or their activity, and to eliminate foreign ("nonself") material and to remember the foreign material if inter to the body again and to eliminate it again (memorize) and in any time".

The immune response: When foreign recognizable macromolecules inter to the body, the body will detect it and signals will send to the immune system to start to get positive response and action against such invader. The immune system cells after getting alarm as signals about a foreigner they start to produce powerful chemicals, regulate their own growth and behaviour and direct themselves to such foreigner. The immune response simply is a bodily defence reaction that recognizes an invading substance (an antigen: such as a virus, fungus, bacteria or transplanted organ) and produces antibodies specific against that antigen.

The specificity and the selectivity of the immune system: Antibodies are the major functional element in the immune system. Antibodies like the enzymes and any active protein are governed by the role of the protein structure/function/specificity. The antibodies can protect the body from foreign recognizable components such as the pathogens or their toxins. However, the equation is not that simple and one should observe some key factors could affect on the antibodies structure/function/specificity. For example, some foreign components have low molecular weight and could not be detected by the immune system but only if linked to larger molecules (e.g. protein). Such low molecular weight molecules or compounds named as "Hapten". For that, molecules must reach certain molecular weight and should have enough antigenicity to be recognizable by the immune system. Other macromolecules as the polysaccharide did not recognized as antigens.

Presence of new foreign recognizable element(s) will induce antibodies production specific for it (Humoral immune response) as well as specific cells (Cellular immune response) able to attack such element(s), to neutralize it, to precipitate it, to inactivate it and to destroy it. In general, to get rid of it to trigger both of the Humoral and the Cellular immune response special conditions are required.

In case of diseases could effect on or attack the immune system itself the equation become different, where the immune system will not be able to work efficiently and to detect a previously proved to be detectable antigens or macromolecules (by healthy individuals). Example about such cases the individual who acquired immunodeficiency or the immune compromised patients. In such cases, the immune system will lose most of its efficacy. Alternatively, it can react non-specifically and attack indigenous components in case of the autoimmune disease. The losing of the immune system specificity and selectivity will lead to the autoimmune diseases. Autoimmune disease can be defined as an immune response of the body against substance normally present in the body. In the autoimmune diseases, the immune system will lose its specificity and selectivity. Alternatively, its memory is changed. Or, for some extend the body itself was changed but our immune system still remembers the old one and resist the change.

The immune system is able to produce very specific antibodies that could differentiate between two close recognizable components such as two close proteins or two close microbial strains. Specificity is very critical factor where if the immune system loss its specificity and selectivity or the body itself do, autoimmune disease will emerged. Presence of antibodies for a certain protein means, that protein is a foreigner for the body,

(except in cases such as the autoimmune diseases). If the immune system attacks its body by mistake, serious degenerative diseases could be emerged. One should observe that different persons in different ages and environments have different immune status.

The personal immune status can be roughly classified to:

1- Healthy and mature (e.g. healthy adults).

2- Healthy but immature (e.g. healthy infants).

3- Each of the above with or without immunological experiences with certain pathogen (e.g. highly sanitized live style).

4- Immunocompromised (e.g. diabetic, alder).

5- Immunodeficiency (e.g. AIDS).

6- Autoimmune diseases.

For each of the above immune status special precautions should be considered. For example, one might be healthy but grown in a high-sanitized environment. Such condition will leave the immune system without any experiences. Such person's immune system could manage some infections while it is anyhow healthy. However, it might react strongly in abnormal way like what is observed nowadays after the infection happened by some influenza viruses.

The Microbes

Pathogenic microbes are the main human enemy and the main source of antigens, which activate the immune system. The microbes will be classified in this review to beneficial, pathogenic and opportunistic. However, beneficial not means that they are compatible with our immune system. All microbes are foreigner to our healthy immune system. Pathogens are able to invade us with different rate and quite number of them are able to crake our immune system defences even one is healthy. For that, treatment with antimicrobial agent is so important especially for fatal microbes where the time needed for the identification of the infection type will be critical. Passive vaccination against such microbe is essential for surviving. In contrast, vaccination should be happened before one becomes infected.

As a microbiologist writes a review about the vaccines, I thought that it might be interesting to give some information about the microbes themselves and the other components which have antigens that could interact with our immune system. There are different kinds of microbes that could be classified in different ways; however a rough classification will be followed based on the topic of this review.

Rough classification for viable microbes

1- Benefit or harmless (in normal amount)

2- Pathogenic microbes (able to cause diseases)

3- Opportunistic "Taking immediate advantage, often unethically, of any circumstance of possible benefit"

Classifying the microbes to benefit (or harmless), pathogenic (or harmful) or opportunistic will help in simplifying our understanding to the microbes and their interactions with our immune system. However, scientific facts must not be neglected. Any microbe is a foreigner to our immune system. In addition, our bodies are designed and created to be able to interact and tolerate in some of their parts with foreign bodies' including microbes (for some extend) such as the digestive system however; some others are forbidden for any foreigner such as the circulating system. In addition, one should memorized that some microbes have low antigenicity but if existed or taken in large quantity that will be enough to change our body response to them. That means some microbes, which are classified as benefit, might be harmful based on their quantities only and vics verse with limitation. Some harmful microbes might not harm us in very low Colony Forming Unit (CFU). Meanwhile, some beneficial microbes can harm if consumed in large quantities. Additionally, some beneficial microbes could acquire extra components such as gene(s) of toxic protein on plasmid and become pathogenic. The line, which separates the three above categories, is so faint and fixable. In addition, that one should understand those factors, which could take a microbe from one category to another. The genetic mobility between different species plays a critical role. Genetic elements can be transfer from microbe to another by the aid of different processes including, transformation, transduction, phage, transposon, in nature hyperdization. In addition, mutagenesis play critical role in giving new traite to the microbes. The most important is the ability to resist antibiotics. For example, mutant in β-lactamase could extend its resistance to different derivatives of it. About 1387 new genes in comparison with the previously sequenced non-pathogenic laboratory strain *E. coli* K-12 were acquired [28]. For more details about how non-pathogenic microbes turn to be pathogenic, refer to Amara [29] and Amara [30]. The third category the opportunistic pathogen is the best example about the instability in the microbe world. Opportunistic pathogens are friendly microbes but with powerful degrading system. Opportunistic pathogen for my believe is the indicator which tell us did our health and of course our immune system is good or not.

We who are invited the opportunistic pathogens to attach us!

Microbes play key role in the balance of the land ecosystem. Perhaps, their most important role is their ability to degrade others and utilizing them as foods. In fact, they are able to degrade us too; but our immune system prevents them. However, they are waiting to do. In fact, they do their work spontaneously, but under the control of the surrounding biological, chemical and physical factors. Opportunistic pathogens are nothing but microbes with extra power for degradation, unable to attack us when our body is healthy and able to do when our immune system is compromised. They are like an active ant and we are like a crystal of sugar either being protected or not.

Opportunistic pathogens prefer simpler substrate [16,29,31]

Opportunistic pathogen lives friendly with us but suddenly they start to attack us. In fact, they not like to attack us but we invited them to do. To understand that a simple Egyptian practice can be mentioned her, where those who have badly odder between their finger due to the fungal infection are advices to put sugar between their fingers. Soon the bad odder will disappear. *Candida albicans* prefers sugar than us and will stop to attack our tissue, and will be happy with the sugar. However, the emergence of the signals that one become susceptible to the opportunistic pathogens is one of the early biological indicators about that the immune system has a problem or more than one problem. This example is given to highlight those extra elements, which is not related to the immune system could be used to help it, particularly those that reduce or inhibit the microbial virulence factors. In addition, one should try those simple tools before going to the complicated ones.

In immunocompetent individuals, immune response generated at mucosal sites is crucial and preferable for effective clearance of the infection and long-term protection such as intramuscular, intradermal or subcutaneous injection [32].

Scientific facts about the microbe's interaction with the immune system

1- All microbes even those who are friendly to us are foreigner to our body and will be attacked by our immune system, but not equally based on where they are existed.

2- Microbes live with/on/in us. Such microbes is named microflora.

3- Friendly microbes and those which existed as part of our microflora and which are safe to us could acquired extra chromosomal elements (plasmid or by transposons or any other components) which could transfer pathogenic element(s) and so they become pathogenic to us.

4- Microflora could inhibit the growth of pathogenic microbes by filling in any suitable place for colonization. For that, it is important to rebuild our microflora after losing it for example by the effect of the antibiotic treatment.

5- Some microbes can go deeper in our cells evens they can live in the mitochondria so they escape from the immune system and become invisible to it such as the Salmonella.

6- Opportunistic pathogens are mostly part of our microflora but if our immune system becomes weak (by any means), they start to attack us.

7- Microbes have virulence factors, some could be so fatal to us, and our immune system could not have enough time and chance to control them if enter to our bodies as viable cells or complete viruses. Besides that, they have virulence factors and can replicate inside our bodies.

8- The type of microbes could affect on our immune system ability to control it. For example *S. pneumoniae* surface polysaccharides prevent correct immune response for their antigens.

9- Some microbes are able to immunize us against others. Perhaps the most famous example is the cowpox, which is able to protect us against the smallpox.

10- The size of the microbe is critical to the efficacy of the immune system.

11- Naturally dead microbes, naturally produced microbial ghost as suggested by Amara [33], are a source of natural immunization.

12- Mild infection such as the common cold could help in our immune system activation. For that, we should give our body the chance to recover from such infection as suggested by Amara [28]. Second infection will be like the second booster.

13- Vaccine using deactivated, live attenuated, genetically modified, similar strains, fragment of the microbes etc., will help us to be ready and protected from fatal diseases.

Passive immunity [34- 36]

In special cases such as antibiotics resistance, to control microbes that are able to produce fatal toxins, immunocompromised patient using antibodies against microbial infection are recommended. In such case previously, prepared antibodies should be available for fast aid (Passive immunity). Passive immunity which is an temporary form of acquired immunity in which antibodies against a disease are acquired naturally (as through the placenta to an unborn child) or artificially (as by injection of antiserum)". Previously prepared antibodies for foreign microbes, proteins and toxins are very important to resist sudden infection with fatal pathogen, where the immune system, even the person is healthy will not be able to control such foreigners and need an immediate help by a previously prepared antibody. If such prepared antibodies not existed so strong or effective antibiotic(s) must be used. Antibiotic(s) can be used alone or simultaneously with the passive antibodies. Human and animals are both sources for antibodies production. The antibodies will react at once upon injection and will find the antigens in the surface of any foreigner. The produced antibodies will do with it an antibody-antigen reaction, and will turn it to non-harmful foreigner. Antibody-antigen reaction is the best currently biomarker for the vaccine efficacy [32,37]. In addition the antibody-antigen reaction will enable the body either to get rid from the harmful antigens or microbes or to destroy them. However, the antibodies itself will be target for cleaning and the body will loss the antibody in few weeks or might be in lesser or longer times. Infant receives from his mother antibodies, which are transported across the placenta during the last 1–2 months of pregnancy and become protected from some diseases for up to a year such as measles, rubella and tetanus. Lactation will extend the mother gift of antibodies to her infant [37,38]. For that, lactation is important. However, unfortunately as if it protects the infant from infection it will protect it from any antigen including the vaccine itself. Some ethnic groups provide infant with two years lactation. For infants, the immune system is immature and less capable of developing memory. In such case, the duration of protection can be very short-lived for polysaccharide antigens and passive immunity will be essential for protection against some infections. Alternatively, lactation might satisfy the infant demand. Vaccines can be given to the mother itself.

Active immunity

The personalization of the immune response: In case of active immunity, the protection is individual-based (personal) and different from individual to another. It based on many aspects include the health condition, the body experience with such foreigner, the power of the immune system, the individual behaviour, age, gender, sleeping hrs., etc. Active immunity is a form of acquired immunity in which the body produces its own antibodies against disease-causing antigens. Such differences could be given a general title "The personalization of the immune response". The active immunity is usually permanent. Vaccines induced immunity may vary among individuals depending on their genetic characteristics [32], the vaccine type, amount, route of administration and the dosage repeated number, etc.

The active immunity is based on individual stimulation by either viable, attenuated microbes or viruses or toxins from some microbes. Such self-interaction will activate the body immune system. Based on the activation power the response either will be translated to both of the humoral and cellular immunity or is limited to one of them.

Other factors also involved. Infection happened in the tissue such as the muscles by a pathogen that will trigger the cellular immunity.

The Memory

Upon the recovery from the disease the individual will be immunized permanently against such disease despite cases where the pathogen can change its surface antigen such as virus C, and the influenza virus. Pathogen has polysaccharide disable correct recognition and/ or correct detection [39] (in further immunized one) such as in case of *S. pneumonia*. Correct exposure to an antigen will produce certain cells (memory B cells) which circulating in the blood and are residue in bone marrow for many years. The second exposure to the same antigen, this memory cell will start to reproduce and to produce antibody. Vaccination (artificially) by viable attenuated or killed microbes or viruses are able to produce similar immune response can do by viable non-attenuated microbes. Many vaccines are able to induce correct immunologic memory. However quite number need more than one dosage. For that, it is recommended to give the same vaccine again after the first exposure. The second dosage was given the name first booster [40]. In most cases, the first booster is not enough and second booster is required to ensure correct immunization. Viable microbes or viruses ensure the full activation, however in some cases attenuated microbes or viruses are not available or too dangerous to be used. Particularly with immune compromised or immune deficient patients However, in contrast killed microbes, microbial ghosts or damaged viruses will not be able in some cases to induce correct immunization such as infants and the presence of maternal antibody, immune compromised and immunodeficiency patients [38]. Host factors such as age, nutritional factors, genetics, and coexisting disease, may also affect the response [41]. Maternal antibodies interfere with vaccine by interacting against it (antigen/ antibody) and dilute the response of the immune system [42]. An alternative strategy is done by vaccine the mother before pregnant. In addition, in modern vaccine, vaccination of pregnant mother becomes available nowadays by safe vaccines.

If antigen is changed – the immune response is changed

One should observe that our immune system provide us a perfect protection against mild infections like the common cold. Common cold will be a good example to show that our immune system could sense that there is a need to do some response and not satisfy by the already existed antibodies. Why not to react naturally? The new microbe is responsible for the new infection or not new but modified old ones due to changes were happened in the surface antigens that will case the immune system to response. There is no need for being so worried from mild infection. We did not know what the common mild infectious (which could be passed by the aid of some help), provide to us. Even a fatal disease, the smallpox can be avoided by the infection with the cowpox virus. The mild infection with the common cold influenza will flourish our immune system and its components including the memory B cells. In addition, might be able to protect us from severe influenza infection or from unknown type of infection.

Old civilization and the vaccines technology [3,15,33,43]

There are evidences that ancient civilizations developed some practice could improve the immune system. They are either knowledgeable more than we expected or they find solutions by chances.

In Egypt, there is a calibration between two seasons ("Sham ennisim' at the beginning of spring). In that day the Egyptian, eat salted sub-rotten fishes usually green-onion also included, which contain of both probiotic and antibiotic. The mild rotten fish is excessively subjecting to salting. Such salting will kill most of the existing microbes including pathogenic and non-pathogenic hence will let them to be as an oral vaccine upon their eat.

Like any creature, each microbe has a life cycle ends with the death. The microbes' cell walls after their death could stay longer and resist decaying. However, due to natural decay, environmental effect, enzymatic activity of other microbes, or any other expected mechanism. That can lead the microbes to loss their cytoplasm or become inactive. For that, cell ghosts and microbial ghosts are produced daily in our bodies (inside, or outside). They are produced in the lung, in the stomach, in the surface of our skin, in our aged food and so on. So dead and microbial ghost are natural phenomenon. They play different roles in natural immunization.

It is also an observation that in Egypt there is another type of food, which might be designed to be used as a food that could immunize. It is the old cheese or the Mish (also can be said

"Mesh). Mish might be the first invented strategy to attenuate microbes and the first known vaccines, but the question is did the ancient Egyptians know that?

Personally, the Mish was produced in my father family house in the village Bardala at Kafer Al Dawar city (~ 1940 - 1985) until nobody existed in this house permanently. One closed fermentation system plugged with rice-straw plug (or other natural tissues) where cheese left to be aged and seeded with a previous old cheese. This system was a static continuous fermentation, which means it used for decades. The taken pieces of cheese are substituted regularly with new ones. Such aged cheese must contain dead and attenuated microbes, which should stimulate the immune system.

Another signal about one of the traditional used technique in Egypt concern with the skin infection particularly in the face, That was simply treated by taking with your finger soap from your mouth when you wake up early in the morning and putting it on the infected area, only one time/day. Such traditional practice might not be documented or neglected because some might found it not scientific. In fact being used by many, being practical, and transferred through generations from unknown time their success makes their food for investigation. Personally, my Aunt advised me to do that. After being a microbiologist, I understand that mouth which contains microflora, saliva contain antibodies and lysozyme, which they all could collaborate to kill mild superficial infection.

The modern vaccine start with simple observation, which was well known within farmer but less explained until a physician explains it. The milkmaid who infected in their hand by cowpox is protected against smallpox and she was happy because her face will be beautiful. She starts to song and a physician heard this song "*I shall never have smallpox for I have had cowpox. I shall never have an ugly pockmarked face.*" The cow name is "Blossom". The physician start to investigate the case, then he concluded that the infection with cowpox will protect against the smallpox. He made manual infection from arm to arm by the lymph node.

By tracing such practice by milkmaid in the old literature there is evidence that such practice are well known. The following text has been found.

'Where are you going, my pretty maid?'

'I'm going a-milking, sir' she said.

'May I go with you, my pretty maid?'

'You're kindly welcome, sir' she said.

'What is your father, my pretty maid?'

'My father's a farmer, sir' she said.

'What is your fortune, my pretty maid?'

'My face is my fortune, sir' she said.

'Then I cannot marry you, my pretty maid.'

'Nobody asked you, sir' she said.

Vaccine

Vaccine type

The word vaccine was derived from the Latin name of the cowpox Variolae vaccine which Edward Jenner prove in 1798 that it is able to prevent smallpox infection in humans. Today the term 'vaccine' applies to all preparation derived from living organisms or viruses, that enhances immunity against disease and are able to prevent (prophylactic vaccines) or, treat disease (therapeutic vaccines). All diseases are not yet prevented by the vaccination. Vaccines could not successfully apply for some elderly and pregnant women [42]. The major reported problems was for preparing vaccines against HIV, TB and malaria and also for elderly, infants and cancer patients [42].

Classification of vaccines

Fatal microbes usually have virulence factors able to bypass the immune system defenses. For that, then virulence factors should be reduced or the microbe itself should be killed.

Variolation: Variolation is the obsolete process of inoculating a susceptible person with material taken from a vesicle of a person who has smallpox. Variolation was known also for the cheep pox and some other viruses.

Live attenuated vaccines [44,45]: Attenuated microbes are living microbes that were weakened or attenuated, usually by cultivating them under suboptimal conditions. Genetic modification also can reduce the microbe's ability to cause disease.

Pathogens, which are attenuated, or weakened, in a laboratory, usually by, repeated culturing. For example, the measles virus used as a vaccine today was isolated from a child with measles disease in 1954. Almost 10 years of serial passage using tissue culture media was required to transform the wild virus into attenuated vaccine virus.

Live attenuated vaccines are viable pathogens but after reducing their virulence to the limit where the body is able to control them. In addition, to produce antibodies against their mother viable cells and to be safe enough to be introduced to a person who has no previous infections or has no antibodies to such live attenuated or the wild pathogens. In another word, if the wild pathogen used it will be harmful but if attenuated microbe is, used one could pass without complication and with correct immunization to it (the vaccine itself) and to the fatal viable one (after infection) and for long time or permanent.

Inactivated vaccines: Whole organisms that were inactivated by chemical, thermal or other means is named inactivated vaccines. Inactivated vaccine is not alive, hence it cannot replicate. Removing the replication ability from a pathogen will turn it to dead microbes, which can be used as vaccine, and the used amount can be adjusted to safe quality. They are deactivated by any safe mean such as physical (heat and gamma radiation), chemical (usually formalin, or microbial ghosts prepared by Sponge-like protocol) or even biological (e.g. Bacterial ghosts by using the *E lysis* gene strategy or ghost

cells prepared by lysozyme) [28,33]. The inactivated pathogens can be used as they are. In case of viruses, one should be sure that their RNA/DNA is well deteriorated. Because viruses are non-live particles but could be turn to be alive and active in the related host cells. So, any of the viruses' genetic elements must be deteriorated. For example, interring a very small fragment of the virus representing the virus promoter this will let the cell to produce the protein downstream to the virus promoter in large amount, which might cause health problems. The attenuated microbe can be subjected to further purification to collect the most antigenic parts such as in case of polysaccharides capsule of pneumococcus. Inactivated vaccines generally used in dosage number more than the attenuated ones. And also need more boosters to insure correct immune response.

Inactivated whole virus influenza vaccine and completely inactivated bacterial vaccines (pertussis, typhoid, cholera and plague) are no longer available in the United States [40].

Polysaccharides-conjugate vaccine: Polysaccharides vaccines for pneumococcal disease, meningococcal disease [41], Haemophilus influenza type b (Hib) [48,49] and *Salmonella Typhi* [50,51] are examples. In the late 1980s, it was discovered that the problems of less antigenic due to the polysaccharide cotes could be overcome through a process called conjugation. The polysaccharides are chemically combined with a protein molecule. Conjugation changes the immune response from T-cell independent to T-cell dependent. Such change increases the vaccine immunogenicity [37,52] in infants. The first conjugated polysaccharides vaccine was for Hib. A conjugate vaccine for pneumococcal disease was licensed in 2000. A meningococcal conjugate vaccine was licensed in 2005 [53,54].

Combining vaccines: Combining several serotypes of a disease-causing microbes in a single vaccine is well established (e.g. 13-valent pneumococcal conjugate vaccine) to provide protection against several different diseases at once. These combined vaccines may contain different types of vaccines. Diphtheria, tetanus, pertussis, Haemophilus influenzae type b, Hepatitis B and polio are commonly used in one combined vaccine. These vaccines incorporate both viral and bacterial vaccines and contain toxoids, purified protein subunit vaccine, conjugated polysaccharides vaccine, recombinant protein vaccine and inactivated viral vaccine respectively.

Vaccines may also contain antigens against several types (or serotypes) of the same disease-causing organism to provide protection against each type. Polio and influenza vaccines each protect against three types of virus and some bacterial vaccines like pneumococcal vaccine protect against up to 23 different serotypes of *S. pneumoniae*.

Toxoid vaccines: [55] A bacterial toxin that was weakened until it is no longer toxic. However, it is still strong enough to induce the formation of antibodies and immunity to the specific disease caused by the toxin and to prevent diseases caused by bacteria that produce toxins. When the immune system receives a vaccine containing a toxoid it produces antibodies against it. As an example of toxoid vaccine, the vaccine containing diphtheria and tetanus toxoids (DTaP). The used quantity usually adjusted to produce correct immune response without harming. Using purified microbial toxin has an advantage over using the weaking or attenuated viable microbe. That will give chance for better immunization and safe usage out of the risk of using attenuated microbe might be turn to fully viable one at any time.

Subunit vaccines: Include only parts of the microbes. The pertussis (whooping cough) which is a part of the DTaP vaccine is an example of subunit vaccine. Fractional vaccines include subunits (hepatitis B, influenza, acellular pertussis, human papilloma virus, and anthrax) and toxoids (diphtheria [56], tetanus. [57-59]) A subunit vaccine for Lyme disease is no longer available in the United States.

Genetically engineered vaccine: Antigens can be produced artificially as it is by the genetic engineering and molecular biology tools. Or, they can be engineered and modified using the mutagenesis different tools. For more details, refer to Amara [30]. These products are sometimes referred to as recombinant vaccines. Four genetically engineered vaccines are currently available in the United States. Hepatitis B and Human Papillomavirus (HPV) vaccines are produced by insertion of a segment of the respective viral gene into the gene of a yeast cell or virus. The modified yeast cell produces pure hepatitis B surface antigen or HPV capsid protein when it grows. Live typhoid vaccine (Ty21a) is *Salmonella Typhi* bacteria that were genetically modified to be safe (no illness). Live attenuated influenza vaccine was engineered to replicate effectively in the mucosa of the nasopharynx but not in the lungs.

Passive immunity is based on preparing antibodies in both of animals and humans for a specific requires and collecting them to be ready requirement. Such production out of the host may lead to some kind of incompatibility after administration including antigenicity and less activity. In cases of the antibody-conjugate, such problems also raised.

Antigens as well are subject for both of the genetic engineering and protein engineering, which can be employed to improve both of the antigen production and their loading on safer surface also as their alteration. With the help of both of the recombinant DNA technology and the protein engineering altering protein either as antigen or as antibodies to match certain structure/function/specificity, become, available. Engineering antibodies can be mediated on the level of genes or proteins. Approximately 350 biotechnology drugs currently undergo development. These include vaccines, gene therapy, antisense technology, and antibodies derived from 'humanized' transgenic mice [56,60].

Protein engineering is important for improving therapeutical pharmaceutics proteins with specified increasing their solubility and stability. Major protein-based drugs practical application problem that they show activity in vitro and have promising role in medicinal practical application but they are primary molecules with suboptimal affinity and poor half-life in vivo, which lead to poor efficaciousness [61].

Antibodies are protein in nature and antigens as well are protein in most cases. As protein but in a few cases is non-human

that caused immune responses against the vaccine itself. Affinity, half-life, and dosing regimen are all inter-related and act as their role in determining the clinical efficacy from different point of view including the production cost. The immunization and the immune response are generally the most crucial issue peculiarly for those which will inter to our body from various rote of administration and will be a subject for the immune system response [1-3].

For example Pulmozyme (Genentech) is human DNAse derived drugs used in managing cystic fibrosis and bovine pancreatic DNAse I, study the immune response for such product is one of the most important issue [62].

The immunogenicity of mouse antibodies to a human protein was a major scientific issue to be dissolved. This major problem raised by the early monoclonal antibodies. Chimaeric antibodies by fusing mouse variable domains to human constant domains improve the body acceptance. This chimaeric hold binding specificity and reduce the amount of mouse sequence in their backbone. In 1998, Remicade (Centocor), a TNF-neutralizing chimaeric monoclonal antibody, was approved for use in treating Crohn's disease and rheumatoid arthritis [63]. A reduction in monoclonal antibody immunogenicity has taken a stage further by complementarily determining region (CDR) grafting, where the CDRs of mouse antibodies were grafted onto human frameworks to further reduce the proportion of mouse sequences in the drug while retaining its binding specificity.

Other forms of vaccines: Such forms include proteins vaccine, nucleic acids vaccine and recombinant vaccines. Innovative technologies currently used in vaccine research and development including adjuvant, vectors, nucleic acid vaccines and structure based antigen design [42].

Microbial Ghosts (MGs) The Cheapest Vaccine Technology

The first shown empty microbial cells might be in the first sample shown under the light microscope by Antonie van Leeuwenhoek. Red blood cells and bacteriophage were evacuated earlier than the bacteria [28,33]. The bacteriophage lysis of gram negative bacteria was also leading to empty cells [64]. However, the first directed evacuation of cells was aimed to isolate the cells DNA, RNA and protein. The cells themselves did not seem to be interesting (adjusting the conditions to keep the cells 3D structure safe). The first attempt to evacuate microbe without damaging their structure was for the bacteriophage. And apparently the first cells were for the Red Blood Cells (RBCs). After the emerging of the genetic engineering and molecular biology tools, the bacteriophage *E lysis* gene was cloned; the *E. coli* was evacuated using the *E lysis* protein. The heat sensitive promoter controlled the *E lysis* gene, which enable producing biomass following by the expression of the *E lysis* protein using different temperature. After that, E lysis protein-based method show many successes. However, it was a weak point that the *E lysis* gene based method is restricted only to the gram negative bacteria. The gram-positive bacteria have an additional layer, which interfere with the mechanism of the lysis process by the

E lysis protein. Some authors upon their describing their work have reported that microbial cell was evacuated due to the effect of some chemical compounds or some physical parameters. Also biological factors are reported [65].

The concept of using MIC (Minimum Inhibition Concentration) and MGC (Minimum Growth Concentration)

Apparently, the first attempt to design a full protocol to introduce pore(s) in bacterial cells using the concept of using critical chemical concentration (both of the MIC and the MGC) of some compounds which did not damage the cells 3D structure under the used concentration were Amara, et al. [66].

Amara [28,33] suggested that this method will be able to evacuate any microbe hence any parasite (or one of its stage) and it might be a critical step to establish a new science given the name "Evacology". It is concerning with evacuating cells from their cytoplasmic content without deforming the cell 3D structure. It is a science emerged from the BGs (bacterial ghosts), which recently after turning viruses and eukaryotic to ghosts was named Microbial Ghosts (MGs) [26,55]. For more details about both of the E lysis gene based protocol and the Sponge-like protocol for MGs preparation application, involved microbes, etc., refer to Amara [26,55]. Using lysozyme from the hen egg white to prepare ghosts from *Bacillus stearothermophilus* might be the first protocol to prepare oral vaccines without chemicals and with components [28,33] from the nature.

Microbial Ghosts (MGs) definition as described recently by Amara [28] is "MGs are empty and dead microbial cells or viruses (envelopes) devoid of cytoplasmic contents or any internal fluidized or any genetic element. One or more than one pore happened in their cell walls lead to direct remove of the genetic components. They have correct 3D structure, morphology and native surface antigens structure and able to induce the immune system of the delivered host to produce specific antibodies that could react correctly with the mother viable cell or viruses. As being empty cells, they can be used as drug delivery system for various drugs, genes and antigen or surface antigen expressed protein from another potent pathogenic bacteria. The critical chemical concentration and enzymes activity methods have extended the spectrum of the BGs from the gram negative bacteria to all types of bacterial strains including gram-positive, Archaea as well as eukaryotes such as yeast ghosts. Viruses were achieved and parasites are expected to join the Ghosts Family. The future of the MGs is bright [33]". This simple method for turning microbes to ghosts has kept free and did not covered by any patent, to be in the hand of any one particularly the developmental countries as well as any country or population in need for it.

Cancer Immunotherapy

Using immunological tactics for cancer treatment is based on inducing an extra-activation for the immune system aiming to produce antibodies that might somehow attack the cancer cells. Such random activation will lead to different forms (in major case) of non-specific antibodies but might also produce in minor case specific ones. Such hopes might produce antibodies that can incapacitate the cancer cells. In experimental animals, success was achieved in which the frequency of induced tumors was reduced after increasing the level of immunological responsiveness by administration of adjuvant (e.g. BCG (Bacillus of Calmette and Guérin), vaccine, *Corynebacterium parvum*, Vaccinia vaccine, or extracts of yeast cells). Drugs such as levamisole that stimulates the immune mechanisms provide another possible means of non-specific stimulation's. Active immunization with malignant cells is achieved by killing tumor cells by irradiating them *in vitro* before re-injecting them. The antigenicity remains and a degree of immunity to living tumor cells is imparted.

Passive immunity may be transmitted by transferring serum, lymphocytes or bone marrow from an animal, which was immunized against proteins of human bladder cancer [2,35,38].

Lymphokines are peptides produced by lymphocytes that regulate the immune system and mobilize defences against foreign invaders including bacterial and viral infections [67,68].

The polyclonal antisera are antibodies resulting from injecting animals with the appropriate antigen for large production of antibodies.

The specificity of the antibody can vary quite markedly, because large molecules and cells are in themselves a collection of different antigens, each of which may elicit an antibody production. Each antibody may be produced in different amounts, and they may bind to their corresponding antigen with different degrees of affinity [68,69].

The other strategy is aiming to produce more specific antibodies such as the monoclonal antibodies, which will attack the cancer cells more specifically. It is a hope to find specific target in the cancer cells that can be attacked by the antibodies. The main problem is that the cancer cells are collectively similar to the human cells. For that the person who have cancer, its immune system did not recognize the presence of subnormal cells, the cancer cells.

The monoclonal antibody has a greater antigen specificity, homogeneity and availability, produced by a clone, or colony of cells that drive from white blood mother cell-B-lymphocyte and so are identical. When an antigen stimulates them to manufacture an antibody, they all make the same one. Their advantage comes from their high specificity for the antigen produced *in vitro*, and their homogeneity. Hybridomas between cancer cells and these immunized lymphocytes can be used as continues source for monoclonal antibodies. Leukopheresis machine, a blood cell separator, takes about four hours to separate red from white blood cells, re-infuse the red cells immediately, and then isolates five to ten billion white blood cells (leukocytes) in a plastic bag. These leukocytes (potential killer cells) were then incubated with monoclonal antibodies that would instruct them to attack cancer cells. This mixture can be intravenously infused into the patient who over one to two hours can go back home on the same day [68,69].

Another promising strategy is to synthesis modified antibodies (antibody-conjugate or chimeric) to be more target and fatal to the cells [70]. Such modified antibodies either by doing modification in its own backbone or by hybrid it with another fatal protein (chimeric) or linking it with bioactive molecules. The tools of the genetic engineering, PE and protein chemistry can do that. Catalytic antibodies are proteins that normally bind to a specific molecule but do not alter the bound molecule in any way.

A catalytic antibody is a variant of an antibody which was changed by mutations to have a novel sequence that folds into a structure, resulting into a specific reaction (such as amide bond formation, ester hydrolysis, and decarboxylation). Catalytic antibodies function like enzymes, and are created to catalyze reactions for which there are no naturally occurring enzymes [71,72]. Fifty or more reactions were made by the action of catalytic antibodies, which were obtained individually by the methods of Protein Engineering (PE) [30,73-76]

What does a vaccine contain?

Vaccines from different types were prepared in different formula. The most important criteria in the vaccine formulathat the final formula must ensure good administration and shelf life. Vaccines are administered mostly in a liquid form. They are injected, either by oral, or by intranasal routes. Vaccines are composed of either the entire disease-causing microorganism (Viable, attenuated, killed or in the form of MGs) or some of its components.

The first unspecific vaccine used by modern tools, the cowpox, breeding centres was established for animal vaccinifers. Cows with cowpox symptoms were transported from place to place. After that development, the primitively preserved vaccinal lymphs were to be transported. Liquid paraffin, lanolin and glycerine were evaluated. Glycerine proved to be the most desirable.

In addition to its antigenic components, vaccines are formulated (mixed) with other fluids (such as water or saline), additives or preservatives and sometimes adjuvant. These ingredients are known as the excipients [77]. These ensure the quality and potency of the vaccine over its shelf life. That because vaccine is mostly contains proteins, which are substrates to proteolytic enzymes, and active proteins of any existed microbes. Vaccines are always formulated to be both safe and immunogenic when injected into humans. Vaccines are usually formulated as liquids, but may be freeze-dried (lyophilized) for reconstitution immediately prior to the time of injection. Preservatives ensure the sterility of the vaccine over the period of its shelf life. When a first dose of vaccine is extracted from a multi-dose container, a preservative will protect the remaining product from any bacteria that may be introduced into the container. Preservatives if needed are added during manufacture to prevent microbial contamination. The used preservatives are non-toxic in the used amounts and do not conflict with the potency of vaccines.

Vaccines and the immune response

The antibody produced by B lymphocytes is the main cell involved in the process [78]. Cytotoxic CD8+ T Lymphocytes (CTL) limits the spread of infectious agents by recognizing and killing infected cells or secreting specific antiviral cytokines. Growth factors and signals provided by CD4+ T helper (Th) lymphocytes are commonly subdivided into T helper 1 (Th1) and T helper 2 (Th2) subtypes, are essential in the process. They are controlled by regulatory T cells (Treg) that are involved in maintaining immune tolerance [79]. CD4+ T cells are required for most antibody responses, while antibodies exert significant influences on T cell responses to intracellular antigen [79].

The immune response to protein free polysaccharides vaccine is typically T-cell independent. Vaccines are able to stimulate B cells without the assistance of T-helper cells. T-cell–independent antigens, including polysaccharides vaccines, are not consistently immunogenic in children younger than 2 years of age probably because of immaturity of the immune system. Repeated doses of most inactivated protein vaccines cause the antibody titer to go progressively higher, or "boost." This does not occur with polysaccharides antigens. Repeat doses of polysaccharides vaccines usually do not cause a booster response. Antibody induced with polysaccharides vaccines has less functional activity than that induced by protein antigens. This is because the predominant antibody produced in response to most polysaccharides vaccines is IgM, and little IgG is produced

The French and English early contact with the Middle East and Africa

It is clear that after the invasion of Nablion to the Middle East particularly to the Egypt; many traditional practices were gained. Egypt, which is one of the old known civilizations and might be the oldest one located in the central of the world, is a pool where experiences are collected from everywhere, improved or modified and new one developed. The French crop uses a strategy to document everything as it is and precisely which was helpful in many fields of sciences. Also, the traditional medicinal Arabic books, which describe many tactics for improving health, protection and curing from different diseases, which in some time contain unexplained tools, all that were in the hand of the French Moreover, they succeeded to reintroduce the importance of the medicinal advices of the ancient Egyptian civilization, the Pharaonic (including some of their medicinal applications). Perhaps the oldest known reservoir, which allows correct and safe fermentation including safe air transfer, was an Egyptian invention for producing old cheese. Such facts will not be explained in detail in this review. However, the progress of the vaccines against times is summarized in Table 1 [3,15].

Smallpox

There is documentary evidence of the use of the inoculation technique against smallpox by nomadic herders in Africa (e.g. Tulani). Somebody is in Africa mentioned inoculation against sheep pox as long ago as the 16th Century. Apparently, the human variolation was attempted in Egypt, Sudan, China, East Africa or

Table 1: The time table of vaccines updated after Lahariya 2014 [3,15,81]

Time	Achievement in the filed of immunology and vaccine production
Ancient time Egypt, India, China and others	Mummification and its process including cuprous drying, the use of resin and wine, proves of the awareness of the presence of the microbes or at least deterioration agents.
7000 BC	Rabies and smallpox are acknowledged in Egypt and Africa.
3000 BC	Smallpox is thought to have originated from India or Egypt.
430 BC	Thucydides describing a plague in Athens, he wrote that only those who had recovered from the plague could nurse the sick because they could not contract the disease a second time.
300 BC	Description of smallpox in Sanskrit literature.
700 BC	Buddhist monks drank snake venom in order to acquire immunity versus snakebites.
910 AD	Smallpox was differentiated from Measles by Abu Bakr El-Razi and its use of rose extract to protect the eye during the infection.
1000 AD	Inoculation documented from China Inoculation was reportedly practicable in India also.
1500 AD	Chinese and Turks reports suggest that the dried crusts derived from smallpox pustules were either inhaled into the nostrils or inserted into small cuts in the skin (a technique called variolation). Also reported by herders in Africa and for human protection in the middle east.
1600	Documented evidences of practice of inoculation (variolation) from India.
1718	Lady Mary Wortley Montagu, the wife of the British ambassador to Constantinople, observed the positive effect of the variolation on the native population and had the technique performed on her own children.
1767	Dr. Holwell gave a description of practice of inoculation in India to College of Physicians in London.
1774	Benjamin Jesty did experiment on his wife and two children by injecting cowpox matter.
1796	Edward Jenner conducted the famous observation on milkmaids.
1798	Jenner's observations were published and smallpox vaccine was discovered.
1802	First documented smallpox vaccination was done in India.
1810	Gennaro Galbiati, an Italian physician, used cows for vaccine production.
1870	Animal vaccine production in USA.
1876	First vaccine farm in Lakeview, New Jersey, USA.
1879	First laboratory vaccine produced by Louis Pasteur for Chicken Cholera.
1885	Louis Pasteur, known for his animal vaccines, injects a rabies vaccine into two people and causes controversy. Few people at the time were comfortable with the idea of introducing a deadly, live virus into a human being.
1897	A killed vaccine for the plague was developed.
1896	Vaccine for cholera and typhoid are developed using killed versions of bacteria.
1901	Noble prize for Emil von Behring – Germany for his work on serum antitoxins.
1905	Noble prize for Robert Koch – Germany for his work on cellular immunity to tuberculosis.
1908	Noble prize for Elie Metchnikoff – Russia for his work on the role of phagocytosis and Paul Ehrlich for his work on antitoxin.
1904-1908	Typhoid vaccine trial was done on British Army officials posted to India and Egypt.
1909	Lucien Camus develops first air-dried smallpox vaccine in Paris.
1913	Noble prize for Charles Richet – France for his work on Anaphylaxis.
1923	A powerful toxin from diphtheria bacteria is chemically inactivated and used as a "toxoid" to kill bacteria. Before the vaccine, as many as 200,000 cases occurred each year, with 15,000 deaths.
1930	Noble prize for Karl Landsteiner – United State for his work and his discovery for the blood group.
1950	Noble prize for Max Theiler – South Africa for his work on the development of yellow fever vaccine.
1960	Noble prize for F. Macfarlane Burnet- Australia and Peter Medawar – Great Britain for their work and their discovery for the acquired immunological tolerance.
Between 1980 and 1995,	Only four children died from diphtheria.
1926	A killed vaccine for pertussis ("whooping cough") is developed, using the whole pertussis organism.
1927	A tetanus "toxoid"is developed. Before the tetanus vaccine, there were about 600 cases a year in the U.S. with 180 deaths, now about 70 cases occur, causing 15 deaths.
Late 1940s	Tetanus was combined with diphtheria and pertussis as the children's vaccine "DTP."

1948	BCG, a vaccine for tuberculosis developed by Albert Calmette and Camille Guérin. BCG Laboratory in Guindy, Madras (now Chennai) set up BCG vaccination was started at pilot level.
1954	Jonas Salk develops a killed polio virus that decreased paralysis cases from 20,000 in 1952 to 1,600 in 1960.
1958	World Health Assembly passed a resolution to eradicate smallpox.
1961	Alfred Sabin develops an oral polio vaccine using a live virus, which is easy to take and was successful at eliminating the spread of polio.
1963	A safe and effective measles vaccine is developed, reducing the number of cases from four million in 1962 and 3,000 deaths, to 309 cases in 1995, with no deaths.
1964	A killed rabies vaccine is developed, but requires up to 30 painful shots in the abdomen. By 1980, a newer version requires only five shots in the arm to protect against this deadly disease.
1967	A vaccine for mumps is licensed, reducing the incidence from about 200,000 cases annually with 20 to 30 deaths to about 600 cases with no deaths.
1970	Several strains of rubella are weakened to make a vaccine. Between 1964 and 1965 there were about 12 million cases leading to birth defects in 20,000 children. Now there are about five cases of birth defects each year.
1972	Noble prize for Rodney R. Porter – Great Britain and Gerald M. Edelman – United State for their work on the chemical structure of the antibodies.
1975	Last case of smallpox was reported.
1977	Noble prize for Rosalyn R. Yalow – United State for their work on the development of radioimmunoassay.
1977	Last case of smallpox was reported from the world India declared smallpox free Source: Refs 3, 5-20.
1971,	Measles, mumps and rubella vaccines were combined into a single shot known as MMR.
1970s & 80s	Meningoccocal, pneumococcal and Haemophilus influenza type b (Hib) vaccines are developed, using a piece of the bacteria cover to provide a safe antigen for the body to react to. These vaccines help protect against life-threatening diseases such as meningitis, blood infections and some pneumonias.
1980	Noble prize for George Snell – United State and Jean Daussct – France and Baruj Benacerraf – United State for their work on the MHC.
1984	Noble prize for Cesar Milstein Great Britain and Georges E. Köhler –Germany for their work on the monoclonal antibodies and Niels K. Jerne – Denmark for his work on immune regulatory theories.
1986	A vaccine for hepatitis B is licensed with an antigen that is cloned rather than grown.
1987	Noble prize for Susumu Tonegawa –Japan for his work on gene rearrangement in antibody production.
1990	A killed vaccine for hepatitis A is developed.
1991	Noble prize for E. Donnall Thomas and Joseph Murray – United States for their work on Transplantation immunology.
1995	A varicella (chicken pox) vaccine is licensed for use in children.
1996	Noble prize for Peter C. Doherty Australia and Rolf M. Zinkernagel – Switzerland for their work on the role of MHC in antigen recognition by T cells.
1996	The first "DTaP"vaccine is approved, using only part of the pertussis organism, combined with diphtheria and tetanus. Annual pertussis deaths have dropped from 8,000 before the vaccine to about 10 today.
2000	Influenza vaccine use reaches 70 million doses. Premature death related to influenza is estimated at 20,000 people annually. While many advances have occurred in the last two centuries, science is poised for even more in the future.

India from undocumented time. In Europe, efforts for immunizing by inoculation were made for the sheeppox. Sheeppox is a virus, which is close to smallpox in humans. Belgian physician Willems made effort with the bovine contagious pleuropneumonia disease, which was inoculated by using the ancient civilization practice at 1853. The inoculation was made in the base of the tail of the animals with a small amount of the isolated infective material [3].

In the French language, a term was used to refer specifically to inoculation with sheep pox, clavélisation, from the French word for the disease, clavelée.

Lady Mary Wortley Montagu, the wife of the British ambassador to Constantinople, observed the positive effect of the variolation on the native population and had the technique performed on her own children.

Inoculation with smallpox was by using only human material, serous matter from pustules (A small-inflamed elevation of skin containing pus; a blister filled with pus) and scabs taken from a subject with a mild form of the disease.

Inoculation with unspecific virus aiming to immunization against the smallpox was made by using the cowpox by an English doctor, Edward Jenner (1749-1823) [3].

The inoculation with unspecific virus but have the same symptoms (but non-fatal) the cowpox which proposed by Jenner in 1798 prove to be more efficient but also safer.

Lombard, et al. [3] wrote "At the beginning of the 19th Century, Jenner's vaccination procedure rapidly spread around the world, supported by governments favourable to a measure that could reduce the devastating effects of epidemics on their populations. The President of the United States of America (USA); the Tsar of Russia; the King of Sweden; the Emperor of France, Napoleon I; and the Pasha of Egypt, Ali Mohammed, to mention but a few, were greatly enthusiastic about the vaccine and actively promulgated it, in some cases, as with Napoleon I in 1812, going as far as to make it compulsory in the army, and even in society as a whole. When it came to putting these plans into action, however, it was of course quite a different story."

The treating the smallpox in such age eradicating it before fully clarifying the origin and behaviour of poxviruses and their vaccines throughout history [3]

The Pasteurian Era

Apparently, the Pasteurian era is depending on the microbial attenuation, which is a trend which attracts many of the scientists before Louis Pasteur. Apparently, the target was the anthrax, which was subjected to many form of attenuation to use it as vaccine [3,15].

Fowl Cholera

In 1876, the French veterinarian Henri Toussaint (1847-1890) cultured a causal bacterium of fowl cholera in neutralized urine, described two years later by Perroncito (and subsequently known as *Pasteurella avicida* or *gallicida*, and now as *P. multocida*). The hen survived inoculation with the 'forgotten' cultures and even became resistant to a subsequent, virulent inoculation. It was in fact an empirical trial to attenuate the culture by re-seeding the medium at longer intervals devised by Emile Roux with the help of a system of continuous oxygenation to accelerate the ageing process [3].

Anthrax

John Burdett-Sanderson and William Greenfield (England, in 1878), by re-seeding the culture at 35°C succeeded to attenuate the virulence of the strain without affecting its immunizing potential. In 1880, Henry Toussaint proposed that if animals were vaccinated with blood heated at 55°C they could then survive an otherwise lethal inoculation. He successfully immunized five ewes using this technique.

Toussaint admires Pasteur, in the year 1879. He isolated the microbe of "cholera des poules" (Pasteurella) and gives Pasteur this new microbe. Pasteur used this microbe for his research about reducing the virulence by successive subculturing.

In 1880 his publishing on July 12 at the Academy of science, Toussaint presents his successful results with an attenuated vaccine against anthrax. He used dog's sheep and vaccine, which he has reduced off virulence by chemical manner [82].

The fowl cholera culture, which might be attenuated chemically and then left as Pasteur, describes longer to be aged. The chickens become ill, but not dead. And after they are given active viable culture, they were survived and become protected.

Following his teacher, in 1881, Louis Pasteur undertook his still famous trial at the farm in Pouilly-le-Fort, near Paris. In the presence of an extensive public consisting of farmers and veterinarians, he compared the behaviour of vaccinated and unvaccinated sheep. Initially, his vaccine had consisted of a culture attenuated simply by heating same as what done by Henry Toussaint. However, Pasteur's disciples persuaded him to take the precaution of using an attenuated culture also containing an antiseptic known to inhibit the formation of spores (this was 'the secret of Pouilly-le-Fort'), [3].

Swine erysipelas

An attenuated vaccine against swine erysipelas, a disease caused by a bacillus that had been discovered by Louis Thuillier, This attenuated vaccine was lapinised, (attenuated by serial passages through rabbits). Such observation made a critical breakthrough.

The observation that an increase in virulence when a disease is passed from one individual to another during an epidemic is common to both physicians and veterinarians in contrast, the *in vivo* attenuation of virulence when microbes affecting one species are passed through another species is critical observation of long date and research. It proved to be a successful source of research for the Pasteurian school [3,70,83,92].

Rabies

Pierre-Victor Galtier (1846-1908) a veterinarian, a student of Chauveau at the Lyons veterinary school (France), who demonstrated rabies to be an affectionateness of the nervous system, with a variable incubation period. In 1879, he evoked that laboratory dogs could be replaced by rabbits, which arise a paralytic form of the disease with a faster course than in dogs, so making them more controllable.

In 1881 and 1882, Louis Pasteur and his students Charles Chamberland, Emile Roux and Louis Thuillier entered the fray and modified Galtier's technique by inoculating nervous tissue from a rabid animal directly into the brain after trephination. By successive passages in dogs, they obtained a virus of maximal virulence coupled with a fixed incubation period of around 10 days. They then were use the strategy of change that host species to attenuate the virulence of the virus indirectly by passages through rabbits. Emile Roux made up the chosen attenuation procedure. It consisted of suspending the spinal cord of a rabid rabbit in a flask, in a warm dry atmosphere, to accomplish slow desiccation. Using animals as alive propagating medium, Pasteur and his group succeeded in producing 'attenuated viruses of different strengths', in short a standardized range of viruses, the weakest of which could be used to prepare a vaccine [3,93,97].

Bovine and human tuberculosis

In 1882, Robert Koch (1843-1910) described the tubercle bacillus responsible for tuberculosis in humans. Tubercular infection was also well known in cattle. However, Theobald Smith in the USA drew attention to differences between the bovine and human bacilli.

Koch suggested inoculating a calf with human tubercle bacilli treated with phenol. Working on behalf of the firm Hoechst, Emil von Behring prepared a bovo-vaccine based on desiccated human bacilli reduced to a powder. At around the same time, a physician in Berlin named Friedmann suggested using a tuberculosis bacillus in humans that was not thought pathogenic since it came from an animal of a distant species, a turtle [3].

Vaccination against tuberculosis is still based on the historic vaccine of Calmette and Guérin whose initials it bears (BCG vaccine [bilious bacillus vaccine of Calmette and Guerin] [98,99]. In 1897, Albert Calmette and Camille Guérin, a student of Nocard, start to working together. A bovine bacillus, isolated by Nocard in a sample taken from the udder of a tuberculous cow, was cultured by passages through glycerinated bile potato medium as being a laboratory strain it eventually resulting in an attenuated form. The tubercular bacillus has a fatty capsule, which makes it difficult to blend. The idea of using bovine bile in the culture medium most likely came from the veterinarian Vallée, who had used dilapidated bacilli in his vaccination trials: at that time. The bacillus, from 1908 to 1921, was subsequently transformed by serial passages (230 passages) without regaining virulence in susceptible animals. The vaccine was called 'BCG' (which stands for 'vaccin bilié de Calmette et Guérin') [3].

Also in France in 1921, the first clinical trial of BCG took place, involving a newborn child in a family with a history of tuberculosis [100].

Adjuvants

Adjuvants are agents which increase the stimulation of the immune system by enhancing antigen presentation (depot formulation, delivery systems) and/or by providing co-stimulatory signals (immunomodulators). Aluminium salts are most often used in today's vaccines.

The discovery of adjuvants of immunity by Gaston Ramon (1886-1963), a veterinarian at the Pasteur Institute who became one of the first Directors General of the World Organization for Animal Health (OIE) (then known as the Office International des Epizooties), following its creation in Paris in 1924.

Gaston Ramon developed an anti tetanus vaccine in 1924, consisting of the tetanus toxin treated with formaldehyde and heat, which he called 'anatoxin' (i.e. toxoid). This discovery was to prove a model for many subsequent applications. He also proposed that the efficacy of this 'anatoxin' could be enhanced by using, beside the specific antigens, substances known as adjuvants of immunity, [3, 101].

Adjuvants can effect on our bodies in different ways include [42]:

1- Improve the vaccine efficacy.

2- Increase in the antibody titers and CD4 T-cell frequencies.

3- Improve duration of protective responses.

4- Increase cross-protection against different microbial strains.

5- Reducing the antigen dose amount and the number of treatment to gain correct titer.

6- Antigen can modulate the quality of the antibody (isotypes) and the T-cell (Th1; Th2; Th17) response.

Modern adjuvants belong to two main groups the vehicles and the immunostimulants. Vehicles are substances that enable optimal presentation of the vaccine antigen to the immune system [32]. They includes

1- Mineral salts (aluminium or calcium phosphate)

2- Emulsions

3- Liposomes

4- Virosomes

5- Biodegradable polymeric microspheres

Immunostimulants differ in that they directly increase the immune response to antigens. Often they are microbial products such as

1- TLR ligands

2- Lipopolysaccharide (LPS)

3- Cytidine-phosphate-guanosine

4- Flagellin [50,102,103]

5- Lipoproteins

6- Zymosan

7- Bacterial DNA

8- Bacterial toxins

9- Cytokines [104]

10- Plant products (Saponins) [32].

Rinderpest

A disease is known from time immemorial in Europe and central Asia. It is fatal and its mortality range from 90 to 100%. Rinderpest or the Cattle plague (also steppe murrain) caused by rinderpest virus, (group V ((-) ssRNA comprises among the great historical besets that cause destroyed human farm animal since centuries [105,106].

The cattle plague is eradicated from Europe along the end of the nineteenth Century by simple program of hygienic criteria; even before the identification of the causative agent.

It is valuable noting the work of Geert Reinders (1737-1815) in the Netherlands was a farmer in the state of Groningen and a self-taught man who remarked that calves from recovered cows were immune to infection. A phenomenon of maternally-derived resistance [38]. His use of three separate inoculations at early age. There were trials to immunize cattle against cattle plague applying the smallpox vaccine. This practice comprised passionately supported in England on the epizootics of 1865 to 1867 [107].

Henri Bouley established the total deficiency of cross-protection between cattle plague, smallpox and cowpox in 1865. For this aim, he sent eight cattle to England, where the cattle plague epizootics were violent. These cattle, which had already been used in France to produce the anti-smallpox vaccine, all got cattle plague [3].

Afterward, Robert Koch, doing work in South Africa, recommended that cows could be saved by subcutaneous injection of blood serum from immunized animal and bile from an infected animal. This extremely unsafe formula was shortly substituted by the employ of immune serum and later on by a mixing of immune serum and virulent virus. Afterward, the method was improved by consecutive passages of the bovine virus through goats, which enabled Edwards to produce a compromised vaccine in India in the 1920s. Runs with inactivated vaccines as well occurred. At last, the successful isolation of the virus in cell culture led to the in vitro developing of a weakened strain and from this the production of a secure and highly efficient vaccine [23,29,30-32]. Robert Koch is the honour of the first publication about the practical method of immunizing cattle against the rinderpest. He injects the uninfected animal with the bile of the animal died by the rinderpest and after that by the serum of an immunized animal.

Contagious Bovine pleuropneumonia

It is a disease of cattle and water buffalo caused by *Mycoplasma mycoides* subsp. Mycoides (*M. mycoides*). The microbe attacks the lungs and the membranes lining the thoracic cavity. It is highly contagious with a mortality rate up to 50%.

The disease was epidemic in Europe on the 19th Century, extending to Belgium in 1828, the Netherlands in 1833, and the United Kingdom (UK) in 1841. Louis Willems, a Belgian doctor, tried to use inoculation to prevent the disease. Willems inoculate cattle at the tail, provoking large abscesses; the animals exhibited common clinical signs but not the characteristic signs of the disease (pleuropneumonia) and became secure when exposed once again [3,108]. It essential be noticeable that a operation alike to it planned by Willems was by trial and error evolved in Western Africa, where cattle were immunized with virulent pleuropneumonia tissues; vaccination occasionally provoked exostosis leading to a horny protrusion on the nasal bone. The skulls of such vaccinated animals even led to false recognition of a new species named "Bos triceros" [10,11,17].

Foot And Mouth Disease (FMD)

Like most of the virus infection it is highly mutated, hence it is difficult to be controlled by vaccine. It has huge variant and even within serotypes. There is no or very week protection between serotype. Two strains within a given serotype may have 30% nucleotide differences. Earlier vaccine is made by dead sample of FMDV to inoculate animals. In 1981, US government announced the first FMD genetically modified vaccines.

Protective herds against the consequences of FMD were a concern for cattle breeders for centuries, believably since antiquity. Vaccination is a new evolution (between the two World Wars) in the history of livestock breeding, and was preceded by various choice measures, all of them orientated to protect the herd from losing brought on by the threatened disease [3,109,110].

Conflict Of Interest

The authors declare that there is no any kind of conflict of interest concerning this review.

Financial Support

The authors declare that there is no financial support for this review.

References

1. Plotkin SA. Vaccines: correlates of vaccine-induced immunity. Clin Infect Dis. 2008;47(3):401-9.doi: 10.1086/589862.

2. Gorter E. [Natural cowpox and vaccine]. Ned Tijdschr Geneeskd. 1951;95(36):2594-2601.

3. Lombard M, Pastoret PP, Moulin AM. A brief history of vaccines and vaccination. Rev Sci Tech. 2007;26(1):29-48.

4. Garlapati S, M. Facci, M. Polewicz, S. Strom, L.A. Babiuk, G. Mutwiri, et al. Strategies to link innate and adaptive immunity when designing vaccine adjuvants. Vet Immunol Immunopathol. 2009;128(1-3):184-91.doi: 10.1016/j.vetimm.2008.10.298.

5. Honda T, Okamura H, Taneno A, Yamada S, Takahashi E. The role of cell-mediated immunity in chickens inoculated with the cell-associated vaccine of attenuated infectious laryngotracheitis virus. J Vet Med Sci. 1994;56(6):1051-5.

6. McGhee JR, K. Fujihashi, J. Xu-Amano, R.J. Jackson, C.O. Elson, K.W. Beagley, et al. New perspectives in mucosal immunity with emphasis on vaccine development. Semin Hematol.1993;30:3-12; discussion 13-5.

7. Zhao ZZ, H.B. Zhang, Q. Chen, D. Su, Z. Xie, Y.Y. Wangl, et al. Promotion of immunity of mice to Pasteurella multocida and hog cholera vaccine by pig interleukin-6 gene and CpG motifs. Comp Immunol Microbiol Infect Dis. 2009;32(3):191-205. doi: 10.1016/j.cimid.2007.10.001.

8. Bentley DW. Pneumococcal vaccine in the institutionalized elderly: review of past and recent studies. Rev Infect Dis. 1981;3 Suppl:S61-70.

9. Girard MP, T. Cherian, Y. Pervikov, and M.P. Kieny. A review of vaccine research and development: human acute respiratory infections. Vaccine.2005;23(50):5708-24. doi: 10.1016/j.vaccine.2005.07.046.

10. Haber M, A. Barskey, W. Baughman, L. Barker, C.G. Whitney, K.M. Shaw, et al. Herd immunity and pneumococcal conjugate vaccine: a quantitative model. Vaccine. 2007; 25(29):5390-8. doi: 10.1016/j.vaccine.2007.04.088.

11. Klein PJ, M. Vierbuchen, B. Roth, J. Fischer, K. Strick, A. Bannach, et al. [Hemolytic anemia in infections caused by neuraminidase-producing bacteria]. Verh Dtsch Ges Pathol. 1983; 67:415-8.

12. Lee CJ. Bacterial capsular polysaccharides--biochemistry, immunity and vaccine. Mol Immunol. 1987;24(10):1005-19.

13. Stanislavskii ES. [Cross-reacting microbial antigens and vaccine prophylaxis (review)].Zh Mikrobiol Epidemiol Immunobiol.1971;48(10):42-7.

14. Greene CJ, Chadwick CM, Mandell LM, Hu JC, O'Hara JM, Brey RN 3rd, et al. LT-IIb(T13I), a non-toxic type II heat-labile enterotoxin, augments the capacity of a ricin toxin subunit vaccine to evoke neutralizing antibodies and protective immunity. PLoS One. 2013;8(8):e69678. doi: 10.1371/journal.pone.0069678.

15. Lahariya C. A brief history of vaccines and vaccination in India. Indian J Med Res. 2014;139(4):491-511.

16. Wuthrich M, Filutowicz HI, Warner T, Deepe GS Jr, Klein BS. Vaccine immunity to pathogenic fungi overcomes the requirement for CD4 help in exogenous antigen presentation to CD8+ T cells: implications for vaccine development in immune-deficient hosts. J Exp Med.2003;197(11):1405-16.doi: 10.1084/jem.20030109.

17. Nader S, Bergen R, Sharp M, Arvin AM. Age-related differences in cell-mediated immunity to varicella-zoster virus among children and adults immunized with live attenuated varicella vaccine. J Infect Dis 1995;171(1):13-17.

18. Qin H, Zhou C, Wang D, Ma W, Liang X, Lin C, et al. Enhancement of antitumour immunity by a novel chemotactic antigen DNA vaccine encoding chemokines and multiepitopes of prostate-tumour-associated antigens. Immunology. 2006;117(3):419–430. doi: 10.1111/j.1365-2567.2006.02322.x.

19. Garcia-Peñarrubia P, Koster FT, Kelley RO, McDowell TD, Bankhurst AD. Antibacterial activity of human natural killer cells. J Exp Med. 1989;169(1):99-113.

20. Chin J and San Gil F. Skin delivery of a hybrid liposome/ISCOM vaccine implicates a role for adjuvants in rapid modulation of inflammatory cells involved in innate immunity before the enhancement of adaptive immune responses. Immunol Cell Biol.1998;76(3):245-55. doi: 10.1046/j.1440-1711.1998.00742.x.

21. Wynn TA, Jankovic D, Hieny S, Cheever AW, Sher A. IL-12 enhances vaccine-induced immunity to Schistosoma mansoni in mice and decreases T helper 2 cytokine expression, IgE production, and tissue eosinophilia. J Immunol. 1995; 154(9):4701-9.

22. Dreskin SC. ATP-dependent activation of phospholipase C by antigen, NECA, Na3VO4, and GTP-gamma-S in permeabilized RBL cell ghosts: differential augmentation by ATP, phosphoenolpyruvate and phosphocreatine. Mol Cell Biochem. 1995;146(2):165-70.

23. Bento D, Staats HF, Goncalves T, Borges O. Development of a novel adjuvanted nasal vaccine: C48/80 associated with chitosan nanoparticles as a path to enhance mucosal immunity. Eur J Pharm Biopharm.2015;93:149-64. doi: 10.1016/j.ejpb.2015.03.024.

24. Cosgrove CA, Castello-Branco LR, Hussell T, Sexton A, Giemza R, Phillips R, et al. Boosting of cellular immunity against Mycobacterium tuberculosis and modulation of skin cytokine responses in healthy human volunteers by Mycobacterium bovis BCG substrain Moreau Rio de Janeiro oral vaccine. Infect Immun. 2006;74(4): 2449–2452. doi: 10.1128/IAI.74.4.2449-2452.2006.

25. Bartholomew WR, Shanahan TC. Complement components and receptors: deficiencies and disease associations. Immunol Ser. 1990; 52:33-51.

26. Edwards SW, Morgan BP, Hoy TG, Luzio JP, Campbell AK. Complement-mediated lysis of pigeon erythrocyte ghosts analysed by flow cytometry. Evidence for the involvement of a 'threshold' phenomenon. Biochem J. 1983; 216(1):195-202.

27. Powers JH, Buster BL, Reist, Martin E, Bridges M, Sutherland WM, et al. Complement-independent binding of microorganisms to primate erythrocytes in vitro by cross-linked monoclonal antibodies via complement receptor 1. Infect Immun. 1995;63(4):1329-35.

28. Amara AA. Lysozymes, Proteinase K, Bacteriophage E Lysis Proteins, and some Chemical Page 16 of 16 Compounds for MGs Preparation: a Review and Food for Thought. SOJ Biochem. 2016;2(1)-16.

29. Amara AA. Opportunistic pathogens and their biofilm Food for thought. Science against microbial pathogens: communicating current research and technological advances. Microbiology Book Series. 2011;3:813-25.

30. Amara AA. Pharmaceutical and industrial protein engineering: where we are? Pak J Pharm Sci. 2013;26(1):217-32.

31. Holst PJ, Bartholdy C, Stryhn A, Thomsen AR, Christensen JP. Rapid and sustained CD4(+) T-cell-independent immunity from adenovirus-encoded vaccine antigens. J Gen Virol. 2007;88(Pt 6):1708-16. doi: 10.1099/vir.0.82727-0.

32. Mortellaro A, Ricciardi-Castagnoli P. From vaccine practice to vaccine science: the contribution of human immunology to the prevention of infectious disease. Immunol Cell Biol.2011;89(3):332-9. doi: 10.1038/icb.2010.152.

33. Amara AA. Kostenlos viral ghosts, bacterial ghosts microbial ghosts and more, Schuling Verlag - Germany, 2015.

34. Bohl EH, Frederick T, Saif LJ. Passive immunity in transmissible gastroenteritis of swine: intramuscular injection of pregnant swine with a modified live-virus vaccine. Am J Vet Res. 1975;36(3):267-71.

35. Kohara J, Hirai T, Mori K, Ishizaki H, Tsunemitsu H. Enhancement of passive immunity with maternal vaccine against newborn calf diarrhea. J Vet Med Sci. 1997;59(11):1023-5.

36. Lemaire M, Hanon E, Schynts F, Meyer G, Thiry E. Specific passive immunity reduces the excretion of glycoprotein E-negative bovine herpesvirus type 1 vaccine strain in calves. Vaccine. 2000;19(9-10):1013-7.

37. Andruskevich SM, Perry P, Houpt K, Houpt TR. The relation of maternal fluid balance to offspring passive immunity. Physiol Behav. 2013;122:155-8. doi: 10.1016/j.physbeh.2013.09.005.

38. Mondal SP, Naqi SA. Maternal antibody to infectious bronchitis virus: its role in protection against infection and development of active immunity to vaccine. Vet Immunol Immunopathol. 2001;79(1-2):31-40.

39. Abraham-Van Parijs B. Review of pneumococcal conjugate vaccine in adults: implications on clinical development. Vaccine. 2004;22(11-12):1362-71. doi: 10.1016/j.vaccine.2004.01.029.

40. Ribero ML, Fara GM, Del Corno G. [Duration of tetanus immunity in relation to the number of doses of vaccine]. Boll Ist Sieroter Milan 1980;59(4):464-475

41. Woodland DL. Jump-starting the immune system: prime-boosting comes of age. Trends Immunol. 2004;25(2):98-104. doi: 10.1016/j.it.2003.11.009.

42. Delany I, Rappuoli R, De Gregorio E. Vaccines for the 21st century. EMBO Mol Med. 2014;6(6):708-20. doi: 10.1002/emmm.201403876.

43. Shchelkunov SN. Emergence and reemergence of smallpox: the need for development of a new generation smallpox vaccine. Vaccine. 2011;29 Suppl 4:D49-53. doi: 10.1016/j.vaccine.2011.05.037.

44. DeLaine BC, Wu T, Grassel CL, Shimanovich A, Pasetti MF, Levine MM, et al. Characterization of a multicomponent live, attenuated Shigella flexneri vaccine. Pathog Dis. 2016;74(5). pii: ftw034. doi: 10.1093/femspd/ftw034.

45. Tamura S, Ainai A, Suzuki T, Kurata T, Hasegawa H. Intranasal Inactivated Influenza Vaccines: a Reasonable Approach to Improve the Efficacy of Influenza Vaccine? Jpn J Infect Dis. 2016;69(3):165-79. doi: 10.7883/yoken.JJID.2015.560.

46. Lambert SM, Markel H. Making history: Thomas Francis, Jr, MD, and

the 1954 Salk Poliomyelitis Vaccine Field Trial. Arch Pediatr Adolesc Med 2000;154(5):512-517.

47. Bjune G. "Herd immunity" and the meningococcal vaccine trial in Norway. Lancet. 1992;340(8814):315.

48. Choudhury SA, Mishreki NK. Subnormal immunity to Hemophilus influenzae type b (Hib) in previously vaccinated human immunodeficiency virus-infected children 59 months of age or older and response to booster doses of the conjugate vaccine. CLIN PEDIATR. 2004;43(9):831-835. doi: 10.1177/000992280404300907.

49. Pichichero ME, Loeb M, Anderson, Smith DH. Do pili play a role in pathogenicity of Haemophilus influenzae type B? Lancet. 1982;2(8305):960-2.

50. Ben-Yedidia T, Arnon R. Effect of pre-existing carrier immunity on the efficacy of synthetic influenza vaccine. Immunol Lett. 1998;64(1):9-15.

51. Tagliabue A, Villa L, De Magistris MT, Romano M, Silvestri S, Boraschi D, et al. IgA-driven T cell-mediated anti-bacterial immunity in man after live oral Ty 21a vaccine. J Immunol. 1986;137(5):1504-10.

52. André F, Van Damme P, Safary A, Banatvala J. Inactivated hepatitis A vaccine: immunogenicity, efficacy, safety and review of official recommendations for use. Expert Rev Vaccines. 2002;1(1):9-23. doi: 10.1586/14760584.1.1.9.

53. Ghanem S, Hassan S, Saad R, Dbaibo GS. Quadrivalent meningococcal serogroups A, C, W, and Y tetanus toxoid conjugate vaccine (MenACWY-TT): a review. Expert Opin Biol Ther. 2013;13(8):1197-205. doi: 10.1517/14712598.2013.812629.

54. Kristiansen PA, Diomande F, Ba AK, Sanou I, Ouédraogo AS, Ouédraogo R, et al. Impact of the serogroup A meningococcal conjugate vaccine, MenAfriVac, on carriage and herd immunity. Clin Infect Dis. 2013;56(3):354-63. doi: 10.1093/cid/cis892.

55. Bla R. [Dynamics of anti-diphtherial immunity following inoculation with whooping cough-diphtheria vaccine in children from various age groups]. Zh Mikrobiol Epidemiol Immunobiol 1962;33:62-66.

56. Aguila A, Donachie AM, Peyre M, McSharry CP, Sesardic D, Mowat AM. Induction of protective and mucosal immunity against diphtheria by a immune stimulating complex (ISCOMS) based vaccine. Vaccine. 2006;24(24):5201-10. doi: 10.1016/j.vaccine.2006.03.081.

57. Altemeier WA 3rd. A pediatrician's view. Some science and history behind our vaccine schedule. Pediatr Ann. 1998;27(7):401-3.

58. Maĭskaia LM, Basova NN, Bolotovskiĭ VM, Tamm OM, Miartin IaK. [Antidiphtheria and antitetanus antitoxic immunity indices in the population of the Estonian SSR (an assessment of the effectiveness of a planned vaccine prophylaxis)]. Zh Mikrobiol Epidemiol Immunobiol. 1979;(8):21-6.

59. Muller RH, Keck CM. Challenges and solutions for the delivery of biotech drugs – a review of drug nanocrystal technology and lipid nanoparticles. J Biotechnol. 2004;113(1-3):151-70.

60. Whittingham JL, Havelund S, Jonassen I. Crystal structure of a prolonged-acting insulin with albumin-binding properties. Biochemistry. 1997;36(10):2826-31. DOI: 10.1021/bi9625105.

61. JP Moore, RW Sweet. The HIV gp120-CD4 interaction: A target for pharmacological or immunological intervention? Perspect. Drug Discov. 1993;(1)235-250. doi:10.1007/BF02171665.

62. Shak S, Capon DJ, Hellmiss R, Marsters SA, Baker CL. Recombinant human DNase I reduces the viscosity of cystic fibrosis sputum. Proc Natl Acad Sci U S A. 1990;87(23):9188-92.

63. Breedveld FC. Therapeutic monoclonal antibodies. Lancet. 2000;355(9205):735-740.

64. French RC, L Siminovitch. The action of T2 bacteriophage ghosts on Escherichia coli B. Can J Microbiol. 1955;1 (9):757-774.

65. Rowe GE, Welch RA. Assays of hemolytic toxins. Methods Enzymol. 1994;235:657-67.

66. Amara AA, Salem-Bekhit MM, Alanazi FK. Sponge-like: a new protocol for preparing bacterial ghosts. ScientificWorldJournal. 2013; 2013:545741. doi: 10.1155/2013/545741.

67. Jimmie C, Holland M, Lean O, Cullen M. New Insight and Attitudes. In Cancer book: Prevention Detection Diagnosis Treatment Rehabilitation and Cure". (Eds.Arthur I Holleb M.D.) USA; Doubleday Co (1986).

68. Amara AA. Methods other than Experimental Animals for Screening Antitumor Compounds. American Journals of Cancer Science. 2013;(1):Article ID 2013010001, 27.

69. Marshall Goldberg M. The Immune system's newest weapons against cancer In "Cell Wars"(1st Ed.). Toronto, Canada; Collins Publishers (1988).

70. Coenraad JH, Hendriksen FM. Refinement of polyclonal antibody production by combining oral immunization of chicken swith harvest of antibodies from the egg yolk. . ILAR J 2005;46(3):294-299.

71. Paul S, Gabibov A, Massey R. Catalytic Antibodies "Conference Overview". Mol Biotechnol. 1994;1(1):109-11. doi: 10.1007/BF02821514.

72. Ali M, Hariharan AG, Mishra N, Jainal S. Catalytic antibodies as potential therapeutics. Indian Journal of Biotechnology. 2009;(8):253-258.

73. Chang CY, Niblack B, Walker B, Bayley H. A photogenerated pore-forming protein. Chem Biol 1995;2(6):391-400.

74. Grossman I, Ilani T, Fleishman SJ, Fass D. Overcoming a species-specificity barrier in development of an inhibitory antibody targeting a modu-lator of tumor stroma. Protein Eng Des Sel. 2016;29(4):135-47. doi: 10.1093/protein/gzv067.

75. Walker B, Braha O, Cheley S, Bayley H. An interme-diate in the assembly of a pore-forming protein trapped with a genetically-engineered switch. Chem Biol 1995;2(2):99-105.

76. Yu DH, Li M, Hu XD, Cai H. A combined DNA vac-cine enhances protective immunity against Mycobacterium tuberculosis and Brucella abortus in the presence of an IL-12 expression vector. Vaccine 2007;25(37):6744-6754.

77. Arya SC, Agarwal N. Stability of smallpox vaccine in field conditions. Vaccine. 2006;24(9):1235. doi: 10.1016/j.vaccine.2005.09.027.

78. Cooper NR, Nemerow GR. The role of antibody and complement in the control of viral infections. J Invest Dermatol. 1984;83(1 Suppl):121s-127s.

79. Bacchetta R, Gregori S, Roncarolo MG. CD4+ regulatory T cells: mechanisms of induction and effector function. Autoimmun Rev. 2005;4(8):491-6. doi: 10.1016/j.autrev.2005.04.005.

80. Igietseme JU, Eko FO, He Q, Black CM. Antibody regulation of Tcell immunity: implications for vaccine strategies against intracellular pathogens. Expert Rev Vaccines. 2004;3(1):23-34.

81. Fenner F, Pastoret PP, Blancou J, Terré J. Historical introduction. In Veterinary vaccinology (P.-P. Pastoret, J. Blancou, P. Vannier & C. Verschueren, eds). Elsevier, Amsterdam (1997).

82. Chevallier-Jussiau N. [Henry Toussaint and Louis Pasteur. Rivalry over a vaccine]. Hist Sci Med. 2010;44(1):55-64.

83. Angelov S, Kuiumdzhiev I. [Investigations on new principles in active immunization against swine erysipelas with special reference to concentrated adsorption vaccine]. Izv Mikrobiol Inst (Sofiia). 1952;3:5-32.

84. D'Iakonov OB, Podlesnykh LA. [A vaccine against swine erysipelas with an oily additive]. Veterinariia. 1967;44(2):50-3.

85. Eto M, Mishima M, Kobori N, Susumi S, Watanabe S. [Study on hog cholera swine erysipelas combined live vaccine. -III. Field trial of vaccination (author's transl)]. Igaku Kenkyu. 1980;50(7):519-24.

86. Eto M, Mishima M, Kobori N, Susumi S, Watanabe S. [Study on hog cholera swine erysipelas combined live vaccine. -II. Indoor test of vaccination (author's transl)]. Igaku Kenkyu 50 (1980) 519-24.

87. Gavrilova N, Stoev I, Mermerski K. [Mixed crystal-violet vaccine against hog cholera and swine erysipelas. II. Expiration date of the bivalent vaccine]. Vet Med Nauki. 1973;10(7):59-64.

88. Henry S, Kelly B. Swine abortion associated with use of live erysipelas vaccine. J Am Vet Med Assoc. 1979;175(5):453-4.

89. Hopper RJ, Miller ML. Efficacy of swine erysipelas vaccine given simultaneously with medicated rations. Vet Med Small Anim Clin. 1981;76(9):1345-7.

90. Lat J, Kostansky K, Ursiny J. [Experiments on the production of adsorbed vaccine for swine erysipelas]. Cas Cesk Vet. 1950;5(14):319-24.

91. Lawson KF, Walker VCR, Crawley JF. Modified Swine Erysipelas Vaccine. Can J Comp Med Vet Sci. 1958; 22(5): 164-168,169-174.

92. Sizaret P, Reculard P, Caraes Y, Virat B. [Etiology of Some Primary Serum Shocks. Iv. Demonstration and Explanation of the Presence of Equine Antigens in an Anti-Swine Erysipelas a Vaccine. Demonstration of These Antigens in 2 Other Vaccines]. Ann Inst Pasteur (Paris). 1964;106:95-107.

93. Banic S. [Mechanism of antirabies immunity and modified application of Hept's vaccine]. Zdravstveni vestnik. 1951;20(9-10):201-6.

94. Bugyaki L, Moons Jh, Blockeel Sr. [Survival of the fixed Pasteur virus in Fermi-Semple type phenicated antirabies vaccine and its relation to the degree of immunity conferred]. Ann Soc Belg Med Trop (1920). 1959;39:275-80.

95. Costaguta R. [Method used at the Pasteur Institute of Avellaneda in preparation of the Semple type of antirabies vaccine]. Dia Med. 1958;30(80):2857.

96. Fallon Hj. Report on immunity failure of avianized rabies vaccine. J Am Vet Med Assoc. 1952;120(903):373.

97. Habel K. Effect on immunity to challenge and antibody response of variation in dosage schedule of rabies vaccine in mice. Bull World Health Organ. 1956;14(4):613-6.

98. Azzopardi P, Bennett CM, Graham SM, Duke T. Bacille Calmette-Guerin vaccine-related disease in HIV-infected children: a systematic review. Int J Tuberc Lung Dis. 2009;13(11):1331-44.

99. Lugosi L. Theoretical and methodological aspects of BCG vaccine from the discovery of Calmette and Guerin to molecular biology. A review. Tuber Lung Dis. 1992;73(5):252-261.

100. Berdah D.. Entre bovins et humains. La lutte contre la tuberculose en France, dans la première moitié du XX e siècle. DEA histoire des sciences (post-graduate degree thesis).), University of Paris VII – Denis Diderot, Paris. (2002).

101. Ramon G. Sur la toxine et l'anatoxine diphthériques. Pouvoir floculant et propriétés immunisantes. Ann. Inst. Pasteur 38 (1924).

102. Levi R, Arnon R. Synthetic recombinant influenza vaccine induces efficient long-term immunity and cross-strain protection. Vaccine. 1996;14(1):85-92.

103. Schrader JW, Nossal GJ. Effector cell blockade. A new mechanism of immune hyporeactivity induced by multivalent antigens. J Exp Med. 1974;139(6):1582-98.

104. Kayamuro H, Yoshioka Y, Abe Y, Arita S, Katayama K, Nomura T, et al. Interleukin-1 family cy-tokines as mucosal vaccine adjuvants for induction of protective immu-nity against influenza virus. J Virol. 2010;84(24):12703-12. doi: 10.1128/JVI.01182-10.

105. Barrett T, Pastoret P, Taylor W. Rinderpest and peste des petits ruminants: virus plagues of large and small ruminants. Monograph Series: Biology of Animal Infections (P.-P. Pastoret, series editor). Elsevier, Academic Press. (2005).

106. Pastoret PP, Jones P. Veterinary vaccines for animal and public health. In Control of infectious animal diseases by vaccination (A. Schudel & M. Lombard, eds). Proc. OIE Conference, Buenos Aires, Argentina, 13-16 April. Dev Biol (Basel). 2004;119:15-29.

107. Dele E. La peste bovine ou typhus contagieux épizootique en Angleterre (1865-1867). Combe et Van de Weghe, Brussels. (1870).

108. Willems L. Mémoire sur la pleuro-pneumonie épizootique du gros bétail. Rec. Méd. vét. pratique, 3rd part, tome IX 14 (1952).

109. Doel TR. Natural and vaccine induced immunity to FMD. Curr Top Microbiol Immunol. 2005;288:103-31.

110. Patil PK, Bayry J, Ramakrishna C, Hugar B, Misra LD, Natarajan C.Immune responses of goats against foot-and-mouth disease quadrivalent vaccine: comparison of double oil emulsion and aluminium hydroxide gel vaccines in eliciting immunity. Vaccine. 2002;20:2781-9.

7

Comparative toxicity of selected naphthenic acids, oil sands processed water and surface waters in rainbow trout hepatocytes: A gene expression study

Gagné F*, André C, Turcotte P, Gagnon C

Aquatic Contaminants Research Division, Environment and Climate Change Canada, 105 McGill Street, Montreal, Quebec, Canada

Corresponding author: F. Gagné, Aquatic Contaminants Research Division, Environment and Climate Change Canada, 105 McGill Street, Montreal, Quebec, Canada, E-mail: francois.gagne@canada.ca

Abstract

This study examined the cytotoxic properties of selected naphthenic Acids (NA), commercial mixtures of NA in rainbow trout hepatocytes and compares the responses with Oil Sand Processed Water (OSPW) and river water extracts. Hepatocytes were exposed to increasing concentrations of selected NAs ($z = 0, 2, 4, 6, 8$ and 10), 2 commercial mixtures of NA, 2 OSPW, and surface waters upstream and downstream from the Athabasca River at an OS development area (Alberta, Canada) for 48 h at 15°C. Cell viability, total RNA levels and the expression of 15 gene transcripts involved in biotransformation (CYP1A1, CYP3A4, GST and MDR), oxidative stress (SOD, CAT, GST), DNA damage/repair (UNG, APEX, LIG, GADD45, OGG), estrogenicity (VTG and ER2), cell growth (PCNA, GADD45), and glycolysis (GAPDH) were determined. Individual NAs induced the expression most of the genes, except for GST, whose expression was reduced. For NA mixtures, the most sensitive genes were those involved in DNA repair (APEX, LIG, GADD45, and OGG) and biotransformation (CYP1A1 and 3A4). The same pattern was observed for OSPW, except that MDR was the most sensitive gene, and GST gene expression was decreased as occurred with NAs. The responses for river water samples were generally lower than for OSPW, NAs and NA mixtures, and involved biotransformation (GST and MDR) genes, DNA repair (OGG, LIG) genes and potential endocrine disruption (ER2). Canonical analysis of gene expression data and cell viability revealed that genes involved in Xenobiotic metabolism, oxidative stress and DNA damage (repair) were strongly correlated with cytotoxic effects. Decision tree analysis revealed that these compounds separated into 4 distinct clusters (OSPW, upstream and downstream river water, NA mixtures), whereas the individual NAs were generally found in the OSPW cluster. OSPW and NA mixtures showed distinct properties from surface waters collected downstream from the OS development area suggesting that OSPW did not overly contaminated river waters.

Keywords: Oil sands; Hepatocytes; Oncorhynchus mykiss; Biotransformation; Genotoxicity; Endocrine Disruption; Oxidative Stress; Cytotoxicity

Introduction

The Athabaska River (Alberta, Canada) stretches 1,200 km from northern Alberta through the Northwest Territories to the Mackenzie River estuary. It drains an area rich in Oil Sands (OS) deposits and is considered one of the world's largest oil reserves with an estimated yield of 27 billion cubic meters of crude oil [1]. OS consist of about 10% bitumen and 5% water, with the remainder being sand and clay. The extraction process involves adding hot caustic water to OS under aeration to separate bitumen from the OS (Clark extraction process). The crude oil fraction, which partitions at the surface, is then further extracted with the addition of solvents (naphtha, toluene) to yield a more concentrated non-polar fraction. The remaining water and sediment deposits constitute the OSPW, which is released into large tailing ponds covering many square kilometers. Numerous contaminants are found in OSPW, including a number of alkali-extractable compounds such as Naphthenic Acids (NAs), polyaromatic and aliphatic hydrocarbons and various inorganic elements such as sodium, vanadium, strontium, calcium, nickel and sulfates [2-4].Recent evidence has shown that OS mining activity was associated with increased releases of dissolved PAHs and heavy metals, which raised concerns about the ecotoxicological consequences of these activities [5-6]. NAs are cyclic aliphatic hydrocarbons that can reach concentrations as high as 50 mg/ L in tailing ponds [7]. NAs follow the $C_nH_{2n-z}O_x$ rule which comprise at least one carboxylic acid and they are suspected to be a major component of OSPW toxicity. However, the presence of NAs has been observed in runoff waters (leachates) from OS storage sites, indicating that OSPW is not the only source of NA in the environment [8]. This highlights the difficulty of identifying specific markers of OSPW in the environment at low levels, although the highest concentrations of NAs are found in OSPW.

The toxicity of OSPW to aquatic organisms has received increasing attention over the last decade. Toxicity has been examined at various levels, including endocrine disruption, DNA damage, immune competence, biotransformation and reproduction. Goldfish exposed to OSPW for 12 weeks showed elevated expression of proinflammatory genes such as interleukin-1β and tumor necrosis factor 2 in the spleen and kidneys [9]. This led to decreased ability of head kidney macrophages to produce reactive oxygen species, which normally follows phagocytosis of bacteria. In a study with fathead minnows exposed to OSPW, lower plasma 11-ketotestosterone levels and gonado-somatic indices were observed [10]. There were also signs of epithelium degeneration in gills of fish exposed to OSPW compared to reference fish, supporting the notion that the toxicity of OSPW is diverse and not specific to a target organ. This was corroborated by another study, which examined the health status of white suckers exposed to aged OS tailings [11]. Fish maintained in aged OSPW had smaller testes and ovaries and reduced growth, which resulted from limited available energy and endocrine disruption, which was associated with increased CYP1A1 activity in the liver. The first evidence of endocrine disruption based on the Vitellogenin (VTG) pathway was reported by Gagné, at al. [12]. VTG gene expression was induced in trout hepatocytes exposed to surface waters located upstream and downstream from the OS development area as well as to OSPW. Induction was higher in cells exposed to OSPW compared to surface waters, which suggests that OSPW is estrogenic. In a previous study, however, it was found that sea water leachates from oil rigs did not increase mosquito fish plasma VTG [13]. OPSW was shown to be genotoxic to rainbow trout hepatocytes based on the alkaline DNA precipitation and comet assays [14,15]. In addition to OSPW and NA commercial mixtures, DNA damaging compounds were ubiquitous in both upstream and downstream surface waters, which suggest that a number of chemicals contributed to the observed genotoxicity of OS. The multi-drug ATP binding cassette transporter (MDR) was also influenced by OS-related compounds and is involved to the efflux of potentially toxic hydrocarbons from cells. The effects of the soluble fraction of OSPW on MDR transport activity were examined in Japanese medaka fry [16]. It was found that the neutral and basic fractions of OSPW inhibited the extrusion of calcein dye in fry, which suggests that MDR activity could be inhibited and limit the elimination of organic contaminants in fish exposed to OS in the environment. These fractions were shown to contain higher levels of oxygen-, sulfur-, and nitrogen-containing hydrocarbons. This study corroborated earlier findings that MDR gene expression was inhibited in hepatocytes exposed to OSPW but not by OS leachates [17]. A better understanding of OSPW-specific toxic effects from the natural leaching of OS compounds would provide more information on the fundamental ecotoxicological impacts of industrial extraction in the Athabasca watershed. Recent evidence suggests that OSPW is more toxic than passive OS water leachates in the river. For example, a study of the phytotoxicity of laboratory-prepared OSPW and of OS leachates revealed that OSPW contained more light PAHs (naphthalene-like), vanadium, aluminum and chromium than did OS leachates [18]. A study compared the toxic properties of OSPW and OS leachates using primary cultures of rainbow trout hepatocytes [17]. It found that some gene transcripts were more specific to OSPW and either did not respond or respond slightly to OS leachates; these were superoxide dismutase (SOD), glutathione S-transferase (GST), CYP3A4, glyceraldehyde-3-phosphate dehydrogenase, a marker of anaerobic glycolysis, and two genes involved in DNA repair (GADD45 and APEX). DNA repair genes responded to OS

leachates, albeit less strongly than they did to OSPW. Moreover, gene transcripts that were associated with cell viability were chosen on the basis of their implication in cell mortality. It was found that genes involved in biotransformation were closely related to cytotoxicity. Although these genes appeared to be specific to OSPW but not to OS leachate, they were not compared with selected NAs or commercial NA mixtures to identify similarities with OSPW and determine whether surface waters upstream and downstream from the OS mining area display similar toxic properties to OSPW and NAs.

The purpose of the study was to compare the response profiles of gene transcripts in rainbow trout hepatocytes exposed to a selection of individual NA-like compounds of increasing molecular size (z value), commercial mixtures of NA, OSPW and surface water samples collected in locations upstream and downstream from the OS mining area. We hypothesized that the upstream and downstream waters are similar (no influence from mining activities) and that OPSW cannot be explained only by NAs or commercial NA mixtures (null hypothesis). The gene transcripts were chosen based on the results of previous studies on OSPW, OS leachates and surface water. Cytotoxicity was examined at the membrane integrity level in addition to biotransformation, oxidative stress, genotoxicity (DNA repair), anaerobic glycolysis, endocrine disruption (estrogenicity) and cell maintenance/growth. Primary cultures of fish hepatocytes are recognized models for toxicity investigations involving biotransformation [19]. An attempt was made to find common trends in gene expression profiles for OSPW, individual NAs, commercial mixtures of NAs and downstream surface waters in the Athabasca River in a region where industrial OS extraction operations are carried out.

Materials and methods

Naphthenic acids, oil sands processed water and surface water collection preparation

A selection of NA-like compounds and commercial NA mixtures were used in this study in addition to OSPW and surface waters (Table 1). Six mono-carboxylic acid chemicals following the $C_nH_{2n-z}O_2$ rule were chosen based on increasing z values (z = 0, 2, 4, 6, 8, 10). They were purchased from Sigma-Aldrich and dissolved in absolute ethanol at a concentration of 10 mg/ml. The exposure concentrations were 2, 10, 50 and 100 ug/ml at a final 0.2% ethanol concentration. Two commercial NA acid mixtures were also tested: one from Sigma-Aldrich (Na mixture 1) and one from Thermo-Fisher Scientific (Na mixture 2, ACROS Organics brand). The stock solution was diluted in absolute ethanol, and the following concentrations were used at a final absolute ethanol concentration of 0.2%: 2, 10, 50 and 100 ug/ml.

OSPW and surface water samples were collected in the Athabasca River near the OS extraction sites. OSPW samples 1 and 2 correspond to two different OS tailing ponds. The surface waters consisted of samples from the upstream site (in the Athabasca River, 10 km downstream from Fort McMurray) and the OS area (along the western shore of the Athabasca

Table 1: Naphthenic acid characteristics

Naphthenic acid	Chemical formula / MW		Z value	Structure
Decanoic acid	$C_{10}H_{20}O_2$	172.26	0	
Cyclohexanepentanoic acid	$C_{11}H_{20}O_2$	184.28	-2	
2-(1-cyclohexenyl)-butyric acid	$C_{10}H_{16}O_2$	168.24	-4	
3-Noradamantanecarboxylic acid	$C_{10}H_{14}O_2$	166.22	-6	
5β-Cholanic acid	$C_{24}H_{40}O_4$	392.57	-8	
Abietic acid	$C_{20}H_{30}O_2$	302.45	-10	

River opposite the Steep bank River outfall) as well as samples taken at the confluence of the Muskeg and Ells rivers. A volume of 120 L was collected, mixed and sent to various laboratories dark containers at 4° C. Upon receipt (within 24 h), the water samples were filtered on a sterile 0.45-um pore polycarbonate filter and a 500-mL volume was passed through a C18 solid-phase mini-column (Sep-Pak C18, 360 mg, Waters Associates Inc.). The columns were pre-conditioned with 2 mL of absolute ethanol and 10 mL of distilled water at 5 psi. After passing the water or OSPW samples, the columns were washed with 10 mL of bidistilled water and eluted with 15 mL of analytical grade absolute ethanol (Sigma Chemical Company, ON, Canada). The ethanol fraction was reduced to 5 mL under a nitrogen stream to obtain a 100× concentration factor and was then kept at -20°C in the dark until analysis. The exposure concentrations were 0.004%, 0.02%, 0.1% and 0.5%, which corresponded to 0.4%, 2%, 10% and 50% of the original water samples. The maximum ethanol concentration was at 0.5%, and constituted the solvent control. The OSPW and surface water extracts were analyzed for polyaromatic hydrocarbons by fixed wavelength fluorescence spectroscopy, as previously described [14]. The light, medium and heavy PAHs were expressed as µg/ L naphthalene, pyrene and benzo (a) pyrene equivalents, respectively.

Preparation and exposure of rainbow trout hepatocytes

Primary cultures of rainbow trout (*Oncorhynchus mykiss*) hepatocytes were prepared using the double perfusion methodology [20]. In summary, young-of-the-year rainbow trout (10–15 cm fork length) were euthanized in 100 mg/ L

tricaine methanesulfonate buffered with $NaHCO_3$ at pH 7 for 5 min, as recommended by the Canadian Council on Animal Care Committee. The livers from 5 individuals were first perfused with 10 mM EDTA in phosphate-buffered saline (PBS: 140 mM NaCl, 5 mM KH_2PO_4 and 1mM glucose, pH 7.4) for 10 min at room temperature. The livers were then minced in 100 U/mL collagenase in PBS (without EDTA) containing 1mM $CaCl_2$ and 0.05% serum bovine albumin for 30 min at room temperature. The samples were then passed through a 50-µm sieve and centrifuged at 200 ×g for 5 min. The cell pellets were then re suspended in Liebovitz (L - 15) cell culture media containing 10% Fetal Bovine Serum (FBS), antibiotics and antimycotics (100 µg/ mL of streptomycin, 100 U/ mL of penicillin, and 0.25 µg/ mL of amphotericin B) for cell counting and viability assessment as described below. The cells were then washed 2 more times (centrifugation-resuspension) without FBS, resuspended in L15 without FBS and plated at a density of 1 million live cells/mL in 48-well micro plates in L-15 media. The initial cell concentration and viability were determined using trypan blue staining (0.004%) and examination on a hematocytometer at 200× under a microscope after 10 min. The hepatocytes were exposed to the compounds and extract samples, as described above, at 15°C for 48 h under saturated humidity in the dark; control cells included ethanol at 0.2% or 0.5%. After the exposure period, the micro plates were centrifuged at 150 ×g for 5 min at 4°C, and the media were removed by aspiration and the cells resuspended in PBS. Cell viability was determined using trypan blue, as described above. A 250-µL portion of the cell suspension was immediately mixed in the same volume of an RNA stabilizing solution (RLT plus buffer; Quiagen). The mixtures were placed on a QIA shredder spin column and centrifuged at 400 ×g for 2 min. The supernatant was then stored at -80°C until analysis for total RNA and the 15 gene transcripts described below.

Gene expression determinations using real-time polymerase chain reaction

Total RNA levels were extracted using commercial extraction kits, as described previously [12]. RNA concentrations (260 nm) and purity (260 nm/ 280 nm ratio < 1.8) were estimated with the NanoDrop 1000 UV–Vis Spectrophotometer (Thermo Fisher Scientific, ON, Canada). RNA integrity was verified using a micro fluidic-based electrophoresis system (Experion Automated Electrophoresis System; Bio-RAD, ON, Canada). Reverse transcription of RNA was performed using the QuantiTect reverse transcription commercial kit from QIAGEN (Toronto, ON, Canada), which ensured the complete removal of genomic DNA. The incubation temperature was 42°C for 15 min followed by 95°C for 3 min. The cDNA samples produced were stored at -80°C until real-time quantitative Polymerase Chain Reaction (qPCR) analysis. The selected genes that were quantified in this study are highlighted in Table 2. The genes were grouped under oxidative stress, biotransformation, genotoxicity (DNA repair), endocrine disruptors and others (proliferation and anaerobic glycolysis). Genes involved in oxidative stress were catalase (CAT), super oxide dismutase (SOD) and 8-oxoguanine glycosylase (OGG), which is also a marker for genotoxicity . DNA repair genes were apurinic/apyrimidinic endonuclease 1 (APEX), growth-arrest DNA damage inducible protein (GADD45), DNA ligase (LIG), uracil DNA glycosylase (UNG) and OGG, as mentioned above. Genes involved in biotransformation were cytochrome P4501A1 (CYP1A1), cytochrome P4503A4 (CYP3A4), glutathione S-transferase (GST) and multidrug P-glycoprotein pump (MDR). For endocrine disruption, vitellogenin (VTG) and estradiol-17β Receptor 2 (ER2) were determined. Cell proliferation and a marker of anaerobic glycolysis were determined by proliferation nuclear antigen (PCNA) and glyceraldehyde-3-phosphate Dehydrogenase (GAPDH), respectively. Reference genes used in this study were elongation factor 1 α (EF1α), RNA polymerase 1 (RPL), prolylpeptidyl isomerase 1 (PP1A) and hypoxanthine phosphoribosyl transferase concentrations (i.e., no significant increase in baseline mortality) was selected for normalization.

Primers were designed using NCBI's Primer-BLAST (Primer3 with Blast) and Net Primer (Bio soft, Palo Alto, CA) to avoid secondary structures. The selected primer sequences were then synthesized by Integrated DNA Technologies (Coralville, Iowa, USA) and reported in Table 3. The qPCR reactions were performed using the iQ SYBR Green Super Mix (Bio-Rad, Mississauga, On, Canada) and a real-time thermocycler (Master cycler ep realplex2; Eppendorf). All reactions were run in duplicate and consisted of 5 uL of cDNA (equivalent to 20 ng cDNA), 12.5 μL of iQ SYBR Green Super Mix, 0.2 mM of each dNTP, 25 U/mL of iTaq DNA polymerase, 3 mM of $MgCl_2$, and 10 nM of SYBR Green I, primer concentrations of 300 nM each, and DEPC-treated water (Ambion), and were completed to 25μL. Temperature cycles

were 95°C for 2 min, then 40 cycles at 95°C for 15 s, 60°C for 15 s, and 68°C for 15 s. A melting curve analysis was performed to check for lack of amplification specificity; the temperatures used were 95°C for 15 s, lowered to 57°C and increased to 95°C after 10 min.

Data analysis

For gene expression data, quantification Cycle (Cq) values were imported from the real-time thermo cycler instrument to the Genex Pro software (v. 5.4.0.512; Multi D Analyses, AB, Goteborg, Sweden). The selected reference gene, EL1α, was analyzed using the Norm finder [21] algorithms included in the Genex software and served to calculate relative expression. The cells were exposed to $n=8$ replicates of the samples at each treatment concentration. The gene expression data relative to reference gene were finally normalized against the ethanol solvent control. Homogeneity of variance and normality were checked using Levene's test and the Shapiro–Wilk test. The data were then subjected to analysis of variance and the critical difference from the controls was determined using Fischer's Least Square Difference test. Pearson-moment correlation analysis was performed between gene expression changes and cytotoxicity data (cell viability and total RNA levels) for all the tested samples. Canonical correlations were also performed to find relationships between cytotoxicity endpoints (cell viability and total RNA levels) and genes involved in biotransformation (CYP1A1, CYP3A4, GST, MDR), oxidative stress (SOD, CAT, OGG), DNA repair/damage (UNG, LIG, APEX, GADD45 and OGG), endocrine disruption (ER2 and VTG), anaerobic glycolysis (GAPDH) and cell proliferation (PCNA). The global properties of the individual NAs, NA mixtures, OSPW, and upstream/downstream surface waters were analyzed using classification and regression (decision) trees (CART) to determine differences (specific rules) between samples. Significance was set at α= 0.05 and all tests were run using the Statistica software (version 8, France).

Results

Rainbow trout hepatocytes were exposed to individual compounds of increasing molecular size (z) based on the general formula for naphthenic acids: $C_nH_{2n-z}O_2$ (Table 2). The z values ranged from 0 to 10; the z=6 was a diamonid (noradamantane carboxylic acid), a fused tricyclic carboxylic acid. The z=8 compound was cholanic acid, a derivative of steroids such as cholesterol that can be found in decomposed matrices derived from living matter. The same observation holds for the z = 10 compound, abietic acid, a phyto steroid derived from plants and trees. Basic physico-chemical characteristics of the surface water and OSPW samples are reported in Table 4. The pH of the water samples was slightly alkaline with pH values between 8.5 and 8.8. The OSPW samples had a pH of 9 – 9.2. Water conductivity varied between 275 and 300 μS*cm^{-1} at both the upstream and downstream sites. The conductivity at the Muskeg River confluence was 410 μS*cm$_{-1}$ which was still below the values for OSPW (2200 μS*cm-1). The suspended matter was between 10 and 25 mg/ L with an increasing upstream/downstream trend and reached values of 300–400 mg/ L in downstream samples. The

Table 2: Selected gene targets for toxicogenomic analysis in rainbow trout hepatocytes

Target genes	Symbol	Function/role
APEX nuclease (multifunctional DNA repair enzyme)	APEX	Genotoxicity, DNA repair
Catalase	CAT	Oxidative stress, hydrolysis of H2O2
Cytochrome P450 1A	CYP1A	Biotranformation, hydroxylation of coplanar polycyclic aromatic compounds
Cytochrome P450 3A	CYP3A	Biotransformation, hydroxylation of polycyclic aliphatic hydrocarbons
Estradiol receptor β2	ER	Endocrine disruption, estrogen receptor
growth arrest and DNA-damage-inducible, alpha	GADD45	Genotoxicity, DNA repair activity
Glyceraldehyde 3P dehydrogenase	GAPDH	Anaerobic glycolysis
Glutathion-S-Transferase-P	GST	Biotransformation, conjugation of polar hydrocarbons
DNA ligase	Ligase	Genotoxicity, DNA repair activity
8-oxoguanine DNA glycosylase	OGG	Genotoxicity and oxidative stress, DNE repair of oxidized nucleotides (8-oxoguanine)
Proliferating cell nuclear antigen	PCNA	Cell division
P-glycoprotein (Abcb1)	pGP	Phase III « biotransformation », involved in the extrusion of polar hydrocarbons
Superoxide dismutase (Cu/Zn cytosolic)	SOD	Oxidative stress, oxygen radicals transfer for H2O2 genesis
uracil-DNA glycosylase	UNG	Genotoxicity, DNA repair activity.
Vitellogenin	Vtg	Endocrine disruption, egg yolk protein precursor under the control of the estrogen receptor
Reference genes		
Prolylpeptidyl isomerase I	PPIA	Significantly affected, not selected for normalization
Hypoxanthine phosphoribosyl transferase I	HPRT	Significantly affected, not selected for normalization
RNA polymerase I	RPL	Significantly affected but fairly. Could be used for normalization
Elongation factor I α	EFIα	Least significantly affected, used for normalization

same was observed for dissolved organic carbon (DOC) content, which was 4 mg/ L for the upstream site and increased to 7–12 mg/ L at downstream sites. The DOC value for the OSPW was one order of magnitude higher than the surface water reaching 140 mg/ L. The proportion of light PAHs in relation to total PAHs was determined by fixed wavelength spectroscopy. The proportion of light PAHs, expressed as naphthalene equivalents of surface waters, varied between 40% and 51%, whereas the downstream Ells River confluence had the highest proportion at 51%. The proportion of light PAHs in the OSPW extract reached 72%, with a mean value of 81% for commercial NA mixtures. This suggests that commercial NA mixtures and OSPW contain high levels of light PAHs such as naphthalene. Based on the above data, there is an upstream-downstream trend in suspended matter, DOC and percent of light PAHs Table 4.

The toxicity of the individual compounds, NA mixtures, OSPW and surface waters were investigated in rainbow trout hepatocytes (Figures 1A-1F). Toxicity was determined by loss of membrane permeability (try pan blue staining) and totals RNA levels were used as a general indicator of cell activity. With regard to the individual compounds, $z = 0$ and $z = 2$ were the most toxic compounds based on cell viability (Figure 1A). Interestingly, the $z = 0$ and $z = 2$ compounds were more toxic than the equivalent amount of both NA commercial mixtures. Based on total RNA levels, all compounds significantly influenced RNA levels but in a different manner (Figure 1B). Compounds $z = 0$ to $z=6$ increased total RNA levels at the lowest concentration (2 mg/ L) compared with a decrease in RNA levels at concentrations > 10 mg/ L. However, compounds $z = 8$ to $z = 10$ and the 2 commercial mixtures of NAs decreased total RNA levels at all concentrations tested. The toxicity of 2 OSPW samples was assessed; they were similarly toxic as they both increased cell mortality at a concentration of 0.5% (Figure 1C). This corresponds to the original concentration of 50% given that a 100× concentrate of the water/OSPW extract was used. At the total RNA level, the two samples both increased RNA levels at the lowest concentration (0.4% of original concentration) with a decreasing effect as the concentration reached 50% dilution (Figure 1D). This suggests that acute lethality is likely to occur at low OSPW dilutions, while changes in total RNA levels can be detected at 0.4%. Exposure to surface water extracts did not produce any appreciable changes in cell viability (Figure 1E). Total RNA levels were generally increased at 0.02% (or 2% of the original sample) with a trend of stronger responses for downstream sites in the OS area (i.e., decreased levels of total RNA) but with increased total RNA levels for the confluence of Ells River site. Correlation analysis between cell viability and RNA levels revealed that total RNA levels were significantly correlated with cell viability ($r = 0.53$; $p < 0.001$).

Gene expression was investigated in hepatocytes exposed to individual compounds, NA mixtures, OSPW and surface water extracts (Table 4). The individual NA compounds from $z = 0$ to $z = 10$ and the 2 commercial NA mixtures affected gene expression levels for the 15 target genes (Table 4). Xenobiotic biotransformation genes were generally up regulated but were down regulated with GST for all NAs ($z = 0$ to $z = 10$). CYP3A4 gene expression, which is involved in the biotransformation of cyclic aliphatic compounds, was down regulated for the $z = 10$ compound. The biggest changes (based on the fold response/concentration ratio) were observed with CYP1A1 for the $z = 10$ compound and commercial NA mixture 1, CYP3A4 for $z = 2$ and 4 compounds, GST for $z = 6$ and 10 compounds and MDR for $z = 10$ compound. Gene expression involved in oxidative stress

Table 3 : Selection of primers for quantitative gene expression analysis.

Target genes	Symbol	Foward primer	Reverse primer	Amplicon size (bp)
APEX nuclease (multifunctional DNA repair enzyme)	APEX	TGACAACGGCACAGCTCCCG	GGCCTCGTCACGCACCCAAT	199
Catalase	CAT	TGATGTCACACAGGTGCGTA	GTGGGCTCAGTGTTGTTGAG	195
Cytochrome P450 1A	CYP1A	GATGTCAGTGGCAGCTTTGA	TCCTGGTCATCATGGCTGTA	104
Cytochrome P450 3A	CYP3A	TACATGCCATTTGGGGCGGGG	ACGGGCCTCCAGCCTCAGTTT	195
Estradiol receptor β2	ER	CTGACCCCAGAACAGCTGATC	TCGGCCAGGTTGGTAAGTG	125
growth arrest and DNA-damage-inducible, alpha	GADD45	CGAGGCAGCCAAGTCGCTCA	CTCGCAGCAGAACGCCTGGA	130
Glyceraldehyde 3P dehydrogenase	GAPDH	CCAACCAAACGCTACCGAAC	CCAGATTCCATCTCACCTT	173
Glutathion-S-Transferase-P	GST	ATTTTGGGACGGGCTGACA	CCTGGTGCTCTGCTCCAGTT	81
DNA ligase	Ligase	TGGTGCGATTTTGAAGTGTG	GGTCCTGTGTCCTTGTGGTT	147
8-oxoguanine DNA glycosylase	OGG	GGCGGGCAATGGGCAGAAGA	CCGAGTGTGCCCAACCAGCA	101
proliferating cell nuclear antigen	PCNA	ACAACGCAGACACACTCGCCC	GGGCAAACTCCCCCGATGGC	156
P-glycoprotein (Abcb1)	pGP	ACGTGCGCTCCCTGAACGTG	GCGTTGGCCTCCCTAGCAGC	157
Superoxyde dismutase (Cu/Zn cytosolic)	SOD	TGGTCCTGTGAAGCTGATTG	TTGTCAGCTCCTGCAGTCAC	201
uracil-DNA glycosylase	UNG	TGTCTACCCACCCCCTCAGCA	CCGTGATATGGGTCCTGGCCG	96
Vitellogenin	Vtg	AGCCCATCCACGAACTTGCTGTT	AGGGCCAAAACTGCATCAGCCT	190
Reference genes				
Elongation factor I α	EFIα	GAATCGGCTATGCCTGGTGAC	GGATGATGACCTGAGCGGTG	141
RNA polymerase I	RPL	ACTATGGCTGTCGAGAAGGTGCT	TGTACTCGAACAGTCGTGGGTCA	120
Prolylpeptidyl isomerase I	**PPIA**	CATCCCAGGTTTCATGTGC	CCGTTCAGCCAGTCAGTGTT	203
Hypoxanthine phosphoribosyl transferase I	**HPRT**	CCGCCTCAAGAGCTACTGTAAT	GTCTGGAACCTCAAACCCTATG	255

Table 4: Physico-chemical properties of OSPW and surface waters of the Athabasca River

	pH (units)	Conductivity (μs/ cm)	Suspended matter (mg/ L)	DOC (mg/ L)	Light PAHs (phenanthrene)	Total PAHs	Light/total
Upstream	8.8	291.6	10.4	4	159	350	47%
OS development area	8.8	280.3	20.9	12	217	473	47%
OSPW	9	2200	360	55	3300	4600	72%
Muskeg River confluence	8.7	413.7	16.3	8	125	375	33%
Ells River confluence	8.6	268.8	23.1	7	208	411	51%
NA Mix 1	---	---	---		6724	7371	91%
NA Mix 2	---	---	---		1483	2043	72%

revealed upregulation for most NA compounds but was down regulated with CAT for $z = 2$, 4 and 6 compounds and OGG for the $z = 0$ compound. The most marked changes were observed for CAT with $z = 8$ compound, and for OGG with $z = 2$ compound and commercial NA mix 2. SOD gene expression responded with the least intensity; responding only with $z = 0$, $z = 4$, $z = 10$ compounds and the 2 commercial NA mixtures. Genes involved in Genotoxicity (mostly DNA repair activity) were generally up regulated, with the exception of UNG for $z = 0$, $z = 2$ and $z = 10$ compounds. The most marked responses were obtained for UNG with $z = 8$ compound, APEX with $z = 6$ compound and the 2 commercial NAs mixtures, LIG with $z = 2$, $z = 4$, $z = 8$ compounds and commercial NA mixture 2. The strongest response for GADD45 gene expression was observed for commercial NA mixture 1. Genes involved in endocrine disruption were generally up regulated; however, down regulation occurred for ER2 with

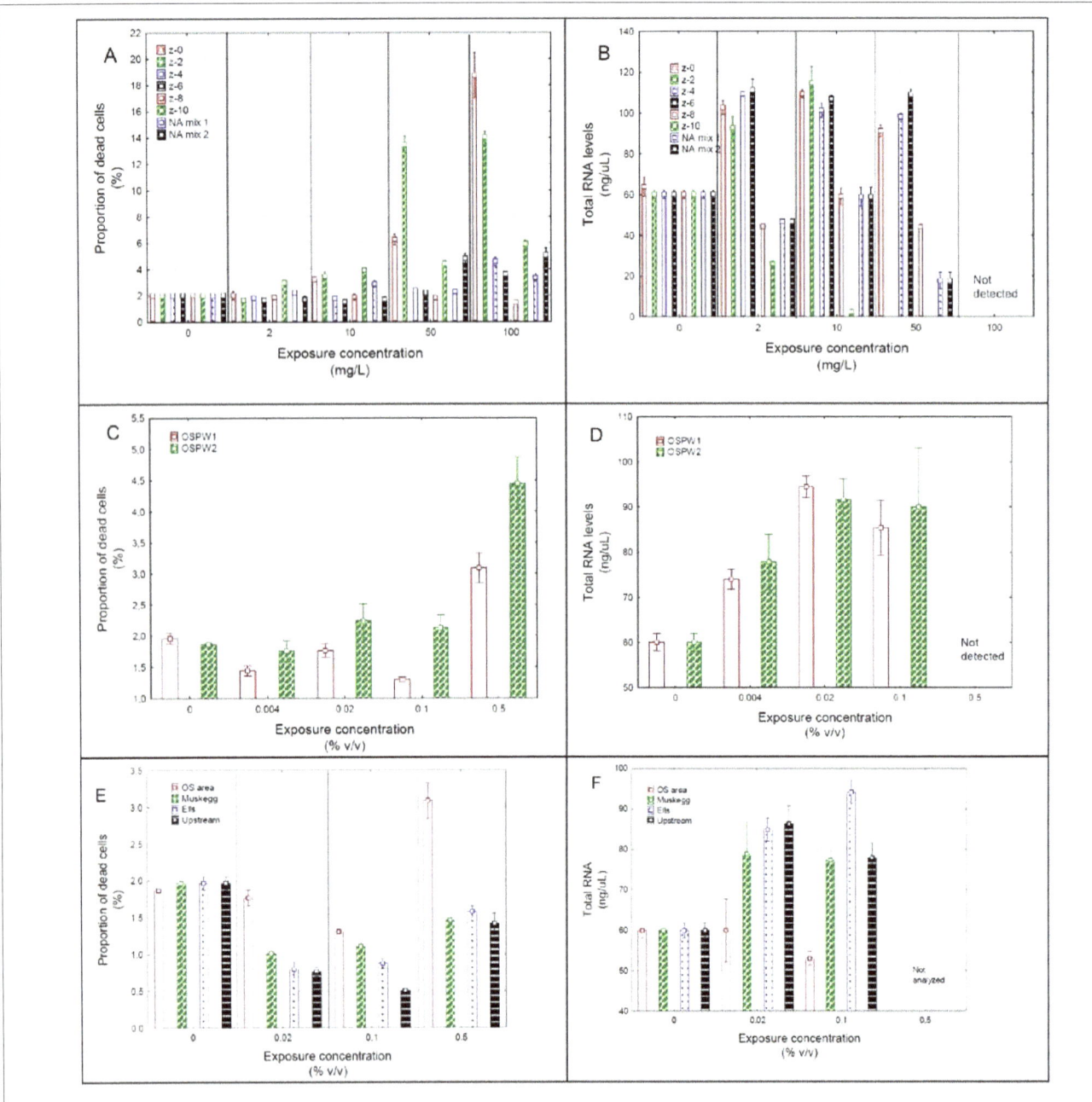

Figure 1: Loss in cell viability and total RNA levels in rainbow trout hepatocytes. Cell viability and total RNA levels in hepatocytes are shown A and B, respectively, for exposure to selected NAs, in C and D, respectively, for OSPW, and in E and F, respectively, for river water.

z=0 and z=4 compounds and for VTG with z=0 compound. The marker gene for anaerobic glycolysis (GAPDH) was significantly up regulated with almost all NA-like single compounds and the mixtures, except for z=6 compound, for which significantly lower levels were observed. The strongest responses for GAPDH were found with z = 8 and z = 10 compounds. The gene marker for cellular proliferation was generally down regulated; however, z = 10 NA and commercial NA mixture 2 produced no changes, and commercial NA mixture 1 induced gene expression of PCNA somewhat (1.3 times the control).

The effect of OSPW and surface water extracts collected upstream and downstream from the OS development area on gene expression was also determined (Table 5). Genes involved in biotransformation were up regulated in most cases, with the exception of GST gene expression, which was always down regulated. The upstream site, located in a high density area of OS deposits(10 km north of Fort McMurray),did not produce changes in all tested genes involved in biotransformation. CYP1A1 gene expression was significantly induced for one of the 2 OSPW samples and somewhat down regulated in the OS extraction

Table 5: Gene expression changes in cells exposed to selected NAs.

Gene category	Gene id	z=0	z=2	z=4	z=6	z=8	z=10	NA Mix 1 / Mix 2
Xenobiotic biotransformation	CYP1A1	1.7 (10 mg/L)[1]	ns[2]	1.9 (50 mg/L)	1.5 (2 mg/L)	ns	**3.4 (2 mg/L)**	**2.8 (2 mg/L)** / 1.5 (2 mg/L)
	CYP3A4	1.6 (10 mg/L)	**2 (2 mg/L)**	**1.7 (2 mg/L)**	1.3 (2 mg/L) 0.65	1.4 (2 mg/L)	0.56 (2 mg/L)	1.9 (2 mg/L) / 1.6 (2 mg/L)
	GST	0.45 (10 mg/L)	0.8 (2 mg/L)	0.7 (2 mg/L)	**0.48 (2 mg/L)**	0.5 (10 mg/L)	**0.44 (2 mg/L)**	ns / ns
	PGP (MDR)	3 (10 mg/L)0.3	4 (10 mg/L)	1.8 (50 mg/L)	1.3 (50 mg/L)	4.4 (10 mg/L)	**7 (2 mg/L)**	1.5 (2 mg/L) / 1.7 (10 mg/L)
Oxidative stress	SOD	2.2 (50 mg/L)	ns	ns	1.4 (2 mg/L)	ns	1.5 (10 mg/L)	1.3 (2 mg/L) / ns
	CAT	1.5 (50 mg/L)	0.5 (2 mg/L)	0.6 (50 mg/L)	0.75(10 mg/L)	**1.7 (2 mg/L)**	1.9 (2 mg/L)	1.7 (2 mg/L) / 1.2 (2 mg/L)
	OGG	0.5 (50 mg/L)	**1.4 (2 mg/L)**	1.4 (2 mg/L)	1.5 (2 mg/L)	1.4 (2 mg/L)	ns	2.2 (2 mg/L) / **1.9 (2 mg/L)**
Genotoxicity	UNG	0.7 (2 mg/L)	0.45(10 mg/L)	1.4 (2 mg/L)	ns	**1.5 (2 mg/L)**	*0.7 (10 mg/L)*	1.6 (2 mg/L) / 2 (10 mg/L)
	APEX	1.8 (10 mg/L)	1.7 (10 mg/L)	2 (10 mg/L)	**2 (2 mg/L)**	2.7 (10 mg/L)	1.3 (2 mg/L)	**3.4 (2 mg/L)** / **3.2 (2 mg/L)**
	Ligase	1.6 (10 mg/L)	**1.6 (2 mg/L)**	**1.5 (2 mg/L)**	1.5 (2 mg/L)	**1.7 (2 mg/L)**	ns	2.2 (2 mg/L) / **2 (2 mg/L)**
	GADD45	2 (10 mg/L)	1.7 (10 mg/L)	ns	2.2 (50 mg/L)	2.4 (10 mg/L)	2 (2 mg/L)	**2.4 (2 mg/L)** / 1.7 (10 mg/L)
Estrogenicity	ER2	**0.4 (2 mg/L)[3]**	ns	1.8 (10 mg/L)	**0.1 (2 mg/L)**	2 (10 mg/L)	1.4 (10 mg/L)	1.9 (2 mg/L) / ns
	VTG	**0.4 (2 mg/L)**	4.8 (10 mg/L)	ns	1.7 (50 mg/L)	1.7 (10 mg/L)	ns	3.6 (10 mg/L) / 1.9 (10 mg/L)
Other	GADPH	1.3 (50 mg/L)	1.3 (10 mg/L)	1.6 (10 mg/L)	0.58 (10 mg/L)	**1.5 (2 mg/L)**	**2.6 (2 mg/L)**	1.8 (2 mg/L) / 1.4 (2 mg/L)
	PCNA	**0.4 (2 mg/L)**	0.5 (2 mg/L)	**0.6 (2 mg/L)**	0.6 (2 mg/L)	**0.66 (2 mg/L)**	ns	1.3 (2 mg/L) / Ns

1. The data are expressed as the significant response factor (normalized to controls) at the lowest concentration. Response factors < 1 are decreased relative to the controls; the concentration is in parentheses.
2. Ns: not significant
3. The 3 most sensitive endpoints are highlighted in **bold**. The endpoints are calculated based on the ratio of the fold change/concentration in mg/L.

area; no effects were observed for the other downstream sites. CYP3A4 was strongly induced by both OSPW samples as well as at the confluence of Ells River but with less intensity. GST gene expression was consistently depressed in OSPW1 and at the confluence of both the Muskeg and Ells rivers. The MDR transcript levels were significantly higher in both OSPW1 and OSPW2 and at the confluence of both the Muskeg and Ells rivers. Genes involved in oxidative stress were generally induced, although some showed a decrease (OSPW1 for SOD). CAT and OGG gene expression was increased at the upstream site but the increase was less marked based on the fold change/ concentration ratio. The strongest responses were observed for the OGG gene as follows: for both OSPW samples, the OS downstream area and at the confluence of Ells River. Strong responses were also observed for CAT at the Muskeg and Ells rivers. Genes involved in DNA repair or Genotoxicity were also examined. The upstream site showed a small increase in DNA repair activity based on APEX and GADD45 gene expression but the responses occurred at higher concentrations compared to those at downstream sites. UNG gene expression was induced by both OSPW samples and by all the other downstream sites, whereas the upstream site did not produce any significant changes. The same was observed for LIG gene expression but the responses were stronger than for UNG. APEX gene expression was strongly induced in both OSPW samples at the lowest concentration tested, but the gene was induced at the upstream site as well. The Estrogenicity evaluation based on ER2 and VTG gene expression revealed some upstream/downstream trends. ER2 transcript levels were significantly higher in OSPW2 and at all downstream sites, but there was no effect at the upstream site. VTG gene expression was only induced in OSPW2, which suggests that OS tailings have modest estrogenic

Table 6: Gene expression changes in cells exposed to OSPW and river water.

Gene category	Gene ID	OSPW1 OSPW2	OS area	Muskeg River confluence	Ells River confluence	Upstream
Xenobiotic biotransformation	CYP1A1	ns 1.5(0.02%)	0.7(0.1%)	ns	ns	ns
	CYP3A4	**1.8 (0.004%)** 1.5(0.02%)	ns	ns	1.3(0.02%)	ns
	GST	ns 0.6 (0.1%)	ns	0.65(0.02%)	0.7(0.1%)	ns
	PGP (MDR)	1.6 (0.1%) **2.1(0.02%)**	ns	1.5(0.02%)	1.3(0.1%)	ns
Oxidative stress	SOD	0.7(0.004%) 1.2(0.02%)	ns	ns	ns	ns
	CAT	1.4 (0.1%) ns	ns	**1.6 (0.02%)**	**1.6(0.02%)**	2(0.1%)
	OGG	**1.6 (0.004%)** **1.5(0.004%)**	**1.4(0.02%)**	1.5(0.02%)	**1.6(0.02%)**	1.9(0.1%)
Genotoxicity	UNG	1.5 (0.004%) 1.7(0.02%)	**1.7 (0.02%)**	1.8(0.1%)	1.2(0.1%)12	ns
	APEX	**2.1 (0.004%)** **2.5 (0.004%)**	ns	**1.7(0.02%)**	**2.2(0.02%)**	1.6(0.02%)
	Ligase	**2.8 (0.004%)** 1.4 (0.004%)	1.2(0.02%)	1.4(0.02%)	**1.5(0.02%)**	ns
	GADD45	ns 0.8 (0.004%)	0.6(0.1%)	ns	0.7(0.1%)	0.7(0.02%)
Estrogenicity	ER2	ns 1.6 (0.02%)	**2(0.02%)**	**1.6(0.02%)**	1.3(0.1%)	ns
	VTG	ns 18(0.1%)	ns	ns	ns	ns
Other	GADPH	1.47 (0.004%) 1.5 (0.02%)	1.9(0.1%)	1.3(0.02%)	1.3(0.02%)	1.4(0.1%)
	PCNA	1.2 (0.02%) 1.5 (0.02%)	ns	ns	1.2(0.1%)	Ns

1. The data are expressed as the significant response factor (normalized to controls) at the lowest concentration. Response factors < 1 are decreased relative to the controls; the concentration is in parentheses.
2. Ns: not significant
3. The 3 most sensitive endpoints are highlighted in **bold**. The endpoints are calculated based on the ratio of the fold change/concentration in % v/v.

activity. The marker gene for anaerobic glycolysis was induced to a significantly greater extent for all sites, including the upstream site. However, the responses were stronger (i.e., higher response relative to the controls at lower concentrations) for the OSPW1, OSPW2 and OS extraction area sites. The gene marker for cell proliferation showed no clear trends, although some induction was observed for both OSPW sites/ samples. Correlation analysis revealed that cell viability was significantly correlated with CAT (r = -0.38), GST (r = -0.43), SOD (r = -0.22), GAPDH (r = -0.3), PCNA (r = -0.32), CYP3A4 (r = -0.41), UNG (r = -0.37), GADD45 (r = 0.25), VTG (r = 0.29), ER2 (r = -0.22), SOD (r = 0.21) and MDR (r = 0.54), suggesting that these genes were associated with a cytotoxic effect. The analysis also revealed that the changes in total RNA levels were correlated with CAT (r = -0.36), GST (r = -0.52), SOD (r = -0.22), GADPH (r = -0.4), PCNA (r = -0.3), UNG (r = -0.27) and ER2 (r = -0.37).We also performed a canonical analysis of gene expression data to identify the gene groups (biotransformation, Genotoxicity, oxidative stress, endocrine

disruption, anaerobic glycolysis and cell proliferation) linked to changes in Cytotoxicity (cell viability and total RNA levels) (Figure 2). The genes involved in biotransformation, oxidative stress and Genotoxicity were found to be highly correlated with Cytotoxicity. Hence, these groups point to possible adverse toxic outcomes for OSPW and OS development area contaminants.

A classification and regression tree (CART) analysis was performed on the gene expression and Cytotoxicity data to get an global view about the cytotoxic properties of individual compounds, commercial NA mixtures, OSPW and surface water samples towards rainbow trout hepatocytes (Figure 3). The analysis revealed 4 distinct clusters based on a number of properties (rules) related to gene expression and Cytotoxicity data. The first cluster consists of the control and upstream surface water extracts, which were classified based on the following rules or observations: no/low cell mortality and LIG gene expression levels at first, followed by UNG levels > 1.1 and GADD45 levels < 0.7, then the sample belongs to the second sub-cluster, and is still considered upstream; however, if GADD45 is > 0.7, the sample is now considered downstream water. Downstream waters are characterized by increased expression of DNA repair genes such as UNG, GADD45, and LIG. When LIG gene expression reaches > 1.7 fold, the sample belongs to the next (third) cluster, which contains NA mixtures 1 and 2. NA mixtures are characterized by high LIG gene expression levels (> 1.7 fold). OSPW 1 and 2 are found in the 4th cluster, which is characterized by high Cytotoxicity (> 1.3 fold relative to the controls). OSPW samples are characterized by high levels of CYP1A1 and PCNA gene expression for OSPW1. OSPW2 is characterized by increased CYP3A4 expression and low GST values (at least > 0.6). The individual NAs were distributed within the fourth cluster (OSPW), with the exception of z=8, which was located in the third NA mixture cluster. No clear pattern of gene expression was found for the individual NAs, but marked changes in CYP1A1 (> 2.8 fold) and CYP3A4 (> 2 fold) and GADD45 (> 1.1 fold) were found for z = 10, z = 2 and z = 8 compounds. The most important biomarkers (< 80% relative importance) permitting classification of the various samples were PCNA, CYP3A4, GADD45, LIG and CYP1A1. These biomarkers were strongly associated with Cytotoxicity responses, with the exception of PCNA.

Discussion

Analysis OSPW, NA mixtures and surface waters revealed that they contained significant amounts of PAHs. The proportion of light PAHs was significantly higher in the OSPW and NAs mixtures, accounting for about 80% of the total PAHs, compared to 33–51% in surface water samples. The increase in the proportion of light PAHs paralleled the increases in total suspended solids and dissolved organic carbon contents observed in OSPW compared to surface water samples; hence, these properties could be considered general markers of OSPW contamination. This is in keeping that the Clark extraction procedure liberates low molecular weight hydrocarbons and carboxylic acids (NAs) during the caustic hot water extraction process. Based on these parameters, there was no clear evidence that surface waters were tainted with OSPW or NA mixtures. This is consistent with the finding reported in a previous study, specifically that only groundwater showed evidence of OSPW-

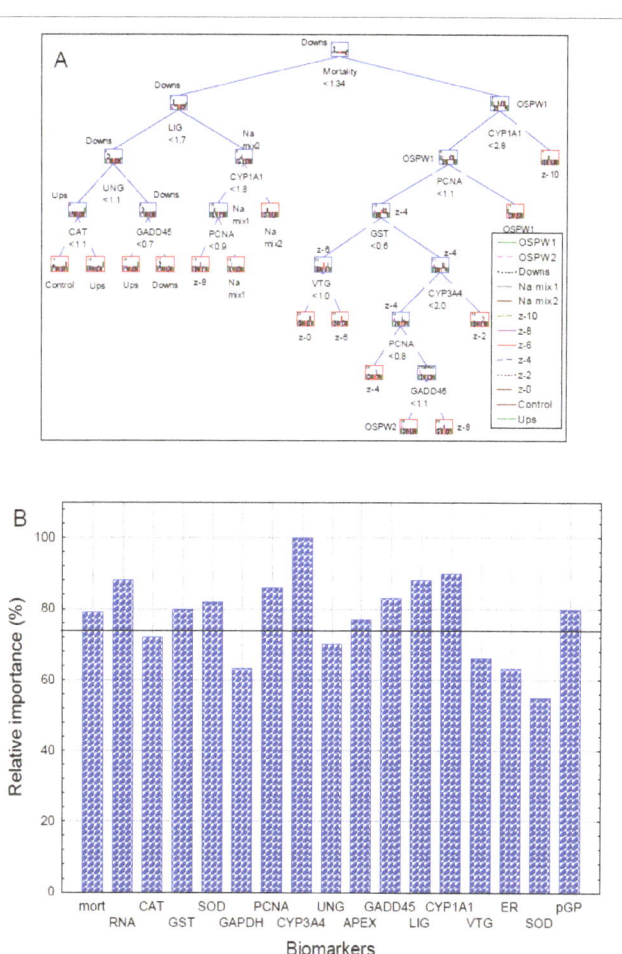

Figure 3: Decision tree classification of OS samples based on gene expression and cell toxicity

Decision tree analysis was performed using the CART algorithm (univariate) to identify the divisions with the best performance. The decision tree and rules for each division are shown in A and the most important biomarkers for sample classification are shown in B. The dashed line in figure 3B represents the 75% threshold.

related contamination [14]. In the present study, CYP3A4 and GST gene expression was up regulated and down regulated, respectively, in hepatocytes exposed to OSPW. Commercial NA mixtures led to increased CYP3A4 gene expression in trout hepatocytes but had no effect on GST gene expression. Increased CYP3A4gene expression, indicative of phase 1 biotransformation of cyclic aliphatic hydrocarbons, was proposed a specific gene marker for OSPW compared to surface waters [12]. No induction inCYP3A4 transcripts in hepatocytes exposed to surface waters, except a small increase (1.3 fold) for the water sample taken at the confluence of Ells River, was detected. A decrease in GST gene expression was observed in previous studies at both gene expression [17] and enzyme activity levels [14]. Decreased GST gene expression was closely associated with OSPW and with surface waters downstream from the OS development area. Although the reason for the decreases in GST activity or gene expression is unclear at present, they seem to be a consistent pattern associated with OS mining activity.

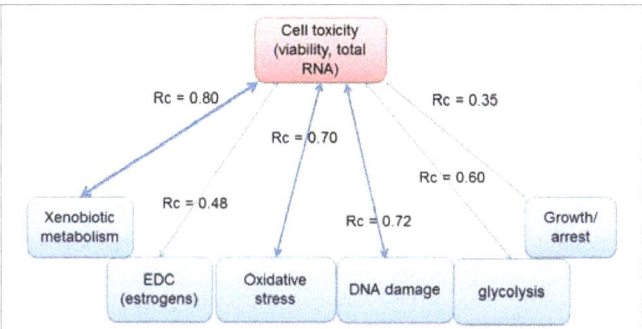

Figure 2: Canonical analysis of gene expression data and cell toxicity

Canonical analysis was performed on the gene expression data and cytoxicitydata, based on loss of cell membrane permeability and changes in total RNA levels, in order to identify pathways of toxicity in hepatocytes exposed to various OS-related products.

OPSW Genotoxicity was found in the present study based on the observation that the most marked responses in DNA repair genes were obtained with the two OSPW samples. To the best of our knowledge, [14] were the first to report DNA damage in a study on rainbow trout hepatocytes exposed to OSPW extracts and surface waters. DNA strand breaks in primary cultures of trout hepatocytes exposed to surface waters and OSPW extracts were found to be the most responsive endpoint; however, they offered the least discrimination between "natural" or background levels in the area rich in OS deposits and the area downstream from OS mining activity. This highlights the challenge involved in discriminating between natural releases of OS contaminants and those associated with mining activities in the region. Moreover, marked DNA repair gene expression was also observed for both upstream and downstream surface waters, with a trend of higher responses for the area downstream from the OS mining operations. This is consistent with the relatively high background values of PAHs found in the area rich in OS deposits [5,13]. PAHs are known to induce CYP1A1 gene expression and produce DNA damage [22]. High molecular weight (heavy) PAHs were found to be responsible for most of the mutagenic activity. The comet assay was used in rainbow trout hepatocytes to examine the contribution of some NAs (including commercial mixtures), OSPW and OS leachates in terms of mimicking the natural background release of genotoxic compounds [15]. Genotoxicity was observed in all samples from the individual compounds ($z = 6$ to $z = 10$), commercial NAs, OSPW and OS leachates. However, genotoxic potential was higher for OSPW than for the OS leachates and it was strongly involved oxidative DNA damage. A recent study revealed that diamondoid naphthenic acids caused *in vivo* DNA damage in the hemocytes and gills of marine mussels [23]. This is keeping with our results and those of the Lacaze, et al. [15] study in which a noradamantane carboxylic acid increased the expression of APEX, LIG and GADD45 genes. Interestingly, the noradamantane carboxylic acid ($z = 6$) also induced the expression of OGG (DNA repair of oxidized guanine) at lower intensities (1.5 fold at 2 mg/ L) than for LIG and APEX, which produced 2-fold (2 mg/ L) and 2.2-fold (50 mg/ L) responses, respectively. This was also found with the comet assay on trout hepatocytes at much lower concentrations (0.2 mg/ L) in the Lacaze, et al. study[15].

Recent evidence suggests that some OS products are capable of inducing VTG gene expression in rainbow trout [12]. To the best of our knowledge, this was the first report VTG gene upregulation in trout hepatocytes exposed to OSPW and to surface waters from the OS development area.VTG gene expression was markedly increased in one of the OSPW samples with a response 18 times greater than for the controls. There was no indication of induction in hepatocytes exposed to surface waters. ER2 gene expression was increased for surface waters from sites downstream from the OS development area and for one of the OSPW extracts. As in the case of Genotoxicity responses, these data suggest that endocrine disrupting substances are present in OSPW and in river water in the area with rich OS deposits; however, the intensity of responses seems to vary from year to year. The commercial NA mixtures also induced VTG and ER2 gene expression, further supporting the hypothesis that OSPW-related material

is estrogenic to fish. In a previous study, VTG gene expression increased about 30-fold when hepatocytes were exposed to OPSW, whereas surface waters were 2X less potent in inducing VTG (i.e., 15-fold increase relative to controls).Exposure to NAs isolated from OSPW and to commercial NA mixtures significantly increased the production of progesterone and estradiol-17β but decreased testosterone levels in H295R cells [24]. This corroborates the present study's findings of increased CYP3A4 gene expression caused by OSPW and commercial NAs mixtures. Indeed, CYP3A4 codes for the corresponding cytochrome P450, which has 6β-testosterone hydroxylase activity and is involved in the elimination of testosterone [25].In another study with young-of-the-year zebra fish (*Danio rerio*), exposures to NAs in OSPW and commercial NA mixtures led to the upregulation of aromatase (CYP19b), Erα and VTG gene expression [26]. These changes in gene expression were associated with delays in embryo hatching, which was followed by developmental lesions such as pericardial edema, yolk sac edema and spinal malformation. These results further support the notion that NAs derived from OS can negatively impact endocrine function in fish and can contribute to long-term toxicity of OSPW. Aromatic NAs were found to weakly induce VTG gene expression in zebra fish larvae [27], suggesting that other types of NAs or other compounds found in OSPW contribute towards estrogenicity or produce effects at the steroid metabolism level, as shown above. It appears that NA endocrine-disrupting effects occur across the brain–gonad–liver axis in fathead minnows exposed to untreated and ozone treated OSPW [28]. The abundances of transcripts of estrogen-responsive genes were greater in livers from male fish exposed to untreated OSPW compared to control male fish. However, the opposite effect was found in female fish, which showed a decrease in estrogen-responsive genes in the liver. The same pattern was observed for the gonads. By contrast, in brain tissue, the abundance of transcripts of genes important for synthesis of gonadotropins was greater in both male and female fish exposed to OSPW than in control fish. These results indicate that the endocrine-disrupting effect goes beyond the liver and affects the entire brain–gonad–liver axis.

The reported toxicity of the individual organic compounds in this study is limited to aquatic species. Chemical characterization of commercial NAs revealed that the majority of NAs in commercial mixtures were composed of 1 to 3 rings [29]. These commercial preparations also contain monoaromatic acids and non-acids (both found in the light PAHs fraction), PAHs and sulfur heterocyclic hydrocarbons. Abietic acid ($z = 10$) was found not to influence VTG gene expression, although a small increase in ER2 gene expression is observed at a relatively high concentration of 10 mg/ L. It was found that abietic acid in combination with estrogenic compounds contributed to male feminization of the roach [30]. This suggests that some NAs could potentiate the effect of environmental estrogens. Abietic acid was found to cause inhibition in GST gene expression. GST inhibition was also observed in mussels exposed to 3 μM (0.9 mg/ L) of abietic acid for up to 24 h [31]. DNA integrity and oxidative stress in mussel hepatopancreas were respectively decreased and increased in exposed mussels, a finding that parallels the present study's 2-fold increase in GADD45 gene expression with increases in SOD and CAT, which is indicative of oxidative stress responses. The Genotoxicity and biotransformation potential of abietic acid was investigated in the eel *Anguilla anguilla L* [32,33]. Low concentrations of abietic acid (0.1 to 0.3 mM) were found to increase EROD activity (CYP1A1) in eels, which is in line with the observed induction in gene expression of CYP1A1 (3.4 fold relative to the controls). Abietic acid also increased erythrocyte nuclear abnormalities and DNA strands in the liver of exposedeels. Thee resin acid, abietic acid, which originates from the decomposition of plants/trees, may be present in bitumen but it also occurs naturally in the environment, as well as in municipal and pulp mill effluent wastewaters; hence, it is not considered specific to OSPW.

In this study, based on canonical analysis, the genes involved in Xenobiotic biotransformation, oxidative stress and DNA repair activity were strongly related to Cytotoxicity. Exposure to OSPW was shown to alter gene expression of male fathead minnows involved in oxidative stress, oxidative metabolism (which involves biotransformation), apoptosis and immune function [34].In another study, fathead minnow embryos exposed to OSPW hatched prematurely and the embryos exhibited higher incidences of hemorrhage, pericardial edema and malformation of the spine [35]. These embryos had elevated reactive oxygen species with a greater abundance of transcripts for CYP3A, GST, SOD and caspase 9. This suggests that OSPW caused oxidative stress and biotransformation of Xenobiotic s which can lead to mitochondrial dysfunction and apoptosis. Hepatocytes exposed to commercial NA mixtures and OSPW had elevated levels of GAPDH, which is involved in anaerobic glycolysis. This could be a consequence of altered mitochondria in the liver due to increased levels of oxidative stress and biotransformation. It is noteworthy that this marker gene was equally expressed in surface waters downstream from the OS extraction area and at the upstream site, suggesting that OS-rich deposits may contain chemicals that affect the aerobic/ anaerobic balance in cells. However, based on canonical analysis, this endpoint was not as strongly correlated to Cytotoxicity compared to the other gene endpoints, indicating that this effect may not be a major driver of toxicity. Nonetheless, this also raises the possibility that disturbance in aerobic/ anaerobic glycolysis could be a contributing factor to the toxicity of OSPW and NAs.

In conclusion, exposure of rainbow trout hepatocytes to OS-related products revealed various effects at the gene expression level, including biotransformation, oxidative stress, DNA repair activity, anaerobic glycolysis and growth arrest. It was found that genes involved in biotransformation, oxidative stress and DNA damage (repair) were the most strongly associated ($r_c \geq 0.70$; $p < 0.001$) with Cytotoxicity based on cell viability and total RNA levels. Decision tree analysis revealed that upstream waters, downstream waters, commercial NA mixtures and OSPW formed 4 distinct groups based on gene expression data. Some trends related to location upstream/downstream from the OS development area were observed, especially with genes involved

in DNA repair (UNG, APEX and LIG) and biotransformation (GST downward expression). Expression of endocrine-disrupting genes was not associated with cell toxicity, VTG was strongly induced by only one of the OSPW samples, and ER2 was expressed in downstream waters. These results collectively suggest that endocrine disruption is not a major effect of exposure to OS-derived products. However, more research is required to determine whether these upstream/downstream trends result from the particular hydrodynamics (confluence of many rivers at downstream sites such as Muskeg and Ells rivers) and mining activities that characterize this region.

Acknowledgements

The authors wish to thank Joanna Kowalczyk for conducting the biochemical assays in the laboratory. This work was funded by the Oil Sands research and monitoring initiative of Environment and Climate Change Canada.

References

1. Gosselin P, Hrudey SE, Naeth A, Plourde A, Therrien R, Van Der Kraak G, et al. Environmental and health impacts of Canada's oil sands industry. The Royal Society of Canada/LaSociété Royale du Canada, Ottawa. 2010.

2. Frank RA, Roy JW, Bickerton G, Rowland SJ, Headley JV, Scarlett AG, et al. Profiling oilsands mixtures from industrial developments and natural groundwaters for source identification. Environ Sci Technol. 2014;48(5):2660-70. doi: 10.1021/es500131k.

3. Bauer AE, Frank RA, Headley JV, Peru KM, Hewitt LM, Dixon DG. Enhanced characterization of oilsands acid-extractable organics fractions using electrospray ionization-high-resolution mass spectrometry and synchronous fluorescence spectroscopy. Environ. Toxicol.Chem. 2015;34(5):1001-1008. doi: 10.1002/etc.2896.

4. Puttaswamy N, Liber K. Influence of inorganic anions on metals release from oil sands coke and on toxicity of nickel and vanadium to Ceriodaphniadubia. Chemosphere. 2012;86(5):521-9. doi: 10.1016/j.chemosphere.2011.10.018.

5. Kelly EN, Short JW, Schindler DW, Hodson PV, Ma M, Kwan AK, et al. Oil sands development contributes polycyclic aromatic compounds to the Athabasca River and its tributaries. Proc Natl Acad Sci U S A. 2009;106(52):22346-51. doi: 10.1073/pnas.0912050106.

6. Kelly EN, Short JW, Schindler DW, Hodson PV, Ma M, Kwan AK, et al. Oil sands development contributes elements toxic at low concentrations to the Athabasca River and its tributaries. Proc Natl Acad Sci U S A. 2009;106(52):22346-51. doi: 10.1073/pnas.0912050106.

7. Martin JW, Han X, Peru KM, Headley JV. Comparison of high- and low-resolution electrospray ionization mass spectrometry for the analysis of naphthenic acid mixtures in oil sands process water. Rapid Commun Mass Spectrom. 2008;22(12):1919-24. doi: 10.1002/rcm.3570.

8. Holowenko FM, MacKinnon MD, Fedorak PM. Characterization of naphthenic acids in oil sands wastewaters by gas chromatography-mass spectrometry. Water Res. 2002;36(11):2843-55.

9. Hagen MO, Katzenback BA, Islam MD, Gamal El-Din M, Belosevic M.The analysis of goldfish (Carassiusauratus L.) innate immune responses after acute and subchronic exposures to oilsands process-affected water. Toxicol Sci. 2014;138(1):59-68. doi: 10.1093/toxsci/kft272.

10. Kavanagh RJ, Frank RA, Solomon KR, Van Der Kraak G. Reproductive and health assessment of fathead minnows (Pimephalespromelas)

inhabiting a pond containing oilsands process-affected water. Aquat Toxicol. 2013;130-131:201-9. doi: 10.1016/j.aquatox.2013.01.007.

11. Arens CJ, Hogan NS, Kavanagh RJ, Mercer AG, Kraak GJ, van den Heuvel MR. Sublethal effects of aged oilsands-affected water on white sucker (*Catostomuscommersonii*). Environ Toxicol Chem. 2015;34(3):589-99. doi: 10.1002/etc.2845.

12. Gagné F, Douville M, André C, Debenest T, Talbot A, Sherry J, et al. Differential changes in gene expression in rainbow trout hepatocytes exposed to extracts of oil sands process-affected water and the Athabasca River. Comp Biochem Physiol C Toxicol Pharmacol. 2012;155(4):551-9. doi: 10.1016/j.cbpc.2012.01.004.

13. Casini S, Marsili L, Fossi MC, Mori G, Bucalossi D, Porcelloni S, et al. Use of biomarkers to investigate toxicological effects of produced water treated with conventional and innovative methods. Mar Environ Res. 2006;62 Suppl:S347-51. doi: 10.1016/j.marenvres.2006.04.060.

14. Gagné F, André C, Douville M, Talbot A, Parrott J, McMaster M, et al. An examination of the toxic properties of water extracts in the vicinity of an oil sand extraction site. J Environ Monit. 2011;13(11):3075-86. doi: 10.1039/c1em10591d.

15. Lacaze E, Devaux A, Bruneau A, Bony S, Sherry J, Gagné F. Genotoxic potential of several naphthenic acids and a synthetic oil sands process affected water in rainbow trout (*Oncorhynchus mykiss*). Aquat Toxicol. 2014;152:291-9. doi: 10.1016/j.aquatox.2014.04.019.

16. Alharbi HA, Saunders DM, Al-Mousa A, Alcorn J, Pereira AS, Martin JW, et al. Inhibition of ABC transport proteins by oilsands process affected water. Aquat Toxicol. 2016;170:81-8. doi: 10.1016/j.aquatox.2015.11.013.

17. Gagné F, André C, Turcotte P, Gagnon C, Sherry J, Talbot A. A comparative toxicogenomic investigation of oil sand water and processed water in rainbow trout hepatocytes. Arch Environ Contam Toxicol. 2013;65(2):309-23. doi: 10.1007/s00244-013-9888-2.

18. Debenest T, Turcotte P, Gagné F, Gagnon C, Blaise C. Ecotoxicological impacts of effluents generated by oil sands bitumen extraction and oil sands lixiviation on Pseudokirchneriellasubcapitata. Aquat Toxicol. 2012;112-113:83-91. doi: 10.1016/j.aquatox.2012.01.021.

19. Baksi SM, Frazier JM. Isolated fish hepatocytes—Model systems for toxicology research. Aquat.Toxicol. 1990;16(4):229–256.

20. Klaunig JE, Ruch RJ, Goldblatt PJ. Trout hepatocyte culture: Isolation and primary culture. In Vitro Cell Dev Biol. 1985;21(4):221-8.

21. Andersen CL, Jensen JL, Ørntoft TF. Normalization of real-time quantitative reverse transcription-PCR data: a model-based variance estimation approach to identify genes suited for normalization, applied to bladder and colon cancer data sets. Cancer Res. 2004;64(15):5245-50. doi: 10.1158/0008-5472.CAN-04-0496.

22. Marvin CH, Lundrigan JA, McCarry BE, Bryant DW. Determination and genotoxicity of high molecular mass polycyclic aromatic hydrocarbons isolated from coal-tar-contaminated sediment. Environ. Toxicol. Chem. 1995;14(12):2059-2066.

23. Dissanayake A, Scarlett AG, Jha AN. Diamondoid naphthenic acids cause *in vivo* genetic damage in gills and haemocytes of marine mussels. Environ Sci Pollut Res Int. 2016;23(7):7060-6. doi: 10.1007/s11356-016-6268-2.

24. Wang J, Cao X, Sun J, Huang Y, Tang X. Disruption of endocrine function in H295R cell in vitro and in zebrafish in vivo by naphthenic acids. J Hazard Mater. 2015;299:1-9. doi: 10.1016/j.jhazmat.2015.06.004.

25. Yamazaki H, Shimada T. Progesterone and testosterone hydroxylation by cytochromes P450 2C19, 2C9, and 3A4 in human liver microsomes. Arch Biochem Biophys. 1997;346(1):161-9. doi: 10.1006/abbi.1997.0302.

26. Wang J, Cao X, Huang Y, Tang X. Developmental toxicity and endocrine disruption of naphthenic acids on the early life stage of zebrafish (Danio rerio). J Appl Toxicol. 2015;35(12):1493-501. doi: 10.1002/jat.3166.

27. Reinardy HC1, Scarlett AG, Henry TB, West CE, Hewitt LM, Frank RA, et al. Aromatic naphthenic acids in oil sands process-affected water, resolved by GCxGC-MS, only weakly induce the gene for vitellogenin production in zebrafish (*Danio rerio*) larvae. Environ Sci Technol. 2013;47(12):6614-20. doi: 10.1021/es304799m.

28. He Y, Wiseman SB, Wang N, Perez-Estrada LA, El-Din MG, Martin JW, et al. Transcriptional responses of the brain-gonad-liver axis of fathead minnows exposed to untreated and ozone-treated oil sands process-affected water. Environ Sci Technol. 2012;46(17):9701-8. doi: 10.1021/es3019258.

29. Swigert JP, Lee C, Wong DC, White R, Scarlett AG, West CE, et al. Aquatic hazard assessment of a commercial sample of naphthenic acids. Chemosphere. 2015;124:1-9. doi: 10.1016/j.chemosphere.2014.10.052.

30. Lange A, Sebire M, Rostkowski P, Mizutani T, Miyagawa S, Iguchi T, et al. Environmental chemicals active as human antiandrogens do not activate a stickleback androgen receptor but enhance a feminising effect of oestrogen in roach. Aquat Toxicol. 2015;168:48-59. doi: 10.1016/j.aquatox.2015.09.014.

31. Gravato C, Oliveira M, Santos MA. Oxidative stress and genotoxic responses to resin acids in Mediterranean mussels. Ecotoxicol Environ Saf. 2005;61(2):221-9. doi: 10.1016/j.ecoenv.2004.12.017.

32. Maria VL, Correia AC, Santos MA. Anguilla anguilla L. genotoxic and liver biotransformation responses to abietic acid exposure. Ecotoxicol Environ Saf. 2004;58(2):202-10. doi: 10.1016/j.ecoenv.2003.12.005.

33. Pacheco M, Santos MA. Induction of EROD activity and genotoxic effects by polycyclic aromatic hydrocarbons and resin acids on the juvenile eel (Anguilla anguilla L.). Ecotoxicol Environ Saf. 1997;38(3):252-9. doi: 10.1006/eesa.1997.1585.

34. Wiseman SB, He Y, Gamal-El Din M, Martin JW, Jones PD, Hecker M, et al. Transcriptional responses of male fathead minnows exposed to oil sands process-affected water. Comp Biochem Physiol C Toxicol Pharmacol. 2013;157(2):227-35. doi: 10.1016/j.cbpc.2012.12.002.

35. He Y, Patterson S, Wang N, Hecker M, Martin JW, El-Din MG, et al. Toxicity of untreated and ozone-treated oil sands process-affected water (OSPW) to early life stages of the fathead minnow (Pimephales promelas). Water Res. 2012;46(19):6359-68. doi: 10.1016/j.watres.2012.09.004.

8

Sublethal Effects of Poly (Amidoamine) Dendrimers in Rainbow Trout Hepatocytes

Auclair J[1], Morel E[2], Wilkinson KJ[2], Gagne F[1]*

[1]Aquatic Contaminants Research Division, Environment and Climate Change Canada, 105 McGill, Montréal, QC, Canada
[2]Department of Chemistry, University of Montreal, C.P. 6128, Succ, Centre-Ville, Montreal, Canada H3C 3J7

*Corresponding author: F. Gagné, Aquatic Contaminants Research Division, Environment and Climate Change Canada, 105 McGill Street, Montreal, Quebec, Canada, E-mail: francois.gagne@canada.ca

Abstract

The purpose of this study was to examine the toxicity of drug vectors—poly (amidoamine) (PAMAM) dendrimers—to rainbow trout hepatocytes. Primary cultures of rainbow trout hepatocytes were exposed to concentrations of G2, G4, G5 PAMAM dendrimers and a representative antibiotic—minocycline—in municipal effluents for 48 h at 15°C. After the exposure period, cells were harvested for the assessment of viability, heat shock protein 70 (HSP70) level and glutathione S-transferase (GST) activity. The results revealed that the PAMAM dendrimers were toxic to rainbow trout hepatocytes, with the G4 and G5 PAMAM dendrimers being 5 times more toxic than the G2 PAMAM dendrimer. In addition, the G4 and G5 PAMAM dendrimers increased HSP70 levels, while the G2 PAMAM dendrimer systematically reduced those levels. The G5 PAMAM dendrimer alone was able to induce GST activity, which is indicative of oxidative stress. Minocycline was found to be toxic to rainbow trout hepatocytes at high concentrations (> 90 µg/ mL) which are not likely to occur in municipal effluents. The antibiotic also systematically reduced HSP70 levels and GST activity. In conclusion, PAMAM dendrimers are cytotoxic to rainbow trout hepatocytes but acute toxicity occurs at concentrations not expected to be found in hospital and municipal effluents. The sublethal effects of these dendrimers on HSP70 levels and GST activity suggest that chronic effects could also occur.

Keywords: Oncorhynchus mykiss; Hepatocytes; PAMAM dendrimers; Cell viability; Heat Shock Proteins; Glutathione S-Transferase

Introduction

Nanotechnology has under gone exponential development which has reached many sectors of our economy. NMs have found many applications, from electronic devices, paints/dyes, cosmetics and personal products, to biomedical uses such as imaging and drug and gene delivery strategies. Any product at the nanoscale with at least one dimension between 1 and 100 nanometers (nm) is considered a NM. Compounds produced at the nanoscale offer new and interesting emerging properties with tremendous potential for commercial applications. For example, the use of nanoparticles or nano-vectors can permit enhanced delivery of a given drug within the body and can target drug release to specific sites in the body. However, the increasing use of NMs has raised concerns about the inadvertent release of such products into the environment and potential impacts on aquatic ecosystems [1]. The toxicity of nanomaterials arises from the cumulative effects of four basic properties associated with colloids: 1) the leaching of low-molecular-weight molecules or ions, 2) the geometry (size and shape) of the NMs including their aggregates, 3) the surface properties (reactivity), and 4) the vector effect. The last property has been extensively studied in connection with the development of drug, gene and peptide delivery systems in therapeutics. Some NMs have the ability to interact with xenobiotics (drugs) and can increase their bioavailability and toxicity by promoting their internalization in tissues/cells [2]. For example, the cytotoxicity of Adriamycin to Chinese hamster cell line DC3F increased when it was associated with cyanoacrylate nanoparticles. In addition, an Adriamycin-resistant hamster cell line became more sensitive to the drug when it was associated with cyanoacrylate nanoparticles, which provides evidence of vector effect. From an environmental risk assessment perspective, it is important to gain a better understanding of the toxicity associated with NMs used as drug delivery "devices" before seeking to determine the vector effect in contaminated environments.

The development and use of poly PAMAM dendrimers for targeted and enhanced drug and gene delivery have been extensively examined [3,4]. The interest in these dendritic NMs stems from their structural properties including uniformity, size, shape, monodispersity and functionalized surfaces [5]. Dendrimers are composed of an initiator amine core (-NH2) with attached amidoamine units that are radially distributed around the core (Figure 1). Each successive branching that forms a surface layer is termed a generation (G). Full-generation dendrimers (G1, G2, G3, etc.) have cationic amine-terminated groups at physiological pH, while half-generation dendrimers (G2.5, G3.5) have anionic carboxylic moieties at physiological pH. Finally, each successive generation has twice the number of terminal groups and increased diameter size. Cationic dendrimers have been shown to exhibit cytotoxicity and haemolysing properties which are dependent on size and surface charge (Zeta potential) [6]. It appears that dendrimers produce small "nanoholes" or "nanopores" in membranes, which can perturb membrane potential integrity and permeability. Thus, the toxicity of dendrimers could be due to their surface properties in addition to their vector properties.

Studies on the toxicity of PAMAM dendrimers to non-target species are relatively scarce at present and the environmental risk of these NMs in not well understood at the present time. Hence the examination of cytotoxicity of PAMAM dendrimers at both the lethal and sublethal levels in fish hepatocytes is of relevance in the understanding of the potential toxicity of these compounds in aquatic ecosytems. G4 PAMAM dendrimers were found to decrease growth and larval development in zebra fish embryos [7]. In an earlier study, G4 PAMAM dendrimers were associated with reduced algal survival, enhanced oxygen production and stimulation of photosystem II reaction centre activity [8], which points to the formation of reactive oxygen species and oxidative stress. Depending on their size and shape, nanoparticles may induce interactions in the protein space domain leading to protein denaturation. The heat shock proteins of the 70 kDa family (HSP70) are stress proteins that are involved in stabilizing protein conformation [9]. This process is clearly energy demanding since these chaperone proteins require ATP to function. For example, it was estimated that one heat shock protein requires up to 100 moles of ATP to re-fold denatured rhodanase protein [10]. Heat shock proteins were also shown to respond to oxidative stress [9]. Rainbow trout yearlings exposed to cadmium-based quantum dots and to dissolved cadmium showed increased HSP70 levels and oxidative damage [11]. However, correction of HSP70 levels against oxidative stress markers (oxidized proportion of metallothioneins or lipid peroxidation) failed to remove the inducing effects of the quantum dots, suggesting that interactions other than oxidative stress were at play. Oxidative stress and xenobiotic conjugation can be conveniently monitored on the basis of glutathione S-transferase (GST) activity. GST requires reduced glutathione (GSH) in order to function, which can be a limiting factor during oxidative stress. The formation of oxygen adducts to molecules during oxidative stress could also be neutralized by conjugation with GSH. For example, GST activity was used as a marker of oxidative stress in marine mussels exposed to cadmium-based quantum dots [12]. Exposure to 10 µg/ L cadmium-based quantum dots increased oxidative stress and GST activity, while dissolved cadmium at the same concentration failed to induce GST activity. Given that PAMAM dendrimers are likely to be released into the environment in wastewater effluent containing many pollutants such as antibiotics, the toxicity of a representative antibiotic to fish liver cells is relevant. Tetracyclines (minocycline) are commonly found in hospital and municipal wastewaters [13]. Minocycline levels were found to range from non-detectable to 530 µg/ L in hospital effluents and from 95 to 920 µg/ L in wastewater treatment plant effluents. In addition, these compounds are continuously released in to the environment from municipal effluents. This could lead to accumulation in non-target organisms if exposure to such compounds exceeds their capacity to eliminate them.

The purpose of this study was to investigate the cellular toxicity of G2, G4 and G5 PAMAM dendrimers and of minocycline in rainbow trout hepatocytes. Cytotoxicity and the levels of stress proteins (HSP70) and GST activity were also determined in order to evaluate the toxicity and mechanisms of action of these NMs in fish hepatocytes.

Methods

Preparation and exposure of rainbow trout hepatocytes

Second, fourth and fifth generation PAMAM dendrimers were purchased from Sigma Chemical Company (Ontario, Canada). They were diluted in High Quality water at 200 mg/ mL to perform dynamic light scattering (DLS) analysis in order to measure particle size distribution and Zeta potential and hepatocyte exposure. The analysis was done using a DLS) instrument with a gel electro mobility option (Wyatt-Instrument Mobius, 532-nm laser). Zeta potential was determined from gel mobility data as described in Domingos et al., 2013 [14]. The measurements were made at 1 mg/mL under identical conditions as in High Quality water. The analytical performance of the instrument was validated with NIST polystyrene standard beads (42 nm diameter) and a Zeta potential standard solution (Ostuka mobility Standard, lot No. 302013). Primary cultures of rainbow trout (Oncorhynchus mykiss) were prepared using a perfusion method with saline citrate and albumin [15]. Briefly, young-of-the-year (8- to 10-cm fork length) rainbow trout (3 livers pooled) were used. After the trout were anesthetized with 25 mg/L tricaine buffered to pH 7.4 with 1 M NaHCO3, the excised livers were perfused with 10 mM citrate in 125 mM NaCl, pH 7.2, at 4°C until the liver tissue acquired a light brown coloration. The livers were then minced and placed in 10 mL of citrate perfusion media containing 0.5% serum bovine albumin. The suspension was stirred slowly with a magnetic stirring bar at 20–40 rpm for 30 min at room temperature. After this period, the suspension was passed through a cell extraction sieve (40-µm diameter mesh, Sigma Chemical Company) and the cells were washed in phosphate-buffered saline (PBS: 140mM NaCl, 5 mM KH2PO4, 5 mM NaHCO3, 1 mM glucose, pH 7.4) containing 0.1% serum bovine albumin, followed by centrifugation (200 ×g for 5 min)/ re suspension 3 to 4 times until a clear supernatant (free of debris) was obtained. A portion of the cell suspension was stained with 0.004% trypan blue in PBS for the determination of cell concentration and viability. The cells were counted and viability was determined (live cells remain transparent and dead ones are blue) using a hematocytometer under a microscope at 200X enlargement. Hepatocytes were plated in 48-well microplates at a density of 0.5 × 106 viable cells/mL (6 replicate wells per treatment) in Liebovitz (L-15) cell culture media containing 10 mM HEPES-NaOH, pH 7.4, 50 units penicillin, 50 µg/mL streptomycin and 0.1 µg/mL amphotericin B. The cells were exposed to increasing concentrations of G2, G4 and G5 PAMAM dendrimers and to minocycline at 1.6, 8, 40 and 200 µg/mLfor 48 h at 15°C in a saturated humidity atmosphere. At the end of the exposure period, the microplates were centrifuged at 250 ×g for 3 to 5 min and the exposure medium was removed by aspiration. Cells were suspended in PBS (without albumin) for cell density and viability assessments. Relative cell density was determined by measuring the absorbance at 600 nm.

Cell viability assessment

Hepatocyte viability was determined by the fluorescein dye retention assay as described elsewhere [15]. A portion (20 µL) of the cell suspension was mixed with 180 µL of 10 µM fluorescein diacetate in PBS containing 1 mM glucose and kept in dark-coloured microplates for 20 min at 20°C. The microplate was centrifuged at 250 ×g for 5 min and the supernatant removed. The cells were then resuspended in 200 uL of phosphate-buffered saline, and fluorescence was measured at 485 nm excitation and 520 nm emission using a microplate reader (Chameleon II, Bioscience, USA). A positive control (100% mortality) was prepared by adding cells to separate wells containing 20% DMSO to completely permeabilize the cells. The data were normalized to controls and expressed as a fold change (reduction) in fluorescence.

HSP70 levels were determined using an enzyme-linked immunoassay as described earlier (Louis et al., 2010) [11]. The hepatocytes were first homogenized using a Teflon pestle tissue/ cell grinder (4 passes at 4°C) and centrifuged at 12,000 ×g for 20 min at 4°C. The supernatant (S12) was diluted to 1 µg total protein in 50 mM sodium carbonate buffer at pH 9.6. Total protein was determined using the Coomassie brilliant blue protein binding assay with serum bovine albumin for calibration [16]. The material was added to high-binding microplate wells (Immulon-4 microplate) and held overnight at 4°C. Afterwards, the wells were rinsed with 200 µL of PBS twice and incubated with PBS containing 1% albumin for 30 min at 20°C to block the remaining sites. The wells were washed with 200 µL of PBS, and 100 µL of HSP72 polyclonal antibody (recombinant human HSP72 IgG SPA-812; Stressgen, USA) diluted 1:1,000 in PBS containing 0.5% albumin was added to each well. The wells were incubated at 37°C for 60 min. The cells were washed 3 times in PBS, and 100 uL of the secondary antibody (rabbit anti-IgG linked with peroxidase) diluted 1:5,000 in PBS containing 0.5% albumin was added and incubated for 30 min at 20°C. The wells were washed 3 times in PBS (200 µL), and peroxidase activity was determined with 1 uM luminol and 10 µM hydrogen peroxide. Luminescence was measured at the initial mixing and monitored for up to 20 min using a luminescence microplate reader (Chameleon II, Bioscience, USA). The data were expressed as peroxidase activity (increase in luminescence)/min. GST activity was determined in the S12 fraction of the supernatants using the colorimetric assay procedure with reduced GSH and 1-chloro-2, 4-dinitrobenzene co-substrates [17]. The data were expressed on the basis of the rate of increase in absorbance at 340 nm / (min × mg proteins).

Data analysis

The hepatocytes were exposed to n = 6 replicates of each concentration of the tested compounds. The toxicity of the PAMAM dendrimers and of minocycline was expressed in terms of toxicity thresholds, which corresponds to the geometric mean of the lowest significant effects concentration (LSEC) and the no-effect concentration (NEC): TT = (LSEC × NEC)1/ 2. The data were checked for homogeneity of variance and normality using Levene's test and the Shapiro-Wilk test, respectively. Analysis of variance was performed, and critical differences were determined using Dunnett's t test. Correlation analysis was also performed using Pearson's product-moment procedure and the tests were performed using the Statistic software package (version 8.).

Results and Discussion

The prepared dendrimers consisted of the G2, G4 and G5 PAMAM dendrimers which have a poly diamine core and described in Figure 1. The G2, G4 and G5 dendrimers have a theoretical diameter of 2.9.4.5 and 5.4 nm, respectively (Table1). Although the size of these dendrimers did not change much, the number of functional amine groups (-NH4+) at their surface readily increased from 16 to 128 for G2 and G5 dendrimers, respectively. This change was accompanied by an increase in molecular weight in such a manner that an equivalent 20 µg/ mL solution consisted of 6, 1.4 and 0.7 µM for G2, G4 and G5 PAMAM dendrimers, respectively. Compared to the same amount of minocycline, the dendrimer concentrations were at least one order of magnitude lower than the antibiotic, i.e., minocycline was in excess compared to the PAMAM dendrimers. Given the

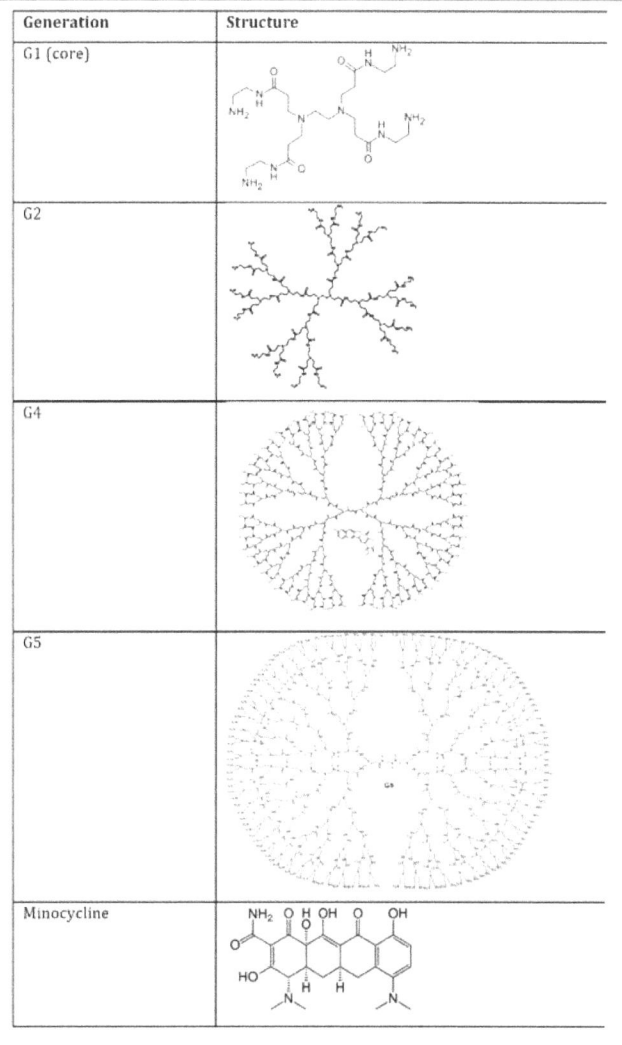

Figure 1: Molecular structure of PAMAM dendrimers and minocycline

pKa values (5 and 9.5) for the 2 amine groups of minocycline, the molecule can be assumed to be cationic at physiological pH as is the case for dendrimers. This is consistent with what is known about these types of drug vectors, which have high surface area chemistries permitting interaction of drugs at the surface. However, dendrimers can induce pore formation, permitting higher diffusion of contaminants into cells [18]. These properties complicate the classical risk assessment paradigm because nanoparticles could change the bioavailability of contaminants in the environment.

The effects of PAMAM dendrimers of increasing size were examined in rainbow trout hepatocytes (Figure 2). G4 and G5 PAMAM dendrimers were found to be more toxic than G2 PAMAM dendrimers, with toxicity thresholds of 3.6 μg/mL compared to 20 μg/ mL. Minocycline was the least toxic test substance; it caused a significant drop in cell viability at 200 μg/ mL, giving a toxicity threshold of 90 μg/ mL. This is in keeping with other studies which showed that dendrimer toxicity is size- and surface charge-dependent [6,19]. The haemolysing potential and the cytotoxicity observed in erythrocytes in creased with higher generation PAMAM dendrimers (G5 and G6). However, the initial positive Zeta potential value in water dropped to a negative value in cell culture media, which points to an interaction with cell culture media components. Increasing cationic charge at the surface of PAMAM dendrimers was proportionally toxic to Daphnia magna and rainbow trout gonad (RTG-2) cell lines [20]. Toxicity was also related to the Zeta potential of G4 to G6 PAMAM dendrimers in the culture media, indicating that toxicity was related to the surface properties (i.e., number of surface groups) of the nanoparticle. Although the Zeta potential decreased in aquarium water, there was no indication of aggregate formation and the dendrimers were shown to influence the innate immunity in zebrafish embryos exposed to G3 and G4 PAMAM dendrimers [21].

The sublethal effects of PAMAM dendrimers were also examined by monitoring HSP70 levels and GST activity (Figures 3 and 4). For protein chaperone HSP70, the dendrimers tended to decrease HSP70 levels, with the exception of G4 dendrimers which increased their levels. The decrease in HSP70 levels occurred at the lowest tested concentrations of the G2 and G5 PAMAM dendrimers. The lowest concentration of minocycline reduced HSP70 levels but with less potency than the dendrimers, however the decrease in HSP70 levels was dampened at 8 and 40 μg/ mL. The activity of GST, a marker enzyme for oxidative stress and xenobiotic conjugation, increased in response to the lowest concentration of G5 PAMAM dendrimer and decreased at the higher dendrimer concentrations. G2 andG4 dendrimers and minocycline reduced GST activity, but G4 PAMAM dendrimer was more potent than the other dendrimers (G2 and G5) in reducing GST activity. Based on correlation analysis, the decrease in HSP70 levels and GST activity was mostly associated with decreased cell viability. Cell viability was significantly correlated with HSP70 (r = 0.52; p < 0.01) and GST activity (r = 0.45; p < 0.05) for G2 PAMAM dendrimer. For G4 PAMAM dendrimer, cell viability was also correlated with HSP70 (r = -0.5; p < 0.01) and GST activity (r = 0.78; p < 0.001). GST activity was significantly correlated with

HSP70 only after correcting against loss of cell viability (residuals) at r = 0.38 (p < 0.05), which suggests that oxidative stress was involved in HSP70 expression, in part at least. For the G5 PAMAM dendrimer, cell viability was significantly correlated with HSP70 level (r = -0.45; p=0.01) and GST activity (r = 0.61; p < 0.001). In the case of minocycline, cell viability was only correlated with GST activity (r=0.46; p=0.01). GST activity (corrected against cell viability) and HSP70 levels were significantly correlated at (r=-0.66 (p < 0.001). Minocycline is recognized as having antioxidant properties in addition to bactericidal activity as shown by reduced lipid peroxidation in brain tissues [22]. In another study, GST induction by cypermethrin was prevented by minocycline inperipheral red blood cells in the rat [23].

To best of our knowledge, this is the first report on the influence of PAMAM dendrimers on HSP70 levels. Decreased expression of HSP70 could render cells less able to defend against changes

Table 1: Physico-chemical characteristics of minocycline and, G2 and G5 dendrimers

Compound	MW g/ mol	Formula	# surface groups	Diameter (nm)	Zeta Potential
Minocycline	494	$C_{23}H_{27}N_3O_7 \cdot HCl$	--		--
G2 PAMAM	3256	$[NH_2(CH_2)_2NH_2]:(G = 2);$ $PAMAM(NH_2)_{16}$	16	2.9	16.6
G4 PAMAM	14215	$[NH_2(CH_2)_2NH_2]:(G = 4);$ $PAMAM(NH_2)_{64}$	64	4.5	32.25
G5 PAMAM	28826	$[NH_2(CH_2)_2NH_2]:(G = 5);$ $PAMAM(NH_2)_{128}$	128	5.4	36

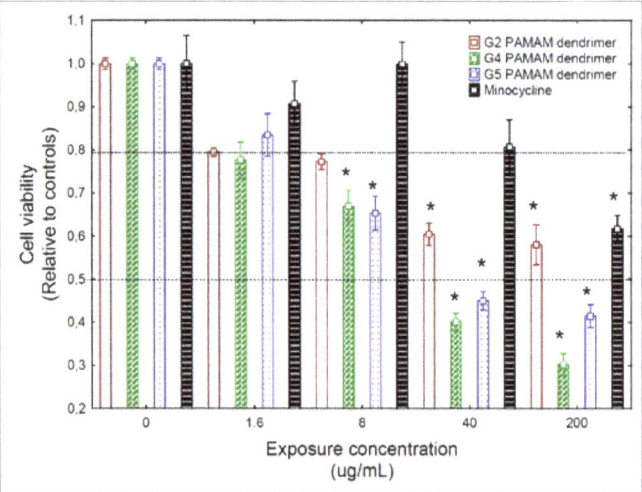

Figure 2: Change in cell viability in trout hepatocytes exposed to PAMAM dendrimers and minocycline. Rainbow trout were exposed to G2, G4 and G5 PAMAM dendrimers and minocycline for 48 h at 15°C. The star * symbol indicates a significant difference from the controls at α<0.05.

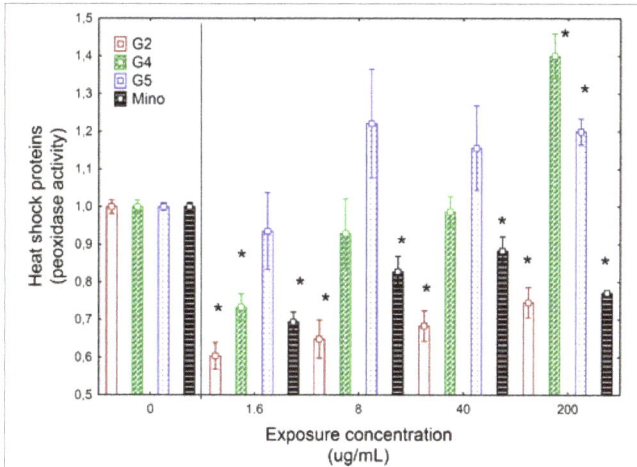

Figure 3: Change in heat shock proteins in trout hepatocytes exposed to dendrimers of increasing sizes. Trout hepatocytes were exposed to increasing concentrations of G2, G4 and G5 PAMAM dendrimers and minocycline for 48 h at 15°C. The star * symbol indicates significance at α=0.05.

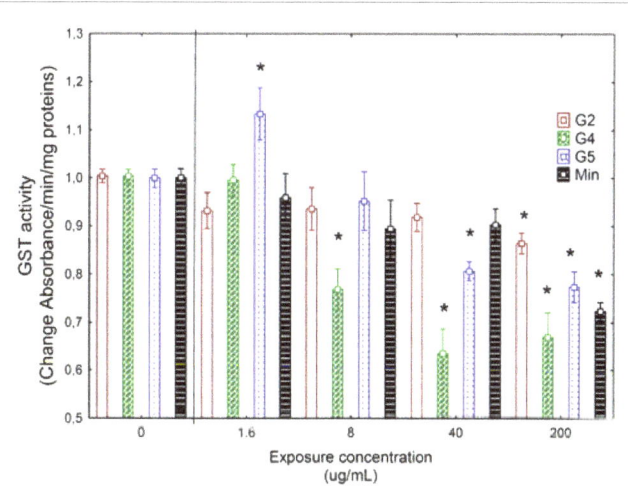

Figure 4: Change in GST activity of trout hepatocytes exposed to dendrimers of increasing size. Rainbow trout were exposed to G2, G4, G5 PAMAM dendrimers and minocycline for 48 h at 15°C. The star * symbol indicates significant difference from the controls at α<0.05.

in protein scaffolding induced by nanoparticles. For example, strong expression of HSP70 occurred in trout hepatocytes exposed to aged cadmium-based quantum dots [24]. Induction of HSP70 was the strongest response to these nanoparticles and involved metallothioneins and labile zinc in cells, suggesting that the release of toxic cadmium ions was at play, at least. It is noteworthy that an increase in HSP70 was also associated with oxidative stress [25]. GST activity is a marker of oxidative stress as well as the conjugation of polar compounds. GST activity was marginally negatively correlated with HSP70 levels (r = -0.31; p=0.09) for the G4 PAMAM dendrimer, the only dendrimer that induced HSP70. Decreased GST activity could also result from the depletion of reduced GSH in cells undergoing oxidative stress [26]. Induction of HSP70 was also associated with oxidative stress in freshwater musselse xposed to zinc oxide nanoparticles

[27] and in zebra fish embryos exposed to C60 fullerene [28]. In a study involving G4, G5 and G6 PAMAM dendrimers, increased production of reactive oxygen radicals and an increase in genotoxicity were observed in fish hepatocellular carcinoma cell lines [29]. The increase in HSP70 levels and the concomitant decrease in GST activity in hepatocytes exposed to G4 and G5 PAMAM dendrimers maybe attributable to oxidative stress. This is consistent with the decrease in HSP70 levels in hepatocytes exposed to minocycline (known to act as an antioxidant), which was not related to decreased cell viability. In conclusion, PAMAM dendrimers were toxic to rainbow trout hepatocytes, with the G4 and G5 dendrimersbeing 5 times more toxic than G2 dendrimer. The G4 and G5 PAMAM dendrimers increased the levels of HSP70, while the G2 dendrimersystematically reduced those levels. Only the G5 dendrimer was able to induce GST activity indicative of oxidative stress. Minocycline was less toxic to rainbow trout hepatocytes than the dendrimers and systematically reduced the levels of HSP70 and GST activity.

Acknowledgements

This study was supported by the Chemical Management Plan of Environment and Climate Change Canada. The technical assistance of Joana Kowalczyk is duly recognized.

References

1. Gagné F, Gagnon C, Blaise C. Aquatic nanotoxicology: A review. Current Top. Toxicol. 2008a;4(1):14.

2. Kubiak C, Couvreur P, Manil L, Clausse B. Increased cytotoxicity of nanoparticle-carried Adriamycin in vitro and potentiation by verapamil and amiodarone. Biomaterials. 1989;10(8):553-6.

3. Yang Y, Sunoqrot S, Stowell C, Ji J, Lee CW, Kim JW, et al. Effect of size, surface charge, and hydrophobicity of poly (amidoamine) dendrimers on their skin penetration. Biomacromolecules. 2012;13(7):2154-2162. doi: 10.1021/bm300545b.

4. Abbasi E, Aval SF, Akbarzadeh A, Milani M, Nasrabadi HT, Joo SW, et al. Dendrimers: synthesis, applications, and properties. Nanoscale Res Lett. 2014; 9(1): 247. doi: 10.1186/1556-276X-9-247.

5. Svenson S, Tomalia DA. Dendrimers in biomedical applications–reflections on the field. Adv Drug Deliv Rev. 2005;57(15):2106-29. doi: 10.1016/j.addr.2005.09.018.

6. Jain K, Mehra NK, Jain NK. Potentials and emerging trends in nano pharmacology. Curr Opin Pharmacol. 2014;15:97-106. doi: 10.1016/j.coph.2014.01.006.

7. King Heiden TC, Dengler E, Kao WJ, Heideman W, Peterson RE. Developmental toxicity of low generation PAMAM dendrimers in zebra fish. Toxicol Appl Pharmacol. 2007;225(1):70-9. doi: 10.1016/j.taap.2007.07.009.

8. Petit AN, Eullaffroy P, Debenest T, Gagné F. Toxicity of PAMAM dendrimers to Chlamydomonas reinhardtii. Aquat Toxicol. 2010;100(2):187-93. doi: 10.1016/j.aquatox.2010.01.019.

9. El Golli-Bennour E, Bacha H. Hsp70 expression as biomarkers of oxidative stress: mycotoxins' exploration. Toxicology. 2011;287(1-3):1-7. doi: 10.1016/j.tox.2011.06.002.

10. Martin, J, Langer T, Boteva R, Schramel A, Horwich AL. Chaperonin-mediated protein folding at the surface of groEL through a 'molten-globule'-like intermediate. Nature. 1991;352(6330):36-42. doi: 10.1038/352036a0.

11. Louis S, Gagné F, Auclair J, Turcotte P, Gagnon C, Émond C. The characterization of the behaviour and gill toxicity of CdS/CdTe quantum dots in rainbow trout (*Oncorhynchus mykiss*). Int.J. Biomed. Nanosci. Nanotech. 2010;1(1): 52-69.

12. Rocha TL, Gomes T, Mestre NC, Cardoso C, Bebianno MJ. Tissue specific responses to cadmium-based quantum dots in the marine mussel Mytilus galloprovincialis. Aquat Toxicol. 2015;169:10-8. doi: 10.1016/j.aquatox.2015.10.001.

13. Pena A, Paulo M, Silva LJ, Seifrtová M, Lino CM, Solich P. Tetracycline antibiotics in hospital and municipal wastewaters: a pilot study in Portugal. Anal Bioanal Chem. 2010;396(8):2929-36. doi: 10.1007/s00216-010-3581-3.

14. Domingos RF, Rafiei ACZ, Monteiro BCE, Khan MAK, Wilkinson KJ. Agglomeration and dissolution of zinc oxide nanoparticles: role of pH, ionic strength and fulvic acid. Environ. Chem. 2013;10(4):306–312.

15. Gagné F. Acute toxicity assessment of liquid samples with primary cultures of rainbow trout hepatocytes. In Small-scale Freshwater Toxicity Investigations,(C.Blaise and J.-F.Férard, Eds.). 2005;1:453-472. Netherlands.

16. Bradford MM. A rapid and sensitive method for the quantitation of microgram quantities of protein utilizing the principle of protein-dye binding. Anal.Biochem. 1976;72:248-254.

17. Boryslawskyj M, Garrood AC, Pearson JT. Elevation of glutathione-S-transferase activity as a stress response to organochlorine compounds in the freshwater mussel *Sphaerium corneum*. Mar. Environ. Res. 1988;24(1-4):101-104. doi:10.1016/0141-1136(88)90263-2.

18. Zhang J, Jing B, Regen SL. Kinetic evidence for the existence and mechanism of formation of a barrel stave structure from pore-forming dendrimers. J. Am. Chem. Soc. 2003;125(46): 13984-13987. doi: 10.1021/ja036390h.

19. Mukherjee SP, Davoren M, Byrne HJ. In vitro mammalian cytotoxicological study of PAMAM dendrimers - towards quantitative structure activity relationships. Toxicol. In Vitro. 2010;24(1):169-177. doi: 10.1016/j.tiv.2009.09.014.

20. Naha PC, Davoren M, Casey A, Byrne HJ. An ecotoxicological study of poly (amidoamine) dendrimers-toward quantitative structure activity relationships. Environ. Sci. Technol. 2009;43(17): 6864-6869. doi: 10.1021/es901017v.

21. Oliveira E, Casado M, Faria M, Soares AM, Navas JM, Barata C, et al. Transcriptomic response of zebrafish embryos to polyaminoamine (PAMAM) dendrimers. Nanotoxicology. 2014;8 Suppl 1:92-9. doi: 10.3109/17435390.2013.858376.

22. Kraus RL, Pasieczny R, Lariosa-Willingham K, Turner MS, Jiang A, Trauger JW. Antioxidant properties of minocycline: neuroprotection in an oxidative stress assay and direct radical-scavenging activity. J. Neurochem. 2005;94(3): 819-827. doi :10.1111/j.1471-4159.2005.03219.x.

23. Tripathi P, Singh A, Agrawal S, Prakash O, Singh MP. Cypermethrin alters the status of oxidative stress in the peripheral blood: relevance to Parkinsonism. J Physiol Biochem. 2014;70(4):915-24. doi: 10.1007/s13105-014-0359-7.

24. Gagné, F, Maysinger D, André C, Blaise C. Cytotoxicity of aged cadmium-telluride quantum dots to rainbow trout hepatocytes. Nanotoxicology. 2009;2(3) :113-120.

25. El Golli-Bennour E, Bacha H. Hsp70 expression as biomarkers of oxidative stress: mycotoxins' exploration. Toxicology. 2011;287(1-3):1-7. doi: 10.1016/j.tox.2011.06.002.

26. Bagnyukova TV, Chahrak OI, Lushchak VI. Coordinated response of goldfish antioxidant defenses to environmental stress. Aquat Toxicol. 2006;78(4):325-31. doi: 10.1016/j.aquatox.2006.04.005.

27. Gagnon C, Pilote M, Turcotte P, André C, Gagné F. Effects of exposure to zinc oxide nanoparticles in freshwater mussels in the presence of municipal effluents. Invert.Surv.J. 2016;13:140-152.

28. Usenko CY, Harper SL, Tanguay RL. 2008. Fullerene C60 exposure elicits an oxidative stress response in embryonic zebrafish. Toxicol Appl Pharmacol. 2008;229(1): 44–55. doi: 10.1016/j.taap.2007.12.030.

29. Naha PC, Byrne HJ. Generation of intracellular reactive oxygen species and genotoxicity effect to exposure of nanosized polyamidoamine (PAMAM) dendrimers in PLHC-1 cells in vitro. Aquat Toxicol. 2013 May 15;132-133:61-72. doi: 10.1016/j.aquatox.2013.01.020.

9

Biological Evaluation of Multivalent-Type N-Acetyl-D-Glucosamine (GlcNAc) Conjugates for Wheat Germ Agglutinin (WGA) by the Surface Plasmon Resonance (SPR) Method

Amrita Kumari[1], Tetsuo Koyama[1], Ken Hatano[1] and Koji Matsuoka*

[1]Division of Material Science, Graduate School of Science and Engineering, Saitama University, Sakura, Saitama 338-8570, Japan.

*Corresponding author: Koji Matsuoka, Professor of Graduate School of Science and Engineering, Division of Functional Material Sciences, Saitama University, Shimookubo, Sakura-Ku, Saitama 338-8570, Japan, E-mail: koji@fms.saitama-u.ac.jp

Abstract

Analysis of the interaction of synthetic avidin-biotin-GlcNAc (ABG) glycocluster complex with a well-known lectin, wheat germ agglutinin (WGA), was performed with a biosensor based on surface plasmon resonance (SPR). In the SPR measurements, WGA was covalently coupled to the gold surface using the amine-coupling method. Artificial glycopolymers of N-acetyl-D-glucosamine, polystyrene-based linear-type glycoclusters with a polymeric backbone of acrylamide, were used as controls. Three glycopolymers, including tetrameric ABG complex, glycopolymer **1** with a monomer and acrylamide ratio of 1:10 and glycopolymer **2** with a ratio of 1:4, were used as analytes. The SPR method was used for the analysis of the interactions that covered a high affinity range; namely, the strong binding of $K_A \sim 6.45 \times 10^7 M^{-1}$ for ABG compared with glycopolymers **1** and **2**, which show binding of $K_A \sim 3.41 \times 10^5 M^{-1}$ and $K_A \sim 3.30 \times 10^5 M^{-1}$ respectively. SPR measurements confirmed that WGA has higher affinity toward the tetrameric ABG complex than toward the linear-type glycopolymers **1** and **2**, and the usefulness of these synthetic glycopolymers as tools in the study of sugar-lectin interactions has been proved due to the very well-known "glycocluster effect".

Keywords: Surface plasmon resonance; Glycocluster effect; Polystyrene; Analyte; Amine coupling; Multivalent

Abbreviations

GlcNAc: N-Acetyl-D-glucosamine, GPC: Gel Permeation Chromatography, WGA: Wheat Germ Agglutinin, SPR: Surface Plasmon Resonance, ABG: Avidin-Biotin-GlcNAc, RU: Resonance Units, PL: Photo Luminescence, PBS: Phosphate Buffer Saline, ELLA: Enzyme-Linked Lectin Assay

Introduction

Carbohydrates present on the cell surface play a vital role in cell recognition, and they help to protect the cell from the outside world and provide biological information. The phenomena of cell adhesion and cell activation prompted by carbohydrate-protein interaction are among the recent topics of active research. Synthetic polyvalent glycoconjugates that imitate the cell surface glycocalyx have been focused on due to their fascinating biological properties [1-2]. In contrast to the weak and poorly specific interactions that arise between individual carbohydrates and proteins, the multivalent display of carbohydrates at the surface of a molecular scaffold is currently used to boost the binding avidity and selectivity toward a target protein [3, 4]. This phenomenon is known as "cluster glycoside effect" [5-7].

Lectins are inherently proteins of a non-immune source that recognize and bind to specific saccharide structural epitopes present on the surface of a cell membrane [8]. Many of the proteins that participate in multivalent interactions are oligomers such as the lectin wheat germ agglutinin (WGA), an N-acetyl-D-glucosamine (GlcNAc)-specific plant lectin that is extensively used in model studies with glycopolymers to monitor their binding selectivity. WGA is a 36-kDa dimer protein with eight specific binding sites for GlcNAc that are separated by a distance of 14 Å [9, 10]. GlcNAc is important in several biological systems. It has been proposed as a treatment for autoimmune diseases, and recent tests have claimed some success [11]. The most important thing about GlcNAc is that it is highly specific towards binding with the plant lectin WGA [12]. This is the reason we have chosen WGA as a model lectin and GlcNAc as a sugar model to do further biological research in this project.

The difficulty in studies on carbohydrate-protein interactions is that binding affinities are very weak, usually with dissociation constants in the millimolar range [13]. This limitation is repeatedly overcome by derivatization of carbohydrates to obtain higher binding values due to the phenomenon of multivalency or polyvalency.

Koivula and co-workers confirmed by using SPR technology that WGA has higher affinity toward self-assembled monolayers

(GlcNAc-SAM) than toward free GlcNAc monosaccharide [14]. Wittman and co-workers developed a library of cycloheptapeptides exhibiting up to six GlcNAc moieties through a urethane spacer. Enzyme-linked lectin assay (ELLA) with WGA has enabled the identification of tetra-, penta-, and hexavalent glycocluster exhibiting higher binding affinity than the monomeric GlcNAc control [15]. On the basis of structural data and recent progress made in the understanding of multivalent effects, a large variety of synthetic glycoclusters based on peptide dendrimers [16], carbosilanes [17], and nanostructures [18] have been developed as ligands for studying biologically relevant targets or providing compounds with anti-pathogenic and anti-tumoral properties.

The availability of radical polymerization approaches, mild reaction conditions and facile purification steps has enabled the synthesis of glycopolymers 1 and 2 in an aqueous solution [19]. Moreover, a tetrameric ABG complex could be prepared by glycosylation steps followed by coupling reaction with biotin and finally avidin-biotin conjugation [20]. Such synthetic multivalent glycopolymers can bind specifically to WGA with enhanced affinity compared to GlcNAc only.

We present here data on the interaction between the plant lectin WGA and synthetic GlcNAc polymers [19, 20] by finding the kinetics/affinity using surface plasmon resonance (SPR). SPR is a highly sensitive and powerful technique that has been used in mechanistic studies of carbohydrate-protein interactions at interfaces in real time and in a quantitative manner [21, 22]. SPR biosensing can provide data with desirable reproducibility and therefore offers the possibility of detailed computational analysis.

Results

Synthesis of glycopolymers

To obtain multivalent carbohydrate-protein interactions, we synthesized a glycopolymer of a tetrameric structure ABG complex by the use of 4-(4,6-dimethoxy-1,3,5-triazin-2-yl)-4-methylmorpholinium chloride (DMT-MM) as a coupling reagent followed by biotin-avidin complexation [Figure 1], and then structurally examined by NMR, IR, MS and elemental analysis [20].

The other two glycopolymers, 1 and 2, were synthesized by the use of 4-(chloromethyl) styrene, and further azidation followed by reduction reactions were able to yield amino styrene. Condensation of amine with GlcNAc monomer was accomplished and it was polymerized with acrylamide to yield the corresponding water-soluble glycopolymers. These glycopolymers were structurally examined by NMR, IR, MS and gel permeation chromatography (GPC) [19].

Surface Plasmon Resonance Studies

Despite the fact that SPR has been used for the analysis of carbohydrate-lectin binding, lectin was often immobilized on the sensor surface instead of the glycan moiety [23] because of the low sensitivity inferred by the low molecular weight

of the glycan. In order to compare the bindings of different carbohydrate derivatives containing N-acetyl-D-glucosamine to WGA, we immobilized lectin on the sensor chip.

pH scouting

The surface concentration of the ligand at pH 5.5 was higher (RU 29783.4) than that at pH 5.0, 4.5 or 4.0, and pH 5.5 therefore appeared to be the best choice [Figure 2]. Thus, for immobilization of WGA, pH 5.5 (a value less than pI 8.7 ± 0.3 of WGA) [24, 25] was chosen for further modification of the sensor chip.

Immobilization of WGA on a CM5 chip

The experimental sensor chips (CM5 chips) were modified with a covalently bound ligand (WGA) *via* amine coupling by 0.4 M [N-ethyl-N'-(3-dimethylaminopropyl)-carbodiimide hydrochloride] (EDC) and 0.1 M N-hydroxysuccinimide (NHS) derivatization. After activation, WGA was immobilized by injection of solutions of 36 μg/ mL and 720 μg/ mL to the respective sensor surfaces, followed by deactivation of residual NHS esters with 1 M ethanolamine. Upon immobilization of

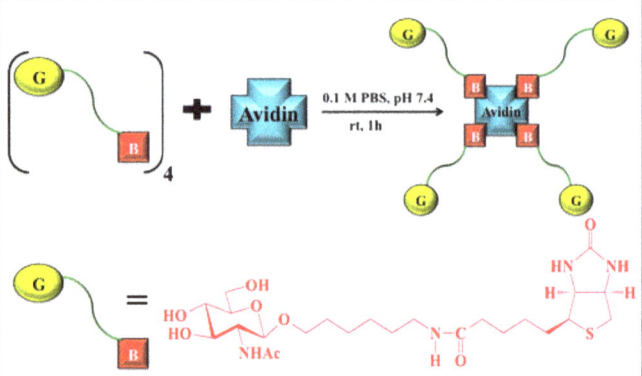

Figure 1: Model representation of the synthetic assembly of avidin-biotin-GlcNAc (ABG) complex [20]. Four units of biotin-GlcNAc conjugate are stirred with avidin for 1 h at room temperature in the presence of PBS buffer

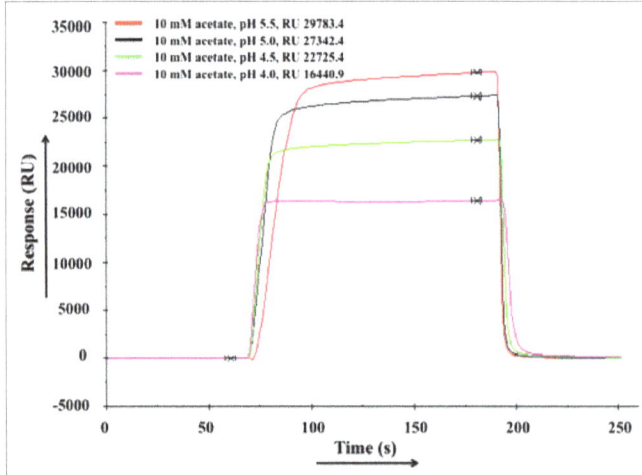

Figure 2: "pH scouting" of WGA Sensorgram showing a relative response-time graph in real time

approximately 50 RU, the final response achieved was 68.50 RU on one chip, and another chip was targeted with 1000 RU, the final response achieved being 695.6 RU [Table 1].

Sensorgrams in Figure 3 show the immobilization of WGA on the flow channel of the CM5 sensor chip *via* amine coupling at pH 5.5 with 10 mM sodium acetate buffer. The procedure includes the following steps. (A) A "pre-concentration" test is carried out to determine the appropriate ligand concentration to inject in order to reach the targeted level of response (RU, response units). Here, RU are 50 in the case of Figure 3a and 1000 in the case of Figure 3b The injection period ends after acquisition and completion of pre-concentration analysis above the baseline of approximately 23,000 [Figure 3a] and 16,500 [Figure 3b]. (B) "Rinse with buffer" involves washing the non-covalently bound ligand with 50 mM NaOH to completely remove the ligand and obtain the baseline. (C) "Activation" involves injection of the coupling reaction mix (NHS/EDC) onto the surface, activating the carboxymethyl group by forming a highly reactive succinimide ester, followed by termination of the injection. (D)There is a slight increase in RU reflecting the activation, compared to the baseline between B and C. (E) Surface activation is followed by injection of a ligand sample diluted in 10 mM sodium acetate pH 5.5 buffer, with continuation of injection resulting in covalent binding of the ligand to the reactive surface to yield the targeted RU of 50 [Figure 3a] and 1000 [Figure 3b]. These values are the difference between the achieved constant level just before F and the baseline at D. (F) The remaining non-reacted activated carboxymethyl groups are blocked by injection of ethanolamine, followed by cessation of injection. (G) Immobilization of the ligand is achieved.

Kinetic analysis

Various sensorgrams can be acquired at different concentrations of the injected compound and they can be simultaneously used to obtain precise kinetic (k) and equilibrium (K) constants. Equilibrium constants can be derived independently from ratios of rate constants or by fitting the

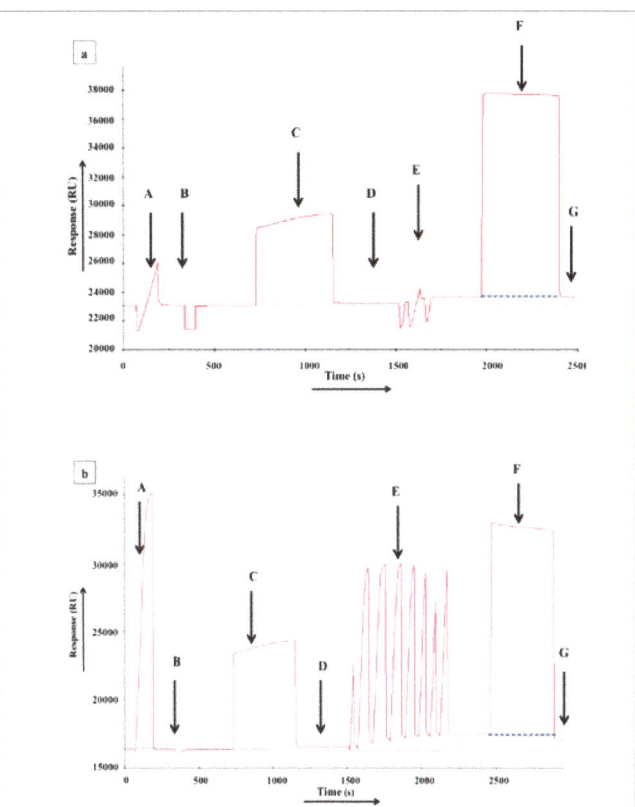

Figure 3: WGA immobilization *via* amine coupling steps: **(A)** pre concentration, **(B)** Washing with 50 mM NaOH, **(C)** NHS/EDC activation, **(D)** RU reflecting the activation, **(E)** amount of ligand bound, **(F)** Ethanolamine injection and **(G)** immobilized ligand

steady-state response versus the concentration of the binding molecule in the flow solution over a range of concentrations. SPR signal alteration is a wonderful method to conclude binding stoichiometry, since the refractive index change in SPR experiments produces basically the identical response for each bound molecule and depends on the molecular weight of the binding molecules. Kinetic analysis was performed by injection of analytes of ABG complex and GlcNAc polymers 1 and 2 dissolved in HEPES buffer at different concentrations on a WGA-immobilized chip. Between binding cycles, the WGA surface was regenerated with a 30 s pulse of 200 mM EDTA [26, 27]. Maximum responses of the glycopolymers to the surface-bound WGA were analyzed and plotted against glycoside concentrations [Figure 4]. All of the three glycopolymers showed binding to WGA. Figure 4a shows that the bound relative response (RU) of ABG complex is high even at low concentrations of (4.30 – 340 nM) because of its strong binding to WGA due to its symmetrically arranged tetrameric structure. However, the binding of glycopolymers 1 and 2 was small even at high concentrations (~100 μM) compared to ABG. The binding of tetravalent ABG complex is approximately 190-times higher than that of glycopolymers **1** and **2**, and this difference is explained by the "glycocluster effect".

Some strange inflections can be seen in Figure 4, and they may have been due to the bulk effect [28, 29]. Such effects basically occur if the running buffer and analyte dilution buffer

Table1: Kinetic parameters obtained from the interactions of lectin (WGA) with artificial GlcNAc polymers by use of a biosensor SPR technique. The closeness of fit is indicated by the value of χ^2

Compound	ABG complex	GlcNAc (1:10)	GlcNAc (1:4)
Molecular weight (Da)	71186.72	1238.37	740.81
Immobilized ligand (RU)	68.5	695.6	695.6
K_D [M]	1.55×10^{-8}	2.99×10^{-6}	3.05×10^{-6}
K_A [M^{-1}]	6.45×10^{7}	3.41×10^{5}	3.30×10^{5}
K_a [M^{-1} s^{-1}]	1.85×10^{5}	0.70×10^{3}	0.80×10^{3}
K_d [s^{-1}]	2.86×10^{-3}	2.05×10^{-3}	2.44×10^{-3}
Theoretical R$_{max}$	135.4	24.0	14.3
Calculated R$_{max}$	118.1	20.2	19.0
χ^2	3.0	2.1	1.1

are not the same. The response in Biacore is the extent of refractive index change at the surface of the sensor chip. The variation in refractive index between the dilution buffer and the baseline buffer (running buffer) is called the "bulk effect" and is induced by the existence of dissolved material including buffer components, biomolecules and salt. Samples should be prepared in running buffer to avoid bulk effects during the injections due to differences in the refractive index between the running buffer and sample. The ABG analyte stock was prepared in 10 mM PBS pH 7.4 containing ~ 137 mM NaCl, while stocks of the other two analytes, glycopolymers **1** and **2**, were prepared in MilliQ and then three-fold dilutions were made in running buffer (10 mM HEPES, pH 7.4 containing 500 mM NaCl), though the dilution buffer and running buffer we used were the same. However, the first higher concentration we used for ABG during the assay contain a mixture of PBS and HEPES buffer and we therefore found some spikes at the end of injection as the buffers contain different salt concentrations.

Surface regeneration

The affinity between immobilized lectin and glycoclusters used in this study seems to be higher ($\sim 10^7$ and $10^5\,\text{M}^{-1}$) than the monosaccharide's affinity ($\sim 10^3\,\text{M}^{-1}$), and this strong binding was dissociated by a special regeneration solution (10 mM HEPES with 200 mM EDTA, pH 7.4 to chelate bivalent Ca^{2+}) [26, 27] to obtain an accurate dissociation rate constant. Regeneration at 30 µL/ min was performed after every cycle of analyte bound to the surface and efficacy of the surface was maintained. The Regeneration process allows the sensor chip to be reused, thus reducing the cost of SPR analysis. However, the lectin surface should be intact without any damage or inactivation of the surface. The regeneration step shortens the analysis procedure as the chip can be used several times without a further immobilization process.

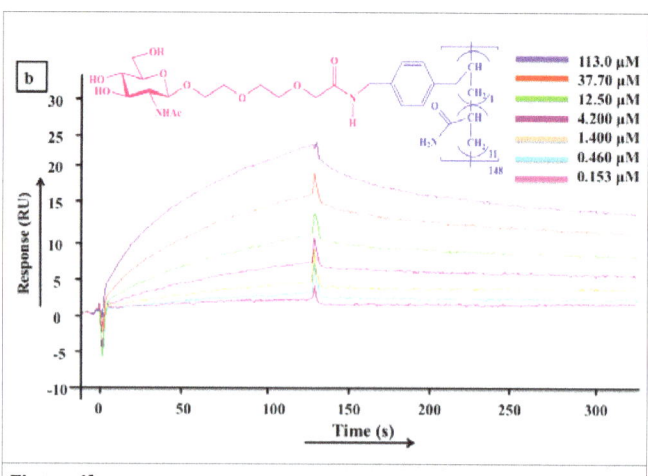

Figure 4b

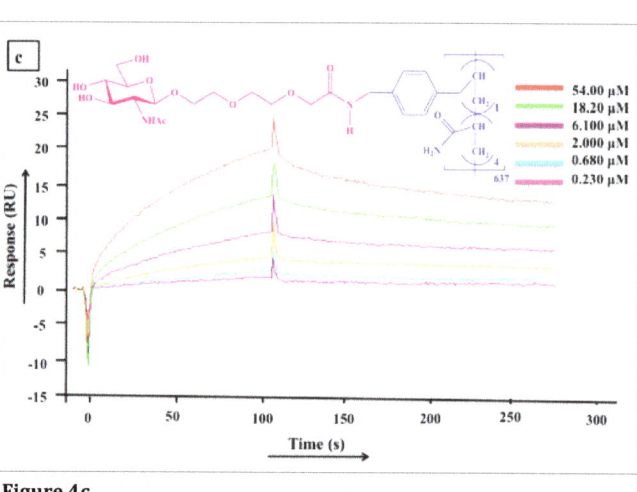

Figure 4c

Kinetic parameters

We calculated dissociation constants K_D, which indicate the concentration of the analyte in the equilibrium state. We performed this calculation by fitting each measured value to a single site interaction model, which is a common procedure used in this type of experiment [30]. The results were analyzed by BIA-evaluation software and are summarized in Table 1. As can be seen, different K_D and K_A values were obtained for different glycol conjugates. Interestingly, higher binding constants were recorded with $K_A \sim 10^7\,\text{M}^{-1}$ in the case of ABG complex, its tetrameric known structure afforded the glycocluster effect in comparison with $K_A \sim 10^5\,\text{M}^{-1}$ for the other two linear-type polymers used as controls.

Discussion

The WGA lectin is a dimer of two identical 18-kDa subunits, each consisting of four homologous domains of 43 amino acids. WGA specifically recognizes GlcNAc. Eight independent carbohydrate-binding sites are present per lectin molecule [9, 10] and it is therefore interesting to use such a lectin model. During the pre-concentration procedure performed for WGA, the optimum pH of sodium acetate buffer was 5.5. This buffer

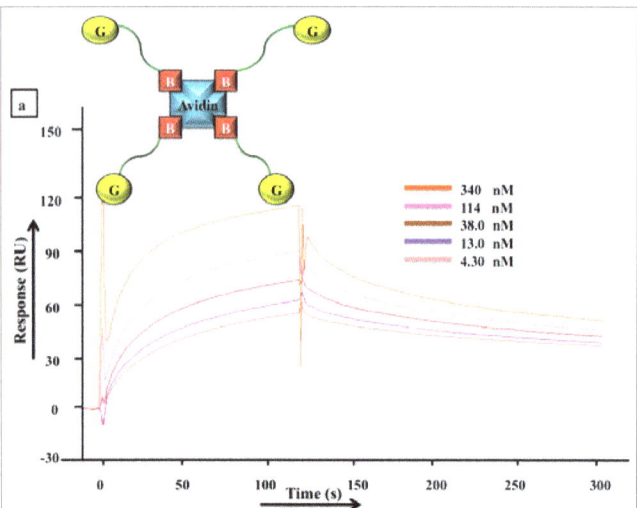

Figure 4: Sensorgrams show interactions of synthetic glycopolymers with lectin WGA. Sensorgrams of the interactions show each association and dissociation phase as a relative response of SPR against time. **(a)** ABG tetramer complex (4.30-340 nm), **(b)** glycopolymer **1** (0.153-113 µM), **(c)** glycopolymer **2** (0.23-54 µM)

was therefore used for immobilization of the lectin in the working channel, resulting in an SPR baseline rise by ~ 700 RU (target of 1000 RU)on one sensor chip and ~ 70 RU (target of 50 RU) on another sensor chip. In the binding experiments with WGA, the ABG glycopolymer was used as an analyte at a concentration of 340 nM, which was diluted three fold, and at such low concentrations, it gave higher binding and higher SPR response (118 RU) than those of the other two glycopolymers. The corresponding glycopolymers, **1** and **2** showed lower SPR responses, which were comparable to the tetrameric ABG complex. Although the valency is four, the avidin-biotin scaffold of the ABG complex enhances binding affinity ~ 6380 times than GlcNAc only and ~ 190 times higher than control glycopolymers **1** and **2** and it is assumed that this exceptionally strong effect is to be due to the phenomenon of chelate binding approach [31-33]. The distance between binding sites of WGA is 13-14 Å (distances between anomeric oxygen's of two bound GlcNAc residues) [34, 35] and the shortest distance between binding sites appears to be as small as 13 Å. For this reason WGA binds strongly to most multivalent analogs that can execute an assured degree of chelation and augment the binding with every valency [36]. The other two glycopolymers (**1** and **2**) are products after radical means of polymerization. These polymers are linear type with pendant top GlcNAc residues and they were used as controls in this study which showed binding affinity towards the WGA ~ 34 times than GlcNAc only.

Conclusions

From the results obtained in this study, it can be concluded that the use of the biosensor BIAcore with SPR as a detection method is a powerful method for investigating the interaction between lectins and glycoproteins. We have already demonstrated that tetravalent ABG complex could be easily prepared using avidin-biotin complexation [20]. Experiments with WGA showed that GlcNAc conjugates after organic modifications had higher protein-carbohydrate binding affinity. However, all the synthetic polymers used in this study showed specific and strong K_A than their monomeric glycosides, but the ABG complex is more in the favor of glycocluster effect due to its tetrameric structure ~ K_A= 6.45 x 10^7 M^{-1}. The binding of tetravalent ABG complex is approximately 190-times higher than that of glycopolymers **1** and **2** and ~ 6380-times higher than its monomer, which can be explained by the glycocluster effect. Glycopolymers **1** and **2** showed ~ K_A= 3.41 x 10^5 $M^{-1}$and ~ K_A= 3.30 x 10^5 M^{-1}, respectively, as measured by SPR. We compared the kinetic affinity calculated by SPR data with previously determined photoluminescence (PL) or fluorospectrophotometry data for the same sugar conjugates and both methods (PL & SPR) are almost in the same agreement for protein-carbohydrate kinetics/affinity. The affinity constant for ABG is 1.39 x 10^7 M^{-1}[20], while those for the other two glycopolymers were in the range of ~10^5 M^{-1} as calculated by PL [19]. Little fluctuation in results of affinity constants calculated by two different methods may occur and it should be emphasized that the PL method measures solution-solution interactions, which are different from the solution-solid interactions measured

by SPR [37]. The analyte-ligand interactions in our case are uniquely determined, and the kinetic affinity (K_A) for synthetic glycopolymers is concentration-dependent.

The SPR method has also been shown to be useful for evaluating the interaction of carbohydrates with proteins in a nano-molar range. In summary, carbohydrate-protein interactions are key steps for many physiological and pathological events. Hence, the development of new carbohydrate conjugates and microarrays is important for detecting these activities by using biophysical methods. Such studies are now concerned to the synthesis of more complex and relevant structures and the study of their biological properties.

Methods

SPR measurement

All SPR measurements were carried out on a BIAcore X 100 instrument (GE Healthcare) by using a CM5 sensor chip at 25 °C at a flow rate of 30 μL/ min. HEPES buffer of pH 7.4 used for all measurements consisted of 10 mM N-(2-hydroxyethyl) piperazine- N'-(2-ethanesulfonic acid) (HEPES), 500 mM NaCl, and 0.02% Tween 20 (p20) detergent. To enhance the lectin-carbohydrate binding, 5.0 mM $CaCl_2$ [38] was also added to the running buffer. SPR experiments were carried out with immobilized WGA using different glycoconjugates as analytes to determine the binding constants. The analytes used in this study were divergent-type ABG glycocluster and linear-type glycopolymers **1** and **2**. The stock of ABG complex (1.028 μM) was prepared in phosphate buffer saline (10 mM PBS, pH 7.4), and 150 μM stock solution of glycopolymer 1 and 100 μM stock solution of glycopolymer 2 were prepared in MilliQ. Then subsequent 3-fold dilutions of these polymers were prepared in 10 mM HBS-P, pH 7.4 buffer in a total amount of 200 μL. An aliquot of the solutions (200 μL) was then injected over the immobilized chip at a flow rate of 30 μL/ min with contact time of 120 s and dissociation time of 180 s. These parameters were the same for all glycopolymers. The chip was regenerated by injection of 30 μL/ min of HBS-EP+ containing 200 mM of EDTA, pH 7.4. The binding assay was performed by a multi-cycle kinetic (MCK) approach. The binding assay also included three startup cycles using running buffer to equilibrate the surface as well as a zero concentration (30% running buffer plus 70% milliQ) cycle of the analyte in order to have a blank response usable for reference subtraction. Each sensorgram was obtained by subtracting the reference cell: buffer only injection and glycopolymer injection without ligand immobilization were performed.

Data analysis

Data analysis was conducted with the software BIAcore X 100 evaluation (version: 2.0.1).

Acknowledgments

The authors would like to thank Dr. Hidenao Arai of Graduate School of Science & Engineering of Saitama University for his valuable advice regarding SPR and BIAcore experiments. The authors are grateful to Rotary Yoneyama Memorial Foundation,

Japan for continuous support and encouragement. This work was partly supported by a grant-in-aid from Saitama Prefecture (K.M.) (Saitama Leading Edge Project).

Author Contributions

A.K., T.K., K.H. and K.M. conceived and designed the experiments; A.K. performed the experiments and analyzed the data; A.K., T.K., K.H. and K.M. contributed to preparation of the paper; and A.K. mainly wrote the paper. This study is a PhD work of A.K. under the supervision of K.M.

Conflicts of Interest

The authors declare no conflict of interest.

References

1. McEver RP, Moore KL, Cummings RD. Leukocyte trafficking mediated by selectin-carbohydrate interactions. J Biol Chem. 1995;270(19):11025-8.

2. Walsh G, Jefferis R. Post translational modifications in the context of therapeutic proteins. Nat Biotechnol. 2006;24(10):1241-52. doi: 10.1038/nbt1252.

3. Mammen M, Choi SK, Whitesides GM. Polyvalent interactions in biological systems: Implications for design and use of multivalent ligands and inhibitors. *Angew. Chem. Int. Ed. Engl.* 1998;37(20):2755-2794.

4. Kiessling LL, Gestwicki JE, Strong LE. Synthetic multivalent ligands as probes of signal transduction. Angew Chem Int Ed Engl. 2006;45(15):2348-68. doi: 10.1002/anie.200502794.

5. Lee YC, Lee RT. Carbohydrate-protein interactions: Basis of glycobiology. Acc. Chem. Res. 1995;28(8):321-327. doi: 10.1021/ar00056a001.

6. Matsuoka K, Nishimura S-I. Synthetic glycoconjugates. 5. Polymeric sugar ligands available for determining the binding specificity of lectins. Macromolecules. 1995;28(8):2961-2968. doi: 10.1021/ma00112a049.

7. Lundquist JJ, Toone EJ. The cluster glycoside effect. Chem Rev. 2002;102(2):555-78.

8. Barondes SH. Bifunctional properties of lectins: lectins redefined. Trends Biochem Sci. 1988;13(12):480-2.

9. Loris R, Hamelryck T, Bouckaert J, Wyns L. Legume lectin structure. Biochim Biophys Acta. 1998;1383(1):9-36.

10. Portillo-Téllez Mdel C, Bello M, Salcedo G, Gutiérrez G, Gómez-Vidales V, García-Hernández E. Folding and homodimerization of wheat germ agglutinin. Biophys J. 2011;101(6):1423-31. doi: 10.1016/j.bpj.2011.07.037.

11. Kamel M, Hanafi M, Bassiouni M. Inhibition of elastase enzyme release from human polymorphonuclear leukocytes by N-acetyl-galactosamine and N-acetyl-glucosamine. Clin Exp Rheumatol. 1991;9(1):17-21.

12. Wright CS. Crystal structure of a wheat germ agglutinin/ glycophorin-sialoglycopeptide receptor complex. J Biol Chem. 1992;267(20):14345-52.

13. Varki A, Cummings RD, Esko JD, Freeze HH, Stanley P, Bertozzi CR, et al. Essentials of glycobiology, 2nd edition. Cold Spring Harbor (NY): Cold Spring Harbor Laboratory Press; 2009.

14. Lienemann M, Paananen A, Boer H, de la Fuente JM, García I, Penadés S, et al. Characterization of the wheat germ agglutinin binding to self-assembled monolayers of neoglycoconjugates by AFM and SPR. Glycobiology. 2009;19(6):633-43. doi: 10.1093/glycob/cwp030.

15. Wittmann V, Seeberger S. Spatial screening of cyclic neoglycopeptides: Identification of polyvalent wheat-germ agglutinin ligands. Angew Chem Int Ed Engl. 2004;43(7):900-3. Doi:10.1002/anie.200352055.

16. Reymond JL, Bergmann M, Darbre T. Glycopeptidedendrimers as Pseudomonas aeruginosabiofilm inhibitors. Chem Soc Rev. 2013;42(11):4814-22. doi: 10.1039/c3cs35504g.

17. Hatano K, Matsuoka K, Terunuma D. Carbosilane glycodendrimers. Chem Soc Rev. 2013;42(11):4574-98. doi: 10.1039/c2cs35421g.

18. Chen Y, Star A, Vidal S. Sweet carbon nanostructures: carbohydrate conjugates with carbon nano tubes and grapheme and their applications. Chem Soc Rev. 2013;42(11):4532-42. doi: 10.1039/c2cs35396b.

19. Hayama R, Koyama T, Matsushita T, Hatano K, Matsuoka K Chloromethylstyrene as a useful starting material for the preparation of glycopolymers and other functional monomers, under preparation.

20. Kumari A, Koyama T, Hatano K, Matsuoka K. Synthetic assembly of novel avidin-biotin-GlcNAc (ABG) complex as an attractive bio-probe and its interaction with wheat germ agglutinin (WGA). Bioorg Chem. 2016;68:219-25. doi: 10.1016/j.bioorg.2016.08.002.

21. Raman R, Raguram S, Venkataraman G, Paulson JC, Sasisekharan R.Glycomics: an integrated systems approach to structure function relationships of glycans. Nat Methods. 2005;2(11):817-24. doi: 10.1038/nmeth807.

22. Ratner DM, Adams EW, Su J, O'Keefe BR, Mrksich M, Seeberger PH. Probing protein- carbohydrate interactions with microarrays of synthetic oligosaccharides. Chembiochem. 2004;5(3):379-82. doi: 10.1002/cbic.200300804.

23. Haseley SR, Talaga P, Kamerling JP, Vliegenthart JF. Characterization of the carbohydrate binding specificity & kinetic parameters of lectins by using surface plasmon resonance. Anal Biochem. 1999;274(2):203-10. doi: 10.1006/abio.1999.4277.

24. Rice RH, Etzler ME. Chemical modification and hybridization of wheat germ agglutinin. Biochemistry. 1975;14(18):4093-4099. doi: 10.1021/bi00689a027.

25. Monsigny M, Sene C, Obrenovitch A, Roche AC, Delmotte F, Boschetti E. Properties of Succinylated Wheat-Germ Agglutinin. Eur J Biochem. 1979;98(1):39-45.

26. http://www.sprpages.nl/sensor-chips-intro-2/biacore/nta

27. Biacore Sensor Surface Handbook BR-1005-71, Edition AB.

28. MJ O'Brien II, SRJ Brueck, VH Perez-Luna, L Tender, GP Lopez. SPR biosensors: simultaneously removing thermal & bulk-composition effects. Biosensors & Bioelectronics. 1999;14(2):145-154.

29. Myszka DG. Improving biosensor analysis. J Mol Recognit. 1999;12(5):279-84. doi: 10.1002/(SICI)1099-1352(199909/10)12:5<279::AID-JMR473>3.0.CO;2-3.

30. D G Myszka, X He, M Dembo, T A Morton, B Goldstein. Extending the range of rate constants available from Biacore: Interpreting mass transport-influenced binding data. Biophys J. 1998;75(2):583–594. doi: 10.1016/S0006-3495(98)77549-6.

31. Lundquist JJ, Debenham SD, Toone EJ. Multivalency effects in protein-carbohydrate interaction: The binding of the Shiga-like Toxin 1

binding subunit to Multivalent *C*-linked glycopeptides. J Org Chem. 2000;65(24):8245-50.

32. Kitov PI, Sadowska JM, Mulvey G, Armstrong GD, Ling H, Pannu NS, et al. Shiga-like toxins are neutralized by tailored multivalent carbohydrate ligands. Nature. 2000;403(6770):669-72. doi: 10.1038/35001095.

33. Wittmann V, Pieters RJ. Bridging lectin binding sites by multivalent carbohydrates. Chem Soc Rev. 2013;42(10):4492-503. doi: 10.1039/c3cs60089k.

34. Wright CS, Kellogg GE. Differences in hydropathic properties of ligand binding at four independent sites in wheat germ agglutinin-oligosaccharide crystal complexes. Protein Sci. 1996;5(8):1466-76. doi: 10.1002/pro.5560050803.

35. Schwefel D, Maierhofer C, Beck JG, Seeberger S, Diederichs K, Möller HM, et al. Structural basis of multivalent binding to wheat germ agglutinin. J. Am. Chem. Soc. 2010;132(25)8704-8719. doi: 10.1021/ja101646k.

36. Beckmann HS, Möller HM, Wittmann V. High-affinity multivalent wheat germ agglutinin ligands by one-pot click reaction. Beilstein. Beilstein J Org Chem. 2012;8:819-26. doi: 10.3762/bjoc.8.91.

37. Kulkarni AA, Weiss AA, Iyer SS. Glycan-based high affinity ligands for toxins and pathogen receptors. Med Res Rev. 2010;30(2):327-93. doi: 10.1002/med.20196.

38. Farajollahi MM, Cook DB, Self CH. Self, Major improvement of lectin-based assays through choice of cation and optimization of cation concentration. Anal Biochem. 1998;261(1):118-21. Doi:10.1006/abio.1998.2563.

Ultra Sensitive Detection of Influenza A Virus Based on Cdse/Zns Quantum Dots Immunoassay

Feng Wu[1,2#], Mao Mao[1#], Qian Liu[2], Lei Shi[3], Yu Cen[2], Zhifeng Qin[3] and Lan Ma[2*]

[1]*Key Laboratory for Special Functional Materials, Henan University, Kaifeng 475004, P. R. China*
[2]*Division of Life Science and Health, Graduate School at Shenzhen, Tsinghua University, Shenzhen, 518055, P. R. China*
[3]*Shenzhen Entry-Exit Inspection and Quarantine Bureau of the People's Republic of China (SZCIQ), Shenzhen, 518045, P.R. China*
#*Authors contributed equally to this article*

Corresponding author: *Lan Ma, Division of Life Science and Health, Graduate School at Shenzhen, Tsinghua University, Shenzhen, 518055, P. R. China, E-mail: malan@sz.tsinghua.edu.cn*

Abstract

An ultrasensitive lateral flow immunoassay system (LFIAS) was established for the detection of *influenza A virus*. In this LFIAS, hydrophilic dihydrazide-modified CdSe/ZnSquantom dots (QDs) were conjugated with specific antibodies and used as fluorescent labels, and a pair of matched anti-nucleoprotein of *influenza A virus* antibodies were used to form a sandwich immunoassay. The QDs were in conjugation with the fragment crystallizable region (Fc region) of specific *influenza A virus* antibodies through aldehyde-hydrazide covalent chemistry, conferring high sensitivity. The antibodies used for detection are specific for the most conserved and popular nucleoprotein of *influenza A virus* and ensure the accuracy and specificity. The QDs-LFIAS can analyze the nasal-pharyngeal swab samples through simple steps and get results within 15 min. Detection of nasal-pharyngeal swab samples makes it more rapid and convenient, and it is highly efficient for identification of influenza infection and improves influenza patients' management. The limit of detection of this QDs-LFIAS for recombinant nucleoprotein of *influenza A virus* was 0.01 ng/ mL, which was 1000-fold higher than the sensitivity of colloidal gold method. The detection of actual patient samples indicated that the QDs-LFIAS had a high compliance with real-time PCR.

Keywords: Lateral flow immunoassay; *Influenza A virus*; CdSe/ZnS; Quantum dots

Introduction

Influenza viruses circulate each year and cause mild to severe illness and even death in humans. Type *A influenza* can outbreaks in seasonal and regional, and the viruses are susceptible to mutation. On the basis of the antigenic nature of the surface glycoproteins, hemagglutinin (HA) and neuraminidase (NA), type *A influenza viruses* are subdivided into several subtypes [1]. At present, the seasonal *influenza A virus* subtypes caused by human infection is *influenza A* (H1N1) and A (H3N2) [2]. Avian influenza viruses such as A (H5N1) A (H5N6), A (H7N9) and A (H9N2) can sometimes spread to domestic poultry and cause severe outbreaks disease, and they can cause serious infections among people.

To date, numerous analytical methods have being used to detect *influenza A virus*. According to the type of detection target, these methods could be categorized into, for example, virus isolation and identification, nucleic acid-based detection, antigen detection, and antibody detection [3-8]. Most of these methods need demanding conditions and professional operations, and some of the detection processes is time-consuming. For these reasons, some new biosensors based on nano materials such as gold nanoparticles, magnetic nano beads and quantum dots (QDs) have been coming to the forefront [9-12].Developing a sensitive, specific and fast biosensor for diagnosing the influenza virus at the early stages of infection is a challenge. Lateral flow immune assay method has been widely used in detection of infectious diseases with the advantages of rapid, easy to operate, and low cost [13]. However, traditional method such as colloidal gold lateral flow tests has limitations when high sensitivity is needed [14].QDs have been developed to replace colloidal gold owing to their excellent optical properties such as high fluorescence efficiency, wide range of excitation wavelength, narrow and symmetric emission spectra, and QDs based lateral flow tests have been proved higher sensitivity and reproducibility for detection of pathogen, characteristic protein and even nucleic acid [15-18].

In this paper, we present an ultrasensitive lateral flow immunoassay (LFIA) for detecting *influenza A virus*. This immune sensor is based on site-specific covalent binding with the Fc end of *influenza A virus* antibodies to CdSe/ ZnS QDs. The carboxyl-functionalized CdSe/ ZnS QDs conjugated with antibodies via an amide bond often result in random links of the antibody structure in the QD conjugates and block certain antigen-binding sites. In order to obtain site-specific linking, we modified carboxyl-functionalized QDs with adipicdihydrazide (ADH) and oxidized the carbohydrate groups on the antibody's Fc region to obtain reactive aldehyde groups, the oxidized antibodies can then conjugated with dihydrazide-modified QDs [19]. The Fc region of the antibody was linked to the QDs surface with the fragment antigen-binding region (Fab region) facing outward as shown in Figure 1A. The antibodies used for detection by the developed sensor are specific for the most conserved and popular nucleoprotein of *influenza A virus*.

Materials and Methods

Materials

Sodium periodates, Ethylene glycol, Adipic acid dihydrazide (ADH), D-(+)-Trehalose dehydrates, 2-(N-morpholino) ethane sulfonic acid (MES), and D-(+)-glucose were purchased from Sigma-Aldrich (Shanghai, China). 1-Ethyl-3-[3-dimethylaminopropyl] carbodiimide hydrochloride (EDC) was purchased from Thermo Fisher Scientific, Inc. (Waltham, MA. U.S.A.). Sodium phosphate dibasic, sodium phosphate monobasic monohydrate, sodium acetate, bovine serum albumin fraction V (BSA), and Tween-20were purchased from Shanghai Sangon Ltd. (Shanghai, China). Goat anti-mouse IgG antibody was purchased from Arista Biologicals, Inc.(Allentown, PA. U.S.A.).Anti-nucleoprotein of *influenza A virus* antibodies (mAb IgG) and recombinant nucleoprotein of *influenza A virus* were obtained from Life Science Division, Tsinghua University. *Influenza A* subtype viruses HI test antigens were purchased from Harbin Weike Biotechnology Development Company (Habin, China). Influenza B subtype viruses, Measles virus, Mumps virus, Rubella virus, *Varicella zoster virus*, *Staphylococcus aureus* and *Pseudomonas aeruginosa* test antigens were purchased from China food and Drug Inspection Institute. All chemical reagents used were of analytical grade without any further purification.

Conjugation of influenza A virus antibodies to QDs via oxidized Fc-carbohydrate groups

Hydrophilic carboxyl-functionalized CdSe/ Zn SQDs were prepared according to the previous work of our group [20-22]. The carboxyl-functionalized QDs were dihydrazide-modified by ADH. Briefly, 5 mg of the QDs were mixed with 5 mM EDC and 5 mM ADH in 0.01 M phosphate-buffered saline (PBS) and then incubated for 4 h at room temperature. After washing and centrifuged three times at 20,000 g for 0.5 h, the QDs were dispersed in 0.01 M PBS buffer. *Influenza A virus* antibodies were oxidized by sodium periodate [19, 23]. Briefly,0.1 M sodium periodate dissolved in 0.1 M sodium acetate buffer (pH=5.5), and 100 µL of the solution was added to 1 ml of the antibody in 0.01 M PBS buffer(1.5 mg/mL), then incubated in the dark for 0.5h at 0°C. The reaction was terminated by adding 100 µL ethylene glycol. Oxidized antibodies were purified by dialysis overnight against 0.01 M PBS buffer (pH=7.2). To form QDs-Ab conjugates, 0.3 mg of oxidized *influenza A virus* antibodies were mixed with5mg dihydrazide-modified QDs, and then incubated for 2 h at room temperature. For blocking residual active coupling sites, 1% glucose solution was added and incubated for 0.5 h at room temperature. The conjugated QDs-Ab were purified by centrifugation at 20,000 g for 3 times, and stored in 0.01 M PBS buffer at 4°C before use.

Preparation of QDs-LFIA strips

In order to prepare the QDs-LFIA strips, the antibodies were dispensed onto the nitrocellulose membrane while the conjugated QDs-Ab was dispensed onto the sample pad using the XYZ Dispensing System (Bio Dot Inc, Irvine, CA). Briefly,

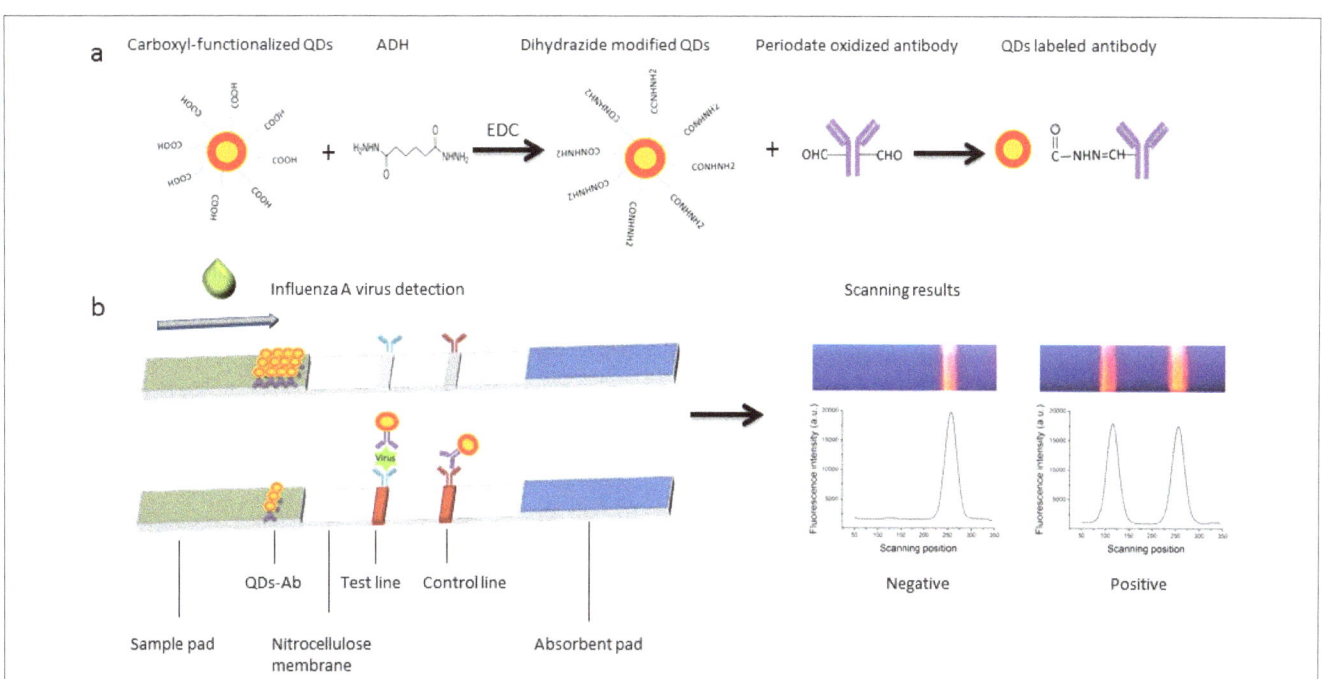

Figure 1: Schematic representation of the QDs-LFIAS (A) Conjugation of antibodies to QDs via oxidized Fc-carbohydrate groups (B) Analytical representation of the QDs-LFIAS

anti-*influenza A virus* coating antibody was dispensed onto the nitrocellulose membrane at 2 mg/ mL as the test line, and the goat anti-mouse IgG was dispensed at 0.5 mg/ mL as the control line. The dispensed nitrocellulose membranes were dried at 37°C in a vacuum oven for 4 h. The sample pad was pretreated by PBS buffer containing D-(+)-trehalose dehydrate (1%, w/ v), BSA (1%, w/v) and Tween-20 (0.1%, w/v), and dried at 37°C for 3 h. The conjugated QDs-Ab was dispensed at a ratio of 10 µL/ cm onto the pretreated sample pad and dried at 37°C for 3 h. The QDs-LFIA strip was assembled in its standard configuration as shown in Figure 1B. The completed QDs-LFIA was cut into individual 3.5 mm strips and stored in sealed and dry condition.

Analytical procedure

Sixty micro liter of analyte was added onto the sample port of QDs-LFIA strip, the captured fluorescence intensity of test line and control line were scanned by a fluorescence test strip scanning device after a 15 min reaction. The captured fluorescence QDs on the test line and control line produced a bright fluorescent band in response to 365 nm ultraviolet excitation, and the fluorescence intensity at 620 nm were detected by the device [22]. As shown in Figure 1B, once *influenza A virus* was in the sample, the QDs-Ab bound the *influenza A virus* specifically and were later captured by the second anti-*influenza A virus* antibody at the test line, and QDs-Ab without *influenza A virus* bound were captured by the goat anti-mouse IgG at the control line, the captured QDs-Ab aggregated into a fluorescent band under 365 nm ultra violet excitation. The fluorescence intensity of test line was closely correlated with the concentration of captured QDs-Ab and virus complex..The cutoff value was obtained by detecting 50 negative samples, the average of fluorescence intensity was calculated, and two times of the average value was defined as the cutoff value. The calculated cut off value is 102 a.u. Fluorescence intensity of 102 a.u. or above was defined as positive, while less than 102 a.u. was negative. The limit of detection (LOD) was estimated by analyzing samples at various concentrations. Briefly, the recombinant nucleoprotein of *influenza A virus* was ten-fold diluted by 20 mM PBS (1000, 100, 10, 1, 0.1, 0.01 and 0.001 ng/ mL) and test separately, 20 mM PBS was test as a negative control. The fluorescence intensity of test line was detected while the LOD was calculated. The LOD was defined as the lowest concentration of recombinant nucleoprotein whose fluorescence intensity was the minimum above the cutoff value. We also test the reproducibility of QDs-LFIA strip by analyzing 20 replicates of recombinant nucleoprotein sample sat different concentrations (1, 10 and 100 ng/ mL). The specificity and cross-reactivity was analyzed by detecting *influenza A virus* subtypes(H1N1, H3N2, H5N1 re-4/6, H7N9, H9N2 re-2,)HI test antigens, *influenza B virus* subtypes(1704 strain, Victoria strain, Yamagata strain),Measles virus, Mumps virus, Rubella virus, *Varicella zoster virus*, *Staphylococcus aureus* and *Pseudomonas aeruginosa*.

Practical field sample tests

Sixty samples (human throat swabs collected into sterile Hanks' balanced salt solution viral transport media)—which were collected and preserved by Shenzhen International Travel Health Care Center, Shenzhen Entry-Exit Inspection and Quarantine Bureau—were assayed using QDs-LFIA strips. Commercial *influenza A* antigen rapid diagnostic test kit (colloidal gold) was conducted in parallel, and real-time PCR assay was used as a reference method to evaluate the accuracy of QDs- LFIAS. A pair of specific primers (Forward Primer 5'-GACCRATCCTGTCACCTCTGAC-3', Reverse Primer 5'-AGGGCATTYTGGACAAAKCGTCTA-3') and a probe (FAM-TGCAGTCCTCGCTCACTGGGCACG-BHQ1) for detecting *influenza A virus* was used in real-time PCR assay.

Results and discussion

Conjugation of antibodies to QDs via oxidized Fc-carbohydrate groups

The reaction mechanism of hydrophilic carboxyl-functionalized CdSe/ ZnS QDs has been previously described in the literature [24, 25]. The dynamic light scattering analysis in Figure 2 shows that the average hydrodynamic size of carboxyl-functionalized CdSe/ Zn SQDs was 42.86 nm with Stdev 18.72 nm, and this size increased to 109.5 nm with Stdev 52.16 nm after dihydrazide modified and conjugation with antibodies, which indicated that antibodies were successful conjugated.

Limit of detection of QDs-LFIA strip

The LOD was estimated by analyzing recombinant nucleoprotein of *influenza A virus* samples at various concentrations, the fluorescence intensity of test line was scanned by fluorescence test strip scanner after 15 min of addition of the samples. The diluted recombinant nucleoprotein of *influenza A virus* samples was also tested by commercial *influenza A* antigen rapid diagnostic test kit (colloidal gold). As demonstrated by Figure3A, with the recombinant nucleoprotein concentration increasing, the fluorescent band of test line at 0.1 ng/ mL was still visible. The commercial *influenza A* antigen rapid diagnostic test kit (colloidal gold) could only detect the recombinant nucleoprotein at 10 ng/ mL concentration through Figure 3B. As show in Figure 3C, employing the fluorescence test strip scanning device, the recombinant nucleoprotein could be detected at 0.01 ng/ mL by QDs-LFIA strips, which was 1000-fold higher than the sensitivity of colloidal gold method.

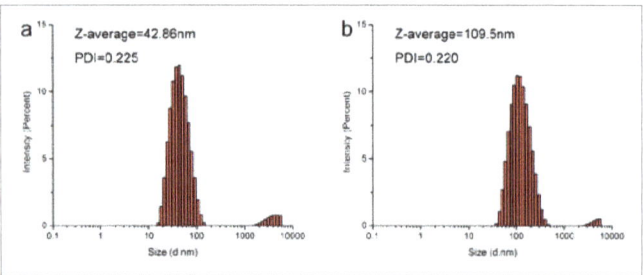

Figure 2: DLS data of QDs and conjugated QDs-Ab. (A) Carboxyl-functionalized QDs. (B) Antibody conjugated QDs. Average hydrodynamic size of CdSe/ZnS QDs was 42.86 nm and this size increased to 109.5 nm after conjugation with antibodies

Reproducibility of QDs-LFIA strip

The reproducibility of this QDs-LFIA strip was analysed by detecting 20 replicates of the recombinant nucleoprotein at various concentrations (1, 10 and 100 ng/mL). The fluorescence intensity of test line was detected while the relative standard deviations (RSD) were later calculated. Table 1 shows the RSD results and the RSD values were below 8%, demonstrating the QDs-LFIA strip had good reproducibility.

Specificity and cross-reactivity of QDs-LFIA strip

The specificity and cross-reactivity of the QDs-LFIA strip was analyzed by detecting *influenza A virus* subtypes (H1N1, H3N2, H5N1 re-4/6, H7N9, H9N2 re-2,) HI test antigens, *influenza B virus* subtypes(1704 strain, Victoria strain, Yamagata strain), Measles virus, Mumps virus, Rubella virus, *Varicella zoster virus*, *Staphylococcus aureus* and *Pseudomonas aeruginosa*. Figure 4A shows that the QDs-LFIA strip could detect all the subtypes of *influenza A virus* used but none of other type antigens. We also tested the *influenza A virus* subtypes with concentration ranging from 1/512 to 16HAU. As demonstrated in Figure4B, the QDs-LFIA could detect *influenza A virus* subtype H1N1 at 1/256 HAU, *influenza A virus* subtype H5N1 at 1/32 HAU, *influenza A virus* subtype H3N2, H7N9 and H9N2 at 1/128 HAU. It demonstrated that the QD-LFIAS could detect *influenza A* subtype viruses with high sensitivity and specificity.

Practical field sample tests

Sixty samples (human throat swabs collected into sterile Hanks' balanced salt solution viral transport media) were detected using QDs-LFIAS. Table 2 shows that all positive samples with low real-time PCR threshold cycle (Ct ≤ 30) were detected by QDs- LFIAS with a high accuracy. For the positive samples with high real-time PCR threshold cycle (Ct > 30), three were detected as negative by QDs-LFIAS, but all were detected as negative by commercial *influenza A* antigen rapid diagnostic test kit (colloidal gold). The results indicated that the QDs-LFIAS had an accuracy of 95%, while that of the commercial *influenza A* antigen rapid diagnostic test kit (colloidal gold) was 56.7% compared with real-time PCR.

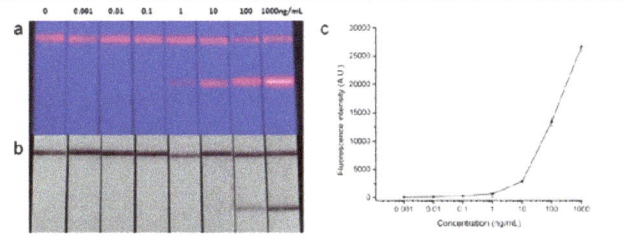

Figure 3: (A) Images of tested QDs-LFIAS in response to excitation with 365 nm ultraviolet light. (B) Images of tested commercial *influenza A* antigen rapid diagnostic test strip (colloidal gold). (C). Fluorescence intensity scans at different concentrations of recombinant nucleoprotein of *influenza A virus* measured by the fluorescence strip scanning device. Shows the fluorescence from test line of QDs-LFIA strips at 0.1ng/mL was still visible. B shows the test line of commercial *influenza A* antigen rapid diagnostic test kit (colloidal gold) at 10 ng/mL was visible. C shows the test line of QDs-LFIA strips at 0.01 ng/mL can be detected by fluorescence test strip scanning device.

Table 1: Reproducibility Tests of QDs-LFIAS. 20 replicates of the recombinant nucleoprotein at various concentrations (1, 10 and 100 ng/mL) were test by QDs-LFIAS. The relative standard deviations (RSD) were then calculated accordingly.

Concentrations (ng/mL)	Fluorescence intensity average (a.u.)	Stdev (a.u.)	RSD/ % n=20
1	650.5	42.5	6.5
10	2957.8	217.8	7.4
100	13827.8	820.7	5.9

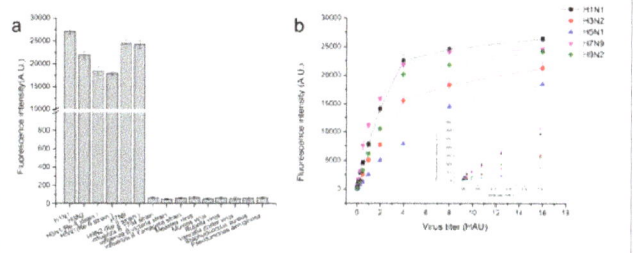

Figure 4: (A) Specificity tests of QDs-LFIAS. (B) Fluorescence intensity scans at different concentrations of influenzaA virus subtypes. Shows QDs-LFIAS could detect all the subtypes of *influenza A virus* used but none of other type antigens. B shows QDs-LFIAS could detect the subtypes of *influenza A virus* with high sensitivity.

Table 2: Practical field sample Tests using QDs-LFIAS. Sixty human throat swab samples were detected using QDs- LFIAS, commercial influenza A antigen rapid diagnostic test kit and real-time PCR. Compared with real-time PCR, the QDs-LFIAS had an accuracy of 95%, while that of the commercial infl uenzaA antigen rapid diagnostic test kit (colloidal gold) was 56.7%.

Samples	Number of samples	QDs-LFIAS result(P/N)	Commercial rapid diagnostic test kit result(P/N)	Real-time PCR result(P/N)
Real-time PCR positive result Ct > 30	20	17/3	0/20	20/0
Real-time PCR positive result Ct ≤ 30	20	20/0	14/6	20/0
Negative sample	20	0/20	0/20	0/20

At the early stage of *influenza A virus* infection, the symptoms of patients were similar to those of common cold and it was difficult and insufficient to use these common symptoms for the diagnosis of type A influenza virus infection [26]. Besides, the *influenza A virus* titer was very low during the early infection. So, high sensitivity is the key for diagnosing the influenza virus at the early stages of infection. Real-time PCR method has ultra-high accuracy but time consuming, as point-of-care testing method, the QDs-LFIAS was proved to detect human throat swab samples more sensitive and accurate than colloidal gold method. Detection of nasal-pharyngeal swab samples makes it more rapid and convenient, and it is highly efficient for monitoring and prevention of influenza outbreak in the hospital emergency, port quarantine, schools and also in-home health care. QDs-LFIAS is a low-cost technique with small amount of immunoassay reagents consumed, and results can be objectively determined by a handheld device. According to our results above, the QDs-LFIAS only requires one step and provides results in 15 min. This simple and less time consuming method could be used in on-site tests, clinical diagnosis and early treatments of influenza virus infection.

Conclusions

The purpose of this study was to develop an ultrasensitive, rapid and low cost lateral flow immune sensor for *influenza A virus* preliminary screening. We developed a QDs-LFIAS method, which rapidly analyzed the sample through one steps. The results were objectively analyzed by an inexpensive, portable device within 15min. The LOD of this QDs-LFIAS for recombinant nucleoprotein of *influenza A virus* was 0.01ng/ mL, which was 1000-fold higher than the sensitivity of colloidal gold method. The QDs-LFIAS could detect *influenza A virus* subtype H1N1 at a concentration of 1/ 256 HAU, *influenza A virus* subtype H5N1 at 1/ 32 HAU, *influenza A virus* subtype H3N2, H7N9 and H9N2 at 1/128 HAU. It demonstrated that the QD-LFIAS could detect *influenza A virus* subtypes with high sensitivity and specificity. This was more sensitive than that of traditional point-of-care testing methods. The specificity and reproducibility were shown to be good. Real patient samples demonstrated that the QDs-LFIAS had high accuracy, and detection of nasal-pharyngeal swab samples makes it more rapid and efficient for identification of influenza infection and improves influenza patients' management.

Acknowledgments

This work was supported by the following sources: the National High Technology Research and Development Program of China (863 Program, NO. 2013AA032204), Science and Technology Planning Project of Guangdong Province (2012B031500003), Shenzhen strategic emerging industry development special funds (JSGG20140716144254155)

Conflict of interest

The authors declare that there is no conflict of interest regarding the publication of this paper.

References

1. Hinshaw VS, Air GM, Gibbs AJ, Graves L, Prescott B, Karunakaran D. Antigenic and Genetic-Characterization of a Novel Hemagglutinin Subtype of *Influenza-a Viruses* from Gulls. Journal of Virology. 1982;42(3):865-872.

2. The World Health Organization Media Center Page,2014.

3. Chen HT, Zhang J, Sun DH, Ma LN, Liu XT, Cai XP, et al. Development of reverse transcription loop-mediated isothermal amplification for rapid detection of H9 avian *influenza virus*. Journal of virological methods. 2008;151(2):200-203. doi:10.1016/j.jviromet.2008.05.009.

4. Alberini I, Del Tordello E, Fasolo A, Temperton NJ, Galli G, Gentile C, et al. Pseudoparticle neutralization is a reliable assay to measure immunity and cross-reactivity to H5N1 *influenza viruses*. Vaccine. 2009;27(43):5998-6003. doi: 10.1016/j.vaccine.2009.07.079.

5. Moore C, Telles JN, Corden S, Gao RB, Vernet G, Van Aarle P, et al. Development and validation of a commercial real-time NASBA assay for the rapid confirmation of *influenza A* H5N1 virus in clinical samples. Journal of virological methods. 2010;170(1-2):173-176. doi:10.1016/j.jviromet.2010.09.014.

6. Yang SY, Chieh JJ, Wang WC, Yu CY, Lan CB, Chen JH, et al. Ultra-highly sensitive and wash-free bio-detection of H5N1 virus by immunomagnetic reduction assays. Journal of virological methods. 2008;153(2):250-252. doi:10.1016/j.jviromet.2008.07.025.

7. Xie Z, Pang YS, Liu J, Deng X, Tang X, Sun J, et al. A multiplex RT-PCR for detection of type *A influenza virus* and differentiation of avian H5, H7, and H9 hemagglutinin subtypes. Molecular and cellular probes. 2006;20(3-4):245-249. doi:10.1016/j.mcp.2006.01.003.

8. Payungporn S, Chutinimitkul S, Chaisingh A, Damrongwantanapokin S, Buranathai C, Amonsin A et al. Single step multiplex real-time RT-PCR for H5N1 *influenza A virus* detection. Journal of virological methods. 2006;131(2):143-147. doi:10.1016/j.jviromet.2005.08.004.

9. Kamikawa TL, Mikolajczyk MG, Kennedy M, Zhang P, Wang W, Scott DE, et al. Nanoparticle-based biosensor for the detection of emerging pandemic *influenza* strains. Biosensors & bioelectronics. 2010;26(4):1346-1352. doi: 10.1016/j.bios.2010.07.047.

10. Krejcova L, Nejdl L, Rodrigo MA, Zurek M, Matousek M, Hynek D, et al. 3D printed chip for electrochemical detection of *influenza virus* labeled with CdS quantum dots. Biosensors & bioelectronics. 2014;54:421-427. doi: 10.1016/j.bios.2013.10.031.

11. Lee C, Gaston MA, Weiss AA, Zhang P. Colorimetric viral detection based on sialic acid stabilized gold nanoparticles. Biosensors & bioelectronics. 2013;42:236-241. doi: 10.1016/j.bios.2012.10.067.

12. Wu Z, Zhou CH, Chen JJ, Xiong C, Chen Z, Pang DW, et al. Bifunctional magnetic nanobeads for sensitive detection of avian *influenza A* (H7N9) virus based on immunomagnetic separation and enzyme-induced metallization. Biosensors & bioelectronics. 2015;68:586-592. doi:10.1016/j.bios.2015.01.051.

13. Ngom B, Guo Y, Wang X, Bi D. Development and application of lateral flow test strip technology for detection of infectious agents and chemical contaminants: a review. Analytical and bioanalytical chemistry. 2010;397(3):1113-1135. doi:10.1007/s00216-010-3661-4.

14. Xie QY, Wu YH, Xiong QR, Xu HY, Xiong YH, Liu K, et al. Advantages of fluorescent microspheres compared with colloidal gold as a label in immunochromatographic lateral flow assays. Biosensors &bioelectronics. 2014;54:262-265. doi:10.1016/j.bios.2013.11.002.

15. Shen H, Yuan H, Niu JZ, Xu S, Zhou C, Ma L, et al. Phosphine-free synthesis of high-quality reverse type-I ZnSe/CdSe core with CdS/Cd(x)Zn(1 - x)S/ZnS multishell nanocrystals and their application for detection of human hepatitis B surface antigen. Nanotechnology. 2011;22(37):375602. doi:10.1088/0957-4484/22/37/375602.

16. Sapountzi EA, Tragoulias SS, Kalogianni DP, Ioannou PC, Christopoulos TK. Lateral flow devices for nucleic acid analysis exploiting quantum dots as reporters. Analytica chimica acta. 2015;864:48-54. doi:10.1016/j.aca.2015.01.020.

17. Li X, Lu D, Sheng Z, Chen K, Guo X, Jin M, et al. A fast and sensitive immunoassay of avian *influenza virus* based on label-free quantum

dot probe and lateral flow test strip. Talanta. 2012;100:1-6. doi: 10.1016/j.talanta.2012.08.041.

18. Zhaohui L, Ying W, Jun W, Zhiwen T, Pounds JG, Yuehe L. Rapid and sensitive detection of protein biomarker using a portable fluorescence biosensor based on quantum dots and a lateral flow test strip. Analytical Chemistry. 2010;82(16):7008-7014. doi: 10.1021/ac101405a.

19. Xing Y, Chaudry Q, Shen C, Kong KY, Zhau HE, Chung LW, et al. Bioconjugated quantum dots for multiplexed and quantitative immunohistochemistry. Nature protocols. 2007;2(5):1152-1165. doi:10.1038/nprot.2007.107.

20. Shen H, Wang H, Tang Z, Niu JZ, Lou S, Du Z, et al. High quality synthesis of monodisperse zinc-blende CdSe and CdSe/ZnS nanocrystals with a phosphine-free method. CrystEngComm. 2009;11(8):1733-1738. doi:10.1039/b909063k.

21. Zhou C, Shen H, Guo Y, Xu L, Niu J, Zhang Z, et al. A versatile method for the preparation of water-soluble amphiphilic oligomer-coated semiconductor quantum dots with high fluorescence and stability. Journal of colloid and interface science. 2010;344(2):279-285. doi: 10.1016/j.jcis.2010.01.015.

22. Wu F, Yuan H, Zhou C, Mao M, Liu Q, Shen H, et al. Multiplexed detection of *influenza A virus* subtype H5 and H9 via quantum dot-based immunoassay. Biosensors & bioelectronics. 2016;77:464-470. doi:10.1016/j.bios.2015.10.002.

23. Gideon Fleminger, Eran Hadas, Tamar Wolf, Solomon B. Oriented Immobilization of Periodate-Oxidized Monoclonal Antibodies on Amino and Hydrazide Derivatives of Eupergit C. Applied Biochemistry and Biotechnology. 1990;23(2):123-137. doi:10.1007/BF02798382.

24. Shen H, Wang H, Zhou C, Niu JZ, Yuan H, Ma L, et al. Large scale synthesis of stable tricolor Zn1– xCdxSe core/multishell nanocrystals via a facile phosphine-free colloidal method. Dalton Transactions. 2011;40(36):9180-9188. doi: 10.1039/C1DT10865D.

25. Shen H, Yuan H, Wu F, Bai X, Zhou C, Wang H, et al. Facile synthesis of high-quality CuInZnxS2+x core/shell nanocrystals and their application for detection of C-reactive protein. Journal of Materials Chemistry. 2012;22(35):18623-18630. doi:10.1039/c2jm33763k.

26. Van dD, C., Hak E, Wallinga J, Van Loon AM, Lammers JWJ, Bonten MJM. Symptoms of *influenza virus* infection in hospitalized patients. Infection Control & Hospital Epidemiology the Official Journal of the Society of Hospital Epidemiologists of America. 2008;29(4):314-319.

11

Biomarker assessment of lanthanum on a freshwater invertebrate, Dreissena polymorpha

Houda Hanana*, Patrice Turcotte, Martin Pilote, Joëlle Auclair, Christian Gagnon and François Gagné*

Aquatic Contaminant Research Division, Environment and Climate Change Canada, 105 McGill, Montreal, Quebec, Canada H2Y 2E7

*Corresponding author: Houda Hanana, Aquatic Contaminant Research Division, Environment and Climate Change Canada, 105 McGill, Montreal, Quebec, Canada, E-mail: houda.hanana@yahoo.fr

François Gagné, Aquatic Contaminant Research Division, Environment and Climate Change Canada, 105 McGill, Montreal, Quebec, Canada, E-mail: francois.gagne@canada.ca

Abstract

The toxicological understanding of rare earth elements (REEs) in the aquatic environment is very limited and concerns are rising about their safety to the environment. This study aimed to determine the bioavailability of Lanthanum (La) in zebra mussel, to evaluate its effects after 14 and 28 days through a multi biomarker approach and to estimate the cumulative effects of exposure to this lanthanide and air-time survival. Results showed that La was bioaccumulated in mussels but did not trigger metallothionein induction. In addition, La caused an antioxidant and prooxidant effects depending on the concentration and the duration of exposure but no genotoxicity was found. A significant decrease in glutathione-S-transferase activity was observed after 14 days but not after 28 days of exposure. Results revealed that low concentration of La enhanced citrate synthase activity, while high concentration trend to decrease its activity. Furthermore, this lanthanide did not significantly affect the air-time survival and mussel weight by comparison to control mussels. We suggested that La could contribute in the resistance of mussels to stress induced by air exposure by improving TCA cycle. This study demonstrates that La has diverse effect on mussel but further experiments are needed to elucidate its exact mechanism of action.

Keywords: Rare earth elements, lanthanum, Zebra mussels, Bioavailability, Oxidative stress, Genotoxicity.

Introduction

Rare earth elements (REEs) are a group of 17 elements including the lanthanide groups and two other elements, scandium and yttrium, closely related to lanthanides. REEs are naturally present in the environment and they have an essential role in the efficient function of the world's economy [1]. Their unique physical and chemical properties make them indispensable for a wide variety of applications in industrial, agricultural and medical sectors [2]. The global production of REEs has increased since 1980/1990 (EPA, 2012) due to their increased use in industry which was estimated about 84,000 tons of oxides in 2003 [3]. However, this growing use of REEs is accompanied by enrichment to the environment [4]. Indeed, REE senter the groundwater and eventually migrate into rivers and lakes [5] and reach humans through trophic transfer [6, 7]. Despite their widespread applications, environmental risks related to REEs have received little attention which explains the paucity of data about their bioavailability and ecotoxicological impact. Thus, more research on the environmental impacts and biological effects of REE sis required [8].

Lanthanum (La), a representative element of light REEs, is widely used in electronics and optoelectronics, lighters, ceramics, battery [9] and also to stimulate the growth of agricultural products [10]. Therefore, this element is considered of specific interest as one of the major lanthanides found in industrial effluents [11]. Although dissolved lanthanum species (La3+) represent a very small proportion of lanthanum compounds in water and sediment, they are bioavailable and cause adverse effects in living biological systems [12]. In fishes and amphibians, several effects of La have been reported particularly on nervous systems [13,14], excretory organs [15] and smooth muscles [16-18]. Previous study demonstrated also that La3+ delayed zebra fish embryo and larval development, decreased survival and hatching rates and causes tail malformation in a concentration –dependent way [19]. Studies performed with crustaceans are focused mainly in evaluating the effect of La on survival, growth and reproduction [20, 21, 22] and showed the highest sensitivity of those organisms towards La-ions dissolved from $LaCl_3$. In our Knowledge, few reports on La effects in marine bivalves have been published before now [23, 24, 25]. Given the lack of data for aquatic organisms, studies with mammals are provided. The trivalent La3+interfere with the immune defense and the function of the liver, spleen, heart and blood vessels and brain [26]. Previous studies indicated that lanthanum chloride accumulated in brain [10] alter the learning capacity and memory in animals

[10,27,28]. Researchers have also shown that La3+entered hepatocytes [29] accumulated and induced oxidative damage in mitochondria [30]. In fish liver, this element induced an inhibitory action on mitochondrial energy turnover [31]. The injury caused by La3+ may be attributed to the occurrence of oxidative stress and to the homeostatic disturbances of essential elements and enzymes [10]. Indeed, the mode of action of La3+on tissues and cells is similar to other lanthanides. *In vitro*, it reacts with various tissue components, like nucleoproteins, amino acids, phospholipids, enzymes, intermediate metabolites and inorganic phosphate and it can precipitate DNA [12]. In general, the biological activities of La3+ on a cellular system is often mediated through the competition for binding sites with calcium ions, since both ions have similar ionic radii [11] and also through its high affinity for phosphate groups of biological molecules.

As indicated above, only a few aquatic toxicological studies addressing environmental effects of La in invertebrates have been published. Therefore, there is clearly a lack of data on the potential Sublethal and long term effects of REEs in aquatic organisms. Biochemical approaches in ecotoxicology have the advantage of finding early warning signals at the molecular level, which can led to effects at higher levels of biological organization [32]. Moreover, the study of biomarkers at different levels of the biological scale allows to elucidate adverse outcome pathways and a better understanding of the mechanisms involved on species [33]. Bivalves are one of animal models that are often used to conduct ecotoxicological studies with biomarkers [34]. Those animals are important members of aquatic invertebrates and have essential ecological functions as food source, calcium mobilization and removal of suspended matter in the water column. Moreover, these organisms are efficient filter feeders which exposed them to numerous contaminants, including REEs. Among bivalves, zebra mussel Dreissena polymorpha, a well-established invasive model organism has been frequently used in Europe as well as in North America for the biomonitoring and for the assessment of environmental quality [35]. Indeed, those mollusks which pervade most environments with enough calcium and zooplankton to sustain their growth and survival are widespread and found in nearly all water bodies in Europe and North America.

Thus, the purpose of this study was to investigate the effect of a sub chronic exposure to $LaCl_3$ in a freshwater invertebrate, the zebra mussel Dreissena polymorpha. La concentrations were determined in mussel tissues to assess its bioavailability. In addition to bioavailability; biomarkers of oxidative and genetic damage were evaluated through a multi biomarker approach. Biomarkers studied in this work were selected given that previous reports revealed cytotoxicity following overproduction of reactive oxidative species (ROS) leading to lipid peroxidation, and DNA damage.

Materials and methods

Sampling

Adult zebra mussels, Dreissena polymorpha, were collected in July 2014 at a reference site in the Saint-Lawrence River near the City of Montréal, Québec, Canada at the following coordinates 45°19′50″N, 73°58′12″W.Mussels were gently cut off from the rocks (they attached themselves to solid substrates through abyssus), quickly transferred to the laboratory in bags filled with river water. For acclimation, mussels were kept in 50 L glass-holding aquaria filled with charcoal and UV-treated tap water (City of Montréal, QC. Canada) in the laboratory at15°C, 16h light/8h dark cycle under constant aeration and were fed three times per week with concentrates of phytoplankton (Phytoplex, Kent Marine, WI) and Pseudokirchneriella subcapitata algal preparations. A total of 240 mussels having similar shell length (1.3 ± 5 mm) were used in this study. Mussels were placed in 4 L containers lined with polyethylene bags and exposed to increasing concentrations (10, 50, 250 and 1250 µg/ L) of lanthanum chloride (262072, Sigma) for 14 and 28 days at 15°C. The control group consisted of mussel exposed to aquarium water only. The number of mussels used was 24 for each group. After exposure periods, tissues from a sub-group of mussel (n=8mussels/treatment) were collected for subsequent biomarker assays and for evaluation of total La accumulation (n=8 mussels/treatment). Another sub-group of survival mussel (n=8mussels/treatment) was removed and used to evaluate the air-time survival i.e., the survival of mussels in air.

Chemical analyses

The bioavailability of La was determined in water and in the soft tissue of mussels. La concentrations in water samples from control and exposure groups were analyzed at the beginning of the exposure period (t=0). Briefly, total La concentrations were determined after acidification with nitric acid (1%) (seastar grade) and analyzed by ion-coupled plasma mass spectrometry (ICP-MS, XSERIES 2 ICP-MS, Thermo Scientific, USA)with a detection limit of 1 ng/L.

Bioaccumulation of La was carried out at the end of exposure periods after keeping mussel in clean water for 24 h to allow them to depurate. Soft tissue was sampled, weighed and frozen at -80°C until analysis. Tissues were acid-digested with 8 ml of concentrated HNO_3, 1 ml of concentrated HCl, and 2 ml of concentrated H_2O_2. The tissues were then digested during 2 h at 170°C using a high pressure microwave oven (Ethos EZ, Milestone Scientific Inc, ON, Canada). The samples were completed to final volume of 12 ml with de ionized water. Total La concentration was afterwards determined by ICP-MS (XSERIES 2 ICP-MS, Thermo Scientific, USA) with a detection limit of 0.02 mg/Kg (wet/weight). The metal bioaccumulation factor (BAF) was calculated by dividing the mean level of metals in soft tissues by the total mean values for metals found in the dissolved and particle-bound phases.

Air time survival

After 14 and 28 days of exposure to La, the survival of mussels in air was evaluated. This experience was performed for up to 7 days in air at 20°C under humidified atmosphere (>80%). Mortality was determined by sustained shell opening and total weight was recorded each day for 7 days. The proportions of weight loss were calculated by the following formula: % weight loss = ((mussel weight at the end of exposure to air / mussel weight before exposure to air) x 100)-100.

Biochemical analyses

Mussels were thawed on ice for approximately 10–15 min. After rapid dissection, soft tissue from each mussel was homogenized in 1:5 (w/v) ratio of buffer solution containing 25 mM Hepes-NaOH buffer, 100 mM NaCl, 1 mM dithiothreitol and 1μg/L of aprotinin with pH adjusted to 7.4. Aliquots of each homogenate were sampled for total protein lipid peroxidation (LPO), DNA damage (DNA). A portion of the homogenates was centrifuged at 1500g for 10 min at 2°C and the resulting supernatant, considered free of unbroken cells or cell debris, was centrifuged at 10000g for 20 min at 2°C and collected for determination of metallothionein (MT) levels and glutathione S-transferase (GST) activity. The pellet was re-suspended in homogenization buffer (200 μL/g of tissue) and used as a mitochondrial fraction to measure the citrate synthase activity. The supernatant (S10), the mitochondrial fraction and the homogenates were stored at -80°C until analysis. Concentrations of studied biomarkers were normalized with the total individual protein concentration according to Bradford methodology [36] with bovine serum albumin as the standard. Total protein concentration was measured with the absorption at 595 nm by a micro plate reader.

Lipid peroxidation: Lipid peroxidation (LPO) was determined according to the thiobarbituric acid (TBARS) methodology (Wills, 1987) which consists to measure the production of malonaldehyde in the mussel homogenate. A volume of 150 μL of 10% TCA containing 1 mM $FeSO_4$ and 75μL of 0.67% TBA was added to 75μL of homogenate, mixed and heated at 70°C for 10 min. To detect thiobarbituric acid reactants, 100 μL of the supernatant was added to a 96 well plate and fluorescence was measured at 540 and 600 nm for excitation and emission, respectively. Blanks and standards of tetramethoxypropane (stabilized form of malonaldehyde) were prepared in the presence of the homogenization buffer. The data were expressed as nmoles of TBARS/ mg of homogenate proteins.

DNA damage: The levels of DNA strand breaks were determined using the alkaline precipitation assay developed by [37] but using fluorescent-based DNA strands for detection. The assay principle is based on the precipitation of protein-containing genomic DNA from protein-free DNA strand breaks left in the supernatant. 200 μL of 2% SDS, 10 mM EDTA, 10 mM Tris and 40 mM NaOH and 200 μL of 0.12 M KCl were added to 25 μL of homogenate than mixed and heated at 60°C for 10 min. Samples were removed, kept at 4°C for 30 min and centrifuged at 8000 g for 5 min at 4°C. A 50 μL sample of the supernatant was added to a 96 well plate and DNA strands were detected using the Hoechst dye at 350 nm excitation and 450 nm emission. DNA quantification was measured with standard solutions of salmon sperm DNA. The data were expressed as μg DNA/ mg proteins in the homogenate.

Glutathione S-transferase activity: Glutathione S-transferase (GST) was measured using 1-chloro-2,4-dichloronitrobenzene (CDNB) as substrate. The reaction mixture containing 10 mM Hepes pH 6.5, 1mM GSH, 1mM 1-chloro-2-4-dinitrobenzene, 125 mM NaCl. The absorbance was measured at 340 nm and standard solutions of freshly prepared GSH were used for calibration. The results were expressed as formation of nmoles of GSH per minute per milligram protein (nmoles GSH/min/mg proteins).

Citrate synthase activity: Citrate synthase (CS) activity was determined using a spectrophotometric assay in the mitochondrial fraction according to the method described by [38] with slight modifications. Briefly, 10 μL of oxaloacetate (final concentration of 0.5 mM) was mixed to 10 μl of mitochondrial preparation and 180 μl of assay buffer (100 mM Tris–HCl pH 8, 0.4 mM Acetyl CoA, 0.2mM 5,5-dithio-bis-(2-nitrobenzoic acid)(DTNB). The assay was performed in a microplate reader for 30 min at 412 nm and 25°C. Activities were determined by the following formula: Δ absorbance /min/ (εx L (cm) x mg mitochondria); ε: is the extinction coefficient of DTNB at 412 nm (13.6 mM cm--1) and L: is the path length for absorbance.

Metallothionein-like proteins (MT): The level of MT like proteins was measured by the silver saturation assay [39] using the modification for non-radioactive silver [40]. A 50 μL sub-sample of the 10000 g supernatant was mixed with one volume of 0.2 M glycine buffer (pH 8.5) and Ag was added to obtain a final concentration of 2 mg/L at pH 8.5 in glycine buffer. After incubation for 15 min at room temperature, 25μL of 2.5% hemoglobin (Hb) was added and incubated for 5 min at 20oC. The sample was then heated at 100°C for 2 min and centrifuged at 10000g for 5 min to remove excess silver bound to denatured proteins and added hemoglobin. The last step was repeated once more to completely remove the excess (loosely bound) silver. The silver concentration in the supernatant was determined by atomic absorption spectrophotometry equipped with Zeeman effect background correction. A ratio of 12 moles of bound Ag to 1 mole of MT was assumed [41]. Results were expressed as nmoles of MT equivalents/ mg of proteins.

Data analysis

Data were expressed as mean ± standard error and normality was checked with Kolomogorov-Smirnov' test. Analysis of variance (ANOVA), was followed by a Fisher LSD post-hoc test to evaluate significant differences (*:$p<0.05$; **: $p<0.01$) between treated samples and controls. The Pearson correlation was carried out to examine the global response patterns of biomarkers and bioaccumulation data between control and exposed groups. All the statistical analyses were conducted with STATISTICA (version 7, statsoft Inc., 1995)

Results

Lanthanum bioavailability

At t=0, La concentrations in water were approximately 15-30% lower than expected in the 10-1250 μg/L range of test concentrations (table1). In the soft tissue, La contents determined after both times of exposure showed a concentration dependent significant increase compared with the control mussels in which La was detected at trace levels (table 1). The concentration of

Table 1: La concentrations measured by ICP MS in water at t = 0and in the soft tissue of zebra mussel exposed to LaCl₃ for 14 and 28 days. Variations of bio concentration factors of La in mussel were also reported. The data represent the mean ± standard error (n=8), *p* values were determined by ANOVA. Fisher LSD post-hoc test was used to identify significant differences between La treatments and control (Ctrl) (* *p <0.05*, ** *p <0.01* level).

La concentration (µg/L)	La biavailability in water (µg/L)	La biavailability in tissue (mg/Kg)		BAF (L/Kg)	
	t = 0	t = 14 days	t = 28 days	t =14 days	t = 28 days
0	0.0074 ±0.0005	0.048 ± 0.012	0.039 ± 0.001	6553,41	5275,75
10	8.68 ± 1.6	1.31 ± 0.17	1.81 ± 0.47	151,73	209,02
50	36.97 ± 5.13	5.19 ± 0.7	5.97 ± 0.79*	140,49	161,43
250	212.57 ± 54.65	21.46 ± 8.5*	15.58 ± 2.47**	100,96	73,30
1250	1042.77 ± 162.97	51.66 ± 6.26**		49,54	

metal accumulated in the tissue after exposure to the highest dose of La was determined only after 14 days of exposure as we have observed mortality in mussels after prolonged exposure of 28 days. The amount of La accumulated after 14 days of exposure to 1250 µg/L reached a value significantly higher than the control (*p<0.0001*), 51.65mg/Kg (wet/weight) and giving a La bioaccumulation factor (BAF) of 49 L/Kg. Whereas, no significant differences could be evidenced between groups exposed to the lower concentrations of La (10 and 50 µg/L) and the control (*p<0.05*).

After 28 days of exposure, results showed no significant differences between groups treated with the lowest dose and the control one (p> 0.05). However, La concentrations were 150 fold (*p<0.05*) and 400 fold higher (*p<0.0001*) than the control mussels for mussels exposed respectively to 50 and 250 µg/L. The BAFs for the mussels exposed to those concentrations were 161 and 73L/Kg respectively. Moreover, comparing to results obtained after 14 days of treatment, concentration of La detected in mussels treated with 250µg/L was lower.

Air-time survival in freshwater mussels

In order to understand the cumulative effects of chronic exposure to La and air exposure, the air-time survival test was determined. The survival of mussels was assessed by evaluating the capacity of mussels to maintain closed shells as it is known that bivalve shells remain closed during stressful situations. Data showed that air time survival was not affected in all groups treated with La by comparison to control (Figure 1A). Overall, mortality occurred after 4 days of exposure to air but after 6 days of exposure mussel shells remain closed. Our results also indicated that mortality in the group treated with the highest concentration of La for 14 days occurs with less body weight loss 30% by comparison to control group 41% (*p<0.05*) (Figure 1B). While after 28 days and for the same dose, body weight loss leading to mortality (31%) was higher than in control group (25%) suggesting that rapid weight loss precedes mortality events. For the other groups of mussels, the body weight loss was not significantly affected (*p>0.05*).

Biochemical effects

Biomarker results are reported in Figure 2. Oxidative damage was determined in mussel tissues by following changes in LPO.

After 14 days, La caused a significantly concentration dependent decrease in LPO levels (Figure 2A).Exposure to La 250 and 1250 µg/L induced a significant decrease of TBARS amount by 31% (*p<0.05*) and 46% (*p<0.001*) compared respectively to the concentration measured in control group. In contrast, a significant increase of LPO was found after 28 days of exposure to 1250 µg/L (*p<0.01*), with values 80% higher than the corresponding control. In contrast, no significant differences were evidenced between the other tested doses and control.

DNA strand breaks (Figure 2B) and MT (Figure 2D) levels in Dreissena polymorpha tissues showed no significant differences between all treatments (*p>0.05*) with respect to the control at either exposure times (14 and 28 days).The increase in LPO observed after 28 days was not correlated with DNA strand breaks(r = 0.2, *p>0.05*) (table 3) suggesting that LPO increase did not affect the DNA repair process and that La is not genotoxic. Our results, showed also that after both times of exposure, no significant variations in GST activity were observed at concentrations up to250 µg/L with respect to the control (*p>0.05*). However, a significant decrease in GST activity (Figure 2C) was observed at 1250 µg/L after14 days but not after 28 days of exposure. A correlation analysis revealed that GST activity after 28 days was correlated to DNA strand breaks (r = 0.4, *p< 0.05*) and MT level (r=0.54, *p<0.05*) (table 3), but no correlation was found after 14 days of exposure (table 2). The citrate synthase activity (Figure 2E) showed a significant increase with the lowest concentration of Lain respect to the control group *(p<0.05)*, after both times of exposure. In contrast, no significant differences were evidenced with the two highest concentrations tested but a trend to decrease could be noticed after 28 days of treatment.

Discussion

The toxicological understanding of REEs in the aquatic environment is very limited but of increasing concern. There are several potential mechanisms by which the REEs may exert toxicity on aquatic organisms; however the exact mode of action remains unknown. In order to elucidate the mechanism of action of La, we have determined its bioavailability in water and mussel tissues and examined the oxidative stress and antioxidant/detoxifying systems status in zebra mussel after chronic exposure.

This study revealed that the amount of La markedly increased in zebra mussel tissues. This finding was in accordance with

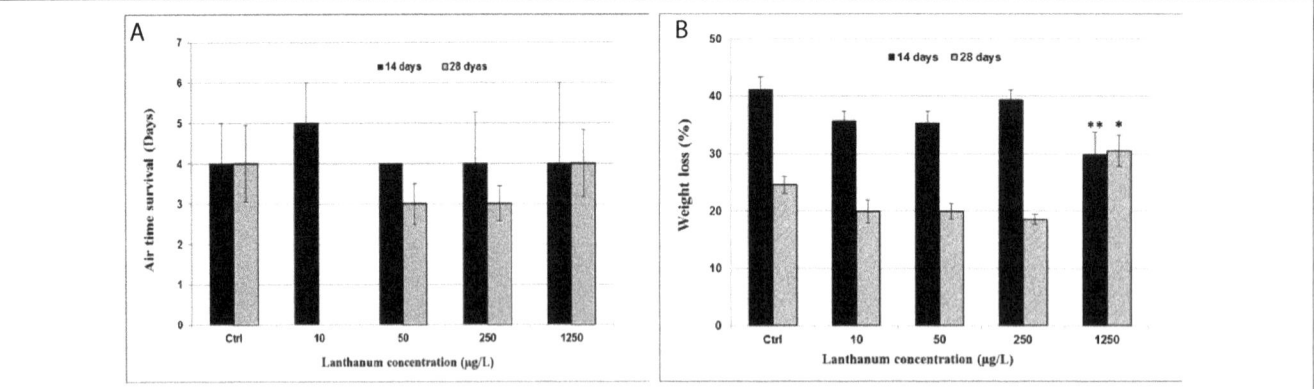

Figure 1: Evaluation of the air time survival (A) and weigh loss of mussels (B) after treatment with La and exposure to air for 7 days. The air-time survival represents the number of days required to die as evidenced by constant shell opening. Data represent the mean with the standard error (n=8). Pvalues were determined by ANOVA followed by a Fisher LSD post-hoc test to identify significant differences between La treatments and control (* $p <0.05$,** $p <0.01$).

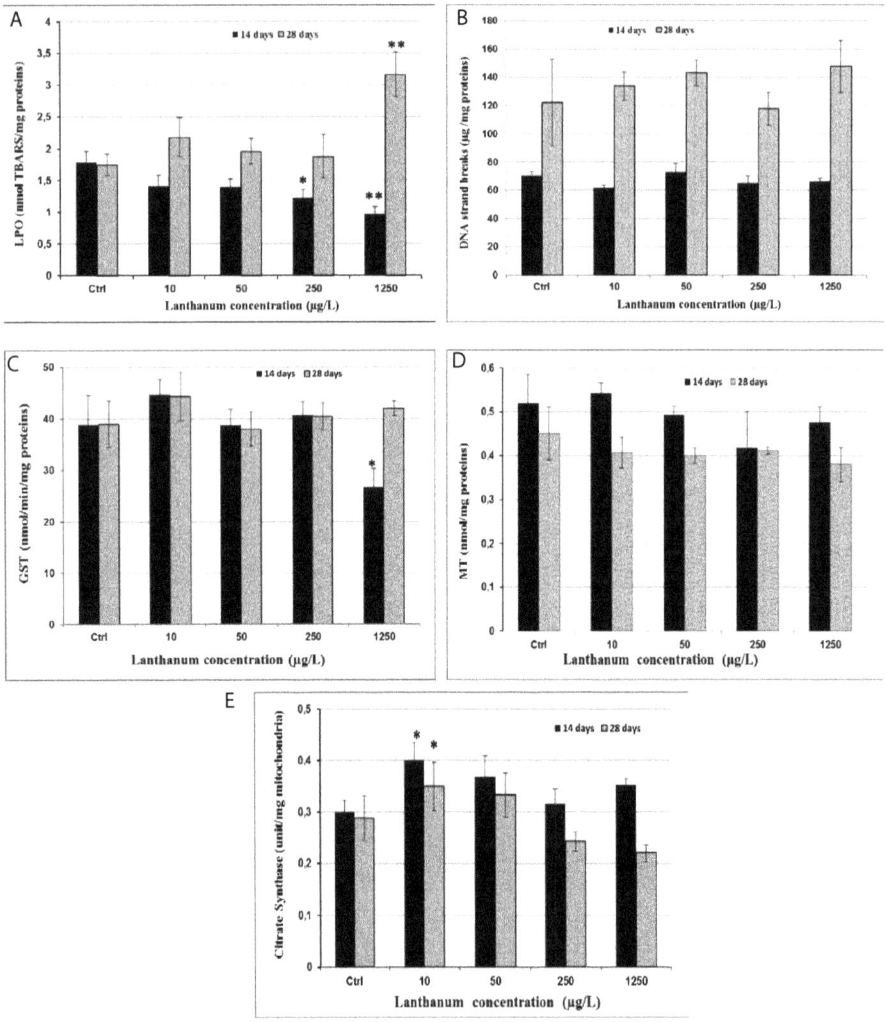

Figure 2: Biomarkers in zebra mussel tissue exposed for 14 and 28 days to different concentrations of LaCl₃. Lipid peroxidation amount (LPO: A), DNA strand breaks level (B), glutathione S-transferase activity (GST: C), metallothionein level (MT: D) and citrate synthase activity (E). Data represent the mean with the standard error (n=8). The asterisks indicate a significant difference from the controls (*$p<0.05$; **$p<0.01$, ANOVA followed by a Fisher LSD post-hoc test)..

Table 2: Pearson correlations calculated between biomarkers analyzed after 14 days of exposure (LPO = lipid peroxidation, APA= DNA strand breaks, GST = glutathione S-transferase, MT = metallothioneins, CS = citrate synthase) and La bioaccumulation (Bio) data.

	LPO	APA	Bio	GST	MT	CS	Weight loss
LPO	1						
APA	0.20	1					
Bio	-0.27	0.32	1				
GST	0.31	0.05	-0.17	1			
MT	0.06	0.24	-0.12	-1.80	1		
CS	-0.16	0.02	-0.01	-1.93	0.09	1	
Weight loss	0.22	0.08	-0.28	-0.02	0.09	0	1

Table 3: Pearson correlations calculated between biomarkers analyzed after 28 days of exposure (LPO = lipid peroxidation, APA= DNA strand breaks, GST = glutathione S-transferase, MT = metallothioneins, CS = citrate synthase) and La bioaccumulation (Bio) data. The asterisks represent a significant difference (*p<0.05; **p <0.01).

	LPO	APA	Bio	GST	MT	CS	Weight loss
LPO	1						
APA	0.18	1					
Bio	-0.26	0.16	1				
GST	0.02	0.40*	0.25	1			
MT	-0.23	0.24	-0.16	0.54**	1		
CS	0.04	0.09	-0.10	0.17	0.22	1	
Weight loss	-0.15	0.02	-0.16	0.05	0.24	-0.10	1

previous field study showing a higher bioaccumulation of La in these aquatic organisms [42]. However, the BAF value (32,000 L/Kg) reported by these authors was greater than that calculated in the current study. This could be explained by differences in the exposure method which may greatly affects the concentration and the bioavailability of La. Moreover, our results showed that BAFs decreased with increased concentration of La and no further accumulation of La was found in mussel exposed to 250µg/L for 28 days compared to those treated with the same dose for 14 days. This may be due to the maintenance of regulatory mechanisms that limit their accumulation (e.g., excretory pathways and/or uptake limitation) [43].

According to biomarkers responses, no tissue damage was observed after 14 days of exposure to La as revealed by LPO levels. However, at the end of 28 days of exposure, this element induced significant increase of LPO suggesting the involvement of the reactive oxygen species (ROS) in the mechanism of La toxicity. Our finding was consistent with previous studies showing that LaCl$_3$ exposure promoted ROS production in the mouse lung [44] and in the brains of the mice [45], which in turn resulted in peroxidation of lipids. However, this also was in agreement with studies indicating that LaCl3 plays a protective role by inhibiting H$_2$O$_2$ induced elevation in ROS level in rat calcifying vascular cells [46]. Thus, it appears that La may have two kinds

of action, an antioxidant and a prooxidant effect depending on the concentration and the duration of exposure. Paiva et al. (2009) indicated that pretreatment of Jurkat cells with Trolox, a known free radical scavenger, leads to a significant decrease of DNA damages in cells incubated with La (NO3)3. Therefore, they suggested that oxidative stress may be involved in the genotoxic process induced by La. However, no genotoxic effect of La could be evidenced in our work after both times of exposure. This was in agreement with studies performed by [47], showing that RE ions at concentration of 1000 µM did not induce genotoxicity. ROS scavenging by LaCl$_3$ observed after 14 days could in turn explain the absence of DNA strand breaks. Indeed, low concentrations of antioxidants drive ROS levels to an optimal "physiological range", which reduces oxidative stress induced DNA damage without impairing the DNA repair system [48].

An antioxidant effect played by cerium was also reported in the literature and authors suggested that this lanthanide could scavenge ROS and act as a catalyst miming and improving the activity of SOD and catalase [49, 50]. The role of La in suppressing H$_2$O$_2$'s effects might be played through calcium signaling as it acts as a Ca^{2+} antagonist [51] and a calcium channel blocker [46]. Indeed, lanthanides can compete with Ca^{2+} in ion and channel binding sites due to their similar properties to Ca^{2+}, since both ions have similar ionic radii but not the same charge. The involvement of calcium in the mechanism of action of LaCl$_3$ is controversial because other studies indicated that lanthanides bound to Ca^{2+} binding proteins decrease the Ca content into the cell and generate an excess of extracellular calcium, leading to oxidative stress production [3, 45]. It was also reported that blocking of intracellular Ca^{2+} influx by GdCl$_3$ induced suppression of GST gene expression [52]. Pre- and post-treatment of freshwater catfish exposed to fluoride with ascorbic acid, which exhibit good antioxidant and free-radical scavenging activities, lead to the decrease of LPO and GST level in the liver and ovary tissues [53]. Thus, the decreased GST activity observed with the highest dose could be explained by gene down regulation resulting from decreased H$_2$O$_2$ contents and the imbalance of calcium level in mussel's tissue. Taken together, it is possible that oxidative stress is mediated by calcium in zebra mussels which is calcium sensitive species for growth and shell generation.

It is well known that MT induction is enhanced by an increase in the availability of divalent essential metals (Cu and Zn) and non-essential metals (Cd, Ag and Hg). In the present study, La bioaccumulation in the tissue of zebra mussel did not trigger MT induction which suggests that La remains trivalent or is strongly bound to other sites. This result is not consistent with those obtained for Tetrahymena thermophila cells showing that La induced the expression of MT [54]. MT induction was also reported by Kobayashi et al. [55] after administration of cerium chloride to mice. In those studies analysis of MT induction was performed under short exposure times to lanthanides. Amirad et al [56]. Indicated that the time periods of contamination affect the induction of MT and Barka et al. [57] confirmed that in the copepod T. brevicornis, cadmium induced MT on the first day of exposure, but not in the following days. Thus, it is possible that

MT induction occurred after short time of exposure to La but it did not play a main role in the detoxification or homeostasis of this trivalent metal after chronic exposure.

CS activity is one of the key regulatory enzymes in the tricarboxylic acid (TCA) cycle [58] and it was extensively used as a metabolic marker in assessing oxidative, mitochondrial abundance and respiratory capacity [59]. Our results showed that low concentration of La enhanced citrate synthase activity which indicated that this lanthanide increase the TCA cycle. A recent study showed that lead exposure caused a substantial increase in the ADP/ATP ratio in the hippocampus of rats and this change is restored by CeO_2 and or Y_2O_3 [60]. They suggested that increase of ATP can be through decrease of oxidative stress and thereby improvement of mitochondrial function and respiratory chain. However, previous studies have indicated that decreases in the activities of CS induced oxidative stress [61, 62]. Those findings are in accordance with our results because with the highest concentration, citrate synthase activity tends to decrease at the end of exposure time. The depletion of citrate synthase activity could be explained by the formation of a complex with certain carboxylic acids as suggested for Gd^{3+} [63].

Our results showed that mussel exposed to prolonged periods to air did not show significant decrease in air-time survival and mussel weight in groups exposed to air after treatment with La compared to a control group. Exposure of freshwater mussels to prolonged periods of air leads to important physiological changes in the attempt to survive outside of water [64]. During this period, mussels expend metabolic energy under low oxygen tension, leading to increased anaerobic metabolism and loss of energy reserves [64]. Occurrence of anaerobic glycolysis was also reported when bivalves are exposed to environment pollutant. It was indicted that phoxim poisoning was reduced in larvae of B. mori via deriving some energy from anaerobic glycolysis. However, added $CeCl_3$ lead to decreased phoxim toxicity by improving TCA cycle, meeting the required energy demands and therefore increase survival rate of larvae under phoxim toxicity [65]. Thus, we suggested that $LaCl_3$ may contribute to the resistance of mussel to air exposure by suppressing anaerobic metabolism and improving TCA cycle which lead to increasing energy supply and survival rate.

Conclusion

Investigations on the bioaccumulation and biochemical effects of lanthanides can be of particular interest to elucidate their effects on aquatic organisms. Data showed that La concentration markedly increased in mussel tissues. However, no further accumulation of La was detected after 28 days of exposure, suggesting the existence of regulatory mechanism that limits its accumulation. Based on the multi biomarker approach, the presence of La in the environment does not seem to present a risk to aquatic ecosystems except at some hotspots or for peak concentrations. Indeed, no strong adverse effects of La were observed after 14 days of exposure as it acts as an antioxidant. By contrast, a prooxidant effect was evidenced after 28 days in mussels exposed to the highest concentration. However,

this study was conducted under controlled condition and the possibility of cumulative effects in more complex environmental conditions could not be excluded. Therefore, this work confirms the relevance of studying the effect of this lanthanide under chronic exposure but complementary studies should be performed under different scenarios to estimate potential future risks in a long term perspective. Investigations of the different pathways involved in the mechanism of action of this lanthanide, particularly those related to calcium are required for better understanding of its effects. Furthermore, co-exposure with pro-oxidant substances should be conducted to confirm the potential antioxidant properties of this lanthanide observed after 14 days.

Acknowledgments

We thank the Natural Sciences and Engineering Research Council of Canada (NSERC) and the Chemical Management Plan (CMP) of Environment Canada for their financial support.

Reference

1. Gonzalez V, Vignati DAL, Leyval C, Giamberini L. Environmental fate and ecotoxicity of lanthanides: Are they a uniform group beyond chemistry? Environ Int. 2014;71:148-157. doi: 10.1016/j.envint.2014.06.019

2. Du X, Graedel TE. Uncovering the Global Life Cycles of the Rare Earth Elements. Sci Rep. 2011;1:145. doi:10.1038/srep00145

3. Wu J, Yang J, Liu Q, Wu S, Ma H, Cai Y. Lanthanum Induced Primary Neuronal Apoptosis Through Mitochondrial Dysfunction Modulated by Ca2+ and Bcl-2 Family. Biol Trace Elem Res. 2013;152(1):125-134. doi: 10.1007/s12011-013-9601-3

4. González V, Vignati DAL, Pons M-N, Montarges-Pelletier E, Bojic C,Giamberini L. Lanthanide ecotoxicity: First attempt to measure environmental risk for aquatic organisms. Environ Pollut. 2015;199:139-147. doi: 10.1016/j.envpol.2015.01.020

5. Kulaksız S, Bau M. Rare earth elements in the Rhine River, Germany: First case of anthropogenic lanthanum as a dissolved micro contaminant in the hydrosphere. Environ Int. 2011;37(5):973-979. doi: 10.1016/j.envint.2011.02.018

6. Li N, Wang S, Liu J, Ma L, Duan Y, Hong F. The Oxidative Damage in Lung of Mice Caused By Lanthanoide. Biol. Trace Elem. Res. 2010; 134(1): 68-78. doi: 10.1007/s12011-009-8448-0

7. Paiva AV, de Oliveira MS, Yunes SN, de Oliveira LG, Cabral-Neto JB, de Almeida CE. Effects of Lanthanum on Human Lymphocytes Viability and DNA Strand Break. Bull. Environ. Contam. Toxicol. 2009; 82(4):423-427. doi: 10.1007/s00128-008-9596-1

8. Yang J, Liu Q, Qi M, Lu S, Wu S, Xi Q, et al. Lanthanum chloride promotes mitochondrial apoptotic pathway in primary cultured rat astrocytes. Environ Toxicol. 2013;28(9):489-497. doi: 10.1002/tox.20738

9. Pałasz A, Czekaj P. Toxicological and cytophysiological aspects of lanthanides action. Acta Biochim Pol. 2000;47(4):1107-1114.

10. Feng L, Xiao H, He X, Li Z, Li F, Liu N, et al. Neurotoxicological consequence of long-term exposure to lanthanum. Toxicol Lett. 2006;165(2):112-120. doi: 10.1016/j.toxlet.2006.02.003

11. Herrmann H, Nolde J, Berger S, Heise S. Aquatic ecotoxicity of lanthanum – A review and an attempt to derive water and sediment quality criteria. Ecotoxicol Environ Saf. 2016;124:213-238. doi: 10.1016/j.ecoenv.2015.09.033

12. Das T, Sharma A, Talukder G. Effects of lanthanum in cellular systems. A review. Biol Trace Elem Res. 1988;18:201-228.

13. Tomlinson G, Mutus B, McLennan I, Mooibroek MJ. Activation and inactivation of purified acetylcholinesterase from Electrophorus electricus by lanthanum (III). Biochim Biophys Acta. 1982; 703(2): 142-148.

14. Tokimasa T, North RA. Effects of barium, lanthanum and gadolinium on endogenous chloride and potassium currents in Xenopus oocytes. J Physiol. 1996;496(Pt 3):677-686.

15. Hardy MA, Balsam P, Bourgoignie JJ. Reversible inhibition by lanthanum of the hydrosmotic response to serosal hypertonicity in toad urinary bladder. J Membr Biol. 1979;48(1): 13-19.

16. Mellanby J, Thompson PA. The interaction of tetanus toxin and lanthanum at the neuromuscular junction in the goldfish. Toxicon. 1981; 19(4):547-554.

17. Miledi R, Molenaar PC, Polak RL. The effect of lanthanum ions on acetylcholine in frog muscle. J Physiol. 1980;309:199-214.

18. Mellanby J, Beaumont MA, Thompson PA. The effect of lanthanum on nerve terminals in goldfish muscle after paralysis with tetanus toxin. Neuroscience. 1988; 25(3):1095-1106.

19. Cui J ZZ, Bai W, Zhang L, He X, Ma Y, Liu Y, et al. Effects of rare earth elements La and Yb on the morphological and functional development of zebrafish embryos. J Environ Sci (China). 2012;24(2): 209-213.

20. Barry MJ, Meehan BJ. The acute and chronic toxicity of lanthanum to Daphnia carinata. Chemosphere. 2000;41(10): 1669-1674.

21. Bogers M. Daphnia Magna, reproduction test with Lanthanum (La). Report No.: 139499. Testing Laboratory: NOTOX. B. V.'s-Hertogenbosch, The netherlands. Owner company : Kemira Pernis B. V, 's-Hertogenbosch, Rotterdam. 1995.

22. Stauber JL, Binet MT. Canning River Phoslock Field Trial – Ecotoxicity Testing Final Report, Report No: ET317R, CSIRO Centre for Advanced Analytical Chemistry Energy Technology, Australia 2000.

23. Stefano GB, Brogan JJ, Aiello E, Hiripi L. Lanthanum blockade of serotonin release from the branchial nerve of the mussel Mytilus edulis. J Exp Zool. 1980;214(1):21-26.

24. Muneoka Y, Twarog BM. Lanthanum block of contraction and of relaxation in response to serotonin and dopamine in molluscan catch muscle. J Pharmacol Exp Ther. 1977;202(3):601-609.

25. Brink PR, Kensler RW, Dewey MM. The effect of lanthanum on the nexus of the anterior byssus retractor muscle of Mytilus edulis L. Am J Anat. 1979;154(1):11-26.

26. Valcheva-Traykova M, Saso L, Kostova I. Involvement of lanthanides in the free radicals homeostasis. Curr Top Med Chem. 2014; 14(22):2508-2519.

27. He X, Zhang Z, Zhang H, Zhao Y, Chai Z. Neurotoxicological Evaluation of Long-Term Lanthanum Chloride Exposure in Rats. Toxicological Sciences. 2008;103(2):354-361. DOI: 10.1093/toxsci/kfn046

28. Che Y, Cui Y, Jiang X. Effects of Lanthanum Chloride Administration in Prenatal Stage on One-Trial Passive Avoidance Learning in Chicks. Biol Trace Elem Res. 2008;127(1):37-44. DOI: 10.1007/s12011-008-8225-5

29. Liu Y, Chen D, Chen A, et al. Study on lanthanum deposit in liver of rats chronically exposed to lanthanum nitrate at low dose. Jou of Health Toxicol. 2003;17(4):203-205.

30. Huang P, Li J, Zhang S, Chen C, Han Y, Liu N, et al. Effects of lanthanum, cerium, and neodymium on the nuclei and mitochondria of hepatocytes: Accumulation and oxidative damage. Environ Toxicol Pharmacol. 2011; 31(1):25-32. DOI: 10.1016/j.etap.2010.09.001

31. Wu M, Gao J-L, Sun M-X, Zhang Y-Z, Liu Y,Dai J. Effects of La(III) and Ca(II) on Isolated Carassius auratus Liver Mitochondria: Heat Production and Mitochondrial Permeability Transition. Biol Trace Elem Res. 2015;163(1-2):217-223. doi: 10.1007/s12011-014-0178-2

32. Gagné F. Biochemical ecotoxicology: principles and methods. Book. 2014: p, 257.

33. Vasseur P, Cossu-Leguille C. Biomarkers and community indices as complementary tools for environmental safety. Environ Int. 2003;28(8):711-7. doi: 10.1016/S0160-4120(02)00116-2

34. Burgeot T,Gagné F. Contaminant exposure and ecotoxicological impacts in estuaries. Environmental Science and Pollution Research. Environ Sci Pollut Res. 2013;20(2):599-600. doi: 10.1007/s11356-012-1324-z

35. Binelli A, Della Torre C, Magni S, Parolini M. Does zebra mussel (Dreissena polymorpha) represent the freshwater counterpart of Mytilus in ecotoxicological studies? A critical review. Environmental Pollution. 2015;196:386-403. doi: 10.1016/j.envpol.2014.10.023

36. Bradford MM. A rapid and sensitive method for the quantitation of microgram quantities of protein utilizing the principle of protein-dye binding. Anal Biochem. 1976;72:248-254.

37. Olive PL. DNA precipitation assay: a rapid and simple method for detecting DNA damage in mammalian cells. Environ Mol Mutagen. 1988;11(4):487-495.

38. Dudognon T, Lambert C, Quere C, Auffret M, Soudant P,Kraffe E. Mitochondrial activity, hemocyte parameters and lipid composition modulation by dietary conditioning in the Pacific oyster Crassostrea gigas. J Comp Physiol B. 2014; 184(3):303-317.doi: 10.1007/s00360-013-0800-1

39. Scheuhammer AM,Cherian MG. Quantification of metallothioneins by a silver-saturation method. Toxicol Appl Pharmacol. 1986;82(3):417-425.

40. Gagné F, Marion M,Denizeau F. Metal homeostasis and metallothionein induction in rainbow trout hepatocytes exposed to cadmium. Fundam. Appl. Toxicol. 1990;14(2):429-437. doi: 10.1016/0272-0590(90)90221-5

41. Kille P, Hemmings A,Lunney EA. Memories of metallothionein. Biochim Biophys Acta. 1994;1205(2):151-161.

42. Weltje L, Heidenreich H, Zhu W, Hubert Th Wolterbeek, Siegfried Korhammer, Jeroen J M de Goeij, et al. Lanthanide concentrations in freshwater plants and molluscs, related to those in surface water, pour water and sediment. A case study in The Netherlands. Sci. Total Environ. 2002;286(1–3):191-214. doi: 10.1016/S0048-9697(01)00978-0

43. Rainbow PS. Trace metal concentrations in aquatic invertebrates: Why and so what? Environ. Pollut. 2002;120(3): 497-507. doi: 10.1016/s0269-7491(02)00238-5

44. Hong J, Pan X, Zhao X, Yu X, Sang X, Sheng L, et al. Molecular mechanism of oxidative damage of lung in mice following exposure to lanthanum chloride. Environ. Toxicol. 2015;30(3):357-365. doi: 10.1002/tox.21913

45. Zhao H, Cheng Z, Hu R, Chen J, Hong M, Zhou M, et al. Oxidative Injury in the Brain of mice Caused by Lanthanid. Biol. Trace Elem. Res. 2011;142(2):174-189. doi: 10.1007/s12011-010-8759-1

46. Shi Y, Gou B-D, Shi Y-L, Zhang T-L,Wang K. Lanthanum chloride suppresses hydrogen peroxide-enhanced calcification in rat calcifying vascular cells. BioMetals. 2009;22(2):317-327. doi: 10.1007/s10534-008-9168-1

47. Wakabayashi T, Ymamoto A, Kazaana A, Nakano Y, Nojiri Y, Kashiwazaki M. Antibacterial, Antifungal and Nematicidal Activities of Rare Earth Ions. Biol. Trace Elem. Res. 2016;174(2):464-470. doi: 10.1007/s12011-016-0727-y

48. Li T-S, Marbán E. Physiological levels of reactive oxygen species are required to maintain genomic stability in stem cells. Stem cells (Dayton, Ohio). 2010;28(7):1178-1185. doi: 10.1002/stem.438

49. Ling Q, Hong F. Antioxidative role of cerium against the toxicity of lead in the liver of silver crucian carp. Fish Physiol. Biochem. 2010;36(3):367-376. doi: 10.1007/s10695-008-9301-7

50. Liu J, Ma L, Yin S, Hong F. Effects of Ce3+ on Conformation and Activity of Superoxide Dismutase. Biol. Trace Elem. Res. 2008;125(2):170-178. doi: 10.1007/s12011-008-8165-0

51. Dong S, Zhao Y, Liu H, Yang X, Wang K. Duality of effect of La3+ on mitochondrial permeability transition pore depending on the concentration. BioMetals. 2009;22(6):917-926. doi: 10.1007/s10534-009-9244-1

52. Kim SG, Choi SH. Gadolinium Chloride Inhibition of Rat Hepatic Microsomal Epoxide Hydrolase and Glutathione S-Transferase Gene Expression Drug Metab Dispos. 1997;25(12):1416-1423.

53. Yadav SS, Kumar R, Khare P, Tripathi M. Oxidative Stress Biomarkers in the Freshwater Fish, Heteropneustes fossilis (Bloch) Exposed to Sodium Fluoride: Antioxidant Defense and Role of Ascorbic Acid. Toxicol. Int. 2015;22(1):71-76. doi: 10.4103/0971-6580.172261

54. Wang Q, Xu J, Zhu Y, Chai B, Liang A, Wang W. Lanthanum(III) Impacts on Metallothionein MTT1 and MTT2 from Tetrahymena thermophila. Biol. Trace Elem. Res. 2011;143(3):1808-1818. doi: 10.1007/s12011-011-9004-2

55. Kobayashi K, Shida R, Hasegawa T, et al. Induction of Hepatic Metallothionein by Trivalent Cerium: Role of Interleukin 6. Biological and Pharmaceutical Bulletin. 2005; 28(10): 1859-63. doi: 10.1248/bpb.28.1859.

56. Amiard JC, Amiard-Triquet C, Barka S, Pellerin J, Rainbow PS. Metallothioneins in aquatic invertebrates: Their role in metal detoxificationand their use as biomarkers. Aquat Toxicol. 2006;76(2):160-202. doi: 10.1016/j.aquatox.2005.08.015

57. Barka S, Pavillon JF, Amiard JC. Influence of different essential and non-essential metals on MTLP levels in the Copepod Tigriopus brevicornis. Comparative Biochemistry and Physiology - C Toxicology and Pharmacology. 2001; 128(4):479-493. doi: 10.1016/s1532-0456(00)00198-8

58. Siu PM, Donley DA, Bryner RW, Alway SE. Citrate synthase expression and enzyme activity after endurance training in cardiac and skeletal muscles. J. Appl. Physiol. 2003;94(2):555-560. doi:10.1152/japplphysiol.00821.2002

59. Rooyackers OE, Adey DB, Ades PA, Nair KS. Effect of age on in vivo rates of mitochondrial protein synthesis in human skeletal muscle. Proceedings of the National Academy of Sciences of the United States of America. Proc Natl Acad Sci U S A. 1996; 93(26):15364-15369.

60. Hosseini A, Sharifi MA, Abdollahi M, Najafi R, Baeeri M, Rayegan S, et al. Cerium and Yttrium Oxide Nanoparticles Against Lead-Induced Oxidative Stress and Apoptosis in Rat Hippocampus. Biol Trace Ele Res. 2015;164(1):80-89. doi: 10.1007/s12011-014-0197-z

61. Ježek P, Hlavatá L. Mitochondria in homeostasis of reactive oxygen species in cell, tissues, and organism. The International Journal of Biochemistry & Cell Biology. 2005;37(12):2478-2503. doi: 10.1016/j.biocel.2005.05.013

62. Feldstein AE, Werneburg NW, Canbay A, Guicciardi ME, Bronk SF, Rydzewski R, et al. Free fatty acids promote hepatic lipotoxicity by stimulating TNF-α expression via a lysosomal pathway. Hepatology. 2004;40(1):185-94. doi: 10.1002/hep.20283

63. Riri M, HM, Kamal O, Eljaddi T, Benjjar A, Hlaïbi M. New gadolinium(III) complexes with simple organic acids (Oxalic, Glycolic and Malic Acid). J Mater Environ Sci. 2011;2(3):303-308.

64. Gagne F, Auclair J, Peyrot C,Wilkinson KJ. The influence of zinc chloride and zinc oxide nanoparticles on air-time survival in freshwater mussels. Comp Biochem Physiol C Toxicol Pharmacol. 2015;173:36-44.

65. Li B, Xie Y, Cheng Z, Cheng J, Hu R, Sang X, et al. Cerium Chloride Improves Protein and Carbohydrate Metabolism of Fifth-Instar Larvae of Bombyx mori Under Phoxim Toxicity. Biol Trace Elem. Res. 2012;150(1):214-220. doi: 10.1007/s12011-012-9465-ys

12

Thermodynamic Studies on the Interaction between Phenylalanine with Some Divalent Metal Ions in Water and Water-Dioxane Mixtures

Ebrahim Ghiamati* and Samieh Oliaei

Chemistry Department, University of Birjand, Birjand, South Khorasan, Iran

Corresponding author: Ebrahim Ghiamati, Chemistry Department, University of Birjand, Birjand, South Khorasan, Iran,
E-mail : eghiamati@birjand.ac.ir

Abstract

A new and simple method was developed to determine the stability constants of phenylalanine complexes of Co (II), Ni (II), Cu(II), Zn (II) and Pb (II) metal ions in water and water-dioxane mixtures at four different temperatures of 25, 37, 45 and 55°C potentiometrically using modified Bjerrum method. Ionic strength of medium was retained at 0.10 M by sodium nitrate. Our results revealed that the stability constant values are greater in water-dioxane mixtures than in water alone. The increasing trend in stability constant values in water and mixture of water-dioxane are the same as follows:

$$K_{f\,Co\,(II)\text{-Phe}} < K_{f\,Zn\,(II)\text{-Phe}} < K_{f\,pb\,(II)\text{-Phe}} < K_{f\,Ni\,(II)\text{-Phe}} < K_{f\,Cu\,(II)\text{-Phe}}$$

Furthermore, by knowing the stability constants at different temperatures, thermodynamic parameters of $\Delta H°$, $\Delta S°$ and $\Delta G°$ for the respective complexes were acquired. $\Delta H°$, $\Delta S°$ values were positive. Negative $\Delta G°$ values conveyed the spontaneity of the complex formation process. Also it is found out that the stability constant of the pertinent complexes increases as the temperature rises meaning that the reactions are endothermic.

Keywords: Stability constant, Amino acid complex, Potentiometric titrations, Thermodynamic parameters.

Introduction

The amino acids have special importance among the other chemical groups since they are building block of proteins. The interactions between metal ions and amino acids have attracted the attention of many biochemists, because they can be used as a model for metal-protein reactions mimicking metal-enzyme mechanism. The explanation of these phenomena in the biological systems requires the determination of the stability constants as a measure of how well the complex of the amino acids with various metal ions in a medium similar to those of biological systems forms.

Among various methods for determining stability constants of complexes, potentiometry has its own advantages. Potentiometric titration of amino acids in the presents of metal ions is generally used as a method for measuring metal complex stability constants. This technique first described by Bjerrum [1] and has been investigated extensively by numerous researches [2-7]. D.J. Perkins examined amino acid structures on the stabilities of complexes formed with metals of group II [8]. A. E. Martell and coworkers have conducted vital studies on amino acid complexes and predicted their stability constants [9-11].

The behavior of the complexes at different temperatures was probed by M.S. Masoud et al. [12]. Thermodynamic parameters for the formation of glycine with metal ions were investigated by S. Sammartano [13]. Formations of binary and ternary complexes were studied by M.M. Shoukry et al. [14]. Cu (II) amino acids complexes are useful antibacterial agents [15]. The stability constants of copper (II) complexes with several amino acids were calculated in dioxane-water mixtures by A. Dogan et al. [16].

The stability of binary complexes of L-aspartic acid in dioxane-water mixture was probed by R.S. Rani et al. [17]. H. Demirelli, et al. have determined the formation constants of phenylalanine complexes of Ni(II), Cu(II), and Zn(II) in water media at 25°C and $\mu = 0.1$ mol L^{-1} KCl [18]. A.A. Mohamed et al. [19] have measured stability constants and thermodynamic parameters for glycine and L-threonine complexes with some rare metal ions in water. The interactions of L- glutamic and L-aspartic acid with some metal ions has been probed by S.A.A Sajadi [20]. Critical survey of formation constants of phenylalanine with metal ions has been reported by L.D. Pettit [21]. A. Eid Fazary, et al. have investigated the protonation equilibria of α- amino acids in water and dioxane mixtures [22]. The stability constants of Ni (II) with some amino acids were probed by N. Turkel [23].

Phenylalanine is a one of the few amino acids that can directly affect brain chemistry by crossing the blood-brain barrier. Phenylalanine is used to cure depression, attention deficit-hyperactivity disorder (ADHD), Parkinson's disease, chronic pain, osteoarthritis, rheumatoid arthritis, alcohol withdrawal symptoms, and a skin disease called vitiligo [24].

In this work, the stability constants of phenylalanine complexes of some divalent metal ions in water and water-dioxane mixtures at four different temperatures have been evaluated. In addition thermodynamic parameters of pertinent complexes have been determined.

Experimental Section

a. Materials and procedure

Phenylalanine with purity of 99%, the nitrate salts of Co(II), Cu(II), Zn(II), Ni(II) and Pb(II) (all pro-analysis), nitric acid (HNO₃), sodium hydroxide (NaOH), hydrochloric acid (HCl), perchloric acid (HClO₄) and sodium nitrate (NaNO₃) all were purchased from Merck and used as received. Deionized water was employed in all of the experiments. The pH potentiometric titrations were performed using Schott pH meter, Thermostat MLW16, glass cell, digital burette, and magnetic stirrer.

A special glass vessel (reactor) for potentiometric titrations was made which had a double wall with entries for combined glass electrode, nitrogen, and base from burette. Temperature inside the reactor was kept constant through circulation of water with an accuracy of ±0.1°C. A 25.00 mL solution mixture prepared so that it was 5.000×10^{-3} M with respect to phenylalanine, 3.000×10^{-3} M with respect to the respective metal ions and 1.690×10^{-2} M with respect to HClO₄. A sufficient amount of 0.10 M NaNO₃ was added to adjust the ionic strength. The solution was thermostatted to desired temperatures of 25, 37, 45 and 55°C and then titrated with an accurately standardized NaOH solution while the titrand constantly was purged. The pH was recorded after each addition of titrant in 0.050 mL increments. The two electrodes used for measuring pH were glass electrode and calomel electrode. The pH meter was calibrated using Merck standard buffer solutions with pH of 4.0, 7.0 and 9.0.

b. Calibration of the Glass Electrode

Calibration of the combined glass electrode and calomel electrode was performed in both acidic and alkaline regions by titrating a solution of 0.01 molL⁻¹ hydrochloric acid with standard sodium hydroxide prior to each titration to read the hydrogen ion concentration directly. The emf values (E) depend on [H⁺] according to $E = E_0 + slog[H^+] + J_H[H^+] + J_{OH}[OH^-]$ where J_H and J_{OH} are fitting parameters in acidic and alkaline media in order to correct experimental errors. These errors arise mainly from the liquid junction and the alkaline and acidic errors of the glass electrode [25].

c. The Method for determination of stability constant

The Bjerrum's pH titration procedure assumes the presence of the reacting species H_2L^+ as amino acid, HL as the monoprotonated amino acid, and L^-

The anion of amino acid

$$H_2L^+ \rightleftharpoons HL_{(aq)} + H^+_{(aq)} \tag{1}$$

$$K_{a_1} = \frac{[HL][H^+]}{[H_2L^+]} \tag{2}$$

$$HL_{(aq)} \rightleftharpoons L^-_{(aq)} + H^+_{(aq)} \tag{3}$$

$$K_{a_2} = \frac{[L^-][H^+]}{[HL]} \tag{4}$$

$$M^{2+} + HL \rightleftharpoons ML^+ + H^+ \tag{5}$$

$$K_{f1} = \frac{[ML^+][H^+]}{[M^{2+}][HL]} \tag{6}$$

$$ML^+ + L^- \rightleftharpoons ML_2 \tag{7}$$

$$K_{f2} = \frac{[ML_2]}{[ML^+][L^-]} \tag{8}$$

Here K_{f1} and K_{f2} is the first and the second stability constants of the complexes. We define $\bar{n}$ as:

$$\bar{n} = \frac{\text{\# of bond ligands}}{\text{total metal ion concentration}} = \frac{L_{bound}}{C_M} = \frac{L_{total} - L_{free}}{C_M} \tag{9}$$

The concentration of free ligand is the sum of concentration of contained ligand species at different form, i.e.

$$L_{free} = [H_2L] + [HL] + [L^-] \tag{10}$$

The bound ligand concentration (L_{bound}) could then be estimated as:

$$L_{bound} = L_{total} - L_{free} \tag{11}$$

After rearrangement and substitutions we have:

$$\bar{n} = \frac{T_{H_2L^+} - [H_2L^+] - [HL] - [L^-]}{T_{M^{2+}}} \tag{12}$$

Then: $$\bar{n} = \frac{[ML^+] + 2[ML_2]}{[M^{2+}] + [ML^+] + [ML_2]} \tag{13}$$

According to mass balance relation we have:

$$T_M = [M^{2+}] + [ML^+] + [ML_2] \tag{14}$$

$$T_{HL} = [HL] + [L^-] + [ML^+] + 2[ML_2] \tag{15}$$

$$\left[ClO_4^- \right] = T_{HClO4} + 2\,T_M \tag{16}$$

$$[ML^+] + 2[ML_2] = [Na^+] - T_{HClO_4} + [H^+] \tag{17}$$

$$\bar{n} = \frac{[Na^+] - [HClO_4] + [H^+]}{T_M} \tag{18}$$

$$[HL] = \frac{K_a(T_{H_2L^+} - \bar{n}T_M)}{k_a + [H^+]} \tag{19}$$

From plot of p_{HL} versus $\bar{n}$ the stability constants could be calculated.

$$K_{f1} = \frac{1}{[HL]_{\bar{n}=\frac{1}{2}}} \tag{20}$$

$$K_{f2} = \frac{1}{[HL]_{\bar{n}=\frac{3}{2}}} \tag{21}$$

All our calculations in this work were executed by GRCβeta computer-program developed in our lab. The software asks for a) initial volume of solution containing the amino acid, metal ion, and perchloric acid, b) the concentration of perchloric acid, c) the concentration of sodium hydroxide, d) the concentration of amino acid, and e) pKa$_1$ and pKa$_2$ of the amino acid in the specified medium and at desired ionic strength which we found them in literature. After insertion of the pertinent values, the software plots calculated pH (corrected pH) of the titrand solution versus the concentration of added standardized NaOH, plus drawing two curves, one for a $\bar{n} = 0.5$ and the other for $\bar{n} = 1.5$. The intersection of the potentiometric titration curve with these two curves produces two points (Figure 1) whose corresponding pHs will be used to evaluate the respective stability constants of the metallic ion-amino acid complexes. Additionally the software

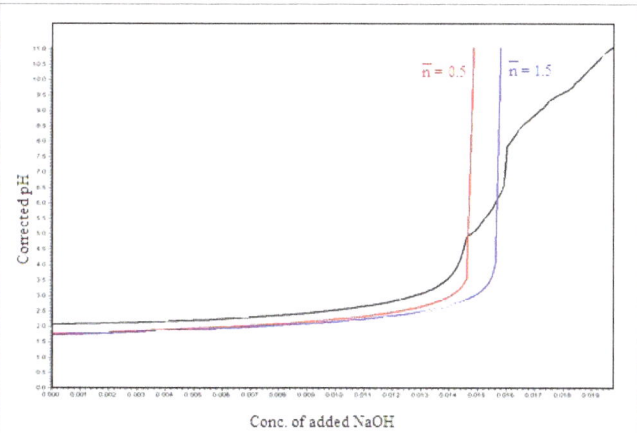

Figure 1: Plot of pH versus concentration of added standardized NaOH for Cu(II)-Phe complex in (70-30) % water - dioxane mixture solution at 25°C

is capable of plotting first and second derivative of d-pH versus d-V$_{NAOH}$ to clarify the end points. For each potentiometric titration approximately 4-7 mL of standardized sodium hydroxide was used.

Thermodynamic calculations were conducted as follows:

The Gibb's free energy change, ΔG°, can be calculated from the equation below:

$$\Delta G^0 = -RT \ln K_f \tag{22}$$

$$\ln K_f = \frac{-\Delta G^0}{RT} \tag{23}$$

By taking the derivative with respect to 1/T from both side of equation (23) we have:

$$\frac{d \ln K_f}{d\frac{1}{T}} = -\frac{1}{R} \left\{ \frac{d}{d\frac{1}{T}}\frac{\Delta G^0}{T} \right\} \tag{24}$$

$$\frac{d \ln K_f}{d\frac{1}{T}} = -\frac{1}{R} \left\{ \Delta G^0 + \frac{d\Delta G^0}{Td\frac{1}{T}} \right\} \tag{25}$$

$$\frac{d \ln K_f}{d\frac{1}{T}} = -\frac{1}{R} \left\{ \Delta H^0 - T\Delta S^0 + T\Delta S^0 \right\} = -\frac{\Delta H}{R} \tag{26}$$

So: $$\frac{d \log K_f}{d\frac{1}{T}} = -\frac{\Delta H^0}{2.303R} \tag{27}$$

Regarding equation (27), the plot of log K$_f$ versus 1/T produces straight line with slop equals:

$$slope = \frac{-\Delta H^0}{2.303R} \tag{28}$$

Using Equation (28) enables us to calculate Enthalpy change. For calculating ΔS^0 we have:

$$\Delta G^0 = \Delta H^0 - T\Delta S^0 \tag{29}$$

Knowing Gibbs free energy and enthalpy changes we can evaluate ΔS^0

$$\Delta S^0 = \frac{\Delta H^0 - \Delta G^0}{T} \tag{30}$$

Results and Discussion

As an example, the output of the software as demonstrated in Figure 1 is a plot of pH versus concentration of added standardized NaOH for Cu (II)-Phe complex in aqueous solution. Figure 2 illustrates the potentiometric titration curves of phenylalanine complexes with respective metal ions. As it is cleared, with increasing the stability of the complex, titration curve for Cu (II) inclines more toward the right. Table 1 represents the stability constants values of the phenylalanine complexes of Co (II), Ni (II), Cu (II), Zn (II) and Pb (II) in temperatures of 25, 37, 45 and 55°C in aqueous solution. The stability constants of the complexes in 70-30% (v/v) water-dioxane mixture have been shown in Table 2. The results indicate that the order of increasing stability constants in both media are the same and as follows:

$$K_{f \, Co \, (II)\text{-}Phe} < K_{f \, Zn \, (II)\text{-} \, Phe} < K_{f \, Pb \, (II)\text{-} \, Phe} < K_{f \, Ni \, (II)\text{-} \, Phe} < K_{f \, Cu \, (II)}^{-} \, _{Phe}$$

This stability trend is in agreement with Irving-William series [26], which is based on ionic potential of metallic ions. The more charge density, the more electrostatic forces appear between ligand and metallic ion causing an increase in stability constant (except Cu^{+2}). Also the stability constant of complexes is related to their stabilization energies. Cu^{+2} with d^9 configuration has the highest stability energy (Jahan-Teller effect) and Co^{+2} possesses the lowest stability energy among the first transition series. Pb^{+2} is located in fourth period and its stability constant cannot be compared with the others.

Potentiometric titration curves for Cu (II)-Phe complexes

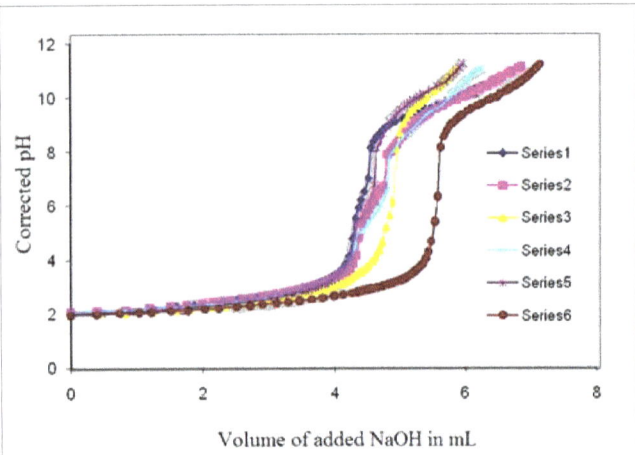

Figure 2: Potentiometric titration curves for the respective complexes at 25°C in water solution; series 1: free of metal ions, series 2: Co (II) ions, series 3: Zn (II) ions, series 4: Ni (II) ions, series 5: Pb (II) ions and series 6: Cu (II) ions.

Table 1: The Log of the stability constants values for the respective metal ion-Phe complexes in aqueous solution at four different temperatures

Complex	Stability constants	25°C	37°C	45°C	55°C
Co(II)-Phe	Log k_1	4.22	4.38	4.56	4.78
	Log k_2	3.61	3.78	3.94	4.18
Ni(II)-Phe	Log k_1	5.80	5.94	6.16	6.27
	Log k_2	4.17	4.39	4.52	4.82
Cu(II)-Phe	Log k_1	7.57	7.72	7.81	7.91
	Log k_2	6.21	6.28	6.38	6.93
Zn(II)-Phe	Log k_1	4.61	5.51	6.12	6.20
	Log k_2	4.56	4.75	5.40	5.53
Pb(II)-Phe	Log k_1	5.69	5.89	7.21	7.48
	Log k_2	3.56	4.28	4.89	5.11

Table 2: The Log of stability constants values for the respective ion metal-Phe complexes in 70-30 % (v/v) water- dioxane mixture at different temperatures

Complex	Stability constants	25°C	37°C	45°C	55°C
Co(II)-Phe	Log k_1	4.80	4.95	5.12	5.28
	Log k_2	3.88	3.97	4.18	4.33
Ni(II)-Phe	Log k_1	6.32	6.37	6.56	6.73
	Log k_2	5.20	5.32	5.50	5.65
Cu(II)-Phe	Log k_1	7.96	8.18	8.38	8.94
	Log k_2	6.43	6.81	7.38	7.67
Zn(II)-Phe	Log K_{f1}	5.80	5.82	5.87	5.89
	Log k_2	5.75	5.86	5.93	6.01
Pb(II)-Phe	Log k_1	6.07	6.18	6.23	6.46
	Log k_2	5.76	5.82	5.95	6.04

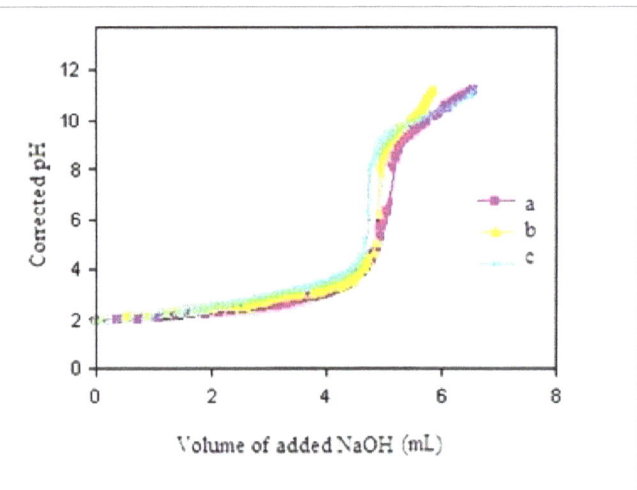

Figure 3: Potentiometric titration curves for Cu (II)-Phe complex at 25°C in a- 50-50% water-dioxane, b-70-30% water-dioxane, c-water alone

for 50-50%, 70-30% (v/v) and water alone have been shown in Figure 3. The more increase in the stability constant of a complex, the more its titration curve is drawn to the right. This means higher stability constant causes more H^+ to be released at lower pH. By changing the solvent, the acidity and basicity of solute varies. The acidic and basic dissociation constant of any species will be measured with respect to the solvent. If a solvent with dissociation constant value of less than water is used, the acidic property of that species increases, therefore the shape of titration curve inclines toward the lower pH with respect to water as solvent.

With increasing the percent of dioxane, the stability constant increases. Because the dissociation constant of amine group of phenylalanine is lower in dioxane than water, so, the stability constant should decrease. This statement is in contrast with the above results. The discrepancy can be explained by solvating ability of ML_2 molecular species, which have more solvating ability in an organic solvent than in water. This is due to lower dielectric constant of dioxane, 2.3 with respect to water, 80. Instead, the solvating ability of M^{+2} molecular ion species is higher in aqueous solution than in organic solvent. It can be expected that the stability constant values are greater in aqueous-organic mixture than in aqueous alone.

The thermodynamic parameters values in Tables 3 and 4 indicate that change in enthalpy for water and water-dioxane mixtures are positive, showing the reactions endothermocity. In all complex reactions with metal ions, the Gibb's free energy changes are negative referring to the reactions spontaneity. The trend has the same pattern for the formation of the complexes in water and in water-dioxane mixtures. It is worthy to note that on increase in the dioxane contents, the free energy becomes more negative, which is an evidence for increasing the stability of the respective complexes.

Conclusions

The stability constants of some divalent metal ion-Phe complexes in water, 70-30% and 50-50% (v/v) water-dioxane

Table 3: Thermodynamic parameters for the pertinent metal ion – Phe complexes in water at 25°C

Complex	ΔH°_1 (KJ/mol)	ΔS°_1 (J/ K)	$-\Delta G^\circ_1$ (KJ/mol)
Co(II)-Phe	6.16	99.2	23.4
Ni(II)- Phe	26.7	198.6	32.5
Cu(II)- Phe	49.2	306.4	42.1
Zn(II)- Phe	44.85	269.2	35.4
Pb(II)- Phe	6.91	100.3	23.0

Table 4: Thermodynamic parameters for the pertinent metal ion- Phe complexes in water dioxane mixture at 25°C

Complex	(70-30% v/v)water-dioxane			(50-50%v/v)water-dioxane		
	ΔH°_1	ΔS°_1	$-\Delta G^\circ_1$	ΔH°_1	ΔS°_1	$-\Delta G^\circ_1$
Co(II)-Phe	20.2	162.6	28.2	12.5	137.7	28.5
Ni(II)-Phe	19.1	177.1	33.7	16.8	170.6	34.1
Cu(II)-Phe	16.8	213.5	46.8	18.3	220.3	47.4
Zn(II)-Phe	55.4	307.0	36.1	12.5	173.7	39.3
Pb(II)-Phe	9.90	125.3	27.4	12.1	134.6	28.0

Table 5: Comparison of the stability constants values for the pertinent metal ion- Phe complexes in water at 25° C in or Lab and in the literature

Cation	Co^{2+}	Ni^{2+}	Cu^{2+}	Zn^{2+}	Pb^{2+}
$\log\beta$	$\log\beta_1, \log\beta_2$	$\log\beta_1, \log\beta_2$	$\log\beta_1, \lambda o\gamma\beta_2$	$\log\beta_1, \lambda o\gamma\beta_2$	$\log\beta_1, \lambda o\gamma\beta_2$
Acquired in our Lab	4.22, 7.83	5.80, 9.97	7.57, 13.78	4.61, 9.17	5.69, 9.25
The literature[21]	4.08, 8.08	5.46, 9.99	7.51, 14.25	4.80, 9.11	4.03, 8.79

mixtures have been determined. The results indicate that the least stable complex is Co (II)-Phe and the most stable one is Cu (II)-Phe. As the percentage of dioxane in the solvent mixture increases, the stability of complexes rises too. This is due to a decrease in dielectric constant of water with respect to dioxane. In fact, co-solvent could affect the protonation-deprotonation equilibria in solution. This will happen by change in dielectric constant of the medium, which alters the relative contribution of electrostatic and non-eletrostatic interactions. Furthermore, thermodynamics parameters of ΔH°, ΔS° and ΔG° were calculated. The data shows that the enthalpy change is positive for all the complexes indicating the reactions are endothermic. The negative ΔG° values for all complexes gives an evidence for spontaneity of the complex reactions.

Acknowledgement

We wish to thank the University of Birjand research council for the finantioal support.

References

1. Bjerrum J. Metal-Ammine Formation in Aqueous Solution. Copenhagen, Denmark: P. Haase and Son. 1941;296p.

2. NormanCL, Doody EJM, WhiteE. Copper (II), nickel and uranyl complexes of some amino acids. J. Am. Chem. Soc. 1958;80(22):5901-5903.

3. Rossotti H. Chemical Applications of Potentiometry. N.J. Van Nostrand. 1969.

4. Rosenberg B, Sigel H, Marcel D. in Marzilli LG (Ed.) Metal Ions in Biological Systems. Wiley-Interscience, New York. 1980.

5. Sovago I, Kiss T, Gergely A. Critical survey of the stability constants of complexes of aliphatic amino acids, Pure Appl. Chem. 1993; 65:1029-1080.

6. Taha M, Khalil MM. Mixed-ligand complex formation equilibria of cobalt (II), nickel (II), and copper (II) with N, N-bis (2-hydroxyethyl) glycine (bicine) and some amino acid. J. Chem. Eng. Data 2005;50(1):157-163.

7. Bastug AA, Goz SE, Talman, Yesim, Gokturk S, Asil E, Caliskan, EJ Coord. Chem. 2011;64: 281-92.

8. Perkins DJ. A study of the effect of amino acid structure on the stabilities of the complexes formed with metals of group II of the periodic classification. Biochem J.1953;55(4):649-652.

9. Martell AE, Smith RM. Critical Stability Constants of Amino Acids. Plenum, New York. 1974.

10. Martell AE. Critical Stability Constants of Metal complexes: Plenum Press. New Yor. 2006;26.

11. Smith RM, Motekaitis RJ, Martell AE. Prediction of stability constants. II. Metal chelates of natural alkyl amino acids and their synthetic analogs. Inorg Chim Acta. 1985;103(1):73-82.

12. Masoud MS, Abdel-Nabby BA. The behavior of some cobalt, nickel and copper amino acid complexes at different temperatures. Thermochim Acta. 1988;128:75-80.

13. Casale A, De Robertis A, De Stefano C, Gianguzza A, Patane G, Riango C, et al. Thermodynamic parameters for the formation of glycine complexes with magnesium (II), calcium (II), lead(II), manganese (II) at different temperatures and ionic strengths with particular reference to natural fluid conditions. Thermochim Acta 1995, 255:109-41.

14. Shoukry MM, Shehata MR, Mohamed MMA. Binary and ternary

complexes of Cd (II) involving triethylenetetramine and selected amino acids and DNA units. Mikrochim Acta. 1998;129(1):107-113.

15. Iqbal MZ, KhurshidS, IqbalMS. Antibacterial activity of copper-amino acid complexes. J Pak Med Assoc. 1990;40(9):221-2.

16. Dogan A, KoseogluF, Kilic E. The stability constants of copper (II) complexes with some α-amino acids in dioxane-water mixtures. Anal Biochem. 2001;295(2):237-9.

17. Rani RS, Rao GN. Stability of binary complexes of L-aspartic acid in dioxane-water mixtures. Bull. Chem.Soc. Ethiop. 2013;27(3):367-376.

18. Demirelli H, Koseoglue F. Equilibrium studies of Schiff bases and their complexes with Ni (II), Cu (II), and Zn (II) derived from salicyldehyde and some α-amino acids. J Solu Chem. 2005;34(5): 561-577.

19. Mohamed AA, Bakr MF, Abd El-Fattah KA. Thermodynamic studies on the interaction between some amino acids with some rare earth metal ions in aqueous solutions. Thermochim Acta. 2003;405(2):235-253.

20. Sajadi SAA. Metal ion-binding properties of L- glutamic acid andL-aspartic acid, a comparative investigation. Natural Sci. 2010;2(2):85-90.

21. Pettit LD. Critical survey   of formation constants of complexes of histidine, phenylalanine, tyrosine, L-DOPA and   tryptophan. Pure Appl. Chem. 1984;56(2): 247-292.

22. Fazary AE, Mohamed AF, Lebedeva NS. Protonation equilibria studies of the standard α-amino acids in NaNO3 solutions in water and in mixtures of water and dioxane. J. Chem.Thermodyn. 2006;38(11):1467-1473.

23. Turkel N. Stability constants of mixed ligand complexes of Ni (II) with adenine and some amino acids. Bioinorg. Chem. Appl. 2015:1-9. doi. org/10.1155/2015/374782

24. Choi TB, Pardridge WM. Phenylalanine transport at the human blood brain barrier: Studies with isolated human brain capillaries. J Biol Chem. 1986;261(4): 6536–6541.

25. Chalmers RA, Chemistry of Complex equilibria: Van Nostrand Reinhold Company. London. UK. 1970.

26. Irving HM, Williams RJP. Stability of Transition Metal Complexes J. Chem. Soc. 1953:3192-3210. doi: 10.1039/JR9530003192

13

The Antisickling Effect of the *Arthrospira platensis* bilins for Liver Protection: a Modeling, Hypothesis, and Food for Thought

Amro Abd Al Fattah Amara*

Protein Research Department, Genetic Engineering and Biotechnology Research Institute, City for Scientific Research and Technological Applications, New Borg Al Arab, Alexandria, Egypt

Corresponding author: Amro Abd Al Fattah Amara, Protein Research Department, Genetic Engineering and Biotechnology Research Institute, City for Scientific Research and Technological Applications, Alexandria, Universities and Research Center District, New Borg EL-Arab, Egypt, E-mail: amroamara@hotmail.com; amroamara@web.de

Abstract

The increase in the number of the liver injury and diabetic patients' worldwide and particularly in sub-Sahara in Africa put a signal about that it might be a link between it and the regional diseases such as Malaria and Sickle Cell Anemia (SCA). SCA is a genetic based disease and is one of the liver injury causative agents. Chronic liver injury patients' are susceptible to more complications such as the infection with viruses (e.g. Virus C), diabetic, immune disease, weakness, anemia etc. *Arthrospira platensis* proves be able to fight and to protect against different diseases at once. It has antiviral, antioxidant, antisickling, nutrient, and edible. In this study their bilins are evaluated for their antisickling effect using molecular modeling aiming to protect patients with SCA particularly the diabetic ones.

In this study molecular modeling for the normal and sickle β-globin, molecules against five bilins [(1) Red bilin, (2) 21H-Bilin-1(22H)-one, (3) 21H-Bilin-1(24H)-one, (4) 1H-Bilin 1 one, and (5) 22H-Biline (21-bilin)] were investigated using protein modeling and docking. MODELLER v 9.8, Hex ver 8.0.0 and Discovery Studio 4.1 Client 4.1.0.14169 (Accelrys software Inc.) were used for modeling and docking. *A. platensis* bilins are evaluated also by comparing the visualized docking results.

The total energy of the system (the molecule) for all of the used bilins with the sickle β-globin molecule particularly in both of the presence or the absence of the porphyrin ring prove to improve such energy to be almost equal to that obtained from the normal β-globin (with or without porphyrin ring). However, porphyrin ring is essential.

A. platensis bilins using molecular modeling prove to be able to stabilize the sickle β-globin molecules particularly in the presence of the porphyrin ring. As being multifunction, it is recommended either as native biomass or as purified bilins to be used for the treatment of the SCA and the protection against further liver deterioration.

Keywords: Liver injury; Bilin; β-globin; protein modeling; structure/function/specificity

Introduction

The proteins structure/function/specificity are governed by their amino acids order, number and arrangement in the protein backbone. Single amino acid change could change the protein property totally. Such change will not effect on the protein family only but will effect on any other macromolecule or micro molecules could gain, or loss affinity to such changed protein. No better example could be described than the SCA. In mammals the O_2 binding protein, myoglobin and hemoglobin are among the most extensively early studied proteins. The single amino acid change in the β-globin (existed on the surface of the protein) cause inefficient O_2 transport. The amino acid change lead to a sticky patch on the β-polypeptide chain of deoxyhaemoglobin, which cause aggregation and precipitation. Such aggregation and precipitation can cause deterioration for the body organs particularly those, which subjected to high blood flow and interaction such as the liver and the spleen. That is not the only type but there are many other variant of hemoglobin mutations. The Fe^{2+} containing heme group is a highly hydrophobic molecule and requires to be placed in a hydrophobic pocket in hemoglobin quaternary structure. It is important to sign that mutants other than the SCA mutant does not necessarily have to be single amino acid substitutions and not have only to affect the heme pocket [1].

Genetic diseases were well known in the old civilization [2]. The World Health Organization (WHO) (1982) estimated that about five percent of the world populations are carriers of genes for clinically important disorders of hemoglobin [3]. Recently Amara highlight some solutions for avoiding different types of genetic diseases including SCA [4]. Each of the β globin genes is represented at least with two copies each in one chromosome gained from the mother and the father to be finally two. So the probability that two incorrect genes find each other will be higher in relatives. Also endemic area with certain incorrect trait or more than one trait of the hemoglobin disease should be considered. Prototype mutants must be detected by using DNA sequencing and mapping the possibility of existing of such type of mutant(s) where one or two bases change can lead to a new mutant which not appears yet. Out group marriage will reduce the disease severity and will give more chances to the correct genes to be existed and the mutated genes to disappear [4].

Normal Hemoglobin Hb A composition is α2Aβ2A with genotype αα/αα β/β. Sickle cell trait hemoglobin Hb A, Hb S its composition α2Aβ2A, α2Aβ2S with genotype αα/αα β/βs. Sickle cell disease hemoglobin HbS its composition α2Aβ2S with genotype αα/αα βs/βs. The hemoglobin four chains 2(α)/2(β) fitted together to form a globular tetramer with a molecular weight of approximate 64000, a structure that for Hb A, is abbreviated as α2β2. The two kinds of chains are almost equal in length, the α chain having 141 amino acids and the β chain 146. α and β chains are encoded by genes at separate loci (the α locus on chromosome 16 and the β locus on chromosome 11). In addition to Hb A, there are five other normal human hemoglobin's, each of which has a tetrameric structure comparable to that of Hb A in consisting of two α or α-like chains and two non-α chains [5].

Liver problems are reported in patients with SCA in 37% cases [6]. Those patients have abdominally meteorism, right upper quadrant pain, or acute painful hepatomegaly. In general different form of liver injury in most cases as cholestatic. Liver infarction has been reported in 34% of autopsies [7]. The SCA associated blood viscosity predisposes to infarction [8,9]. The liver enzymes activity increased abnormally [10,11]. The causing agent in case of SCA patients thought to be obstruction of sinusoidal flow of masses of sickled erythrocytes, trapping them in the liver [12].

There is an increasing in the number of publications, which introduce *A. platensis* as a proposed candidate could be used to reduce the antisickling agent.

This study investigates *A. platensis* bilins structures, which might be able to reduce the sickling process as well as to show that bioinformatics tools must give more concern to the DNA sequences. *A. platensis* bilins have proved to have many useful activities especially as antiviral, antitumor, antioxidant and antisickling. The heme and hemoglobin proteins were docked against five bilins from *A. platensis*. This study refreshes the scientific aim to control and to treat such illness happened by a one nucleotide change. And can be only avoided by avoiding the marriage from the same group. Encourage the marriage from out group is the correct and the simplest solution. And also to protect from further deterioration in organs such as the liver, where diabetic SCA patients will subject to sever liver injury.

Material and Methods

Bilins

The three-diminution structures for five *A. platensis* bilins were used in this study to investigate their abilities to dock with sickle hemoglobin and β- globin protein models. The chemical formula and name of the used bilins are:

1. **Red bilin** [Also known as: CPD-7063, (7S,8S, 101R)-8-(2-carboxyethyl)-17-ethyl-19-formyl-101-(methoxycarbonyl)-3,7,13,18-tetramethyl-2-vinyl-8,23-dihydro-7H-10,12-ethanobiladiene-ab-1,102(21H)-dione]. Which have molecular formula: C35H38N4O7-2; molecular weight: 626.69882 g/mol; InChI Key: HMDDKKOMBDRDIA-DSJLEYPNSA-L. Its IUPAC name is: [3-[(2Z,3S,4S,5Z)-5-[(4-ethenyl-3-methyl-5-oxopyrrol-2-yl)methylidene]-2-[2-[(3-ethyl-5-formyl-4-methyl-1H-pyrrol-2-yl)methyl]-5-methoxycarbonyl-3-methyl-4-oxido-2,3-dihydro-1H-cyclopenta[b]pyrrol-6-ylidene]-4-methylpyrrolidin-3-yl]propanoate].

2. **2.2 1H-Bilin-1(22H)-one,**2,3,7,8,12,13,17,18-octaethyl-19-methoxy-, 113435-10-2, MolecularFormula: C36H48N4O2, Molecular Weight: 568.79192 g/mol, InChI Key: QCYVKTHGTVWFSV-UHFFFAOYSA-N. Its IUPAC name is: [5-[[5-[[5-[(3,4-diethyl-5-methoxypyrrol-2-ylidene)methyl]-3,4-diethyl-1H-pyrrol-2-yl]methylidene]-3,4-diethylpyrrol-2-ylidene]methyl]-3,4-diethylpyrrol-2-one].

3.AGN-PC-0O2GJ1,**21H-Bilin-1(24H)-one,**19-hydroxy-, 21H-Biline-1,19-dione,22,24-dihydro-,142550-15-0, 58828-89-0, Molecular Formula: C19H14N4O2, Molecular Weight: 330.34006 g/mol, InChI Key: MQHWQQCOXHUNCS-UHFFFAOYSA-N, Its IUPAC name is: 5-[[5-[[5-[(5-oxopyrrol-2-ylidene)methyl]-1H-pyrrol-2-yl]methylidene]pyrrol-2-ylidene]methyl]pyrrol-2-one.]

4. **1H-Bilin-1-one** [Also known as: AGN-PC-0OFTAO, 66560-67-6].Which have molecular formula: C19H12N4O; molecular weight: 312.32478 g/mol; InChI Key: VGJBOZZPXZVBBI-UHFFFAOYSA-N. Its IUPAC name is: 5-[[5-[[5-(pyrrol-2-ylidenemethyl)pyrrol-2-ylidene]methyl]pyrrol-2-ylidene]methyl]pyrrol-2-one

5. **22H-Biline,** 21H-Bilin, 22H-Bilin, AC1OAGP5, SureCN139406, AGN-PC-02LS4D

Molecular Formula: C19H14N4, Molecular Weight: 298.34126 g/mol, InChI Key: PPRBOEHFGAHFGC-UHFFFAOYSA-N. Its IUPAC name is: 2-(pyrrol-2-ylidenemethyl)-5-[[5-(pyrrol-2-ylidenemethyl)-1H-pyrrol-2yl] methylidene] pyrrole

Another bilins are existed but will not included in this study

The five bilin molecules were downloaded from PubChem (www.ncbi.nlm.nih.gov/pccompound) and saved as SDF format files [13]. The chemical structure of the molecules is given in (Table 1).

The β-globin sequences and software used in this study

The amino acids sequences of both normal and sickle β-globin are represented by the following amino acid sequences.

A: The normal β-globin

VHLTPEEKSAVTALWGKVNVDEVGGEALGRLLVVYP-WTQRFFESFGDLSTPDAVMGNPKVKAHGKKVLGAFS-DGLAHLDNLKGTFATLSELHCDKLHVDPENFRLLGNVLVCVLAHH-FGKEFTPPVQAAYQKVVAGVANALAHKYH

B: The Sickle β-globin

VHLTPVEKSAVTALWGKVNVDEVGGEALGRLLVVYP-
WTQRFFESFGDLSTPDAVMGNPKVKAHGKKVLGAFS-
DGLAHLDNLKGTFATLSELHCDKLHVDPENFRLLGNVLVCVLAHH-
FGKEFTPPVQAAYQKVVAGVANALAHKYH

Normal and sickle β-globin amino acids sequences were the start point of this study. The sequence was obtained from the www.ncbi.nlm.nih.gov (Blast.ncbi.nlm.nih.gov/Blast.cgi) [14].

The software used in this study

Software for Modeling

One published sickle hemoglobin model was used [15]. The β-globin protein model was generated using the software MODELLER v 9.8 [16].

Software for docking

"Hex" is a Molecular Graphic Program (Hex's Home Page: http://www.loria.fr/~ritchied/hex/) for calculating and displaying feasible docking modes of pairs of protein and DNA molecules [17,18,19]. Hex software can also calculate Protein-Ligand Docking, assuming the ligand is rigid, and it can superpose pairs of molecules using only knowledge of their 3D shapes [20]. It uses Spherical Polar Fourier (SPF) correlations to accelerate the calculations and its one of the few docking programs which has built in graphics to view the result [18]. Simply, the protein pdb is loaded from the "File > open > receptor" and the bilin or the porphyrin ring loaded from "File > open > ligand" and then

from the control option docking is selected and the parameter in is used (Figure 1). The binding energy result is normally negative, stating that a better binding affinity is established from the highest negative result. Low (negative) energy indicates a stable system.

To determine the behavior of both of the protein molecules under study and whether we need high negative energy or lower ones; porphyrin ring has been docked firstly against both of the normal and the sickle β-globin molecules which obtained from the Modeller software and the published sickle hemoglobin model as above.

The five used bilins and one porphyrin ring which used in this study are summarized in (Table 2). The docked molecules' 3D structures have been saved as pdb files and visualized in to show the different interactions (Table 3,4).

Software for the molecules study

The software Discovery Studio 4.1. Client 4.1.0.14169 (Accelrys software Inc.) was used to visualize the docking of the bilins with the proteins models and to show ligands binding sites and the other analysis for the docked molecules [21].

For better 3D structure, the background of the images have been converted to white and the 3D image have been adjusted and saved. All of the docking images have been putted in tables to enable better comparisons between the interaction between the β-globin and the bilins (Table 3,4).

Table 1: Total energy of the system (ETotal) from the docking of the five bilins with different β-globin molecules

Molecules name	Etotal (total energy of the system)*$				Differences					
	Normal β-globin (without porphyrin ring)	Normal β-globin (with porphyrin ring)	Sickle β-globin (without porphyrin ring)	Sickle β-globin (with porphyrin ring)	between column 1 and 2	between column 1 and 3	between column 1 and 4	between column 2 and 3	between column 2 and 4	between column 3 and 4
Red bilin	-380	-322.5	-318.2	-305	-57.5	-61.8	-75	-4.3	-17.5	-13.2
21H-Bilin-1(22H)-one	-352.1	-162	-163.9	-149	-190.1	-188.2	-203.1	1.9	-13	-14.9
21H-Bilin-1(24H)-one	-279	-122	-111	-123.1	-157	-168	-155.9	-11	1.1	12.1
1H-Bilin 1 one, and	-277.7	-137.1	-140	-131.4	-140.6	-137.7	-146.3	2.9	-5.7	-8.6
22H-Biline(21-bilin)	-349	-164.1	-161.2	-159.4	-184.9	-187.8	-189.6	-2.9	-4.7	-1.8

*The EShape (energy of shape only approach) is same as the ETotal.
$ the ETotal and EShape of the normal and sickle β-globin with porphyrin ring are -821 and -516 respectively

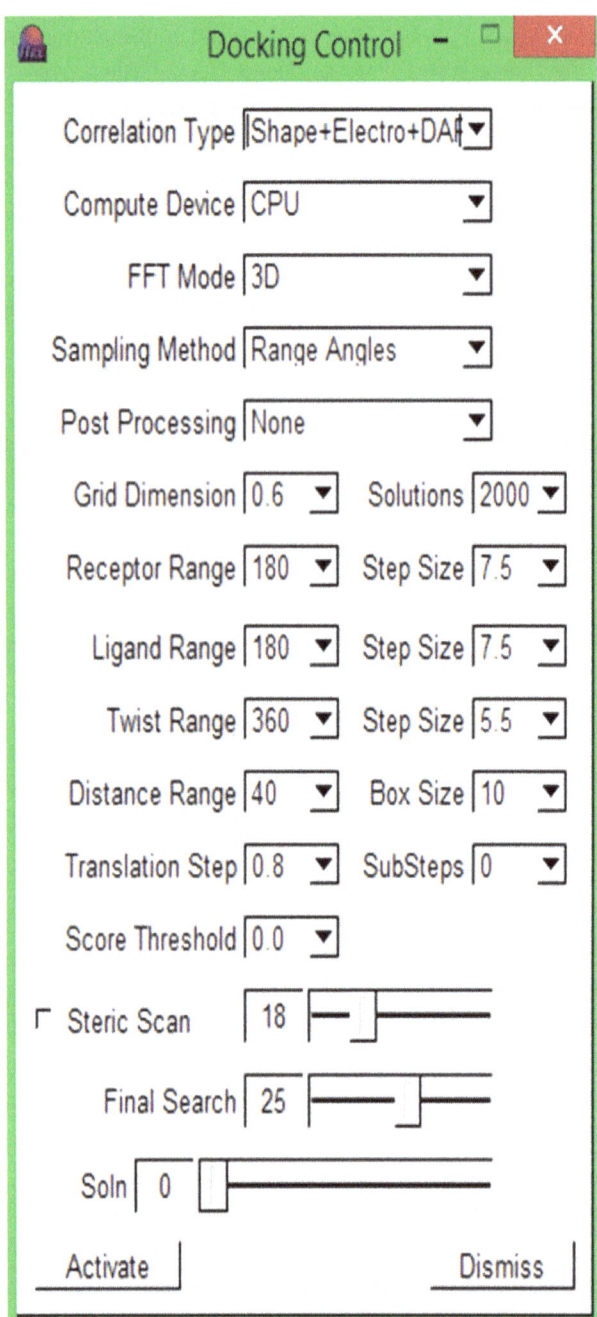

Figure 1: The Hex software docking parameters

Results and Discussion

Our genes are the codes for the proteins in our bodies. Understanding our genes and our proteins will help us to avoid different illness and to design new drugs. Such drugs can be so simple and can be supplied as natural products in the form of food or edible plants. Such natural products can provide us with what the defected genes could not do. Vitamin C is the most famous example. Others forms might can improve certain function like the structure of the SCA defect protein. We are in need to do complicated research to find solutions for some degenerative diseases which alter our macromolecules structure hence their functions and specificity. However, avoiding such type of diseases is so simple. It is just by avoiding the marriage from the same group and from those which have the same disease trite (should not marry from each other). SCA is a known genetic disease in West sub-Sahara in Africa, in the Mediterranean region and other places worldwide [4].

SCA which also named as hemoglobin S disease or hemoglobin SS diseases. After losing the oxygen SCA cells soon stimulate holly leaves or their crescents. Soon they become filamentous and spirculated. Single gene of hemoglobin S and the other is A named sickle cell trait. Individuals have hemoglobin S and β- thalassemea (β-thal) are both prevalent (Greeks and Italians) having a high incidence of S-thalssemia (S-β-thalassemia) [22].

Additionally, scientists, particularly, those from the SCA endemic regions and countries, have summarized their experiences as well the experience gained from their communities in controlling the disease side effect.

Smith and Wood in their book about the Biological Molecules (1991) have written: 'The present-day distribution of effective hemoglobin's has arisen from the accumulation of harmless mutations, early death of individuals with harmful mutation, this confers a selective survival advantage such as increased resistance to malaria, as is the case which sickle cell disease' [1].

For that, lethal mutant are unable the transform their genotype to the second generation. While mutant which gives the minimum survives until the appearance of the seconded generation will do and will be transferred from generation to generation. However there is a 50 % chance that the correct chromosome transfers instead of the one which has the mutant. Alternatively, existing of two globin mutant (on α and β which existed in two chromosomes) will increase the chance of the transfer of globin disease.

Hemoglobin is the oxygen carrier tetrameric molecule and can be found in vertebrate red blood cells, in some invertebrates and in the root nodules of legumes [23]. Each subunit is composed of a polypeptide chain, globin, and a prosthetic group, heme, which is an iron-containing pigment that combine with oxygen and gives the molecules its oxygen-transporting ability. SCA is a global disease and for the Mediterranean and the Africans communities is a local disease [24]. Livingstone, has described in detailed the roles which affect the percentage and the distribution of the SCA in West Africa. From the time of specifying the role of the heredity (the most critical one) till producing artificial blood and artificial oxygen carrier, the scientific progress and the scientist effort did not stop [25].

The biological system is sensitive for the chemical structure. Enzymes could be so specifics. Other protein forms could be also be very sensitive. Red Blood Cells (RBCs) could differentiate between O_2 and CO_2. The conditions and the structure draw the

Table 2: Six macromolecules and six molecules (five bilins and one porphyrin ring) used in this   study

Normal β globin without porphyrine ring	Normal β globin with docked porphyrine ring	Normal β globin with porphyrine ring
Sickle β- globin without porphyrine ring	Sickle β- globin with porphyrine ring	1HHO (One α and one β- globin plus 2 porphyrin rings
Red bilin	21H-Bilin-1(22H)-one	21H-Bilin-1(24H)-one
1H-Bilin 1 one	22H-Biline (21H-Bilin)	Porphyrin ring

Table 3: The different docking results between the five β-globin macromolecules and the five bilins

Names and molecules	Normal β-globin without porphyrin ring against bilin	Sickle β-globin without porphyrin ring against bilin	Normal β-globin with porphyrin ring (with original orientation) against bilin	Normal β-globin with porphyrin ring (with docked orientation) against bilin	Sickle β-globin with porphyrin ring (with docked orientation) against bilin
Red bilin					
21H-Bilin-1(22H)-one					
21H-Bilin-1(24H)-one					
1H-Bilin 1 one					
22H-Biline (21H-Bilin)					

Table 4: Different molecules surface interaction with the β-globin macromolecules

Normal porphyrin ring (surface original)	Normal porphyrin ring (surface original)	Normal β-globin against docked porphyrin ring
Sickle β-globin against porphyrin ring	Red bilin against normal β-globin alone (surface original)	Red bilin against normal β-globin alone (surface modified)
Red bilin Sikle β-globin (surface original)	Red bilin sickle β-globin (surface modified)	21H-Bilin-1(22H)-one against normal β-globin alone (surface original)
21H-Bilin-1(22H)-one sickle β-globin (surface original)	21H-Bilin-1(22H)-one sickle β-globin (surface modified)	21H-Bilin-1(24H)-one against normal β-globin alone (surface original)

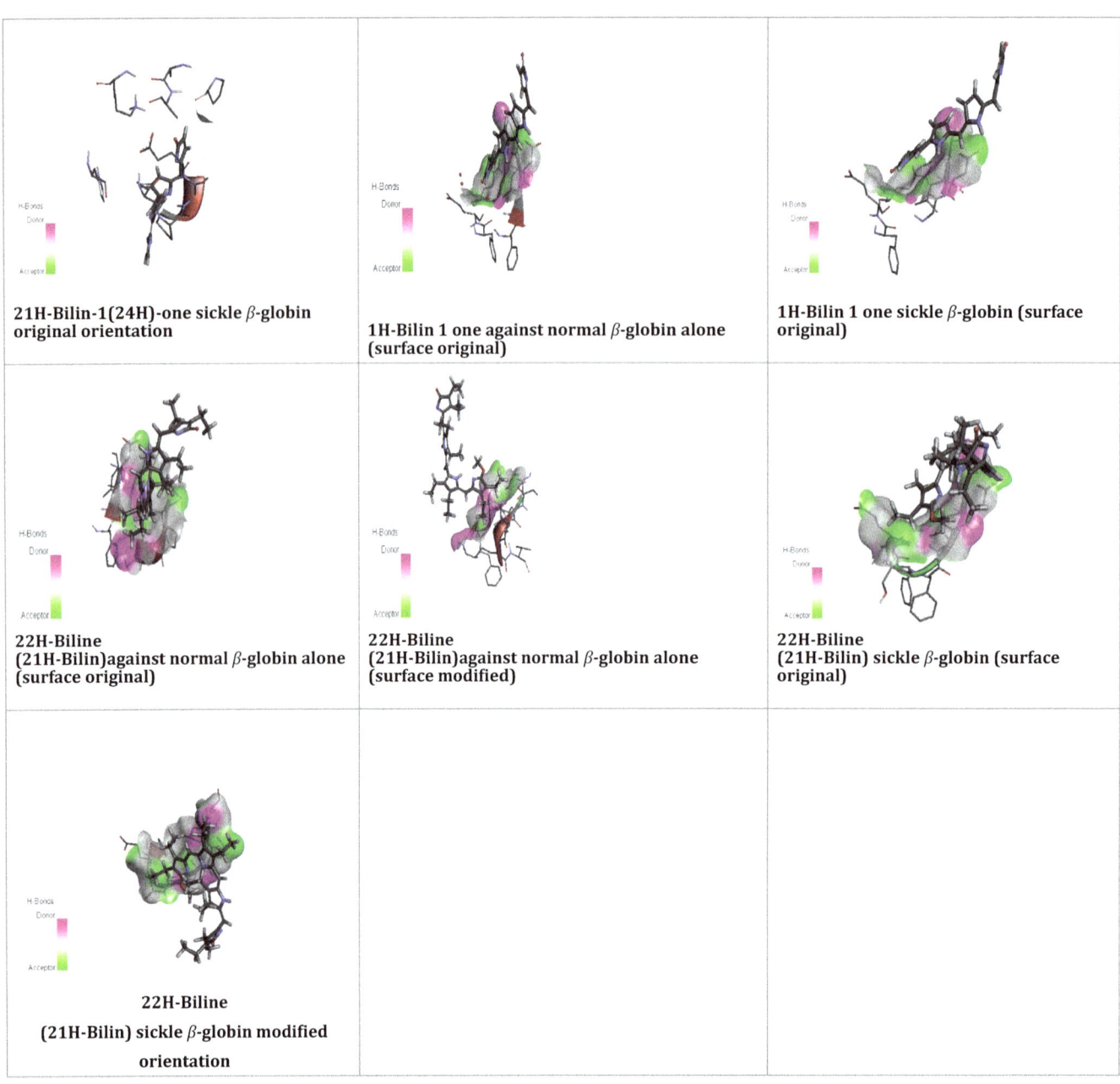

21H-Bilin-1(24H)-one sickle β-globin original orientation	1H-Bilin 1 one against normal β-globin alone (surface original)	1H-Bilin 1 one sickle β-globin (surface original)
22H-Biline (21H-Bilin)against normal β-globin alone (surface original)	22H-Biline (21H-Bilin)against normal β-globin alone (surface modified)	22H-Biline (21H-Bilin) sickle β-globin (surface original)
22H-Biline (21H-Bilin) sickle β-globin modified orientation		

function of the RBCs; where they gain O_2 and where they lose it; where they gain CO_2 and where they lose it. Such specificity is not only in the heme active amino acids but also in the atoms of O_2 and CO_2. For that it is important to investigate molecules could be able to stabilize the sickle hemoglobin particularly during the CO_2 stage which is the aim of this study.

The HBB gene provides instructions for making β-globin. Various versions of β-globin result from different mutations in the HBB gene. One HBB gene mutation produces an abnormal version of β-globin known as Hemoglobin S (HbS). Other mutations in the HBB gene lead to additional abnormal versions of β-globin such as Hemoglobin C (HbC) and Hemoglobin E (HbE). HBB gene mutations can also result in an unusually low level of β-globin; this abnormality is called β-thalassemia. When oxygen is removed

from sickle hemoglobin, those molecules change their shape and combine with one another. The Red blood cell structure changes from ring to sickle shape in the absence of the oxygen. This causes blood to clots and deprives vital organs from their supply of blood, resulting in pain, intermittent illness, and in many cases, a shortened life span. The only difference between normal and sickle cell hemoglobin is that in each β- chain, one glutamic acid is replaced by one valine. Valine, unlike glutamic acid, contains a nonpolar group. The result is a hydrophobic "sticky" region that can interact with hydrophobic region on neighboring molecules, producing the observed clumping. A slight change in the β-globin 3D structure will induce a change in the configuration of it when it interacts with its neighboring subunits of the hemoglobin.

The macromolecules are very sensitive to any effect could

(-821). That might explain the source of the deterioration in case of presence of SCA disease. The purpose might not in the structure only but in the energy as well.

Not all of the pdb files obtained in this study are represented. However, the included images explain also for some extent that it might be used to prove or disprove some facts. Images might be describe in better way where the molecules are interacts and which place are preferable on the protein backbone (to be stabilize or destabilize it). Such docking might not represent the absolute fact, but it will give somehow preliminary judgment and an overall view about what could be happened if such molecules are present in our bodies. Such molecules are derived from edible source so they might be safe if used in an adequate amount. A fact; even seem to be correct in case of using the *A. platensis* native biomass but, need to be proved in case of purified compounds.

For the five bilins it is clear from their models as in that the bilin have good interaction with both of the β-globin molecules (Normal and sickle) (Table 3,4). Apparently, the smaller the molecules the better it become inside the 3D structure. The lager the molecule it tends to attach on the surface of the β-globin molecules. The existence of porphyrin ring or their absence shows different-results. That proves the sensitivity of the process and the importance of the porphyrin ring in the structure of the β-globin, the hemoglobin and the RBCs overall structure.

Based on the five bilin docking data with the β-globin; the docking process is very sensitive.

1. In case of Red bilin the bilin bind to the β-globin in different cases nearly in the same site, except in case of sickle β-globin without porphyrin ring.

2. 21H-Bilin-1(22H)-one show different binding property for all of the five molecules. 21H-Bilin-1(22H)-one is sensitive to any change in the β-globin different molecules.

3. 21H-Bilin-1(24H)-one show different binding sites between normal and sickle molecules but nearly the same in either normal or to the sickle molecules.

4. H1-Bilin-one is bind nearly to the same in place in all of the five macromolecules and show competition against the porphyrin ring.

5. 22H-Bilin (21H Bilin) bind in the same place from the five macromolecules and show competition against the porphyrin ring.

Supposing that β-globin with normal porphyrin ring which obtained without modification from the 1HHO model with native porphyrin ring orientation (no docking) is the most correct macromolecules, in such case Red bilin, 1H-Bilin 1 one and 22H-Bilin (21H Bilin) will be the best molecules which expected to support the normal and the sickle β-globin without the interferes with the porphyrin ring or the macromolecules' 3D structure based on the models obtained as in (Table 3 and 4).

Porphyrin rings attached differently for both of normal and sickle β-globin. Additionally the porphyrin ring orientation is different.

Red bilin against normal and sickle β-globin show nearly full surface fitting. 21H-Bilin-1(22H)-one against normal β-globin alone show good surface fitting but not in case of sickle β-globin.

21H-Bilin-1(24H)-one fitting is totally different in case of normal and sickle globin.1H-Bilin 1 one show the same fitting profile if the different between the normal and the sickle β-globin is considered. 22H-Biline (21H-Bilin) is fit partially to normal β-globin but not the case in the sickle β-globin.

From the models and the results obtained from this study, some facts can be highlighted:

1. Molecular docking is a sensitive process. Docking single protein existed in tetrameric form give different result from that if it studied in its monomeric form.

2. The quaternary structure of the macromolecules such as the hemoglobin is very important where any change or reduce in the number of the protein unit will change its 3D conformation hence change the overall ability to bind to its specific molecules such as in case of hemoglobin and porphyrin ring.

3. Molecules could compete each other if they have similar binding sites such as in case of porphyrin ring and the bilins.

4. Big molecules tend to attach to the surface such as in the porphyrin ring and the big bilin.

5. Small molecules tend to penetrate the protein 3D structure such as in case of small bilin and even so they still sensitive to those molecules bind on the protein surface.

6. The in Silico or in computer modeling could find many useful information however in vitro experiment should be the final judgment. Where the molecules under investigation might affect other macromolecules. And in vivo conditions must be some how different and must be conducted for better evaluation.

7. Natural products must take more interest while they are product prove to be safe, hence they are chemically harmless at least if used in the correct amount and dosage.

8. One amino acid change causes such disease which proves the importance of the protein structure/function/specificity.

9. Avoiding such illness condition can be by avoiding the marriage from the same group or the marriage from individuals who have defected trite.

Such avoidance will lead finally to the disappear of such disease after correct generation.

Our understanding to each condition could effect on our macromolecules will let us to normalize the line between our hope and our bodies for the better of our macromolecules.

Conclusion

A. platensis could support the patients' with SCA from different points where it's well known for their antiviral, antisickling, vitamin rich, antioxidants, high protein content etc. There is an increasing interest for using *A. platensis* in the SCA research. This study suggests a role for five bilins of the *Arthospira*. in treating the SCA. For that the requested molecules have been obtained and generated using different software. Six macro molecules and five bilins have been evaluated best on the interaction between each bilin plus he porphyrin ring and each macromolecules. Two bilins show competition with the docked porphyrin ring while the other three bilins did not with correct interaction with the β-globin 3D structures. It is suggested that some bilins might be used as drugs for treating SCA. However this study did not include any of the *in vivo* study and such study must be done for more perfect judgment. Additionally, the study includes discussions about the conditions, which might affect on our healthy or modified macromolecules which evaluated in this study and represented by the β-globin molecules. As antisickling agent bilins will protect the liver from the liver injury due to the effect of the sickle RBCs which will be indirect way to prevent different diabetic diseases including the viruses' infection, fibrosis, etc.

Reference

1. Smith CA ,Wood EJ. Biological molecule. Springer Netherlands. 1991;1-205.

2. Amara AA. The inevitability of balanced lives: genes – foods- Action-interactions. 2013.

3. Weatherall DJ, Clegg JB, Higgs DR, Wood WG. The hemoglobinopathies In: Scriver CR, Beaudet AL, Sly WS, Valle D (eds). The metabolic basis of inherited disease, 6th ed, MaGraw-Hill:New York;1989.

4. Amara AA. The Need for Early Detection of the Prototype Mutants: Sickle Cell Anemia as a Case Study. doi:104172/jpbS8-006

5. Thompson MW, McInnes RR, Willard HF. Genetics in medicine. Chapter 11: The hemoglobinopathies: Models of molecular disease, W B Saunders Company,5th ed,1991:247-270.

6. Koskinas J, Manesis EK, Zacharakis GH, Galiatsatos N, Sevastos N, Archimandritis AJ. Liver involvement in acute vaso-occlusive crisis of sickle cell disease: prevalence and predisposing factors. Scand J Gastroenterol. 2007;42(4):499-507. doi: 10.1080/00365520600988212

7. Bauer TW, Moore GW, Hutchins GM. The liver in sickle cell disease: a clinicopathologic study of 70 patients. Am J Med. 1980; 69(6):833-837.

8. Mengel CE, Schauble JF, Hammond CB. Infarct-necrosis of the liver in a patient with S-A hemoglobin. Arch Intern Med. 1963; 111(1):93-98. doi:10.1001/archinte.1963.03620250097013

9. Fishbone G, Nunez D Jr, Leon R, Paz G, Isturiz P, McLoughlin C. Massive splenic infarction in sickle cell-hemoglobin C disease: angiographic findings. AJR Am J Roentgenol. 1977;129(5):927-928. doi: 10.2214/ajr.129.5.927

10. Schubert TT. Hepatobiliary system in sickle cell disease. Gastroenterology. 1986;90(6):2013-2021.

11. Rosenblate HJ, Eisenstein R, Holmes AW. The liver in sickle cell anemia: a clinical-pathologic study. Arch Pathol. 1970;90(3):235–245.

12. Ebert EC, Nagar M, Hagspiel KD. Gastrointestinal and Hepatic Complications of Sickle Cell Disease. Clin Gastroenterol Hepatol. 2010; 8(6):483-489. doi: 10.1016/j.cgh.2010.02.016

13. PubChem. PubChem open Chemistry Database; NIH; US National library of Medicine; National Venter for Biotechnology information.

14. Blast Madden T. The NCBI Handbook. The BLAST Sequence Analysis Tool. 2002; McEntyre J, Ostell J, Interne Bethesda MD: National Center for Biotechnology Information US.

15. Shaanan B. 1HHOPDB (Structure of human oxyhaemoglobin at 21 angstroms resolution) Protein Data Bank in Europe EMBL-EBI. J. Mol. Biol. 1983; 31-59.

16. Sali A, Blundell TL. Comparative protein structure modeling by satisfaction of spatial restraints. J Mol Biol. 1993;234(3):779-815.

17. Ritchie DW. Evaluation of Protein Docking Predictions Using Hex 3.1 in CAPRI Rounds 1 and 2. proteins. 2003;52(1):98-106. doi: 10.1002/prot.10379

18. Ritchie DW, Kemp GJ. Protein Docking Using Spherical Polar Fourier Correlations. PROTEINS. 2000;39(2):178-194.

19. Ritchie DW, Kemp GJL. Fast Computation, Rotation, and Comparison of Low Resolution Spherical Harmonic Molecular Surfaces. J Comp Chem. 1999;20:383-395.

20. Ritchie DW. Evaluation of Protein Docking Predictions Using Hex 3.1 in CAPRI Rounds 1 and 2. Proteins. 2003;52(1):98-106. doi: 10.1002/prot.10379

21. Discovery Studio Visualizer V41014169. 2005-2014; Accelrys Software Inc.

22. Kapff CT, Jandl JH. Blood, atlas and sourcebook of hematology. 1st ed. Little Brown and Company Boston USA. 1981;56-57.

23. Aguileta G, Bielawski JP, Yang Z. Evolutionary rate variation among vertebrate beta globins genes: implications for dating gene family duplication events. Gene. 2006;380(1):21-29. doi: 10.1016/j.gene.2006.04.019

24. Kumar R, Sagar C, Sharma D, Kishor P. β-globin genes: mutation hot-spots in the global thalassemia belt. Hemoglobin. 2015; 39(1):1-8. doi: 10.3109/03630269.2014.985831

25. Livingstone FB. Simulation of the diffusion of the beta-globin variants in the Old World. Hum Biol. 1989;61(3):297-309.

26. Sharaf M, Amara A, Aboul-Enein A, Helmi S, Ballot A, Astani A, et al. Molecular authentication and characterization of the antiherpetic activity of the cyanobacteriumA. fusiformis. Pharmazie. 2010;65(2):132-136.

27. Amara AA, Steinbüchel A. New Medium for Pharmaceutical Grade *Arthospira* International Journal of Bacteriology 2013;2013:9. doi: 101155/2013/203432

14

Characterization of Bacterial Lysates by Use of Matrix-Assisted Laser Desorption–Ionization Time of Flight Mass Spectrometry Fingerprinting

Suarez N[1*], Ferrara F[1], Pirez M[2], Rial M[1] and Chabalgoity A[1]

[1]Department of Biotechnology Development, Institute of Hygiene, Faculty of Medicine, Udelar, Uruguay.

[2]Chair of Immunology, Faculty of Chemistry, Udelar, Uruguay.

*Corresponding author: Norma Suárez, Department of Biotechnology Development, Institute of Hygiene, Faculty of Medicine, Udelar, Uruguay, E-mail: nsuarez@higiene.edu.uy

Abstract

Bacterial lysates have for long been used to boost the immunological response to Respiratory Tract Infections (RTI) both in children and adults. They are prepared by growing bacteria usually associated with RTI, followed by chemical or mechanical disruption to prepare single bacterial lysates that are combined in the final product.

Despite the wide range of applications, one drawback for their universal use is the difficulty to assure consistency in their composition given their particular form of preparation; thus there is a need for alternative analytical methods that ensures batch composition consistency.

Here, we demonstrate that MALDI-TOF MS provides reliable and reproducible mass spectral fingerprints for bacterial lysates of S. pneumoniae, H. influenzae, K. pneumoniae, Staphylococcus spp. We also found that mechanical disruption provides markedly better-defined fingerprints. Analysis of the formulated Polyvalent Bacterial Mechanical Lysate (PBML) also showed a characteristic spectrum.

Overall, we found that mechanical lysis coupled to MS analysis allowed for accurate and highly sensible detection of key proteins in each bacterial lysate, a method that can be used to standardize batch-to-batch product composition. Applying this methodology to the production pipeline shall result in better products expanding the acceptance of these cost-effective tools to prevent respiratory tract pathologies.

Keywords: Respiratory infections; bacterial lysates; mass spectrometry.

Abbreviations: MALDI-TOF MS: Matrix-Assisted Laser Desorption–Ionization Time of Flight Mass Spectrometry Mass; RTI: Respiratory Tract Infections; PBML: Polyvalent Bacterial Mechanical Lysate; PBL: Polyvalent Bacterial Lysates; SDS-PAGE: Sodium dodecyl sulfate polyacrylamide gel electrophoresis; OD: Optical Density; TFA: Trifluoroacetic acid.

Introduction

Polyvalent Bacterial Lysates (PBL) has been widely used for decades for prevention of respiratory diseases [1]. They contain whole inactivated microorganisms or defined cellular components from different bacterial strains frequently implicated in upper and lower RTIs i.e. *Staphylococcus aureus, Streptococcus viridans, Streptococcus pneumoniae, Streptococcus pyogenes, Klebsiella pneumoniae, Klebsiella ozaenae, Moraxella catarrhalis, Hemophylus influenzae* [2]. The emergence and up rise of antibiotic resistant bacteria, has posed new problems to the burden and effective control of these pathogens, and there is a need for new treatment modalities that can respond to this situation. Many infectious diseases that were once easily treatable with antibiotics are now a serious health threat because use of antibiotics may lead to even more resistant infectious strains [3].

Because of the clinical relevance of respiratory infections and the importance of the development of new immunization antigens, bacterial lysates used individually or as combination of them (polyvalent) have recently gained new interest due to their capacity to induce a range of effects on the immune system by reducing the level of colonization [4,5]. Although their safety levels are considered acceptable, as are their anti-infective properties, there have been claims that fuller understanding of their mechanism of action is needed, as well as a detailed description of their effects. Consequently an extensive chemical characterization of the antigens to obtain high reproducibility between the preparations and an understanding of their chemical properties may provide better comprehension of their mechanism of action and therefore enhance their safety [2,6].

Bacterial lysates can be obtained by progressive alkaline hydrolysis or by high pressure cell disruption (mechanical disruption). The latter leads to well-conserved antigenic structures avoiding the denaturalization produced by the use of chemical products [7]. There are many different brands of bacterial lysates currently being marketed and several patents describing their preparation and their chemical characterization, that mainly rely on classic laboratory methods such as SDS-PAGE electrophoresis, protein and carbohydrate content among others [7,8,9]. At present considerable effort is being devoted to improve available methods for bacterial lysates characterizations. The introduction of matrix-assisted laser desorption/ionisation

(MALDI-TOF) has proven a significant progress in this field [10,11,12]. The technique allows the acquisition of unique mass signatures for each microorganism and is thus ideal for the characterization of bacterial lysates [13]. MALDI-TOF technique is used to produce a single charged ion (the "soft ionization" method) which preserves proteins integrity. Briefly, the analyte is deposited on a target plate embedded in a crystalline matrix that absorbs laser energy. Once ionized and desorbed sample molecules are analyzed by a mass analyzer (a component of mass spectrometer) and the mass spectrum of the sample sequence can be achieved based on the ratio of molecular weight to charge (m/z) [12,13,14].

Furthermore the technique can quickly identify few bacterial cells present in unknown cellular suspensions or from complex mixtures [13,14]. The ability of technique to rapidly and accurately discriminate between bacterial species has become a revolution in the clinical microbiology and has demonstrated the power of this tool [15,16,17].

Nevertheless, it has been reported that only a small fraction of the proteins primarily separated by SDS PAGE are identified during the MALDI TOF analysis. The reason for such protein behavior is has been explained due to the variation in the ionization properties of the proteins, the limited energy available for proteins ionization process or the presence of non ionizable impurities [18]. However the robustness of the technique has been demonstrated for a wide range of organisms [15,19,20].

The aim of this work was to combine high pressure lysis (mechanical disruption) of bacteria with the MALDI-TOF MS technique, to obtain spectral protein profiles of bacterial lysates and to develop a standard and reproducible protocol for the characterization of these bacterial lysates.

Currently, the work in this field is focused on investigations about the reproducibility of these spectra [19]. The technique can also be used in conjunction with other chemical and biochemical techniques to detect key proteins in complex mixtures as bacterial lysates given that proteins are considered to be the most reliable mass spectrometry biomarkers [15,19,21].

Material and Methods

Monovalent bacterial lysates of Streptococcus pneumoniae (ATCC® BAA334™), Haemophilus influenza (ATCC® 19418™), Klebsiella pneumoniae subsp. pneumoniae (ATCC® 10031™), and Staphylococcus aureus subsp. aureus (ATCC® 25923™), were used for the experiments described in the present work. Hartman serum, inorganic salts glucose, amino acids and vitamins were from Sigma–Aldrich.

MALDI matrix 3, 5-Dimethoxy-4-hydroxy-cinnamic acid (sinapinic acid) and peptide calibration standard mix were purchased from Bruker Daltonics (Billerica, MS, USA).

The total protein concentration of the lysate was estimated by the Bicinchoninic Acid Method [22]. The carbohydrate content was estimated by the Phenol-Sulfuric acid method and SDS-PAGE electrophoresis analysis was done according to [23,24].

The cultivation media for Staphylococcus aureus, Klebsiella pneumoniae, Haemophilus influenza was performed in a standard prepared medium containing: sodium chloride (2g/L), dibasic sodium phosphate (2g/L), sodium acetate (0.5g/L), vegetable soy peptone (40g/L), inosine 0,1 g/L) and glucose(6g/L). Both Hemin and nicotinamide adenine dinucleotide (NADH) in a final concentration of 25mg/ml each was added for Haemophlius influenzae cultivation. Streptococcus pneumoniae was grown in a chemically defined medium previously reported by Texeira [25]. Furthermore the cultivation of S. pneumoniae and H. influenzae was performed at 37°C in 5% CO_2 atmosphere.

Preparation of bacterial lysates

Overnight cultures of each individual bacterium were grown for 12 h at 37°C at 200 rpm. Turbidity was verified and the optical density (O.D) at 600 nm was monitored. 1/100 dilution was made and the culture was reincubated for 4 hours at 37° C until O.D values between 1,0 to 1,2 were reached. For Haemophilus influenzae the dilution after 12 h cultivation was 1/50. At the end of the cultivation time the bacterial suspension was centrifuged, afterwards the pellet was resuspended in 100 mL of Hartmann serum and the lysis was performed.

Mechanical lysis: The bacterial suspension (100 mL) was applied twice to a homogenizer EmulsiFlex-C3 with constant flow-through capacity of 3L/hr. The homogenizing pressure was adjusted between 500 and 30000psi or 35 and 2000bar. The polyvalent bacterial lysates (PMBL) were prepared by mixing 250 µg/mL protein of each monovalent bacterial lysates.

MALDI-TOF analysis

MALDI-TOF measurements were conducted on a Microflex LR MALDI-TOF (Bruker Daltonics, Billerica, MS, USA) with a 337 nm nitrogen laser operated in positive ion lineal mode with delayed extraction and optimized in the m/z range of 0 to 20 kDa. Calibrations were performed with a peptide calibration standard mix (Bruker Daltonics). The laser was fired 100 times at each of ten locations for each sample well on a 96 well plate for a cumulative 1000 shots per sample well taken at 30% intensity.

At the time of analysis, 1 µl of each of the bacterial suspension was purified through a Zip Tip containing C18 and immediately after mixed with 1 µl of matrix solution (sinapinic acid 10 mg/ml in sterile H2O with 1% TFA) at ratios of 1:1. The sample and matrix mix was spotted onto a 96 well stainless steel plate and allowed to air dry for 15 minutes at room temperature.

Results and Discussion

There is a need to better characterize, standardize and control the bacterial lysates extracts in order to make them safer, more effective and longer lasting.

Polyvalent Mechanical Bacterial Lysates (PMBL) containing bacterial lysates derived from pathogenic bacterial strains namely Klebsiella pneumoniae, Haemophilus influenzae, Staphyilococcus aureus and Streptococcus pneumoniae as well as each monovalent lysate were subjected to MALDI TOF-MS analysis in order to identify proteins markers to be used for characterization.

Several research groups have demonstrated that MALDI-TOF is a useful technique to produce protein profiles after cellular extraction and purification [26,27,28].

In the present work each individual bacteria were grown in specific growth media, vegetable-based medium or in the case of S. pneumoniae a chemically defined medium afterwards filtered to remove larger cellular debris [25]. Thus, the resulting bacterial lysates only contained soluble molecular components.

Mechanical lysis of bacteria was selected as the method to lyse the cells, because it has the advantage to preserving the antigenic structure, as opposed to chemical methods like alkaline lysis which produce changes in protein structure [29]. All the lysates (individual or polyvalent) antigenic patterns were analyzed by SDS gel electrophoresis, protein and carbohydrate content and by a mass spectrometry. The matrix selected for the identification of the bacterial lysates proteins was Sinapinic Acid (SA), one of the most predominantly used organic matrices, which demonstrated to provide an effective ionization of the bacterial lysates proteins [30]. The spectra patterns obtained were evaluated on the basis of signal strength, resolution and reproducibility. These criteria are all essential for effective fingerprinting [19,28].

Figure 1a, b, c and d show representative spectral fingerprints of Staphylococcus aureus, Haemophilus influenzae, Streptococcus pneumoniae and Klebsiella pneumoniae by MALDI-TOF MS. The individual spectra profiles observed are the result of mechanical disruption which favors well-conserved protein structure which can be used as ideal markers for the characterization. The signals in the range between 4000 and 20000 Da were previously attributed to be highly cellular conserved housekeeping proteins or cell wall associated proteins and consequently being used as protein markers for identification of bacterial species [19]. Furthermore Krishnamurthy and collaborators reported that disrupted intact cells mixed with sinapinic acid (MALDI matrix) and subjected to direct matrix assisted laser desorption ionization (MALDI-MS) analysis released biomarkers in greater abundance and provide their molecular masses [28]. The protein pattern information is rapidly generated and the technique has clear advantages over the laborious SDS-PAGE technique.

Under the current experimental conditions, most spectral peaks observed were in the mass range of m/z 5000-12000

Da. Figure 1 a showed Haemophilus influenzae three possible biomarkers centered on around m/z 7181, 9123 and 12336 respectively with a high intensity in a reproducible manner. Hagg, et al. reported the use of MALDI/TOF-MS as a technique for the rapid identification and speciation of Haemophilus influenzae [31]. The mass spectral fingerprints reported by the authors differ between whole cells and those that have been lysed. In the present work the m/z protein values are coincident but not identical with the m/z protein values found by the authors for their lysates, a possible explanation for the differences could be due to the lysis method used [31]. Furthermore, in our experiments we found consistency in the patterns obtained from three independent grown samples, suggesting that MS spectrum profile of Haemophilus influenzae is reproducible and reliable and thus may be useful for characterization purposes.

In the case of Staphylococcus aureus Figure 1 b two peaks centered on around m/z 5500 and 7000 showed high intensity, a result similar to that reported by Bernardo and colleagues who reported two specific peaks m/z 5500 and m/z 7000 for S. aureus bacterial lysate obtained by MALDI TOF MS in a reflector mode [32]. Another study reported that nine of S. aureus clinical isolates shared m/z values very near those m/z values found by Bernardo [33]. In the present work these results were reproducible for S. aureus preparations supporting the idea that these proteins could be used as biomarkers for bacterial lysates characterization.

Figure 1 c shows the Klebsiella pneumoniae protein profile. Klebsiella pneumoniae bacterial lysate showed reliable protein pattern accuracy with m/z 7332, m/z 8621 and m/z 9142. All three peaks were previously found in Klebsiella pneumonia clinical isolates [33]. In that study there were reported stable fingerprint spectra for Klebsiella pneumoniae with a few more proteins peaks than the ones observed here, the difference could be explained due the extraction procedure carried out or the protonization efficiency of the proteins.

The bacterial lysate of Streptococcus pneumoniae Figure 1 d gave rise to ions near m/z 6200, m/z 6900 and m/z 8800 in MALDI-TOF mass spectra. Unfortunately there are not many reports in the literature using S. pneumoniae mechanically disrupted lysates for MALDI analysis, to be compared with the protein pattern found here. However there are some groups which used modified approaches for sample preparation of S. pneumoniae isolates to investigate differentiation of species [34]. The spectrum acquisition of S pneumoniae bacterial lysate presented here consistently showed presence of three peaks which proved to be specific for the strain used.

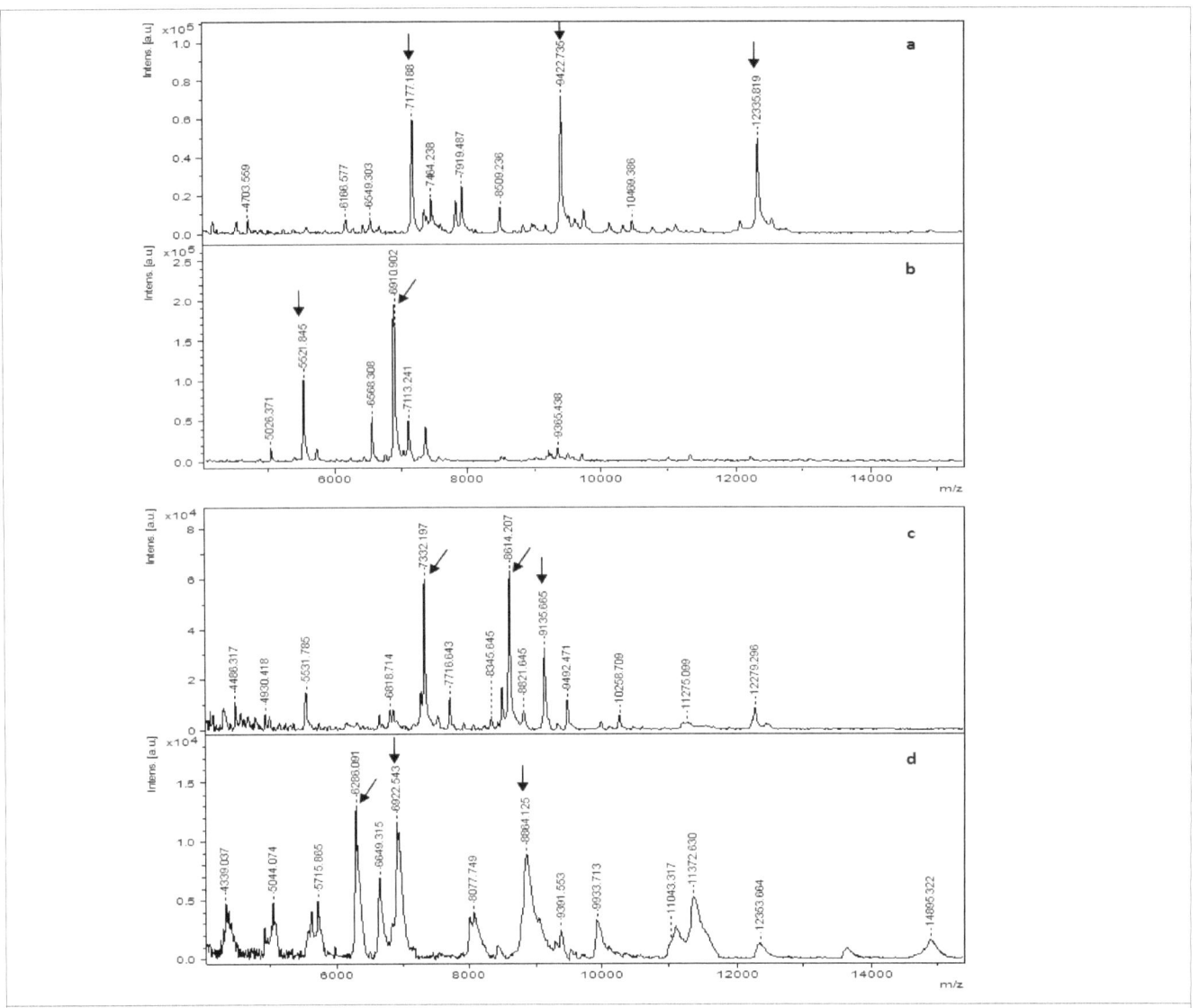

Figure 1: Representative MALDI-TOF mass spectra high intensity peaks are shown in a reproducible manner of; a) *Haemophilus Influenzae* bacterial lysate, m/z 7177, 9422, 12335; b) *Staphylococcus aureus* bacterial lysate, m/z 5521, 6910; c) *Klebsiella pneumoniae* bacterial bacterial lysate, m/z 7332, 8614, 9135; d) Streptococcus pneumoniae bacterial lysate, m/z 6286, 6922, 8864.

One of the factors that affect the obtained spectra is reproducibility. Some authors reported early that reproducibility is dependent on the MALDI-TOF standardization for experimental conditions such as the matrix used the sample: matrix ratio, the culture medium and growth conditions among others, although other studies have shown that a subset of peaks was conserved in the spectra obtained despite different experimental conditions [33,35]. The spectra pattern obtained in the present study for all the strains has shown high reproducibility. As shown in Figure 2, the mass spectral comparison of replicates of Klebsiella pneumoniae grown under same conditions on different days exhibited a high degree of similarity, generates a stable spectrum and provided evidence that these protein signals might serve as signature markers as previously reported [19,28].

Furthermore these ribosomal proteins are well represented and can explain the feasibility to use MALDI-TOF for bacterial

identification and bacterial lysates characterization even without the standardization of experimental conditions. However, it is worth underlining that to optimize the reproducibility, a standardization of sample preparation should be established.

On comparing and analyzing the results obtained by MALDI-TOF MS with those from the conventional techniques SDS-PAGE Figure 3, it was noticed more proteins abundance that the respective MALDI spectra pattern, the discrepant results may be explained due to the deficiency of proton transfer. Ion suppression can arise from impurities that are less ionizable such as salts which can affect the quantity of charged ions hitting the detector [18]. A small pipette tip "Zip-Tip" filled with reverse phase sorbents (C18) was used to overcome this partial ionization suppression and improve the fingerprints obtained.

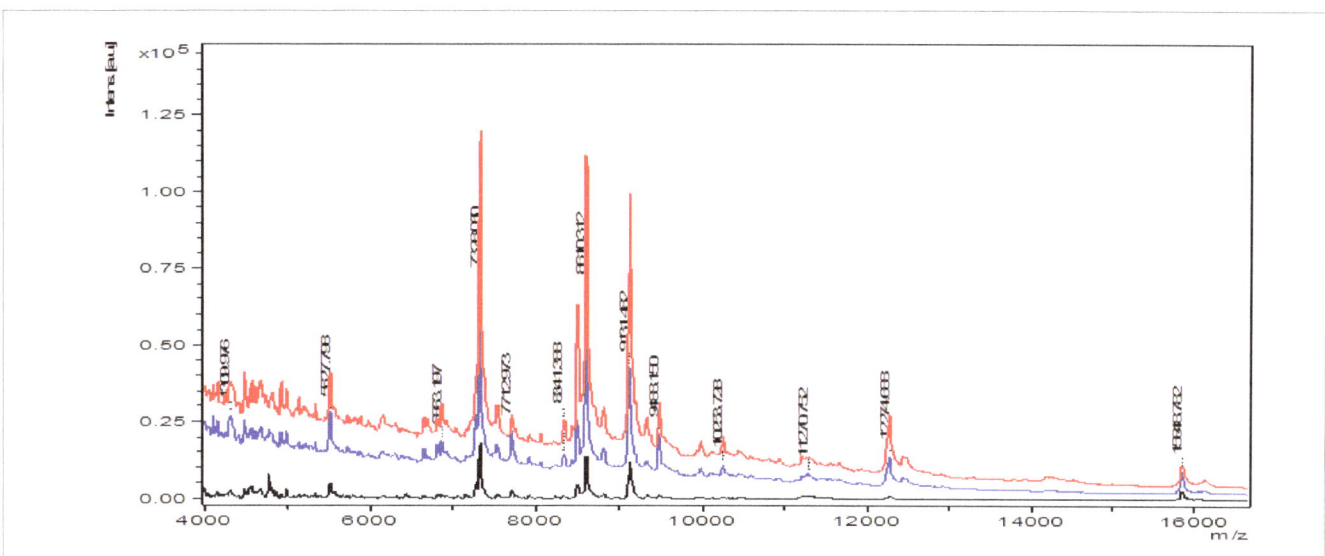

Figure 2: MALDI–TOF MS spectra of *Klebsiella pneumoniae* showing the reproducibility of spectra. Each spectrum represents an independently generated data set of freshly prepared bacterial lysates.

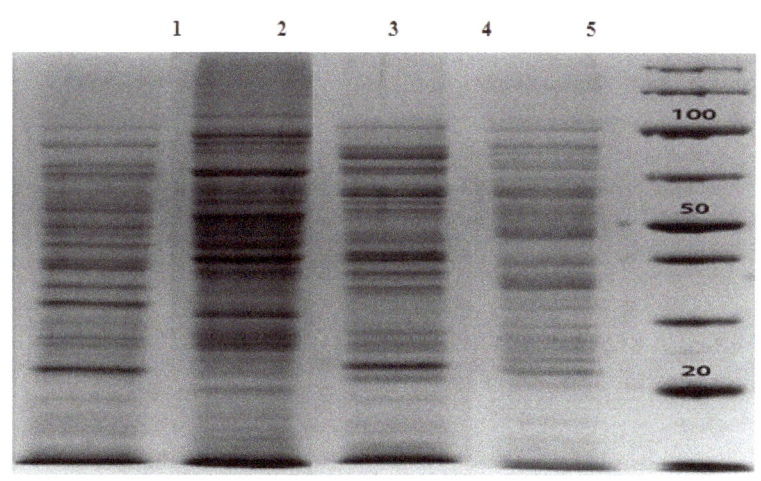

Figure 3: Representative bacterial lysates SDS-PAGE pattern.
SDS-PAGE analysis of mechanically prepared bacterial lysates. **1)** *S. aureus* bacterial lysate **2)** *S. pneumoniae* bacterial lysate **3)** *K. pneumoniae* bacterial lysate **4)** *H. influenzae* bacterial lysate. **5)** Thermo protein standard Molecular Weight 5-250 kDa

In Figure 4 the protein profile of the mechanically polyvalent bacterial lysate is shown. The instrument detected 89 protein peaks in the range of m/z 4000 and 20000 of which the most outstanding were at around on at m/z 7360, m/z 8800, m/z 9540 and m/z 12340. The remarkable reproducibility of the method allowed the measurement of proteins consistently expressed at a high level of abundance, and these biomarkers could be used for the characterization of the bacterial lysates.

To extend the bacterial lysate characterization both protein and carbohydrate content were measured. The average protein and carbohydrate content for all the bacterial lysates were in the range of 0.5 ± 0.019 mg/mL to 1.3 ± 0.019 mg/mL for protein and in 0.4 to 0.5 mg/mL for carbohydrates respectively. These values are in agreement to those found in the literature mainly in the patents which previously reported preparation and characterization of bacterial lysates [7,8,36].

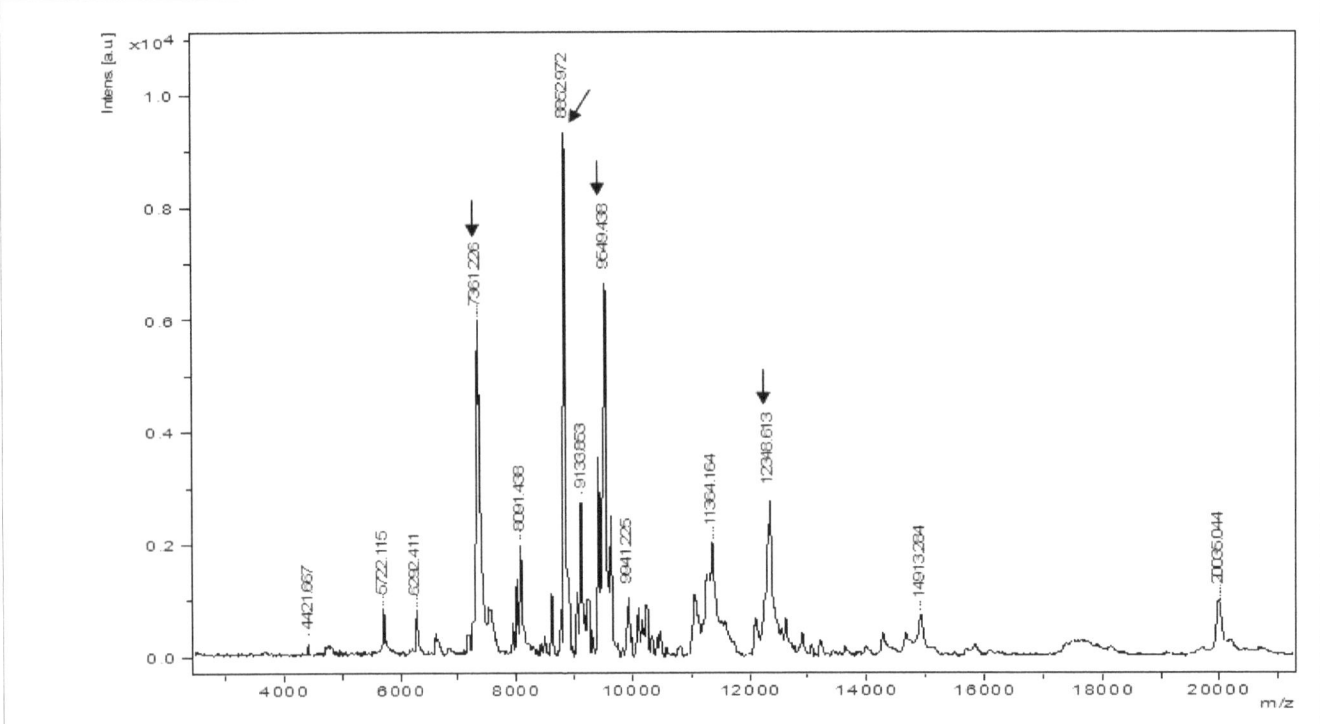

Figure 4: Representative MALDI-TOF spectra of Polyvalent Mechanic Bacterial Lysate (PMBL). MALDI-TOF mass spectra of Polyvalent Mechanical Bacterial Lysate (PMBL) showing four outstanding peaks around m/z 7361, m/z 8852, m/z 9549 and m/z 12348

Conclusions

MALDI-TOF MS analysis of bacterial lysates provided unique spectral patterns for monovalent as well as polyvalent bacterial lysates with high reproducibility. These characteristic fingerprints can thus be used to ensure consistency and reproducibility of bacterial lysate production. Mechanical disruption of bacterial cells using high-pressure lysis, resulted in more precisely defined fingerprints as compared with MS fingerprints obtained from lysates prepared by alkaline lysis, suggesting that mechanical disruption better conserves protein structure in the preparation.

MALDI-TOF MS analysis is simple, rapid and cost-effective, thus the methodology described has shown strong potential to be used as a complementary tool for bacterial lysate characterization. It should be further explored to achieve standard and reproducible protocols in bacterial lysates production, a critical step in the quality control of the final product.

Acknowledgement

This work was partially funded by the National Agency of Research and Innovation (ANII, Uruguay) and PEDECIBA-Química, Uruguay. We would like to thank Lic. Aracĺ Martĺnez for her technical assistance. We are indebted to Dr.Valerie Dee for her valuable revision of this work.

References

1. Braido F, Schenone G, Pallestrini E, Reggiardo G, Cangemi G, Canonica GW, et al. The relationship between mucosal immunoresponse and clinical outcome in patients with recurrent upper respiratory tract infections treated with a mechanical bacterial lysate. J Biol Regul Homeost Agents. 2011;25(3):477- 485.

2. Cazzola M, Capuano A, Rogliani P, Matera MG. Bacterial lysates as a potentially effective approach in preventing acute exacerbation of COPD. Curr Opin Pharmacol. 2012;12(3):300–308. doi: 10.1016/j.coph.2012.01.019

3. Fowler T, Walker D, Davies SC. The risk/benefit of predicting a post-antibiotic era: is the alarm working? Ann N Y Acad Sci. 2014;1323:1–10. doi: 10.1111/nyas.12399

4. Rial A, Ferrara F, Suárez N, Scavone P, Marqués JM, Chabalgoity JA. Intranasal administration of a polyvalent bacterial lysate induces self-restricted inflammation in the lungs and a Th1/Th17 memory signature. Microbes Infect. 2016;18(12):747–757. doi: 10.1016/j.micinf.2016.10.006

5. Braido F, Melioli G, Candoli P, Cavalot A, Di Gioacchino M, Ferrero V, et al. The bacterial lysate Lantigen B reduces the number of acute episodes in patients with recurrent infections of the respiratory tract: The results of a double blind, placebo controlled, multicenter clinical trial. Immunol Lett. 2014;162(2 Pt B):185–193. doi: 10.1016/j.imlet.2014.10.026

6. Rozy A, Chorostowska-Wynimko J. Bacterial immunostimulants--mechanism of action and clinical application in respiratory diseases. Pneumonol Alergol Pol. 2008;76(5):353–359.

7. Bauer J, Hirt P, Schulthess A. Extract of bacterial macromolecules, a process for its preparation and a pharmaceutical composition containing the same. 1995;US5424287 A.

8. Bauer JA, Salvagni M, Vigroox J-PL, Chalvet L, Chiavaroli C. Bacterial extract for respiratory disorders and process for its preparation. 2010;US 2010/0227013 A1.

9. Pillich J, Balcarek JC. Compositions and methods for treatment of microbial infections. US8920815 B2 2014.

10. Lay JO Jr. MALDI-TOF mass spectrometry of bacteria. Mass Spectrom Rev. 2001;20(4):172–194. doi: 10.1002/mas.10003

11. Ryzhov V, Fenselau C. Characterization of the protein subset desorbed by MALDI from whole bacterial cells. Anal Chem. 2001;73(4):746–750.

12. Singhal N, Kumar M, Kanaujia PK, Virdi JS. MALDI-TOF mass spectrometry: An emerging technology for microbial identification and diagnosis. Front Microbiol. 2015;6:791. doi: 10.3389/fmicb.2015.00791

13. Liang X, Zheng K, Qian MG, Lubman DM. Determination of bacterial protein profiles by matrix-assisted laser desorption/ionization mass spectrometry with high-performance liquid chromatography. Rapid Commun Mass Spectrom. 1996;10(10):1219–1226.

14. Pineda FJ, Lin JS, Fenselau C, Demirev PA. Testing the significance of microorganism identification by mass spectrometry and proteome database search. Anal Chem. 2000;72(16):3739–3744.

15. Demirev PA, Fenselau C. Mass Spectrometry for Rapid Characterization of Microorganisms. Annu Rev Anal Chem. 2008;1:71–93.

16. Karimi A, Amanati A. Matrix-assisted laser desorption/ionization time of flight mass spectrometry: A new guide to infectious disease. Arch Pediatr Infect Dis. 2016;4(2):e31816.

17. Hrabák J, Chudáčková E, Walková R. Matrix-assisted laser desorption ionization-time of flight (maldi-tof) mass spectrometry for detection of antibiotic resistance mechanisms: from research to routine diagnosis. Clin Microbiol Rev. 2013;26(1):103–114. doi: 10.1128/CMR.00058-12

18. Annesley TM. Ion suppression in mass spectrometry. Clin Chem. 2003;49(7):1041–1044.

19. Fenselau CC. Rapid characterization of microorganisms by mass spectrometry - What can be learned and how? J Am Soc Mass Spectrom. 2013;24(8):1161–1166. doi: 10.1007/s13361-013-0660-7

20. Qian J, Cutler JE, Cole RB, Cai Y. MALDI-TOF mass signatures for differentiation of yeast species, strain grouping and monitoring of morphogenesis markers. Anal Bioanal Chem. 2008;392(3):439–449. doi: 10.1007/s00216-008-2288-1

21. Murray PR. What is new in clinical microbiology-microbial identification by MALDI-TOF mass spectrometry: a paper from the 2011 William Beaumont Hospital Symposium on molecular pathology. J Mol Diagn. 2012;14(5):419–423. doi: 10.1016/j.jmoldx.2012.03.007

22. Smith PK, Krohn RI, Hermanson GT, Mallia AK, Gartner FH, Provenzano MD, et al. Measurement of protein using bicinchoninic acid. Anal Biochem. 1985;150(1):76–85.

23. Cuesta G, Suarez N, Bessio MI, Ferreira F, Massaldi H. Quantitative determination of pneumococcal capsular polysaccharide serotype 14 using a modification of phenol-sulfuric acid method. J Microbiol Methods. 2003;52(1):69–73.

24. Laemmli UK. Cleavage of structural proteins during the assembly of the head of bacteriophage T4. Nature. 1970;227(5259):680–685.

25. Texeira E, Checa J, Rĺal A, Chabalgoity JA, Suárez N. A new chemically defined medium for cultivation of Streptococcus pneumoniae Serotype 1. Journal of Biotech Research. 2015;6:54–62.

26. Croxatto A, Prod'hom G, Greub G. Applications of MALDI-TOF mass spectrometry in clinical diagnostic microbiology. FEMS Microbiol Rev. 2012;36(2):380–407. doi: 10.1111/j.1574-6976.2011.00298

27. Santos C, Paterson RR, Venancio A, Lima N. Filamentous fungal characterizations by matrix-assisted laser desorption/ionization time-of-flight mass spectrometry. J Appl Microbiol. 2010;108(2):375–385. doi: 10.1111/j.1365-2672.2009.04448

28. Krishnamurthy T, Rajamani U, Ross PL, Jabbour R, Nair H, Eng J, et al. Mass Spectral Investigations on Microorganisms. Toxin Rev. 2000;19:95–117.

29. Islam MS, Aryasomayajula A, Selvaganapathy PR. A review on macroscale and microscale cell lysis methods. Micromachines. 2017;8(3):83. doi:10.3390/mi8030083

30. Fagerquist CK, Garbus BR, Williams KE, Bates AH, Harden LA. Covalent attachment and dissociative loss of sinapinic acid to/from cysteine-containing proteins from bacterial cell lysates analyzed by MALDI-TOF-TOF mass spectrometry. J Am Soc Mass Spectrom. 2010;21(5):819–832. doi: 10.1016/j.jasms.2010.01.013

31. Haag AM, Taylor SN, Johnston KH, Cole RB. Rapid identification and speciation of Haemophilus bacteria by matrix-assisted laser desorption/ionization time-of-flight mass spectrometry. J Mass Spectrom. 1998;33(8):750–756.

32. Bernardo K, Pakulat N, Macht M, Krut O, Seifert H, Fleer S, et al. Identification and discrimination of Staphylococcus aureus strains using matrix-assisted laser desorp- tion / ionization-time of flight mass spectrometry. Proteomics. 2002;2(6):747–753.

33. Panda A, Kurapati S, Samantaray JC, Srinivasan A, Khalil S. MALDI-TOF mass spectrometry proteomic based identification of clinical bacterial isolates. Indian J Med Res. 2014;140(6):770–777.

34. Werno AM, Christner M, Anderson TP, Murdoch DR. Differentiation of Streptococcus pneumoniae from nonpneumococcal Streptococci of the Streptococcus mitis group by matrix-assisted laser desorption ionization - Time of flight mass spectrometry. J Clin Microbiol. 2012;50(9):2863–2867. doi: 10.1128/JCM.00508-12

35. Liu H, Du Z, Wang J, Yang R. Universal sample preparation method for characterization of bacteria by matrix-assisted laser desorption ionization-time of flight mass spectrometry. Appl Environ Microbiol. 2007;73(6):1899–1907. doi: 10.1128/AEM.02391-06

36. Wilson R, Murphy J. Process and compositions for protection of nucleic acids. 2002;US20020197637 A.

15

Development and Validation of Stability Indicating RP-HPLC Method for Rivaroxaban and Its Impurities

Yashpalsinh N Girase[1], Srinivasrao V[2], Dipti Soni[3]

[1] Research Scholar, Pacific Academy of higher Education and Research University, Udaipur, India

[2]Department of Research and Development, Pacific University, Udaipur, India

[2,3]Department of Chemistry, Pacific Academy of higher Education and Research University, Udaipur, India

Corresponding author: Srinivasrao V, Department of Chemistry, Department of Research and Development, Macleods Pharma Ltd, Udaipur, India, E-mail: drvsraoemail@gmail.com

Abstract

Rivaroxaban is oxazolidinone derivative having anticoagulant activity. In literature few analytical methods are discuss about estimation of rivaroxaban; but rarer discussion is available for rivaroxaban impurity profile. The objective of this study is to develop and validate RP-HPLC method for the qualitative analysis of Rivaroxaban. The chromatographic separation was achieved on ZorbaxSB C18 (250 mm X 4.6 mm, 3.5 μ) HPLC column using buffer (0.02M mono basic potassium di hydrogen phosphate) and solvent mixture (acetonitrile: methanol mixture) ingradient programme. The developed methods were validated as per ICH guideline and found to be specific, precise, sensitive and robust.

Keywords: Rivaroxaban; oxazolidinone derivative; Anticoagulant drug, RP-HPLC method; related substance; impurity profile;

Introduction

Rivaroxabanis5-chloro-N-({(5S)-2-oxo-3-[4-(3-oxomorpholin-4-yl)phenyl]-1,3-oxazolidin-5yl}methyl)thiophene-2-carbinoxamide Figure 1.

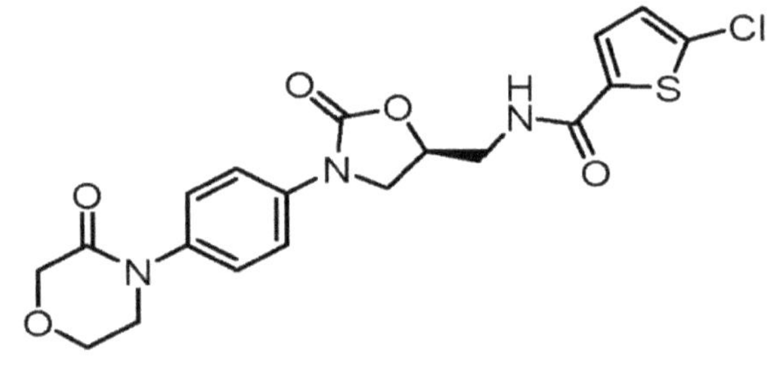

Figure 1: Rivaroxaban

Molecular mass: 435.88

Molecular formula: $C_{19}H_{18}ClN_3O_5S$

In November 2008 the Therapeutic Goods Administration approved new oral anticoagulant drug Rivaroxaban for the prevention of venous thrombosis in patients having elective knee or hip replacement [1, 2]. Rivaroxaban is an oxazolidinone derivative anticoagulant that competitive reversible antagonist of activated factor X (Xa). Factor Xa is the active component of the prothrombinase complex that catalyses conversion of prothrombin (factor II) to thrombin (factor IIa). It is a highly selective direct Factor Xa inhibitor with oral bioavailability and rapid onset of action. Rivaroxabin does not inhibit thrombin (activated Factor II), and no effect on platelets have been demonstrated .

There is no official monograph available for Rivaroxaban or drug product in any pharmacopiea. A preliminary survey of literature for suitable method development for Rivaroxaban has been made [3, 4]. Review of literature suggests that no extensive work has been carried out for the routine analysis of Rivaroxaban, Which can address all process impurities and degradation profiles [5-7]. Monitoring of impurity profiling is very important for quality of drug and patient safety purpose. Also literature survey shows few analytical methods were published for the estimation of Rivaroxaban during formulation and bio availability study for the assay purpose. But rare discussion is available for Rivaroxaban impurity profile study. This study shows detail discussion on monitoring of commercial rout of synthesis and impurity profiling Figure 2, 3. Hence the aim of the present work was to develop accurate and robust routine HPLC method.

Experimental

Material and Reagents

Pure Rivaroxaban was obtained using commercial route of synthesis as per the process described in the Figure-2 [9]. The related impurities including process impurities and degradant impurities (as described in Figure-3) were synthesised in house. The rivaroxaban standard and impurities were characterized using proton nuclear magnetic resonance and mass spectrometry equipped with HPLC. HPLC grade acetonitrile, methanol was procured from J T Baker. Analytical grade potassium dihydrogen phosphate and orthophosphoric acid obtained from Merck chemicals. HPLC grade water obtained from Millipore system was used throughout the analysis. Ion pair reagent Octane sulfonic acid was purchased of Ranchem make.

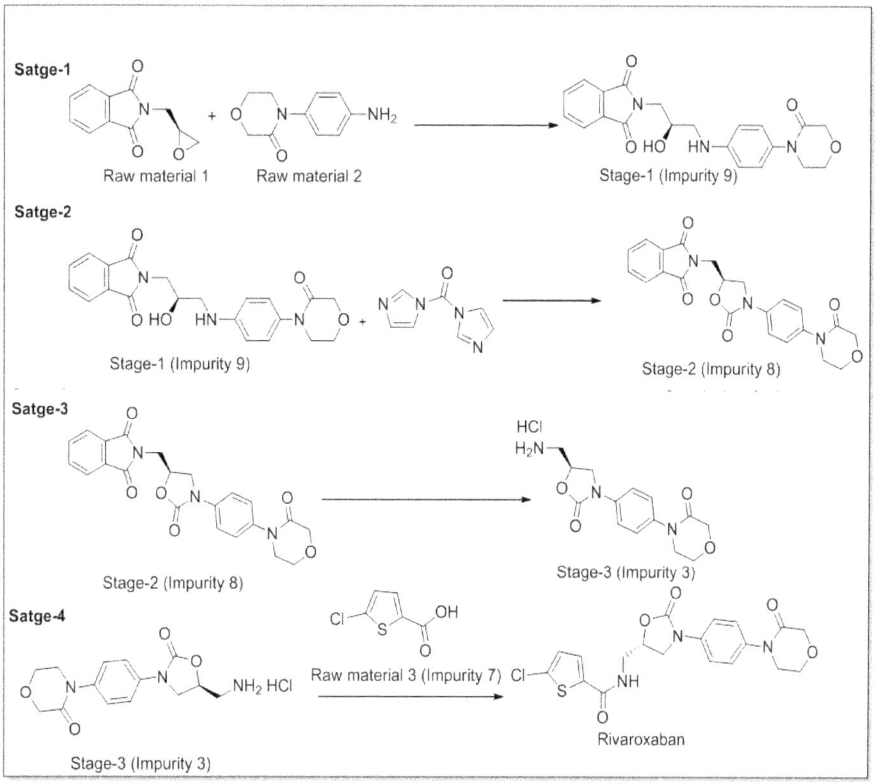

Figure 2: Commercial route of synthesis for rivaroxaban

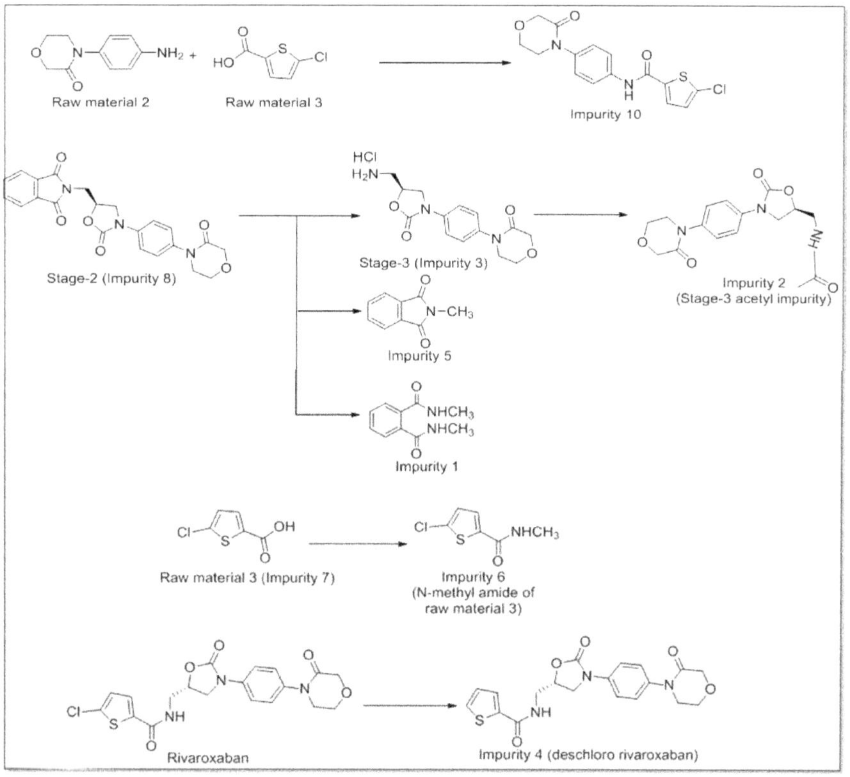

Figure 3: Impurity profiling of rivaroxaban

RP-HPLC Method

Optimization Experiments

In the process of developing RPHPLC method three key parameters were studied which influence the selectivity such as chemistry of stationery phase, pH of the buffer, and organic modifiers. Phophate buffer with Octane sulphonicacid, pH was screened as pH3.0,4.0 and 5.0. HPLC columns were used for development of method were Inertsil ODS 3V, Zorbax phenyl, Kromasil C18 and ZorbaxSBC18. Methanol and Acetonitrile were chosen individual and in different ratio as organic modifier. The impurities spiked solution in Rivaroxaban and sample of all stressed condition were studied and recorded and method was optimized with satisfactory resolutions among all impurities on ZorbaxSBC18 column in Figure 4.

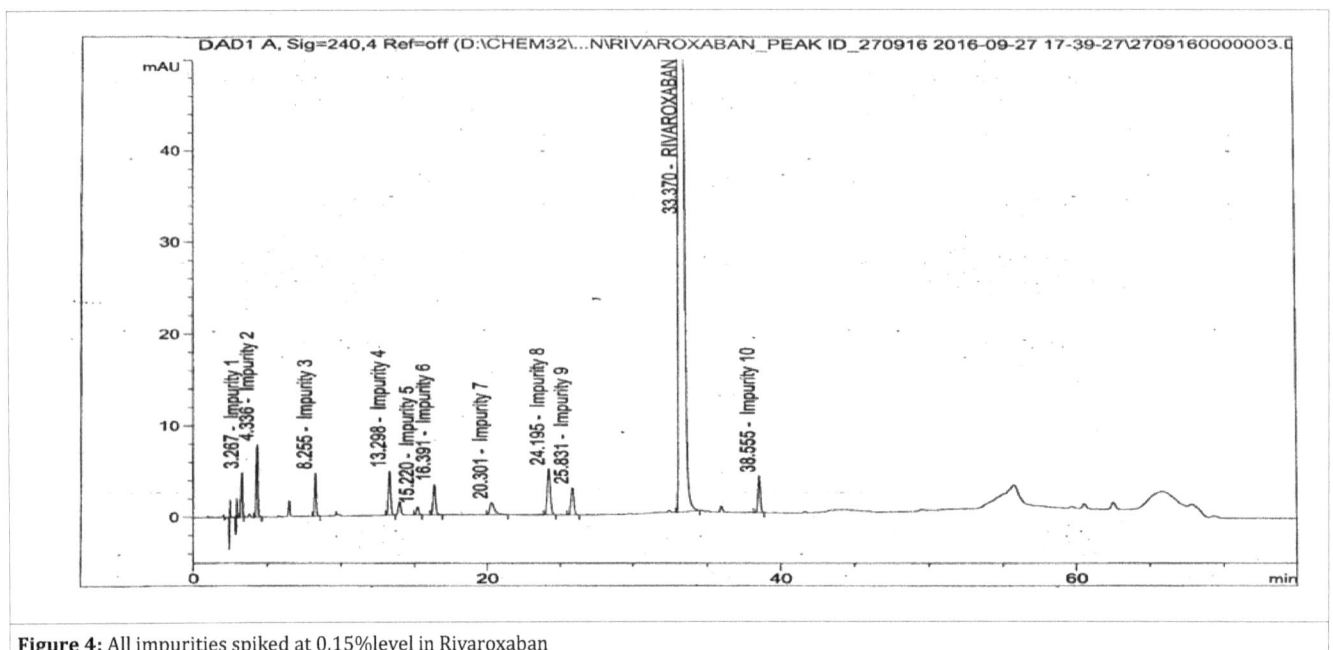

Figure 4: All impurities spiked at 0.15%level in Rivaroxaban

Instrumentation and Chromatographic Conditions

Agilent HPLC 1200 (Agilent Technologies, Germany) equipped with photodiode array detector was used for method development, forced degradation studies and method validation. Zorbax SB C18 (250mmX4.6 mm 3.5 µ) HPLC column. Column thermostat at 45°C was used for the impurities separation. Buffer was prepared using 0.02M of anhydrous potassium dihydrogen phosphate and 1 gm of Octane sulphonic acid solution was adjusted to pH 3.0 with orthophosphoricacid. Solvent mixture was prepared Acetonitrile: Methanol in ratio 820:180v/v. Mobile phase A was prepared by mixing Buffer and solvent mixture in ratio 800 :200 v/v. Mobile phase-B was prepared by mixing Buffer and solvent mixture in ratio of 200:800 v/v. The flow rate and injection volumes were 1.0ml/min and 10µlrespectively. The analysis was carried out under the gradient condition as time(min)/A(v/v):B (v/v); $T_{0.01}$/85:15, $T_{22.0}$/75:25, $T_{35.0}$/55:45, $T_{50.0}$/40:60 ,T65.0/40:60, $T_{66.0}$/85:15 and $T_{75.0}$/85:15. The data was acquired at 240nm for Impurity 1, to Impurity 10; Run time kept 75 min. Chromatographic data processed by using chemstation and chromline HPLC software. The photodiode array detector was used to determine the peak purity of stressed sample.

Preparation of Solutions and Analytical Procedure (System Suitability)

Diluent was prepared by mixing solution A and water in the ratio of 500:500 v/v. Solution A was prepared by mixing methanol and Acetonitrile in the ratio of 500:500 v/v. The test sample solution having concentration of 1000 µg/ml was prepared for the determination of related substances. The stock solution was prepared by dissolving each impurity (Impurity1, Impurity2, Impurity3, Impurity4, Impurity5, Impurity6, Impurity7, Impurity8, Impurity9 and Impurity10) at concentration about 15 µg/mL in diluent and further diluted up to 1.5µg/ml along with Rivaroxaban standard at 1000 µg/ml to prepare the system suitable solution. Inject diluted standard solution in six replicates into the chromatograph and record the chromatogram% RSD for area response of Rivaroxaban peak from six replicate injections of diluted standard solution should not be more than 5.0. The blank, system suitability solution and sample solution of 1000µg/ml, were injected separately and chromatographed under the optimized chromatographic conditions. The resolution NLT 2.0, between Impurity 8peak and Impurity 9 were set as system suitability criteria. All impurities were quantified against 0.1% rivaroxaban diluted standard solution applying the derived Relative Response Factor (RRF). The relative retention time with respect to Rivaroxaban peak and RRF of all impurities are as shown in Table-1, 2.

Table-1:

System suitability results	Retention Time of Rivaroxaban (minutes)	Resolution between Imp-8 and Imp-9 peak	Mean Area	% RSD
Unaltered (Mean Repeatability)	34.529	3.87	27411.35167	0.40

Table-2: Rivaroxaban and its impurities elution order and relative response factor

S. No	Name	RRT	RRF
1	Impurity-1	0.1	0.7
2	Impurity-2	0.13	0.74
3	Impurity-3	0.26	1
4	Impurity-4	0.39	1.06
5	Impurity-5	0.45	1.53
6	Impurity-6	0.48	0.66
7	Impurity-7	0.58	0.57
8	Impurity-8	0.72	1.25
9	Impurity-9	0.77	1.05
10	Rivaroxaban	1	1
11	Impurity-10	1.16	0.57

Validation

Specificity (Selectivity)

Specificity is the ability of method to measure the analyte in presence of its potential impurities. Stress testing of the drug substance performed to identify likely degradation impurities, which intern help to establish the degradation pathways and intrinsic stability of the molecule and validate the stability-indicating power of the analytical procedures used [8,9].

The specificity of developed RP-HPLC method for Rivaroxaban was determined in presence of its impurities (Impurity 1 to Impurity10) and degradation products. Forced degradation studies were also performed on Rivaroxaban to provide an indication of the stability-indicating property and specificity of the proposed method. The dry degradation study was performed by exposing the sample to different stress conditions such as light (1.2 million lux hours), heat (80°C for 12hours), hydrolytic condition (45°C, 75%RH for 48 Hrs.). Wet degradation was performed as acid hydrolysis (1 M HCl for 4hrs at 80°C), base hydrolysis (1 M NaOH for 4hrs at 80°C) and oxidation (5%v/v H_2O_2 for 4 hrs at 40°C). Rivaroxaban was found to degrade significantly in Acid condition, but impurities are found well separated and found method is specific. Mass balance was observed during degradation for all the stressed samples.

Linearity

Linearity solutions were prepared by quantitative dilutions of the stock solution of impurity standard and main drug standard to obtain solutions at LOQ to 250% of the specification limited. Known impurity at 0.15% level and unknown impurity at 0.1% level. A series of solutions were prepared by quantitative dilutions of the stock solution of main drug to obtain solutions at 80% to 120% of the sample concentration.

Each solution was injected and areas were recorded. The linearity of peak areas versus different concentrations was evaluated for Rivaroxaban and its related impurities. The linear regression data for all the impurities plotted and correlation coefficient for all impurities was above 0.99.Linearity results are shown in Table 3.

Table-3: Linearity results

Validation parameter	Imp-1	Imp-2	Imp-3	Imp-4	Imp-5	Imp-6	Imp-7	Imp-8	Imp-9	API	Imp-10
Slope	19622.48	27967.39	27129.84	32929.07	42741.57	20551.2	18186.55	35780.23	30839.05	28544.57	16287.76
Intercept	220.3	440.56	420.31	539.82	1106.65	459.7	146.05	648.37	130.27	466.41	231.29
Correlation coefficient	1	1	1	1	1	1	1	1	1	1	1

Limits of Detection and Quantification (LOD and LOQ)

According to ICH Q2 (R1) recommendations the Limits Of Detection (LOD) and the Limit Of Quantification (LOQ) for Rivaroxaban and its process related impurities (Impurity 1 to Impurity 10) were estimated by calibration curve method [standard deviation of the response (σ) and the slope (S)], by injecting the series of dilute solutions of known concentration. The values of LOD and LOQ found are as depicted in Table-4.

Precision was studied at the LOQ level by injecting six individual preparations of Rivaroxaban and its impurities, followed by the calculation of % RSD of the peaks areas. The %RSD of LOQ precision was below 10%.

Table-4: LOQ and LOD values for Rivaroxaban and its impurity			
S. No	**Name**	**LOQ**	**LOD**
1	Impurity-1	0.033	0.019
2	Impurity-2	0.025	0.012
3	Impurity-3	0.026	0.011
4	Impurity-4	0.021	0.006
5	Impurity-5	0.037	0.015
6	Impurity-6	0.043	0.012
7	Impurity-7	0.048	0.021
8	Impurity-8	0.027	0.009
9	Impurity-9	0.034	0.015
10	Rivaroxaban	0.03	0.018
11	Impurity-10	0.037	0.022

Precision

The precision of method is degree of agreement between the results. Precision of the method was studied for system precision, method precision and intermediate precision. A standard solution of Rivaroxaban at 0.1% was injected for six time to determine the system precision of the method and %RSD was calculated for Rivaroxaban. The %RSD of system precision was found about 0.69%.

Six separate test sample solutions of Rivaroxaban were prepared by spiking the related impurities (Impurity 1 to Impurity 10) at limit level (i.e.0.15% for known and 0.1% for unknown). The % RSD (n = 6) for each related impurities was evaluated and found in between 0.72% to 2.44%.The similar procedure of method precision was carried out by a different analyst, using different mobile phase and diluent preparations and instrument on a different day with different lot of same brand column for intermediate precision study. The %RSD of results for intermediate precision study was calculated and compared with the method precision results.

Accuracy (Recovery)

Accuracy of the method for all the impurities was determined by analyzing Rivaroxaban sample solutions spiked with all the impurities at four different concentration levels of LOQ, 50 %,100 % and 250% of each at the specified limit in both methods. The recovery of all these impurities were found to be in-between the predefined acceptance criterion of 80.0% - 120.0%.

Stability of Analytical Solution

Rivaroxaban spiked with all impurities at specified level were prepared and analyzed immediately and after different time intervals up to 24 hrs to determine the stability of sample solution in both methods. The sample cooler temperature was maintained at about 25^0C and at about refrigerator temperature (2–8^0C). The results from these studies indicated that the sample solution was stable at room temperature and at 2 -8^0C.

Robustness

The chromatographic conditions were deliberately altered to evaluate the robustness of developed method. The resolution between closely eluting peaks was evaluated on altered chromatographic condition. To study the effect of flow rate on the resolution, the flow rate of mobile phase was altered by ± 0.1 mL/min (0.9 to 1.1 mL/min from 1.0 mL/min). The effect of column temperature on resolution was studied at 40^0C and 50^0C instead of 45^0C. whereas all other mobile phase components were held constant similarly to study the pH effect, pH of buffer was altered by ± 0.2keeping rest parameters same. Buffer: solvent mixture was studied by changing composition ±2% absolute of solvent mixture. All these parameters were studied by changing one parameter only at a time. The resolution between all known and unknown peaks present in sample was greater than 1.5 in all the deliberate varied chromatographic conditions indicating the robustness of the method.

Conclusion

A simple accurate and precise HPLC method has been developed for determination of Rivaroxaban and its impurities in bulk drug and dosage form. The method was successfully validated in accordance with ICH guidelines. It can be conveniently used for routine quality control analysis of Rivaroxaban and its impurities in bulk drug and dosage form. Degradation impurities not interference with Rivaroxaban and their impurities, thus the method is stability indicating.

References

1. Roehrig S, Straub A, Pohlmann J, Lampe T, Pernerstorfer J, Schlemmer KH, et al. Discovery of the novel antithrombotic agent 5-chloro-N-({{(5S)-2-oxo-3- [4-(3- oxomorpholin-4-yl)phenyl]-1,3-oxazolidin-5-yl}methyl)thiophene- 2-carboxamide (BAY 59-7939): an oral direct factor Xa inhibitor. J Med Chem. 2005;48(19):5900-5908. doi: 10.1021/jm050101d

2. Celebier M, Recber T, Kocak E and Altınoz S. RP-HPLC method development and validation for estimation of rivaroxaban in pharmaceutical dosage forms. Braz J Pharm. 2013;49(2):359-366.

3. Chandrasekhar K, SatyavaniP, Dhanalakshmi A, Devi C, Barik A and Devanaboyina N. A new method development and validation for analysis of rivaroxaban in formulation by RP-HPLC. Research Desk. 2012;1(1): 24-33.

4. Vaghela D, Patel P. High performance thin layer chromatographic method with densitometry analysis for determination of Rivaroxaban from its tablet dosage form. International Journal of Pharmacy and Pharmaceutical Sciences. 2014;6(6):383-386.

5. Jebaliya H, Dabhi B, Patel M, Jadeja Y, Shah A. Stress study and estimation of a potent anticoagulant drug rivaroxaban by a validated HPLC method: Technology transfer to UPLC.Journal of Chemical and Pharmaceutical Research. 2015;7(10):749-765.

6. Pinaz A Kasad, Murali krishna K.S. Design and Validation of Dissolution Profile of Rivaroxaban by Using RP-HPLC Method in Dosage Form. Asian Journal of Pharmaceutical Sciences. 2013; 3(3):75-78.

16

Differential Antimicrobial Effectiveness of Camel Lactoferrin-Oleic Acid and Bovine Lactoferrin-Oleic Acid Complexes against Several Pathogens

Nawal Abd El-Baky*

Protective and Therapeutic Protein Laboratory, Protein Research Department, Genetic Engineering and Biotechnology Research Institute, City for Scientific Research and Technology Applications, New Borg El Arab, Alexandria 21934, Egypt

*Corresponding author: Nawal Abd El-Baky, Protein Research Department, Genetic Engineering and Biotechnology Research Institute, City for Scientific Research and Technology Applications, Universities and Research Centre District, Ahmed Zewail St, New Borg El Arab, Alexandria 21934, Egypt, E-mail: nawalabdelbaky83@gmail.com

Abstract

Considering the superior biological activities of Camel Lactoferrin (cLf) over lactoferrin from other animal species; which we previously confirmed and continuing the analysis of antimicrobial effectiveness of cLf; we started in previous studies, the current study aimed to formulate a protein-fatty acid complex of cLf and Oleic Acid (OA) and to compare it's in vitro antimicrobial activities against different pathogens with those of a similar Bovine Lactoferrin (bLf)-OA complex. Antimicrobial activity of these complexes was evaluated by agar disc diffusion method, broth microdilution assay, and ELISA-estimating Lf and its complexes binding to bacterial outer membrane proteins. Agar disc diffusion assay results revealed that inhibitory activity of both free cLf and cLf-OA against 13 test pathogens (Methicillin-Resistant *Staphylococcus Aureus* (MRSA), *Staphylococcus aureus*, *Bacillus cereus*, *Escherichia coli*, *Salmonella typhi*, *Shigella sonnei*, *Klebsiella pneumonia*, *Pseudomonas aeruginosa*, *Proteus vulgaris*, *Serratia marcescens*, *Candida albicans*, *Aspergillus niger*, and *Aspergillus flavus*) noticeably exceeded that of corresponding bLf and bLf-OA. Additionally, free OA exhibited antimicrobial activity against MRSA, *S. aureus*, *B. cereus*, and *C. albicans* and to a lesser extent against *E. coli*, *K. pneumonia* as well as *A. niger* and *A. flavus*. Consequently, synergy was evident between cLf/bLf and OA (mostly higher in case of cLf) in prepared complexes against MRSA, *S. aureus*, *B. cereus*, and *C. albicans*. cLf-OA demonstrated 4 times lower Minimum Inhibitory Concentration (MIC) values against MRSA, *B. cereus*, and *C. albicans* than bLf-OA; indicating more superiority in case of cLf-OA than free cLf that showed only twice the activity of bLf. ELISA signals confirmed binding of biotinylated cLf/bLf and cLf/bLf-OA to bacterial membrane proteins. This study proves that cLf obtains enhanced antimicrobial activities after complex formation with fatty acids such as OA even than its free form which has already superior activity than other Lf species; thus this complex may be used as a cure of various microbial infections.

Keywords: Antimicrobial; bovine lactoferrin; camel lactoferrin; oleic acid; protein-fatty acid complexes

Introduction

Lactoferrin is an 80 kDa iron-binding glycoprotein, which is found in numerous secretory fluids, for example milk [1]. Lf displays antimicrobial activity against a variety of pathogenic microorganisms besides modulating the immune system. Most published studies aimed to reveal therapeutic utilization of bLf and Human Lactoferrin (hLf) for treatment of various infectious and inflammatory diseases, but fewer studies have been reported on cLf. Evidences point to the differences in the biological (antimicrobial) activities between cLf and other lactoferrins. It was found that cLf was the most active lactoferrin against many pathogens [2, 3].

Lactoferrin has a variety of biological functions, many of which not related to its iron-binding capability [4]. Lf is an important part of the innate immune system. Besides its major biological function, which is binding and transport of iron ions, lactoferrin has various other functions such as antibacterial, antifungal, antiparasitic, antiviral, antiallergic, catalytic, and anticancer functions. Lf exhibits diverse inhibitory effects against microorganisms, including stasis, cidal, synergistic, bacterial adhesion blocking, opsonic, and cationic mechanisms. Due to broad-spectrum activities of Lf against various bacteria, fungi, parasites, and viruses, along with its immunomodulatory and anti-inflammatory functions, lactoferrin seems to have great impact on practical medicine [5].

OA; the key monounsaturated fatty acid of olive oil has the ability of in vitro killing of different bacterial and fungal pathogens. Long-chain unsaturated fatty acids such as OA and linoleic acid are bactericidal to various pathogens including MRSA and *Helicobacter pylori* [6,7]. Also, they have in vitro killing activity against *C. albicans* [8].

Both LF and α-Lactalbumin (α-LA) have similar iron-binding region structure [9]. It was confirmed that these proteins release the bound ions at acid pH to produce a more open structure; a property that favors OA binding [10-12].

Therefore, in this study, complexes of cLf and bLf with OA were obtained. Then for the first time, their differential antimicrobial activity against several pathogens was evaluated by agar disc diffusion method and broth micro dilution assay, meanwhile, estimating Lf binding to bacterial outer membrane proteins by ELISA.

Materials and Methods

Antimicrobial Agents

cLF and bLf were purified from camel and bovine milk after processing at our lab according to the protocol described by Redwan & Tabll (2007) [13]. Both purified cLf and bLf preparations were sterilized by filtration through 0.22 μm syringe filter (TPP, St. Louis, Mo., USA) and stored at −20°C until use. The iron saturation of both lactoferrins, checked by spectrophotometry, was of approximately 10% in case of bLf and 35% in case of cLf (partially iron-saturated) [14]. The protein content was tested by the Folin phenol reagent [15]. OA (C18:1:9 cis, ≥ 99.0% purity, cell culture tested) was purchased from Sigma-Aldrich (St. Louis, MO, USA).

Discs of 5 antibacterial agents, including carbenicillin (10 μg), vancomycin (30 μg), fucidic acid (10 μg), gentamicin (10 μg), and chloramphenicol (30 μg) were purchased from Mast Diagnostics (Merseyside, UK). Nystatin and amphotericin-B antifungal standards at concentration of 100 μg/ml were obtained from Sigma-Aldrich.

Determination of cLf and bLf Activity

Activity of both lactoferrins was assayed according to the procedure described by Ye et al. with slight modifications [16]. Samples of lactoferrin (50 μl) were added to the mixture containing 15 μl of 50 μM Tris-HCl buffer (pH 8.0), and 75 μl of 300 μM dihydro-nicotinamide-adenine-dinucleotide-phosphate, 300 μM Nitroblue Tetrazolium (NBT), and 30 μM of Phenazin Methosulfate (PMS). The absorbance was checked at 0 and 5 min of reaction at 580 nm. L-ascorbic acid was used as a control. The calculations were based on standard curve prepared with different concentrations of NBT. LF activity was expressed in IU per milligram of protein.

Preparation of cLf-OA and bLf-OA Complexes

Both lactoferrins were dissolved in 10 mM phosphate buffered saline (PBS, pH 8.0) to different concentrations. OA was directly added to each protein solution at 50 molar equivalents (OA: Lf). After vortexing for 30 s, the mixtures were incubated for 20 min at 45°C in a water bath. Finally, excess fatty acid in the complexes was removed by centrifugation at 4°C followed by ultra filtration using a 3000 kDa cut-off membrane.

Oleic Acid Determination in the Prepared Complexes

OA concentration in the prepared complexes was determined according to the colorimetric method of Duncombe [17]. In brief, protein complexes were shaken with chloroform and copper solution. Copper amount in chloroform is corresponding to the amount of OA in the test samples, which is examined by adding sodium diethylthiocarbamate as a color developer. Copper reagent (2.5 ml) consisting of 9 volumes of 1 M triethanolamine, 1 volume of 1 N acetic acid, and 10 volumes of 10% (w/v) copper sulphate was added to either 500 μl test samples or 500 μl standard OA. After shaking vigorously with a vortex mixer, 5 ml chloroform were added to the solution and shaken vigorously for 1 min. Then, 3 ml of the lower layer were carefully transferred to another test tube containing 500 μl of 0.1% sodium diethylthiocarbamate in butanol and absorbance values were recorded at 440 nm. Experiment was done three times, each in triplicate and the results were presented as mean ± SEM.

Test Microorganisms and Growth Conditions

MRSA clinical isolate was obtained from blood of a patient at Almery University Hospital (Alexandria, Egypt) and subjected to the confirming BD GeneOhm™ MRSA assay. The *Staphylococcus aureus* ATCC 25923 and *Candida albicans* ATCC 10231 strains were purchased from Becton Dickinson (France). *Escherichia coli* ATCC 25922, *Salmonella typhi* ATCC 19430, and *Shigella sonnei* ATCC 25931 were obtained from American type culture collection (ATCC, USA). *Bacillus cereus, Klebsiella pneumonia, Pseudomonas aeruginosa, Proteus vulgaris, Serratia marcescens, Aspergillus niger, and Aspergillus flavus* were collected from Al-Azhar University Mycology Center (Cairo, Egypt), and Botany and Microbiology Department, Faculty of Science, Al-Azhar University, Assiut Branch (Egypt).

A 100 μl aliquot culture of each bacterial strain was added to Luria Bertani (LB) broth, incubated at 37 °C for 24 h, and then stored at -80 °C after addition of 20% glycerol to be used as seeds stock. Yeasts such as *C. albicans* and fungi such as *A. niger and A. flavus* were maintained on Sabouraud's dextrose agar at 4 °C. To determine in vitro antibacterial activity, Cation-Adjusted Mueller-Hinton (CAMH) broth and Mueller-Hinton agar were used. While, in vitro antifungal activity was determined using Sabouraud's dextrose broth and Sabouraud's dextrose agar.

Agar Disc Diffusion Assay

Susceptibility screening of test microorganisms to cLf, bLf, OA, cLf-OA, bLf-OA, and different antibacterial and antifungal standards was carried out using the agar disc diffusion technique on Mueller-Hinton agar for bacteria and Sabouraud's dextrose agar for fungi. Plates were overlaid with 100 μl of standardized inoculum suspension using McFarland standard, then wells were bored into the agar media by using a sterile 6 mm cork borer, and about 100 μl of solutions containing different concentrations of cLf, bLf, OA, cLf-OA, and bLf-OA were added into each well. Carbenicillin, vancomycin, fucidic acid, gentamicin, chloramphenicol, nystatin, or amphotericin-B was used as positive control and sterile water was used as negative control. The culture plates were incubated at 4 °C for 2 h to allow proper diffusion of tested antimicrobials before being incubated at 25 °C and 37 °C for 24 h in case of *C. albicans* and bacterial cultures, respectively, and for 5 days at 25 °C in case of fungal cultures. Then plates were examined for the presence of the inhibition zones. The zone of inhibition was measured and interpreted using the Clinical and Laboratory Standards Institute (CLSI) zone diameter interpretative standards [18].

Broth Micro dilution Susceptibility Assay

The minimum inhibitory concentrations of cLf, bLf, OA, cLf-OA, and bLf-OA against test microorganisms were determined by broth micro dilution method as recommended by CLSI [19]. Each 96-well micro titer plate (Greiner, Frickenhausen, Germany) was inoculated with test microorganisms, and then 100 µl of CAMH broth or Sabouraud's dextrose broth containing antimicrobial agents in serial dilution were added. The concentrations of cLf, bLf, cLf-OA, and bLf-OA ranged from 0.0312 mg/ml to 2 mg/ml, while for OA were from 1.25 mM to 10 mM. Plates inoculated with bacteria and *C. albicans* were incubated at 37 °C and 25 °C, respectively for 24 h, whereas plates inoculated with fungi were incubated at 25 °C for 5 days. The MICs were determined by measuring the absorbance at 600 nm for test bacterial strains and *C. albicans* and calculating fungal sporulation using hemocytometer. The MIC was defined as the lowest concentration at which growth was completely inhibited. All MIC determinations were performed in duplicate. Bacteria in CAMH broth and fungi in Sabouraud's dextrose broth were used as control of growth.

Detection of Lf and its Complexes Binding to Bacterial Membrane Proteins by ELISA

Bacterial membrane fractions were prepared from Gram-positive and Gram-negative bacteria under investigation. Test bacterial cells were harvested via centrifugation of their cultures and then washed with PBS at pH 7.4. Cells were suspended in 0.5 mg/ml herbimycin A, 0.1 mM sodium vanadate, 25 mg/ml leupeptin, 50 mg/ml aprotinin, 750 mg/ml benzamidine and 1 mM phenylmethylsulfonyl fluoride in PBS pH 7.1. After sonication at 130 W and 20 kHz for 20 min, the cytosolic fraction (supernatant) and the membrane fraction (precipitate) were separated by centrifugation at 11.000 xg for 10 min. The precipitated pellet was washed then suspended in lysis buffer (0.5 mg/ml herbimycin A, 0.1 mM sodium vanadate, 25 mg/ml leupeptin, 50 mg/ml aprotinin, 750 mg/ml benzamidine,1 mM phenylmethylsulfonyl fluoride, 1% Triton X-100 and 1% CHAPS in PBS pH 7.1). The membrane fraction was obtained by centrifugation at 13.000 xg for 15 min.

cLf, bLf, cLf-OA, and bLf-OA were biotinylated using N-hydroxysuccinimidebiotin (Sigma Chemicals Co., St. Louis, Mo.) according to the protocol described by the manufacturer.

ELISA micro titer plate (Costar, Cambridge, USA) was coated with carbonate/bicarbonate pH 9.6 buffer as a blank or 50 µl of biotinylated cLf/bLf at a concentration of 2 mg/ml as positive control and 50 µl of bacterial membrane fractions preparations at a concentration of 100 µg/ml in carbonate/bicarbonate pH 9.6 buffer for 24 h at 4°C. After washing 5 times with PBS at pH 7.2, the plate was blocked by adding 100 µl of blocking buffer (2% w/v gelatin in PBS) for 1 h at 37°C. Then the plate was washed 5 times with PBS and 50 µl of biotin-labeled cLf, bLf, cLf-OA, or bLf-OA at a concentration of 2 mg/ml were added to blank and test organisms wells. After 2 h of incubation at 37°C, the plate was washed 5 times with PBS, and 50 µl of alkaline phosphatase-conjugated streptavidin (BIO-RAD, Alfred Nobel, Hercules, USA) diluted 1:1000 was added, followed by an incubation of 1 h at 37°C. After washing five times, p-Nitrophenyl phosphate (p-NPP) was added for color development and optical density was calculated at 405 nm using an ELISA micro titer plate reader (Micro Plate Reader, BIO-RAD, USA). Results were represented as mean ± SD of three replicates.

Results and Discussion

Purification of cLf and bLf

Both lactoferrins were purified from skimmed milk by one-step affinity chromatography using heparin-Sepharose column and eluted by 0.0-1.0 M NaCl gradient. Purified fractions were analyzed by SDS-PAGE and then by ELISA (data not shown).

Single discrete band was obtained on 12% SDS polyacrylamide gel of the two proteins and estimated to be 80 kDa as shown in Figure 1.

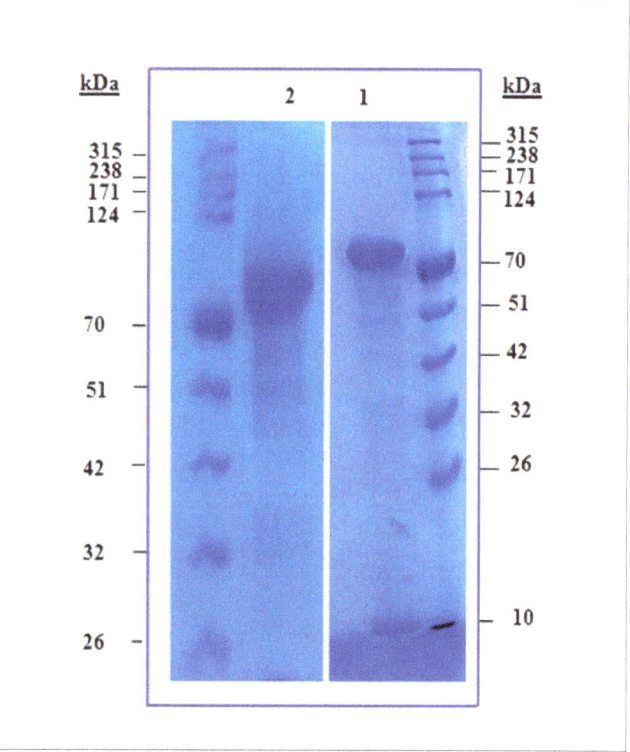

Figure 1: 12% SDS-PAGE of purified cLf and bLf. Lane 1 represents purified cLf and lane 2 represents purified bLf.

cLf and bLf Activity Determination

In the present work, cLf and bLf activity was measured by their ability to inhibit superoxide radical generation. Their mean activity was 6.43 ± 1.71 U/mg of protein for bLf and 9.84±1.18 U/mg of protein for cLf.

Oleic Acid Evaluation in the Prepared Complexes

Concentration values of OA in cLf-OA and bLf-OA complexes were presented as mean ± SEM; the mean concentration value for cLf-OA was 0.241±0.02 mM and for bLf-OA was 0.224±0.025 mM.

Agar Disc Diffusion Assay

cLf and bLf possessed antibacterial besides antifungal activities against a total of 13 test microorganisms (10 bacteria, 1 yeast, and 2 fungi; Tables 1, 2, and 3) and produced concentration dependent inhibition zones Figure 2. The mean diameter of inhibition zone of cLf, evaluated against the test organisms ranged between 7.3 mm against *A. flavus* at a concentration of 0.5 mg/ml and 39.7 mm against *P. aeruginosa* at a concentration of 1 mg/ml. On the other hand, the mean diameter of inhibition zone of bLf, evaluated against the test organisms ranged between 10.0 mm against *A. niger* at a concentration of 1 mg/ml and 42.0 mm against *P. aeruginosa* at a concentration of 1 mg/ml.

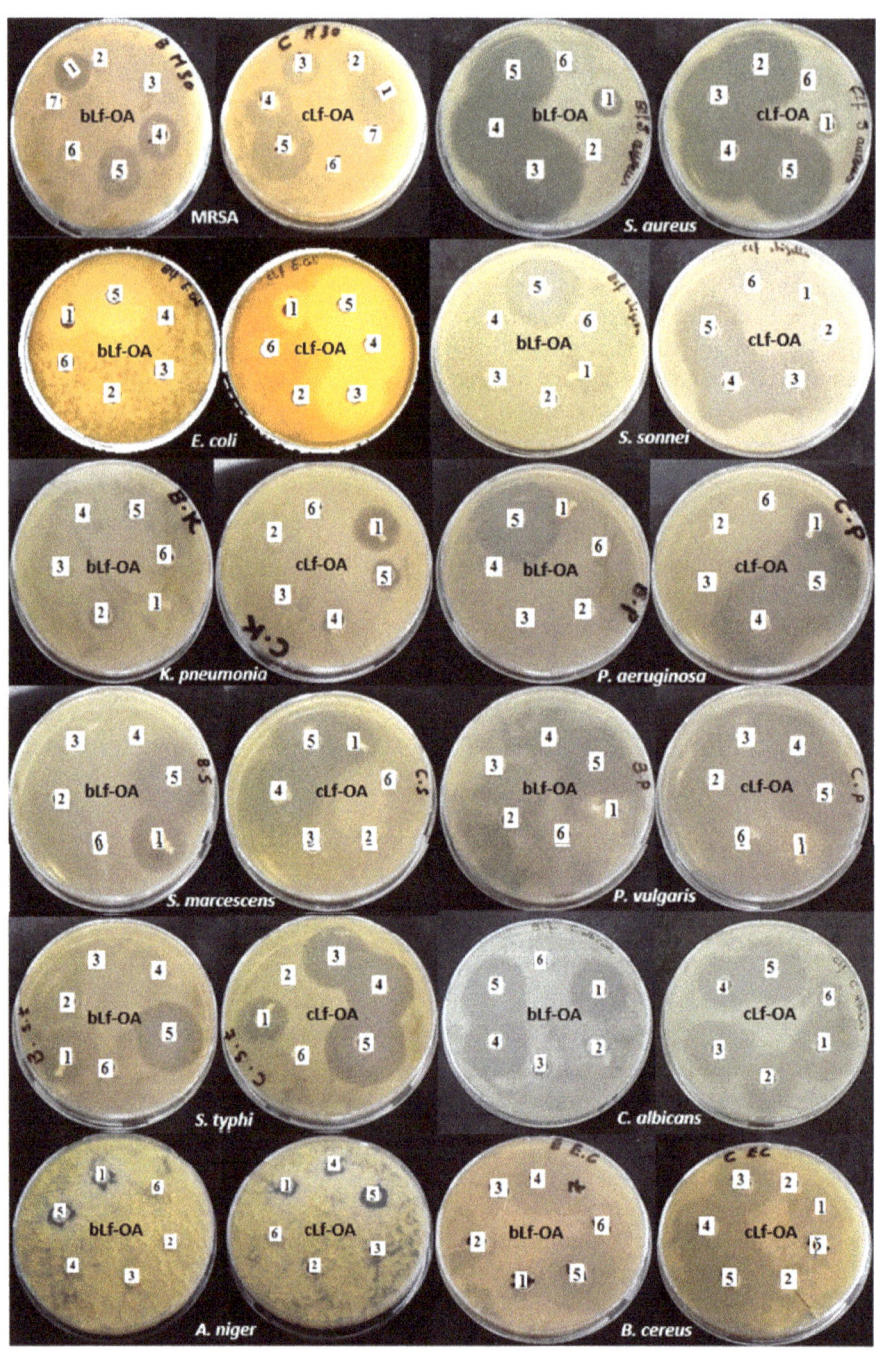

Figure 2: Differential antimicrobial effectiveness of cLf-OA and bLf-OA complexes against several pathogens. 1: represents different antibacterial or antifungal standards; 2: represents cLf/bLf-OA at concentration of 0.0625 mg/ml in case of Gram-positive bacteria and 0.125 mg/ml in case of Gram-negative bacteria and fungi; 3: represents cLf/bLf-OA at concentration of 0.125 mg/ml in case of Gram-positive bacteria and 0.25 mg/ml in case of Gram-negative bacteria and fungi; 4: represents cLf/bLf-OA at concentration of 0.25 mg/ml in case of Gram-positive bacteria and 0.5 mg/ml in case of Gram-negative bacteria and fungi; 5: represents cLf/bLf-OA at concentration of 0.5 mg/ml in case of Gram-positive bacteria and 1 mg/ml in case of Gram-negative bacteria and fungi; 6: represents sterile water used as negative control.

It was observed that neither cLf nor bLf showed any inhibition zones at concentration of 0.125 mg/ml against any of tested bacterial and fungal pathogens except against *S. aureus* in case of cLf which gave an inhibition zone of 19.3 mm mean diameter.

Table 1 showed that growth of MRSA, *S. aureus*, and *B. cereus* was inhibited by cLf at concentrations of 0.25-1 mg/ml, bLf at concentrations of 0.5-1 mg/ml (except for *S. aureus* that was inhibited by bLf at 0.25-1 mg/ml), cLf-OA at 0.0625-0.5 mg/ml, and bLf-OA at 0.25-0.5 mg/ml (except for S. aureus that was inhibited by bLf-OA at 0.125-0.5 mg/ml). This indicates a noticeable increase (this increase was mostly higher in case of cLf) in antibacterial activity against test Gram-positive bacteria of both cLf and bLf by binding to OA in the complexes prepared; a suggestion correlates with the previously published studies, which confirmed that oleic acid was active against many Gram-positive bacteria thus synergy was evident between cLf/bLf and OA [20, 21]. OA exhibited antibacterial activity against the Gram-positive bacteria MRSA, *S. aureus,* and *B. cereus* (zones of inhibition ranging from 10.0-15.7 mm were observed after overnight incubation Table 1).

As presented in Table 2, growth of *E. coli, S. sonnei,* and *S. typhi* was inhibited by cLf and cLf-OA at concentrations of 0.25-1 mg/ml, in addition to bLf and bLf-OA at concentration of 1 mg/ml. *Klebsiella pneumonia* was sensitive to cLf and cLf-OA at concentration of 1 mg/ml but was not sensitive to bLf at 0.25-1 mg/ml or bLf-OA at 0.125-1 mg/ml and only showed sensitivity towards 2 mg/ml bLf (12.0 mm zone of inhibition) and 2 mg/ml bLf-OA complex (15.3 mm zone of inhibition). cLf/cLf-OA and bLf/bLf-OA inhibited growth of *P. aeruginosa, P. vulgaris,* and *S. marcescens* at concentrations of 0.5-1 mg/ml and 1 mg/ml, respectively. The inhibitory activity of OA was lower against Gram-negative *E. coli* and *K. pneumoniae* compared to test Gram-positive bacteria (zones of inhibition of 3.7 mm and 3.3 mm, respectively were observed after overnight incubation) while it has no effect on growth of other test Gram-negative bacteria Table 2. Thus, no change occurred in antibacterial activity against test Gram-negative bacteria of both cLf and bLf by binding to OA in the complexes prepared. This agrees with results obtained by Dilika et al. who reported that OA was inactive against the Gram-negative species they tested [20].

Table 1: Antibacterial activity of cLf-OA and bLf-OA complexes against tested Gram-positive bacteria

	Mean diameter of inhibition zone[a] (±1 mm)																	
Strains	cLf (mg/ml)			bLf (mg/ml)			**OA**	cLf-OA (mg/ml)				bLf-OA (mg/ml)				**CB**	**FC**	**VA**
	0.25	0.5	1	0.25	0.5	1		0.0625	0.125	0.25	0.5	0.0625	0.125	0.25	0.5			
MRSA	19.3	21	30.7	R	16.0	19.7	10.0	6.3	18.0	22.0	31.0	R	R	18.3	22.7	R	26.0	16.0
S. aureus	27.3	30.7	36	20.3	27.0	34.3	15.7	37.3	39.0	40.7	42.3	R	35	35.7	37.0	12.7	NT	NT
B. cereus	20	23.7	32.7	R	19.0	20.3	11.7	7.0	19.3	24.0	32.3	R	R	19.0	24.7	11.3	NT	NT

a - Mean of three assays; OA - oleic acid at concentration of 10 mM; CB – Carbenicillin; FC - Fucidic acid; VA - Vancomycin antibacterial standard discs at concentrations of 10, 10, and 30 µg/ml, respectively; R - Resistant (no inhibition zone); NT – Not tested.

Table 2: Antibacterial activity of cLf-OA and bLf-OA complexes against tested Gram-negative bacteria

	Mean diameter of inhibition zone[a] (±1 mm)															
Strains	cLf (mg/ml)			bLf (mg/ml)			**OA**	cLf-OA (mg/ml)				bLf-OA (mg/ml)				**GM**
	0.25	0.5	1	0.25	0.5	1		0.125	0.25	0.5	1	0.125	0.25	0.5	1	
E. coli	22.3	26.3	35.7	R	R	31.0	3.7	R	26.7	28.3	36.0	R	R	R	37.0	18.3
S. sonnei	21.7	29.3	31.0	R	R	35.3	R	R	22.3	31.0	32.7	R	R	R	36.3	23.7
S. typhi	31.0	31.3	32.7	R	R	30.7	R	R	31.7	32.7	33.7	R	R	R	31.3	22.0
K. pneumonia	R	R	12.7	R	R	R	3.3	R	R	R	14.3	R	R	R	R	21.3
P. aeruginosab	R	36.3	39.7	R	R	42.0	R	R	R	37.0	40.3	R	R	R	42.7	R
S. marcescens	R	35.7	37.3	R	R	34.7	R	R	R	36.3	38.7	R	R	R	36.3	23.0
P. vulgarisb	R	22.7	25.0	R	R	30.7	R	R	R	23.0	26.7	R	R	R	31.3	R

a - Mean of three assays; OA - oleic acid at concentration of 10 mM; GM-Gentamicin antibacterial standard disc at concentration of 10 µg/ml; R -Resistant (no inhibition zone); b-*P. aeruginosa* and *P. vulgaris* were not sensitive to GM but were sensitive to chloramphenicol disc (30 µg) and showed inhibition zones of 19.0 mm and 8.3 mm mean diameter, respectively.

Table 3 showed that cLf/cLf-OA and bLf/bLf-OA inhibited growth of *A. niger* and *A. flavus* at 0.5-1 mg/ml and 1 mg/ml, respectively. Sensitivity of *C. albicans* to cLf, cLf-OA, bLf, and bLf-OA was observed at concentrations of 0.5-1, 0.125-1, 1, and 0.5-1 mg/ml, respectively. Additionally, oleic acid was inhibitory to *C. albicans* after 18 h of incubation and to a lesser extent against *A. niger* and *A. flavus* Table 3; causing a noticeable elevation (this elevation was higher in case of cLf) in antifungal activity against *C. albicans* of both cLf and bLf by binding to OA in the complexes prepared. These synergy results are in agreement with those of Kabara et al., who found OA to be inhibitory to *C. albicans*[22].

Overall results revealed that inhibitory activity of cLf and cLf-OA against test microorganisms noticeably exceeded that of bLf and bLf-OA as previously confirmed [2, 3].

Table 3: Antifungal activity of cLf-OA and bLf-OA complexes

Strains	Mean diameter of inhibition zone[a] (±1 mm)																	
	cLf (mg/ml)			bLf (mg/ml)			OA	cLf-OA (mg/ml)				bLf-OA (mg/ml)				AMP
	0.25	0.5	1	0.25	0.5	1		0.125	0.25	0.5	1	0.125	0.25	0.5	1	
C. albicans	R	20.7	29.3	R	R	28.0	12.3	24.7	27.7	29.0	31.0	R	R	29.3	30.7	29.3
A. niger[b]	R	8.7	10.7	R	R	10.0	3.3	R	R	11.3	13.7	R	R	R	13.3	R
A. flavus[b]	R	7.3	8.3	R	R	10.7	3.7	R	R	9.3	10.0	R	R	R	11.0	R

a - Mean of three assays; OA - oleic acid at concentration of 10 mM; AMP – Amphotericin-B antifungal standard at concentration of 100 µg/ml; R - Resistant (no inhibition zone); b - *A. niger* and *A. flavus* were not sensitive to amphotericin-B but were sensitive to nystatin at concentration of 100 µg/ml and showed inhibition zones of 9.0 mm and 8.0 mm mean diameter, respectively.

The MIC Values of Antimicrobial Agents

cLf showed MIC values of 0.25, 0.125, 0.25, 1, 0.5, 0.5, 0.5, 0.5, 0.5, and 0.5 mg/ml for MRSA, S. aureus, B. cereus, K. pneumonia, P. aeruginosa, S. marcescens, P. vulgaris, C. albicans, A. niger, and A. flavus, respectively, while 0.25 mg/ml for E. coli, S. sonnei, and S. typhi, indicating that it achieved twice and 4 times, respectively higher inhibitory activity than that of bLf against these pathogens Table 4. These data provided further support to the notion that cLf exerted higher antimicrobial activity than bLf.

In view of the MIC results, synergy between cLf or bLf and OA in the prepared complexes was observed against MRSA, S. aureus, B. cereus, and C. albicans causing a 4, 2, 4, and 4 times, respectively increase in cLf antimicrobial activity and a 2 times increase in bLf antimicrobial activity against all of these pathogens. Whereas, the combinations of cLf or bLf and OA in the complexes displayed no synergistic effect against E. coli, K. pneumoniae, A. niger and A. flavus.

Interestingly, OA concentrations in the prepared complexes were significantly (P<0.05) lower than its MICs against sensitive test pathogens, thus the higher antimicrobial activity of cLf/bLf-OA than free forms was not due to a higher OA concentration in the prepared complexes but confirmed the differential participation of lactoferrin proteins in this elevated complex antimicrobial activity.

Table 4: MIC values of cLf-OA and bLf-OA complexes against various pathogens

Test organism	MIC values				
	cLf (mg/ml)	bLf (mg/ml)	OA (mM)	cLf-OA(mg/ml)	bLf-OA(mg/ml)
MRSA	0.25	0.5	2.5	0.0625	0.25
S. aureus	0.125	0.25	2.5	0.0625	0.125
B. cereus	0.25	0.5	2.5	0.0625	0.25
E. coli	0.25	1	5	0.25	1
S. sonnei	0.25	1	R	0.25	1
S. typhi	0.25	1	R	0.25	1
K. pneumonia	1	2	5	1	2
P. aeruginosa	0.5	1	R	0.5	1
S. marcescens	0.5	1	R	0.5	1
P. vulgaris	0.5	1	R	0.5	1
C. albicans	0.5	1	2.5	0.125	0.5
A. niger	0.5	1	5	0.5	1
A. flavus	0.5	1	5	0.5	1

R -Resistant (no antimicrobial activity).

Detection of Lf and its Complexes Binding to Bacterial Membrane Proteins by ELISA

Biotinylated cLf, bLf, cLf-OA, or bLf-OA reacted significantly (P<0.05) with bacterial membrane fractions preparations of all test Gram-positive and Gram-negative bacteria compared to blank; carbonate/bicarbonate pH 9.6 buffer Table 5. Wells of ELISA microtiter plate coated with biotinylated cLf or bLf as positive control gave mean ±SD of 0.829±0.012 and 0.75±0.03, respectively.

We previously confirmed that biotinylated cLf was recognized by two membrane proteins of MRSA [3]. Additionally, bacterial outer membrane protein OmpC of *E. coli and S. typhi* was found to complex with the antibacterial eukaryotic protein camel lactoferrin [23].

Our data agree with various studies in the support of the concept that the ferrochelating properties of both lactoferrins are not the only causing factor of their antibacterial activity [3, 23, 24-27]. In fact, the antimicrobial efficiencies of cLf and bLf samples used in this work were not directly correlated with the levels of their partial iron saturation. Here, the less efficient in antimicrobial activity; bLf had higher partial iron saturation (10%) than the more potent cLf which was 35% iron-saturated. This suggests that some antimicrobial mechanisms for lactoferrins not related to ferrochelation are present such as their binding to bacterial membrane proteins. Also, this proves the fact that OA binding has no effect on Lf binding to bacterial membrane proteins. Even a membrane protein; FadL porin that has highest specific fatty acid binding affinity for oleic acid was characterized in *E. coli* [28].

Table 5: Detection of Lf and its complexes binding to bacterial membrane proteins by ELISA

Test sample/ organism	OD at 405 nm (mean±SD)			
	cLf a	bLf a	cLf-OAa	bLf-OAa
Blank	0.058±0.007	0.04±0.011	0.042±0.005	0.056±0.01
MRSA	0.314±0.012	0.265±0.02	0.291±0.007	0.249±0.041
S. aureus	0.261±0.02	0.35±0.17	0.337±0.018	0.217±0.011
B. cereus	0.306±0.01	0.281±0.04	0.319±0.027	0.229±0.003
E. coli	0.376±0.07	0.314±0.03	0.487±0.016	0.454±0.026
S. sonnei	0.341±0.12	0.217±0.012	0.397±0.008	0.31±0.007
S. typhi	0.322±0.05	0.346±0.03	0.338±0.025	0.347±0.018
K. pneumonia	0.35±0.073	0.276±0.014	0.292±0.023	0.299±0.012
P. aeruginosa	0.311±0.12	0.327±0.01	0.275±0.006	0.306±0.008
S. marcescens	0.279±0.02	0.24±0.015	0.254±0.026	0.197±0.003
P. vulgaris	0.291±0.12	0.254±0.07	0.344±0.014	0.188±0.01

[a]Biotinylated cLf, bLf, cLf-OA, or bLf-OA reacted significantly (*P<0.05*) with bacterial membrane fractions preparations compared to carbonate/bicarbonate pH 9.6 buffer used as a blank.

The observed differences in biological activities between cLf and other species of Lf are likely due to the variance in some of its structure-related characteristics. cLf was found to comprise 689 amino acids and contain 17 disulfide bridges and 4 predicted glycosylation sites, one of them in the N-lobe and three found in the C-lobe. The pattern of disulfide bonds in cLf is the same as that found in hLf, but the positions of predicted glycosylation sites are totally different in cLf. Besides, the amino acid sequence of cLf is 70% identical to the sequences of other lactoferrins, but the first 50 residues or so of cLf N-termini show an identity of less than 40%. Some residues associated with movement of domains in the protein are different in cLf from those found in other lactoferrins, revealing the likelihood of specific structural differences [29, 30]. We also previously proved that elevated levels of intrinsic disorder in the N-terminal region of cLf can influence the functionality of this region [3].

All the tested microorganisms in this study are of significance as human pathogens, mostly showing resistance to many antibiotics, and chosen to be both Gram-negative and Gram-positive bacteria as well as fungi to indicate broad spectrum

activity of the formulated cLf/bLf-OA complexes. *S. aureus* can cause infection in tissues and sites with lowered host resistance as in case of damaged skin or mucous membranes. It is a very common cause of infection in hospitals, mostly capable of infecting newborn babies and surgical patients. Strains of *S. aureus* differ in their degree of susceptibility to particular antibiotics [31]. Moreover, methicillin-resistant strains; MRSA have emerged which complicate the treatment of staphylococci infections because methicillin is considered as the first option in treatment of *S. aureus* infection and also because resistance to methicillin means resistance to all β-lactam antibiotics. The epidemic of MRSA infections occurs mostly in hospitals. MRSA has become one of the leading causes of death in hospitalized patients around the world [32].

B. cereus causes a minority of food borne illnesses (2–5%), resulting in severe nausea, diarrhea, and vomiting. *Bacillus* food borne infections arise because of survival of the bacterial endospores when food is improperly cooked. It produces beta-lactamases, thus it is resistant to beta-lactam antibiotics [33].

E. coli is a pathogen associated with acute gastroenteritis in infants up to 2 years old and rarely in adults with lowered resistance besides infections of urinary tract. Outbreaks of gastroenteritis can cause high fatality rates in maternity nurseries and institutions caring for young children. Antibiotics have insignificant role in treatment of acute stage in severe cases of gastroenteritis as they are not fast enough to stop further body fluid loss [34]. *Shigella sonnei* causes shigellosis (bacillary dysentery) and produces Shiga toxins that target the vascular endothelium, inhibiting protein synthesis within target cells by a mechanism similar to that of ricin [35]. On the other hand, *S. typhi* spreads by food or water contaminated with feces resulting in typhoid fever, with a risk of death of about 20% without treatment [36]. While antibiotics are capable of shortening the span of a diarrheal infection, particularly if administered early, pathogenic *Shigella* and *Salmonella* species are often resisting the effects of common antibiotics, including ampicillin, trimethoprim-sulfamethoxazole, and third generation cephalosporins.

K. pneumonia is a rare cause of bacterial pneumonia but its significance lies in high case mortality in such cases. It is resistant to multiple antibiotics and can produce extended-spectrum beta-lactamases against all beta-lactam antibiotics, except carbapenems [37]. *P. aeruginosa* is a multidrug resistant pathogen associated with serious diseases-hospital-acquired infections such as ventilator-associated pneumonia and sepsis syndromes [38]. *S. marcescens* causes an opportunistic infection in respiratory tract, urinary tract, the eye (keratitis, conjunctivitis, endophthalmitis, and tear duct infections), and wounds. Most *S. marcescens* strains are resistant to numerous antibiotics because of the presence of R-factors; intrinsically resistant to macrolides, ampicillin, and first-generation cephalosporins (such as cephalexin) [39]. *P. vulgaris* is found in individuals in long-term care facilities and hospitals and those with compromised immune systems [40].

Candida infects immunocompromised patients diagnosed with serious diseases such as HIV and cancer. *Candida* commonly causes nosocomial infections. It affects high risk patients who recently undergone surgery, a transplant or are in the Intensive Care Units, leading to malnutrition and interference with the absorption of medication. *A. niger* causes fungal ear infections while *A. flavus* is a common cause of fungal sinusitis and cutaneous infections and noninvasive fungal pneumonia.

Conclusion

This study revealed that cLf and bLf could bind OA and exhibited a much stronger antimicrobial activity than their free forms, especially in case of cLf. Additionally, inhibitory activity of cLf and cLf-OA against test microorganisms noticeably exceeded that of corresponding bLf and bLf-OA. Our study undoubtedly confirmed the presence of no effect by OA binding to lactoferrins on Lf binding to bacterial membrane proteins.

Competing Interests

The author declares having no competing interests.

References

1. Sánchez L, Calvo M, Brock JH. Biological role of lactoferrin. Arch Dis Child. 1992;67(5):657-661.

2. Conesa C, Sanchez L, Rota C, Perez MD, Calvo M, Farnaud S, Evans RW. Isolation of lactoferrin from milk of different species: calorimetric and antimicrobial studies. Comp Biochem Physiol B Biochem Mol Biol. 2008;150(1):131–139.

3. Redwan EM, El-Baky NA, Al-Hejin AM, Baeshen MN, Almehdar HA, Elsaway A, et al. Significant antibacterial activity and synergistic effects of camel lactoferrin with antibiotics against methicillin-resistant Staphylococcus aureus (MRSA). Res Microbiol. 2016;167(6):480-491.

4. Brock JH. The physiology of lactoferrin. Biochem Cell Biol. 2002; 80(1): 1–6.

5. Naidu AS. Lactoferrin: natural, multifunctional, antimicrobial. Boca Raton: CRC Press. 2000; pp. 1–2.

6. Farrington M, Brenwald N, Haines D, Walpole E. Resistance to desiccation and skin fatty acids in outbreak strains of methicillin-resistant Staphylococcus aureus. J Med Microbiol. 1992;36(1):56-60.

7. Hazell SL, Graham DY. Unsaturated fatty acids and viability of Helicobacter (Campylobacter) pylori. J Clin Microbiol. 1990;28(5):1060-1061.

8. Bergsson G, Arnfinnsson J, Steingrimsson O, Thormar H. In vitro killing of Candida albicans by fatty acids and monoglycerides. Antimicrob Agents Chemother. 2001;45(11):3209-3212.

9. Nuijens JH, van Berkel PH, Schanbacher FL. Structure and biological actions of Lactoferrin. J Mammary Gland Biol Neoplasia. 1996;1(3):285-295.

10. Ohashi A, Murata E, Yamamoto K, Majima E, Sano E, Katunuma N, et al. New functions of lactoferrin and β-casein in mammalian milk as cysteine protease inhibitors. Biochem Biophys Res Commun. 2003;306(1):98-103.

11. Baker EN, Baker HM. A structural framework for understanding the multifunctional character of lactoferrin. Biochimie. 2009;91(1):3-10.

12. Fast J, Mossberg AK, Nilsson H. Compact oleic acid in HAMLET. FEBS Lett. 2005;579(27):6095-6100.

13. Redwan EM, Tabll A. Camel lactoferrin markedly inhibits hepatitis C virus genotype 4 infection of human peripheral blood leukocytes. J Immunoassay Immunochem. 2007;28(3):267-277.

14. Parry RM, Brown EM. Lactoferrin conformation and metal binding properties. Advances in experimental medicine and biology. 1974;48:141-160.

15. Lowry OH, Rosebrough NJ, Farr AL, Randall RJ. Protein measurement with the Folin phenol reagent. J Biol Chem. 1951;193(1):265-275.

16. Ye X, Wang H, Liu F, Ng T. Ribonuclease, cell-free translation-inhibitory and superoxide radical scavenging activities of the iron-binding protein lactoferrin from bovine milk. Int J Biochem Cell Biol. 2000;32(2):235-241.

17. Duncombe WG. The colorimetric micro-determination of long-chain fatty acids. Biochem J. 1963;88(1):7-10.

18. Wikler MA. Performance Standards for Antimicrobial Susceptibility Testing. Eighteenth Informational Supplement. M100-S18. ed. C.L.S.I. (Clinical and Laboratory Standard Institute), Pennsylvania, PA, USA. 2008.

19. Cockerill FR, Wikler MA, Alder J, Dudley MN, Eliopoulos GM, Ferraro MJ, Hardy DJ, Hecht DW, Hindler JA, Patel JB, Powell M, Swenson JM, Thomson RB, Traczewski MM, Turnidge JD, Weinstein MP, Zimmer BL. Methods for Dilution Antimicrobial Susceptibility Tests for Bacteria That Grow Aerobically; Approved Standard. CLSI document M07-A9.9. 9 ed. C.L.S.I. (Clinical and Laboratory Standard Institute), Pennsylvania, PA, USA. 2012.

20. Dilika F, Bremner PD, Meyer JJ. Antibacterial activity of linoleic and oleic acids isolated from Helichrysum pedunculatum: a plant used during circumcision rites. Fitoterapia. 2000;71(4):450-452.

21. Choi JS, Park NH, Hwang SY, Sohn JH, Kwak I, Cho KK, et al. The antibacterial activity of various saturated and unsaturated fatty acids against several oral pathogens. J Environ Biol. 2013;34(4):673-676.

22. Kabara JJ, Swieczkowski DM, Conley AJ, Truant JP. Fatty acids and derivatives as antimicrobial agents. Antimicrob Agents Chemother. 1972;2(1):23–28.

23. Sundara Baalaji N, Ravi Acharya K, Singh TP, Krishnaswamy S. High-resolution diffraction from crystals of a membrane-protein complex: bacterial outer membrane protein OmpC complexed with the antibacterial eukaryotic protein lactoferrin. Acta Cryst. 2005;F61:773–775.

24. Arnold RR, Cole MF, Mcghee JR. A bactericidal effect for human lactoferrin. Science. 1977;197(4300):263-265.

25. Ellison RT III, Giehl TJ, LaForce FM. Damage of the outer membrane of enteric gram negative bacteria by lactoferrin and transferrin. Infect Immun. 1988;56(11): 2774-2781.

26. Ellison RT III, LaForce FM, Giehl TJ, Boose DS, Dunn BE. Lactoferrin and transferring damage of the Gram negative outer membrane is modulated by Ca2+ and Mg2+. J Gen Microbiol. 1990;136(7):1437-1446.

27. Visca P, Dalmastri C, Verzili D, Antonini G, Chiancone E, Valenti P. Interaction of lactoferrin with Escherichia coli cells and correlation with antibacterial activity. Med Microbiol Immunol. 1990;179(6):323-333.

28. Black PN. Characterization of FadL-SPECIFIC fatty acid binding in Escherichia coli. Biochim Biophys Acta. 1990;1046(1):97-105.

29. Khan JA, Kumar P, Paramasivam M, Yadav RS, Sahani MS, Singh TP, et al. Camel lactoferrin, a transferrin-cum-lactoferrin: crystal structure of camel apolactoferrin at 2.6 A resolution and structural basis of its dual role. J Mol Biol. 2001;309(3):751-761.

30. Redwan EM, El-Fakharany EM, Uversky VN, Linjawi MH. Screening the anti- infectivity potentials of native N- and C-lobes derived from the camel lactoferrin against hepatitis C virus. BMC complementary and alternative medicine. 2014;14: 219.

31. Mele T, Madrenas J. TLR2 signalling: at the crossroads of commensalism, invasive infections and toxic shock syndrome by Staphylococcus aureus. Int J Biochem Cell Biol. 2010;42(7):1066–1071.

32. Nguyen GC, Patel H, Chong RY. Increased prevalence of and associated mortality with methicillin-resistant Staphylococcus aureus among hospitalized IBD patients. Am J Gastroenterol. 2010;105:371-377. Doi:10.1038/ajg.2009.581

33. Kotiranta A, Lounatmaa K, Haapasalo M. Epidemiology and pathogenesis of Bacillus cereus infections. Microbes Infect. 2000;2(2):189–198.

34. Todar K. Pathogenic E. coli. Online Textbook of Bacteriology. University of Wisconsin–Madison Department of Bacteriology. 2007

35. Torres AG. Current aspects of Shigella pathogenesis. Rev Latinoam Microbiol. 2004;46(3-4):89-97.

36. WHO/Typhoid fever. www.who.int. Archived from the original on 2017-07-27.

37. Hudson C, Bent Z, Meagher R, Williams K. Resistance Determinants and Mobile Genetic Elements of an NDM-1-Encoding Klebsiella pneumoniae Strain. PLOS ONE. 2014;9(6):e99209.

38. Balcht A, Smith R. Pseudomonas aeruginosa: Infections and Treatment. Infectious Disease and Therapy Series. 1994(9); 615.

39. Pathogen Safety Data Sheets: Infectious Substances – Serratia spp. Public Health Agency of Canada. 30 April 2012.

40. Sydnor E. Hospital Epidemiology and Infection Control in Acute-Care Settings. Clinical Microbiology Reviews. 2011;24(1):141–173.

17

Protein and DNA Isolation from *Aspergillus Niger* as well as Ghost Cells Formation

Nawal Abd El-Baky[1#] , Mona M. Sharaf[1], Eman Amer[1], Hoda Reda Kholef[1],
Mohamed Zakaria Hussain[2] and Amro A. Amara[1#*]

[1]*Protein Research Department, Genetic Engineering and Biotechnology Research Institute, City for Scientific Research and Technological Applications, Universities and Research Center District, New Borg El-Arab, Alexandria, Egypt*

[2]*Department of Microbiology and Immunology, Faculty of Medicine, Tanta University, Tanta*

[#]*Both authors contributed equally to this study.*

Corresponding author: *Amro Abd Al Fattah Amara, The head of the Protein Research Department, Genetic Engineering, and Biotechnology Research Institute, City for Scientific Research and Technological Applications, Universities and Research Center District, New Borg El-Arab, Egypt, E-mail: amroamara@web.de*

Abstract

Recently, a protocol for ghost cells preparations was introduced. It was given the name sponge-like protocol: Procaryotes, eucaryotes and virus were turned to ghost cells using such protocol. In this study, with slight modifications, Aspergillus niger ghost cells were prepared using the same protocol. Both the Minimum Inhibitory Concentration (MIC) and the minimum growth concentration (MGC) values for H_2O_2, NaOH, $NaHCO_3$ and SDS against A. niger were determined. Five different randomization experiments were conducted instead of the full Plackett–Burman design. During the ghost preparation steps, the released Protein and DNA were measured spectrophotometrically at 280nm and 260nm, respectively. The quality of the prepared ghost cells were evaluated during the preparation steps using light microscope. Transmission electron microscope was used for evaluating the final steps. Protein and DNA electrophoresis were conducted to evaluate the quality of the released protein and DNA after each randomization experiment. The data obtained prove correct evacuation of the fungal cells from their cytoplasmic content during the successive steps. The study not only introduces a protocol for preparing ghost cells from Aspergillus niger but also enables the isolation of both of protein and DNA. The idea, the concept and the tools used in this study could establish a more sensitive method for protein and DNA isolation using any of four utilized chemical compounds. This proposes the same concept of enzyme-induced cell lysis which is based on minimizing the effect of used chemicals or enzymes. The study recommended extending the benefit of the sponge-like protocol from being a protocol for ghost cells preparation to DNA and protein isolation technique using the same concept.

Keywords: Aspergillus niger; ghost cells; DNA isolation; Protein isolation

Introduction

Aspergillus niger is one of the most commonly found fungi either in in/outdoor environments. It causes black mould diseases in foods as vegetables, grapes, apricots, onions, peanuts, etc. Some reports prove that it is responsible for producing some dangerous mycotoxins. *A. niger* is less likely to cause human disease than some other *Aspergillus sp.* One of the most famous diseases it causes to human is the Aspergillosis; a lung disease. It is one of the otomycosis causing fungi, resulting in ear pain, temporary hearing loss, and head pain etc. In severe cases, it causes damage to ear canal and tympanic membrane. In fact, most people breath in Aspergillus spores including *A. niger* every day. In most cases, it causes nothing for healthy individuals if entered in a little quantity [1, 2]. However, it turns to be pathogenic in cases such as immunocomprimized patients, illness, lung disease, and if inhaled in large amount. One could conclude that weak or compromised immune system is the most claimed causative agent for *A. niger* pathogenesis as well as exposure to large amount of spores or continuous source that enable their accumulation in the body particularly the lung. *A. niger* spores could be found everywhere but one of the most dangerous sources is the dry grinded seeds where the spore will not be germinated and could by one or another way reach to our bodies as spore. Spores are more dangerous than vegetative cells except in case of mycotoxin-producing strains. *A. niger* is not harmful in all cases, it is also beneficial and can be used in different biotechnological applications such as in the citric acid production [3].

Evacuating microbes from their cytoplasmic content is a natural phenomenon [4]. Pores could be introduced to the microbial cells as a result of different mechanisms [1 8]. Such as the evacuation of the gram-negative bacteria by the bacteriophage infections [4]. The bacteriophage E lysis gene is used for evacuating the cells and turning them to ghosts by controlling its expression using heat sensitive promoter [4, 5, 8-11]. Recently, the Sponge Like protocol was introduced [12-15]. Its main concept is using active chemical compounds that could introduce pores in the microbes and degrade the DNA at concentrations which did not change the surface antigens or the 3D structure [13]. This enables evacuating gram-negative and gram-positive bacteria, eukaryotes and viruses [4, 12-14, 16-27].

In this study, *A. niger* representing the fungi was prepared as ghosts. In addition, the idea for using the Sponge Like protocol for releasing both of DNA and protein for various applications was established. Using such protocol based on using chemical compounds or enzymes, not only the ghost cells are produced but also both of DNA and protein are isolated.

Materials and Methods

Fungal strain

A. niger strain used in this study was kindly identified and obtained from Al-Azhar University Mycology Center (Cairo, Egypt).

Cultivation Conditions

The *A. niger* strain was cultivated in one litre flask contains 500 mL Sabouraud's dextrose broth at 28°C for 7 days at 150 rpm.

Determination of the MIC and MGC Values for NaoH, SDS, NaHCO₃ And H₂O₂

Standard broth microdilution susceptibility assay for determining the MIC values for each of NaOH, SDS, NaHCO$_3$ and H$_2$O$_2$ was conducted [13, 14]. The MIC value for each compound was calculated as well as the concentration which allows first fungal growth which is abbreviated as MGC (the concentration showing first growth after the MIC).

Determination of the MIC and MGC values for NaOH, SDS, NaHCO₃ and H₂O₂

Standard broth microdilution susceptibility assay for determining the MIC values for each of NaOH, SDS, NaHCO$_3$ and H$_2$O$_2$ was conducted [13, 14]. The MIC value for each compound was calculated as well as the concentration which allows first fungal growth which is abbreviated as MGC (the concentration showing first growth after the MIC).

Randomization experiment

Five experiments were conducted to map the best preparation conditions either during the cells cultivation or during the ghost cells preparation. The different variables were four chemical compounds representing SDS, H2O2, NaHCO3, and NaOH, and three physical parameters represented as cultivation temperature, shaking rate during treatment, and the cultivation time. The seven variables were randomized according to the design in Table 1.

Table 1: The randomization experiments for ghost cells preparation including 7 variables in +1 or -1.

Experiment Number	NaOH	SDS	NaHCO₃	H₂O₂	Shaking rate during treatment	Cultivation time	Cultivation temperature
1	+1	-1	+1	+1	+1	-1	-1
2	+1	+1	+1	-1	-1	-1	-1
3	-1	+1	-1	+1	-1	-1	+1
4	+1	-1	-1	+1	-1	+1	-1
5	+1	-1	-1	-1	+1	-1	+1

Each variable of the four used chemical compounds was represented at two levels (high and low), which are donated by +1 (MIC) and -1 (MGC) as in Table 1. The +1 in shaking conditions refers to shaking at 150 rpm during the different treatments while -1 refers to static incubation during treatment, +1 in growth time represents cultivation for 10 days while -1 represents cultivation for 5 days, +1 in growth temperature refers to cultivation at 28°C while -1 represents cultivation at 25°C.

The biomass of the different cultivation conditions was collected and washed gently by 0.5% saline and recentrifuged at 6000 rpm for 10 min. The supernatant was then discarded. 5X stock for each of NaOH, SDS, NaHCO$_3$ and 2X stock for H$_2$O$_2$ were prepared from both +1 (MIC) and -1 (MGC), which were determined as mentioned above.

All the experiments were conducted in three steps. The first step contains NaOH, SDS, NaHCO$_3$ where 1 ml from each was added to 1 ml water and finally 1 ml of the fungal suspension (0.5 gm of the fungus mat/ml) was added to get a final concentration of each of NaOH, SDS, NaHCO$_3$ equal to 1X. The second step contains H$_2$O$_2$. The H$_2$O$_2$ was used as 2X and one ml of the fungal suspension was added to reach final concentration of H$_2$O$_2$ equal to 1X.

After each treatment step, the supernatant was collected by centrifuging the fungus at 6000 rpm. After that, the fungal pellet was washed using 1X Phosphate Buffered Saline (PBS) (alternatively common saline solution can be used). In the third step, the cell pellets were washed using 60% Ethanol and left at room temperature. After each of the above washing steps and centrifugation the supernatant was preserved to determine the amount of the released protein and DNA.

Fungal Cells Evaluation Using Light Microscope

Fungal sample from each randomization experiment was examined by light microscope. The quality of the cells from each experiment has been determined based on the cellular structure as being either intact or deformed and then the overall fungal ghost quality is given as %.

Determination of the DNA Concentration

The concentration of DNA in the supernatant after each step for each randomization experiment was determined by measuring the absorbance at 260 nm. Quartz cuvette was used. An extinction 260 = 1 corresponds to 50 µg dsDNA mL^{-1} [28].

Determination of the Protein Concentration

Concentration of released protein from each step in each randomization experiment (the different supernatants) was determined using the spectrophotometer at 280 nm. Quartz cuvette was used. The protein concentration was derived from Bovine Serum Albumin (BSA) standard curve (Figure 1) [15].

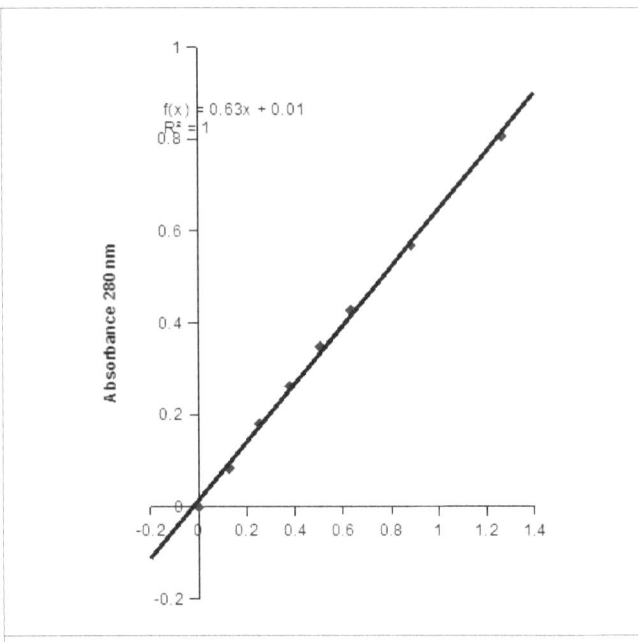

Figure 1: Bovine serum albumin protein standard curve

Determination of the Fungal Viability

The various fungal ghost preparations were investigated for the possibility of the presence of any viable cells by subjecting them to growth on Sabouraud's dextrose agar plates at 28°C for 7 days.

Transmission Electron Microscope for Examination of *A. niger* Gs

Transmission electron microscope (JEOL TEM 100 CX) was used for the examination of *A. niger* Gs.

Agarose Gel Electrophoresis

DNA in the supernatant for each randomization experiment was examined using agarose gel electrophoresis. 1% agarose gel containing 0.5 μg/mL ethidium bromide was run in a horizontal gel electrophoresis unit (Mini-Sub DNA cell, BioRad). The running buffer was TAE (40 mM Tris, 20 mM acetic acid, 1 mM EDTA, pH 8.0). Electrophoresis was carried out at 100 V for 1 h on an Amersham-Pharmacia Biotech (Uppsala, Sweden) power supplier unit ECPS3000/150. The stained bands were visualized with UV light (309 nm) using a transilluminator, and the gel was recorded as digital image using a gel documentation system (UVI-Tech).

Polyacrylamide Gel Electrophoresis

Protein in the supernatant for each randomization experiment was examined on 12% sodium dodecyl sulfate-polyacrylamide gel electrophoresis (SDS-PAGE). About 50 μl of each sample were heated for 2-10 min at 100°C after the addition of 5 μl of 5× loading buffer. Spin down for 1 second to remove debris and subjected to SDS-PAGE with a 0.5-mm-thick gel in Invitrogen device Novex Minicell. Prestained marker heated at 95°C was used to calibrate protein mobility.

Results and Discussion

The production of microbial ghost cells is an aim for scientists as well as the isolation of DNA and protein. There are a large number of protocols for isolating and purifying DNA and protein. Each protocol is designed to match certain criteria and uses certain compounds in defined steps. Adding new idea concerning isolation of both DNA and protein will add new possibility and choices for different situations. The production of ghost cells is achieved by removing the cytoplasm from the microbial cells. The process is very sensitive but easy to conduct after determining both MIC and MGC. From the first step in the Sponge-like protocol, both DNA and protein were given special interest as parameters could be easily monitored spectrophotometrically and indicate microbial cells loss of them during cell cytoplasm loss. *A. niger* was used for first time to evaluate the Sponge-like protocol as a tool which is able to produce microbial ghosts and also could be adjusted to isolate DNA and protein.

As described earlier in the first protocol, the main feature that distinguishes this protocol from others was the determination of both MIC and MGC. An early work was done by introducing modifications in the plasmid alkaline lysis protocol via omitting the last step and adding the phenol extract step to turn it from plasmid isolation to DNA isolation protocol [29]. Such non-enzymatic DNA isolation protocol was used with different microbes and proves efficiency [29]. However, that study revealed variation in the prepared DNA quality as a fixed protocol was applied to different microbes that showed different responses to the used chemical compounds.

After improving the Sponge-Like protocol and proving its efficacy [15], it becomes clear that different microbes have different responses to the used chemicals and thus they show different MIC and MGC values. One of the benefits of this protocol that it can be used for strain differentiation rather than ghost cells preparation. Additionally, enzymes and proteins are able to introduce pores in the microbes when used at their MIC and MGC values [4, 22]. The white egg lysozyme was used to induce an emergence protocol for ghost cells preparation that could be used globally [22].

Our scientific group are working extensively to establish the Sponge-like protocol and to investigate the possibility of its use in DNA and protein preparation. This study is the first trial to prepare fungal ghost cells and investigate the possibility of DNA and protein isolation from fungal cells via Sponge-Like protocol. *A. niger* was selected as a representative fungus. The MIC and MGC values of different used chemicals were determined. NaOH, SDS, H_2O_2, and $NaHCO_3$ showed MIC values of 0.01g/mL, 0.0001g/mL, 0.3mL/mL, and 0.01g/mL, respectively, while gave

MGC values of 0.001 g/mL, 0.00001 g/mL, 0.15mL/mL, and 0.001 g/mL, respectively.

After determining MIC and MGC for such compounds against *A. niger,* five randomization experiments were conducted.Results of these randomization experiments are summarized in Table 2. Data from experiments number 2 and 3 in step one show high release for the protein due to the presence of SDS as +1 in both. In experiment number 3, the culture cultivation temperature was 28°C which might be responsible for the slight increase in the released protein than in experiment number 2.

In step two (H_2O_2) in experiment number 3, results still show high protein release and the same was in step 3. Experiments 4 and 5 in step one, there are no treatment differences but the amount of the released protein in experiment 4 is higher than experiment 5. The apparent explanation that in experiment 4, the cells are grown for a longer time (10 days) and for that they contain more protein. In experiments 4 and 5 in the second step still experiment number 4 has higher release to the protein for the same reason in addition to the high amount of the H_2O_2 might be responsible for the increase in the released protein. In experiment number one SDS was -1 for that it shows lower release to the protein than in experiments number 2 and 3, however and due to the presence of shaking, its release is continuous in step 2 and 3. Considering the effect of the shaking rate as in experiments number 1 and 5 where the SDS was -1, apparently it induces the release of the protein from cells in the successive steps. H_2O_2 apparently not highly effective in the protein release as it was in its +1 at experiments number 1, 3 and 4 but only experiment 3 shows the highest released protein due to the presence of the +1 SDS. Simultaneously, the effect of SDS on the released amount of DNA is clear in step 1 as experiments number 2 and 3 show the highest DNA release. The presence of NaOH and $NaHCO_3$ in +1, decreases the amount of the released DNA because they might cause degradation for it, that can be observed in step

1 at experiment 2 and 3. H_2O_2 in experiments number 2 and 5 shows the lower release which might be due to the degradation of the DNA where NaOH is present in its +1 which might have a synergistic effect with H_2O_2 to degrade DNA and in the absence of NaOH in experiment 3. The amount of the released DNA is more than the experiments number 2 and 5. So in the same experiment (No_3) step 2 where H_2O_2 is -1 shows decrease in the amount of the DNA than step 1 which proves that H_2O_2 is a DNA-degrader. The Ethanol step still show variable release to each of the protein and the DNA according to the pores formed and the protein and the DNA still existe.

In experiments 1 and 5 where the shaking was +1 it might have negative effect on the cells quality. That might be explained by the fact that cells upon their losing to their cytoplasmic content also loss their rigidity.

The different experiments show different ghost cells quality as presented in Table 2. One should put in his consideration that the representative images are taken after the third step for each experiment which represent the final preparation. The quality was given as a percentage and calculated based on 20% for the appearance of clear nuclei in the hypha, 20% for the correct 3D structure of the hypha, 20% for the absence of the spores in the tested sample, 20% for the absence of the sporangia, and 20% for the general appearance of the sample.

Figure 2 summarizes the results of light microscope examination to prepared fungal ghost cells after each experiment. The wild type shows clear sporangia with well-differentiated spores and hyphae. The prepared fungal ghost cells in the five experiments lost the ability to produce the sporangia including the spores. Hyphae elongation was evident. The loss of the nuclei was observed clearly but in variable degrees. The samples were shaked during ghost cells preparation showed a sort of hyphal intertwining and shrinkage (Figure 2 images d and h).

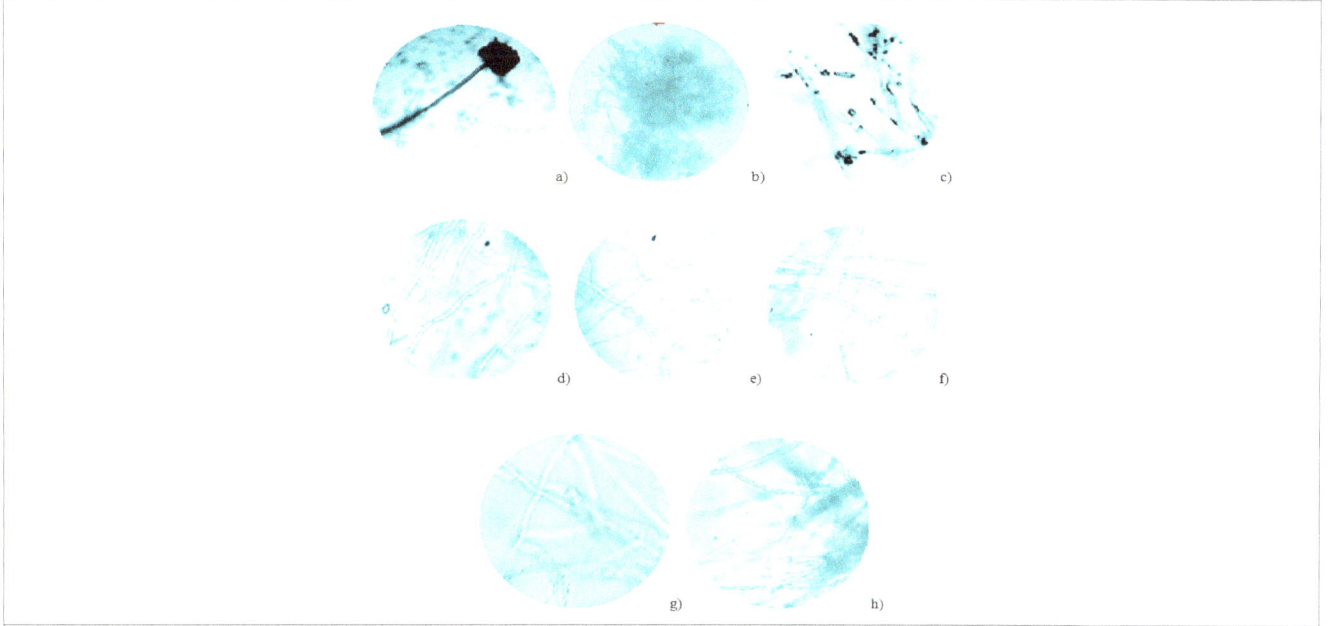

Figure 2: a) A. niger wild type untreated b) A. niger sporangia untreated c) A. niger hyhae untreated d) ghost cells from experiment 1 e) ghost cells from experiment 2 f) ghost cells from experiment 3 g) ghost cells from experiment 4 h) ghost cells from experiment 5

No growth was obtained on Sabouraud's dextrose agar plates after ghost cells cultivation from any of the five experiments which prove the loss of the cytoplasmic content during the process and the death of the fungus. However, the hyphae elongation was not definitely known to occur after which of the three steps of ghost cells preparation.

In conclusion, this study succeeded in fungal cell evacuation and ghost cells preparation releasing both DNA and protein content.

The examination by transmission electron microscope results in a clear prove for the evacuation of the analyzed samples. By comparing experiment 1 with experiment 3 as in Figure 3, the image 3a from experiment number 1 magnification of the section of *A. niger* cells at 1200x represents less loss of A. niger cytoplasm while the image 3b in experiment number 3 magnification of the section of *A. niger* cells at 1200x the sample lost most of

its cytoplasmic content. This does not agree with the results in Table 1 where experiment 1 gives 70% quality while experiment 3 gives 60% quality. Even if they are both near in their quality % however, electron microscope gives better judgement. Image 3b shows better loss for the cytoplasmic content which proves the power of the Sponge-Like protocol in evacuating the microbes. Comparing image 3c which represents sample from experiment number 1 magnied at 10000x with image 3d which represents sample from experiment number 5 magnied at 10000x, the result agrees with that in Table 2 where the quality of the experiment 1 was 70% and of experiment 5 was 55%. The transmission electron microscope proves that the different preparation based on each experiment give different results, however, *A. niger* cells were turned to ghost cells. Some cells did not lose their cytoplasmic content however, they become nonviable as proved by the viability test. Those samples which show existence of the cytoplasm need more washing processes.

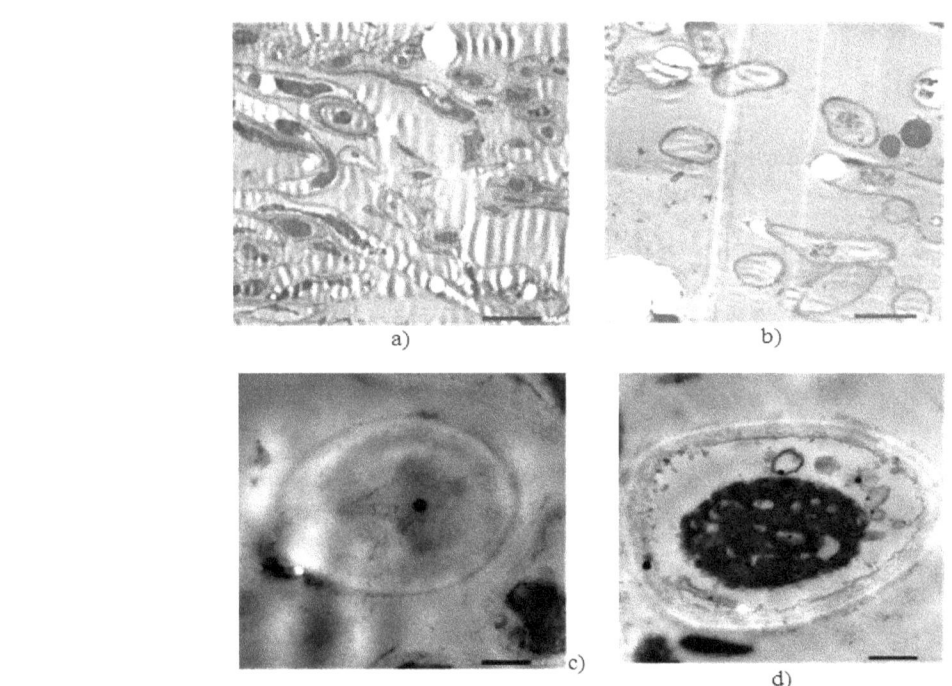

Figure 3: The results of transmission electron microscope for *A. niger* different treatments and magnifications: a) Experiment number 1 magnification of the section of *A. niger* cells at 1200x; b) Experiment number 3 magnification of the section of *A. niger* cells at 1200x; c) Experiment number 1 magnification of the section of *A. niger* cells at 10000x; d) Experiment number 5 magnification of the section of *A. niger* cells at 10000x

Table 2 : The results of randomization experiments for ghost cells preparation.

Experiment Number	NaOH/SDS/ NaHCO₃		H₂O₂		Ethanol (60%)		Cell quality %
	DNA (µg/ml)	Protein (mg/ml)	DNA (µg/ml)	Protein (mg/ml)	DNA (µg/ml)	Protein (mg/ml)	
1	4.25	0.092	17.1	0.181	2.15	0.29	70%
2	28.25	0.616	0.15	0.001	4.35	0.003	80%
3	48.5	0.752	31.45	0.416	1.9	0.138	60%
4	6.2	0.233	11.6	0.119	0.15	0.12	50%
5	3.1	0.061	0.15	0.024	0.3	0.126	55%

The SDS-PAGE analysis of the different samples supernatant prove the existence of protein in all supernatants of the five conducted experiments [Figure 4]. However, the samples represent the experiments end result which is another prove for the loss of the protein during the experiments steps. For obtaining better protein preparation, one should collect the protein of each step during each experiment and follow that with protein precipitation to concentrate the protein samples. The DNA agarose gel results which represent the end point of each of the five conducted experiments are the same [Figure 5, Figure 6]. The DNA from the samples were nearly removed. In addition, during the preparation steps more DNA in each sample is shown. The overall protocol succeeded to prepare *A. niger* as ghost cells and shows the first trial to prepare protein and DNA using the critical chemical concentration of the used compounds. One should not neglect the side effect of H_2O_2 on DNA which has been used in all of the experiments either with MIC or MGC but still effective. The data reveal more optimization is in need. The concept is clear, if MIC and MGC of the used chemical compounds are able to turn *A. niger* to ghost cells by inducing pores so the released DNA and cytoplasm can be collected. By improving such protocol, it will be a universal protocol for preparing both DNA and protein as well as ghost cells. In addition, other micro and macromolecules as well as different biological elements could be prepared.

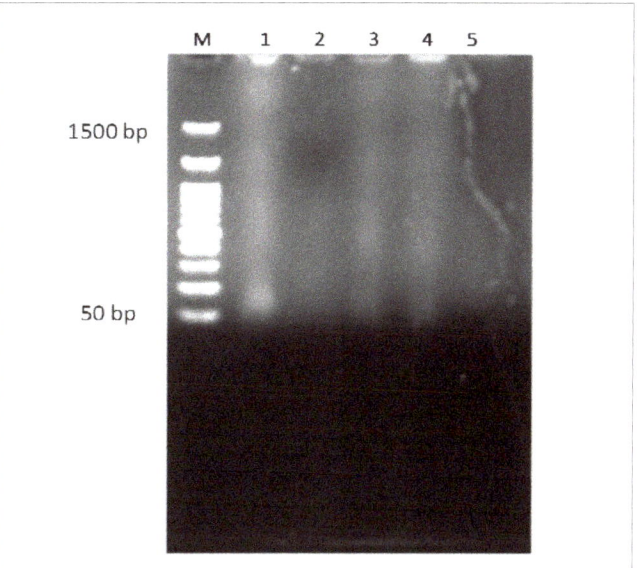

Figure 5: Agarose gel electrophoresis for *A. niger* ghost cells prepared in different experiments. DNA marker (lane M), *A. niger* during ghost preparation (lanes 1-5).

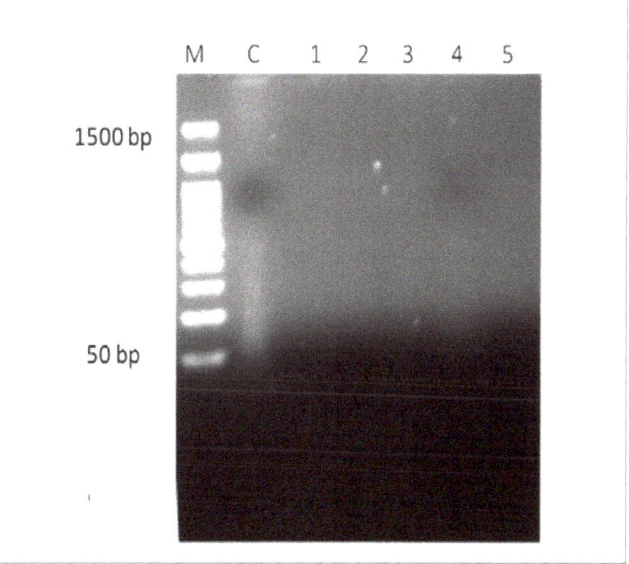

Figure 6: Agarose gel electrophoresis for *A. niger* ghost cells prepared at the end of different experiments. DNA marker (lane M); *A. niger* control (lane C); *A. niger* after ghost preparation (lanes 1-5).

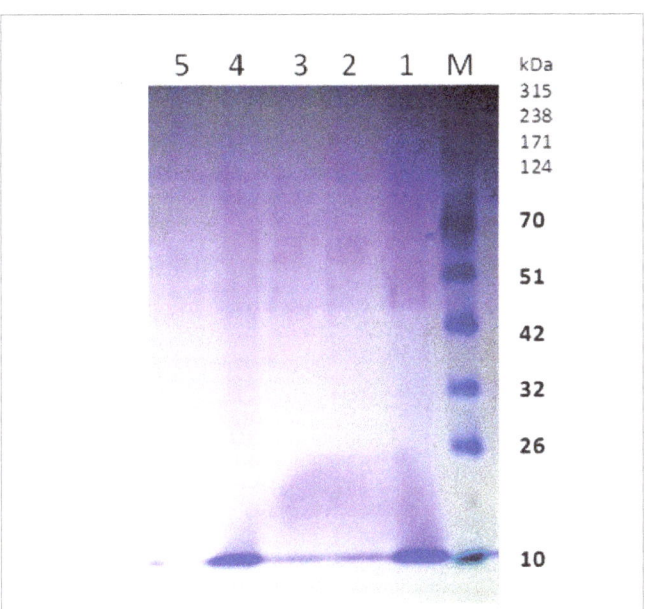

Figure 4: 12% SDS-PAGE analysis of different experiments for *A. niger* ghost cells preparation. Lanes 1-5 represent supernatant from experiments 1 to 5 respectively for *A. niger* ghost cells preparation. M, represent protein marker

Conflict of Interests

The authors have no conflict of interests to declare.

References

1. Egbuta MA, Mwanza M, Babalola OO. A review of the ubiquity of Ascomycetes filamentous fungi in relation to their economic and medical importance. Advanced in Microbiology. 2016;6:1140-1158. DOI: 10.4236/aim.2016.614103

2. Feldmesser M. Prospects of vaccines for medically important fungi. Medical Mycology. 2005;43(7):571-587.

3. Schuster E, D-N Coleman, Frisvad J, van Dijck P.On the safety of Aspergillus niger- A review. Applied microbiology and Biotechnology. 2002;59(4-5).426-435,DOI: 10.1007/s00253-002-1032-6

4. Amara AA. Lysozymes, Proteinase K, Bacteriophage E Lysis Proteins, and some ChemicalPage 16 of 16 Compounds for MGs Preparation: a Review and Food for Thought. SOJ Biochem. 2016;2(1):1-16.

5. Dong H, Han X, Bai H, He L, Liu L, Liu R, et al. Mutation of lambdapL/pR-cI857 system for production of bacterial ghost in Escherichia coli. Sheng wu gong cheng xue bao. Chinese journal of biotechnology. 2012;28(12):1423-1430.

6. Laemmli UK. Cleavage of structural proteins during the assembly of the head of bacteriophage T4. Nature. 1970;227:680-685.

7. Makino K, Yokoyama K, Kubota Y, Yutsudo CH, Kimura S, Kurokawa K, et al.Complete nucleotide sequence of the prophage VT2-Sakai carrying the verotoxin 2 genes of the enterohemorrhagic Escherichia coli O157: H7 derived from the Sakai outbreak. Genes & genetic systems. 1999;74(5):227-239.

8. Panthel K, Jechlinger W, Matis A, Rohde M, Szostak M, Lubitz W.Generation of Helicobacter pylori ghosts by PhiX protein E-mediated inactivation and their evaluation as vaccine candidates. Infect Immun. 2003;71:109-116.

9. Hensel A, Huter V, Katinger A, Raza P, Strnistschie C, Roesler U,et al.Intramuscular immunization with genetically inactivated (ghosts) Actinobacillus pleuropneumoniae serotype 9 protects pigs against homologous aerosol challenge and prevents carrier state. Vaccine. 2000;18(26):2945-2955.

10. Weibull C. The nature of the ghosts obtained by lysozyme lysis of Bacillus megaterium. Exp Cell Res. 1956;10(1):214-221.

11. Witte A, Wanner G, Sulzner M, Lubitz W. Dynamics of PhiX174 protein E-mediated lysis of Escherichia coli. Arch Microbiol. 1992;157:381-388.

12. Amara AA, Salem-Bekhit, M.M., Alanazi, FK. Preparation of bacterial ghosts for E. coli JM 109 using: sponge-like reduced protocol. Asian J Biol Sci.2013; 6(8):363-369.

13. Amara AA, Salem-Bekhit,M. M., Alanazi,F K. 2013. Sponge-like: a new protocol for preparing Bacterial Ghosts. TSWJ.2013; Article ID 545741:1-8.

14. Amara AA. Saccharomyces cerevisiae Ghosts Using the Sponge-Like Re-Reduced Protocol SOJ Biochem. 2015:1-4.

15. Amara AA, Salem-Bekhit MM, Alanazi FK.Sponge-like: a new protocol for preparing bacterial ghosts. The Scientific World Journal. 2013;1-8.

16 . Amara AA, Neama AJ, Hussein A, Hashish EA, Sheweita SA.Evaluation the surface antigen of the Salmonella typhimurium ATCC 14028 ghosts prepared by "SLRP". Scientific World Journal .2014:840863. DOI: 10.1155/2014/840863.

17. El-Baky NA, Amara AA. Newcastle disease virus (LaSota strain) as a model for virus Ghosts preparation using H2O2 bio-critical concentration. International Science and Investigation journal. 2014;3:38-50.

18. Vinod N, Oh S, Kim S, Choi CW, Kim SC, Jang CH. Chemically Induced Salmonella Enteritidis Ghosts as A novel Vaccine Candidate Against Virulent Challenge In Arat Model. Vaccine. 2014;32:3249-3255.

19. Amara AA. Bacterial and Yeast Ghosts: E. coli and Saccharomyces cerevisiae preparation as drug delivery model ISIJ Biochemistry.2015;4:11-22.

20. Amara AA. Kostenlos viral ghosts, bacterial ghosts microbial ghosts and more. Schuling Verlag - Germany. 2015

21. Vinod N, Oh S, Park HJ, Koo JM, Choi CW, Kim SC. Generation of a Novel Staphylococcus aureus Ghost Vaccine and Examination of Its Immunogenicity against Virulent Challenge in Rats. Infect Immun 2015; 83(7):2957-2965. DOI: 10.1128/IAI.00009-15

22. Amara AA. The critical activity for the cell all degrading enzymes: Could the use of the lysozyme for microbial ghosts preparation establish emergance oral vacccination protocol?. International Science and Investigation Journal 2016;5:351-369.

23. Amara AA. Vaccine against pathogens: A review and food for thought . SOJ Biochemistry. 2016;2:1-20.

24. Hussain ZM, Amra AA. Case-by-case study using antibiotic-EDTA combination to control Pseudomonas aeruginosa. Pak J Pharm Sci, 2016;19(3):236-243.

25. Park HJ, Oh S, Vinod N, Ji S,Noh HB, Koo JM et al. Characterization of Chemically-Induced Bacterial Ghosts (BGs) Using Sodium Hydroxide-Induced Vibrio parahaemolyticus Ghosts (VPGs). International Journal of Molecular Sciences. 2016;17(11):1904.

26. Wu X, Ju X, Du L, Yuan j, Wang L, He R, et al. Production of Bacterial Ghosts from Gram-Positive Pathogen Listeria monocytogenes. Foodborne Pathogens and Disease. 2017; 14(1):1-7.DOI: 10.1089/fpd.2016.2184

27. Menisy MM,HA,Ghazy A, Sheweita S, Amara A A. Klebsiella pneumoniae Ghosts as Vaccine Using Sponge Like Reduced Protocol. Cellular and Molecular Medicine.2017;3(2):1-8. DOI: 10.21767/2573-5365.100034

28. Sambrook J , Fristch EF, Mainiatis T. Molecular Cloning a Laboratory Manual. Molecular cloning: a laboratory manual. Cold Spring Harbor. 1989. 2nd edition.

29. Amara AA, Afifi IK, Younis MM, Sharaf MM, Shabeb MS. Non-Enzymatic method for DNA preparation from different microbial strains. Egyptian Journal of Biotechnology. 2005;21:339-349.

18

On the Origin of Life: A Possible Way from Fox's Microspheres into Primitive Life

Zhu Hua*

*Center for Integrative Conservation, Xishuangbanna Tropical Botanical Garden, Chinese Academy of Sciences, Mengla, Yunnan 666303, P. R. China

*Corresponding author: H. Zhu, Center for Integrative Conservation, Xishuangbanna Tropical Botanical Garden, Chinese Academy of Sciences, Mengla, Yunnan 666303, P. R. China, E-mail : zhuh@xtbg.ac.cn

Abstract

The microspheres constituted by proteinoids synthesized from Fox's simulation experiments. They had peptide bond structure and weak catalysis, as well as proliferated themselves. Such microspheres were believed the models for primitive life. Due to lack of metabolism and self-reproduction, the microspheres could not meet requirements of life. Thus, how microspheres could evolve into primitive life remain unsolved mysteries. The microspheres were supposed a dissipative structure and the processes of absorption and hydrolysis could be balanced to maintain their stability by consuming proteinoids. Proteinoid molecules differed in their life spans, which were mainly determined by their multi-space structures. Consequently, molecule selection and retention could occur spontaneously in microspheres and lead to a more organized and stabilized structure of the whole microsphere with time through dissipative process. More complex chain network of chemical reactions could happen in microspheres because the proteinoid with complex, ordered multi-space structure and relatively high catalytic activity would retain. In such microspheres, nucleotides could produce and further aggregate into RNA. The synthesis of real proteins could take place with RNA as the template catalyzed by proteinoids or RNA inside microspheres. When template-based protein molecules replaced the proteinoid inside the microspheres, a protein-based self-catalyzed network of chemical reactions could take place. It is plausible if Fox's proteinoids microspheres is to dawn on a dissipative structure, then molecule selection could occur spontaneously by "dissipative" proteinoids, and the microspheres would acquire catalytic activity due to preserved the proteinoid with a large molecular weight and relatively complex and ordered multi-space structure, and relatively high catalytic activity. Thus the microspheres would spontaneously go to self-organizing, and evolve into primitive life.

Keywords: Proteinoids; microspheres; dissipative structure; molecule selection; origin of life;

Introduction

The question about the origin of life has always been a core issue of philosophy and a long-standing debate between science and religion as well as between materialism and idealism. From the perspective of materialism, life is a special form of material motion, and an inevitable product of the development and evolution of Earth at a certain stage. Though the origin of life could be achieved through chemical pathways, it was not until the 1950s that the proposition was confirmed when the American scholar Stanley L. Miller successfully synthesized amino acids by simulating primitive Earth conditions [1].

Later, Sidney W. Fox, another American scholar, simulated primitive Earth conditions [2]. When a mixture of various amino acids was heated, polymerization occurred among amino acids and a protein-like hyperpolymer with a molecular weight as high as approximately 8,000 to 20,000 Dalton was produced. Fox called this product a "proteinoid," and further speculated that heat polymerization of amino acids could occur in some "hot zones" on primitive Earth. When these proteinoids were dissolved in water, they would automatically aggregate into a microspheric multi-molecular system, which he termed a microsphere. Microspheres resemble bacteria, have a double-layer boundary, and internal structure. They proliferated themselves through budding and division, and had weak catalysis. Fox believed that such microspheres were the models for primitive bacteria.

Due to the success of simulation experiments by Miller, Fox, and other researchers, a hypothesis on the origin of life was tentatively unveiled. As investigations went deeper, however, new problems emerged. Though proteinoids synthesized during the simulation experiments had peptide bond structure and weak catalytic action, and Fox even regarded them as protoenzymes, a significant difference was observed when compared with proteinase. Since proteinoids were not synthesized based on a template, the multi-space structure of the molecule had relatively low complexity, poor order, and irregular primary structure. For instance, acidic and basic amino acids in the backbone structure usually formed branched chains at residues. At some positions of the secondary structure, normal α-helices could be formed, while at other locations, they could not be formed. Thus, in the secondary and tertiary structures, proteinoids could not develop highly ordered and accurate multi-space structures like proteinase can, and the proteinoid molecules, therefore, failed to obtain high catalytic activity or transitivity. In general, such molecules had weak catalytic activity, and the microspheres constituted by such proteinoids, consequently, also had weak catalytic activity. As we know, metabolism and self-reproduction are two main characteristics of life. Metabolism occurs on the basis of an extremely complex and accurate network of chemical reactions, which require highly accurate and high-speed chemical reactions. Such protein-like molecules, however, failed to meet such requirements. At the same time, biological self-reproduction is jointly achieved with nucleic acids and proteins. In the simulation experiment, it is difficult to synthesize nucleic acids. So far, only oligonucleotides with a small molecular weight can be synthesized in the laboratory. Thus, how microspheres synthesized by the limited catalytic activity of proteinoids evolved into primitive life and achieved self-reproduction, and how nucleic acids and proteins established relationships with each other still remain unsolved mysteries.

Proposals for the origin of life remain as hypotheses, meaning they are working assumptions for scientists researching how life began. However, Fox et al. suggested that proteinoid microspheres of appropriate sorts promote the conversion of ATP to adenine dinucleotide and adenine trinucleotide [3]. Other microparticles composed of basic proteinoid and enzymically synthesized poly A cause the conversion of ATP and phenylalanine to various peptides of phenylalanine. When viewed in the context with the origin and properties of proteinoid microspheres, these results model the origin from a protocell of a more contemporary type of cell able to synthesize its own polyamino acids and polynucleotides. Obviously, the hypothesis from Fox et al. is of the most significance to the origin of life [3]. In this article, I make a further inference about how the microspheres could possibly evolve into primitive life.

From Fox's Microspheres into Primitive Life

I put forward the following hypothesis to investigate how microspheres constituted by proteinoids with the limited catalytic activity synthesized from Fox's simulation experiments evolved into primitive life and achieved self-reproduction, and how nucleic acids and proteins established relationships with each other [2, 3].

I speculate that microspheres need to go through several processes before evolving into primitive life. First, microspheres acquired relatively high catalytic activity and became activated microspheres, which made it possible for complex networks of reactions to occur inside the microspheres. Then, activated microspheres provided a special micro-environment, which facilitated the occurrence of many complex reactions inside the microspheres that would have been difficult in primitive oceans such as the synthesis of nucleic acids. When a certain amount of nucleic acids accumulated inside activated microspheres, proteinoid-catalyzed synthesis of proteins started to occur based on nucleic acid template, then the relationship between nucleic acids and proteins was also established. As template-based proteins gradually replaced proteinoid molecules and the self-catalyzed system for the synthesis of proteins, a nucleic acid template was used, and a qualitative change occurred in activated microspheres. The basic metabolic network of reactions was formed and primitive life emerged.

Given the assumption that under suitable conditions, proteinoid microspheres could have absorbed proteinoid molecules to offset the inevitable natural hydrolysis of proteinoid molecules inside themselves. By consuming proteinoid molecules in the solution, microspheres could grow and maintain the stability of their structure. Thus, as long as there was a sufficient supply of proteinoid molecules in the solution, processes could continue within the microsphere for a long time.

According to the dissipative structure theory proposed by Prigogine & Nicolis [4], I supposed that such characteristics of microspheres were just features of a dissipative structure. Microspheres are not in a state of equilibrium. Instead, they continuously absorbed proteinoids from the surrounding solution. Since the microsphere could be maintained for a long time, the processes of absorption and hydrolysis could be balanced to maintain the stability of the microsphere structure. Gradually, the microsphere structure could change from disorder to order, evolving towards complication and organization.

How did microspheres evolve into primitive life? On primitive Earth, there were many pathways and approaches that produced proteinoids. Additionally, the proteinoid molecules differed in their life spans, which were mainly determined by their multi-space structures. In general, if the molecules were large and their multi-space structure was complex, accurate, and ordered, they would have had a strong capacity of self-protection, leading to resistance to hydrolysis and a long life span. On primitive Earth, the non-template synthesized proteinoid molecules usually had low complexity and order in their multi-space structure. The primary structure of some molecules, however, had relatively long, normal α-helices, which led to a relatively complex and ordered multi-space structure, resistance to hydrolysis, and a long life span. Consequently, molecule selection and retention occurred spontaneously in primitive oceans, which was based on the relative difference between the complexity and order of the multi-space structure of proteinoid molecules. Proteinoid molecules with a large molecular weight and relatively complex and ordered multi-space structure would be preserved and accumulated in the primitive oceans due to their resistance to hydrolysis and long life span. Proteinoid molecules without such features would be hydrolyzed. Thus, such preserved and accumulated proteinoid molecules would evolve continuously in the primitive oceans.

Molecule selection inside microspheres was the most significant factor in the origin of life. During the dissipative process in microspheres, proteinoid molecules with a large molecular weight and complex and ordered multi-space structure were absorbed by microspheres, and such molecules would be preserved and accumulated in microspheres due to their stability and long life span. Unstable molecules with a short life span would be decomposed and consumed. Through such processes, simple and unstable molecules would be dissipated, and complex and stable ones would be preserved. Thus, the proteinoid molecules that constituted microspheres became more and more complex and more and more ordered, leading to a more ordered, organized, and stabilized structure of the whole microsphere.

It is possible that the catalytic function of large molecules was related to the state of their multi-space structures. The acquisition of catalytic function was determined by a certain highly ordered multi-space structure. Proteins have enzymatic activity due to their folded, highly accurate, and ordered multi-space structure, of which certain locations can form interactions with substrates. This is the same for proteinoid molecules in that proteinoid molecules with a large molecular weight and complex and ordered multi-space structure would have relatively high catalytic activity. Thus, the structure of the microspheres became increasingly complex and ordered during the dissipative process, and microspheres would acquire catalytic activity due to the molecules with high catalytic activity that constituted them.

When the microsphere originally forms, it should be simple and lack catalytic activity, or perhaps be in a near-equilibrium state. Through dissipation of proteinoid molecules, however, it becomes more and more complex and ordered, and its catalytic activity increasingly becomes enhanced. As it moves away from equilibrium, it eventually becomes an activated microsphere. The dissipative structural features of the microsphere, the length of the life span, and the strength of catalytic activity determined by the state of the multi-space structure of proteinoid molecules were compatible and consistent; therefore, it is an inevitable process for the microsphere to evolve into an activated microsphere.

I supposed that inside the activated microsphere, the speed of chemical reactions increased, the direction and the order of chemical reactions were enhanced, and a more complex chain network of chemical reactions even took place. Thus, the activated microsphere created a special micro-environment where many complex chemical reactions could take place that might have been unlikely to occur naturally in primitive oceans.

For different activated microspheres, their compositions, structures, degrees of organization, and activation could differ greatly, and the forms and content of chemical reactions could also differ from one microsphere to another, thus creating numerous microenvironments for various chemical reactions to take place. In certain activated microspheres, nucleotides were produced and energy chain reactions took place, which was coupled with the decomposition of organic matter. Nucleotides further aggregated into RNA, short molecules of which could combine with amino acids to produce aminoacyl-RNA. It is possible that inside certain activated microspheres, synthesis of real proteins took place with RNA as the template, aminoacyl-RNA as the amino donor, and proteinoids as the catalyst (perhaps RNA was also involved in catalysis). Thus, the relationship between nucleic acids and proteins might be established before the emergence of primitive life. Template-based proteins had regular peptide bonds, normal α-helix secondary structure, and could be folded into a highly ordered multi-space structure, thus had high catalytic activity and specificity. Newly synthesized proteins were continuously integrated into the network of chemical reactions, supplemented and enriched chemical reactions inside microspheres, and gradually replaced proteinoid molecules. When template-based protein molecules replaced all the proteinoid molecules inside the activated microspheres, a highly efficient, protein-based self-catalyzed network of chemical

reactions took place, the basic contents of metabolism formed, activated microspheres experienced a qualitative change, and new life eventually emerged.

Discussion

Matveev suggested that any protocell at the dawn of life on Earth should be a phase system because this kind of physical system has the potential to create special internal conditions necessary for the origin of life and for the first steps of molecular evolution [5]. Conditions for formation of proteinoid biophase and its fundamental physical properties are priorities for the protophysiology. Peptide synthesis rate in potassium ion medium is 3 to 10 times faster than that in the same concentration sodium ion medium, and prebiotic peptides could have formed with K+ as the driving force, not Na+ [6]. Life could originated from an environment with higher potassium content. Ishima et al. studied the ion distribution between proteinoid microspheres and media, and found that the concentration of K + in microspheres is much higher than that in medium, and the microspheres have the chemical conditions for life evolution [7]. The key of origin of life is to create a non-equilibrium physical process. The microspheres are able to absorb new proteinoid molecules from the surrounding solution constantly, making up for the inevitable natural hydrolysis of proteinoid molecules inside. Through the process, growth and structural stability of proteinoid microspheres are maintained. Thus, a non-equilibrium physical process could take place in the microspheres. The dissipative structure feature of the microsphere makes it possible for evolving to primitive life. Retention selection (molecular selection) caused by proteoid molecule's own multidimensional spatial structures could occur spontaneously in microspheres and lead to a more organized and stabilized structure of the whole microsphere with time through dissipative process. The proteoid microspheres could gradually come to self-organization.

The main idea of my inferencing hypothesis was originally published in Chinese, and was similar to several scientists' hypotheses in some aspects and considerations [8]. For example, Huber and Wächtershäuser, proposed the possible formation of peptides by activation of amino acids in simulation experiments [9]. Huber, et al . further put forth a possible primordial peptide cycle [10]. In my hypothesis, due to the dissipative process and molecule selection inside microspheres, more complex chained networks of chemical reactions, including the formation of peptides, could have even took place in microspheres.

Two fundamentally different ideas in the origin of life are: the genetics or replication first scenario [11, 12], and the metabolism first scenario [13] L. E. Orgel, et al. stated that a plausible scenario for the origin of life must, therefore, await the discovery of a genetic polymer simpler than RNA and an efficient, potentially prebiotic, synthetic route to the component monomers [12]. His suggestion that relatively pure, complex organic molecules might be made available in large amounts via a self-organizing, autocatalytic cycle might, in principle, help to explain the origin of the component monomers. In my inferencing hypothesis, in certain activated microspheres, nucleotides could be produced, and nucleotides could be further aggregated into RNA, because inside the activated microsphere, the speed of chemical reactions increased, and a more complex chained network of chemical reactions even took place.

RNA molecules had the dual role of catalysts and information storage systems, which provides support for objections to the genetics first scenario [14]. Existing replicators can serve as templates for the synthesis of additional copies of themselves, but they cannot be used for the preparation of the very first such molecule, which must arise spontaneously from an unorganized mixture [15]. Thus, greater attention should be given to metabolism-first theories, which avoid this conflict [15]. F. A. Anet also stated that the reactions were either spontaneous or were catalyzed by inorganic molecules, or by oligo-peptides or proteinoids formed either in a random manner or by mutual catalysis [13]. C. De Duve clarified that protometabolism must have been dependent on a network of fairly complex chemical processes and suggested that it was likely that ATP and the other NTPs served as precursors for the synthesis of RNA before participating in the formation of the first RNA molecules [16]. C. De Duve also stated that another property of protometabolism is that it must have relied on a set of robust reactions capable of being maintained over the time needed for enzyme-catalyzed metabolism to arise, which could be as long as several millennia [16]. In my inference, the activated microsphere created a special micro-environment where many complex chemical reactions could take place. It is possible that the synthesis of real proteins could occur with RNA as the template catalyzed by proteinoids or RNA inside certain activated microspheres. My inference is compatible with both Orgel's and Shapiro's ideas [11,12,15]. The microspheres were just a dissipative structure and could maintain the stability of their structure to create a special micro-environment where many complex chemical reactions could take place. It is possible that the synthesis of real proteins took place with RNA as the template catalyzed by proteinoids or RNA inside certain activated microspheres. This idea is in the line with De Duve's idea as well.

Carter Jr. & Wolfenden suggested that genetic coding of 3D protein structures evolved in distinct stages, based initially on the size of the amino acid and later on its compatibility with globular folding in water through their experiments [17]. Wolfenden et al. also suggested that the possible situation could be peptide bond catalyzing formation of RNA, and the interaction between amino acids and nucleotides could be existent before origin of life [18]. These new findings support my idea to some extent.

Baum & Vetsigian stated that the core of the origin of life problem is to explain the emergence of chemical systems that exhibit the capacity for heritable change and open-ended evolution [19]. Once such systems arose, adaptive evolution could take over. They suggested that conducting prebiotic selection experiments should be a priority for the origin of life and hope that scientists from diverse disciplinary backgrounds will design concrete experiments to search life-like chemical systems that show evidence of self-propagation and adaptive evolution. My hypothesis is considerable along these lines.

Conclusion

The simulation experiments conducted by Fox and Nakashima & Fox produced microspheres that formed automatically by aggregated proteinoids and these proteinoid microspheres promoted the conversion of ATP to adenine dinucleotide and adenine trinucleotide [2,3,20]. However, how microspheres synthesized by the limited catalytic activity of proteinoids evolved into primitive life and achieved self-reproduction remain unsolved. Based on the dissipative structure theory proposed by I. Prigogine, I supposed that the microspheres were features of a dissipative structure [8]. Molecule selection could occur spontaneously in microspheres. Proteinoid molecules with a large molecular weight and relatively complex and ordered multi-space structure would be preserved and accumulated due to their resistance to hydrolysis and relatively long life span. Thus, the microspheres would acquire catalytic activity due to the molecules with relatively high catalytic activity that were retained through the dissipative process, eventually becoming "activated microspheres." The activated microsphere created a special micro-environment. Inside the activated microsphere, the speed of chemical reactions increased, and a more complex chain network of chemical reactions even took place. In certain activated microspheres, nucleotides could be produced, and nucleotides could further aggregate into RNA. It is possible that the synthesis of real proteins took place with RNA as the template and proteinoids as the catalyst (perhaps RNA was also involved in catalysis) inside certain activated microspheres. Thus, the relationship between nucleic acids and proteins might be established before the emergence of primitive life.

Acknowledgments

This project was funded by The National Natural Science Foundation of China (41471051, 31170195, 41071040). English editing was completed by Top Edit (www.topedit.cn). I would like to the thank reviewers' constructive suggestions on this article.

Declarations

Conflict of interest statement:

Nothing declared:

References

1. Miller SL. Production of amino acids under possible primitive earth conditions. Science. 1953;117(3046):528–529. DOI: 10.1126/science.117.3046.528

2. Fox SW. The origins of prebiological systems and of their molecular matrices. 1st Edition. New York: Acad. Pr. 1965.

3. Fox SW, Jungck JR, Nakashima T. From proteinoid microsphere to contemporary cell: formation of internucleotide and peptide bonds by proteinoid particles. Origins of Life. 1974; 5(1): 227–237.

4. Prigogine I, Nicolis G. Self-Organisation in Nonequilibrium Systems: Towards A Dynamics of Complexity.Bifurcation Analysis. D. Reidel Publishing Company. ISBN 0-471-02401-5.1977.

5. Matveev VV. Comparison of fundamental physical properties of the model cells (protocells) and the living cells reveals the need in protophysiology. International Journal of Astrobiology. 2017; 16(1): 97–104.

6. Dubina MV, Vyazmin SY, Boitsov VM, Nikolaev EN, Popov IA, Kononikhin AS, et al. Potassium ions are more effective than sodium ions in salt induced peptide formation. Origins Life Evol Biosphere. 2013; 43(2): 109–117. Doi: 10.1007/s11084-013-9326-5.

7. Ishima Y, Przybylski AT, Fox SW. Electrical membrane phenomena in spherules from proteinoid and lecithin. BioSystems. 1981; 13(4): 243–251.

8. Zhu H. A hypothesis on the origin of life. Science (KEXUE). 1991;43(1):52–52,11 (in Chinese).

9. Huber C, Wächtershäuser G. Peptides by activation of Amino Acids with CO on (Ni,Fe) S Surfaces: Implications for the Origin of Life. Science. 1998; 281:5377 670–672. Doi: 10.1126/science.281.5377.670

10. Huber C, Eisenreich W, Hecht S, Wächtershäuser G. A possible primordial peptide cycle. Science. 2003;VOL:301(5635): 938–940. Doi: 10.1126/science.1086501.

11. Orgel LE. Prebiotic chemistry and the origin of the RNA world. Crit Rev Biochem Mol Biol. 2004, 39(2): 99-123. Doi: 10.1080/10409230490460765.

12. Orgel LE. Self-organizing biochemical cycles. PNAS. 2000; 97(23): 12507. doi: 10.1073/pnas.220406697.

13. Anet FA. The place of metabolism in the origin of life. Curr Opin Chem Bio. 2004; 18: 654–659.

14. Gilbert W. Origin of life: The RNA world. Nature. 1986; 319: 618.

15. Shapiro R. A replicator was not involved in the origin of life. IUBMB Life. 2000, 49: 173–176. Doi: 10.1080/713803621.

16. De Duve C. A research proposal on the origin of life. Origins Life Evol Biosphere. 2003; 33(6): 559–574.

17. Carter CW, Wolfenden R. tRNA acceptor stem and anticodon bases form independent codes related to protein folding. PNAS. 2015; 112(24): 7489–7494.

18. Wolfenden R, Lewis CA, Yuan Y, Carter CW. Temperature dependence of amino acid hydrophobicities. PNAS. 2015; 112 (24):7484-7488.

19. Baum D, Vetsigian K. An experimental framework for generating evolvabel chemical systems in the laboratory. Orig Life Evol. Biosph. 2017,47(4): 481-497. doi: 10.1007/s11084-016-9526-x.

20. Nakashima T, Fox SW.Formation of peptides from amino acids by single or multiple additions of ATP to suspensions of nucleoproteinoid microparticles. Biosystems. 1981; 14(2) 151–161.

19

Mitochondria and Neurodegeneration "Could Mitochondrial Organelle Transfer be a Cellular Biotherapy for Neurodegenerative Diseases?"

RL Elliott*, XP Jiang and JF Head

Elliott-Baucom-Head Breast Cancer Research and Treatment Center, Baton Rouge, LA 70806, USA

Corresponding author: R.L. Elliott, Elliott-Baucom-Head Breast Cancer Research and Treatment Center, Baton Rouge, LA 70806, USA, Email: relliott@eehbreastca.com

Abstract

It has been known for some time the abnormal function of mitochondria is associated with neurodegenerative diseases. Mitochondrial dysfunction has been implicated in the pathogenesis of Parkinson', Alzheimer's, amyotrophic lateral sclerosis, and Huntington's diseases. Researchers have postulated the therapeutic efficacy of mitochondrially targeted antioxidants, and some have shown encouraging results. We have demonstrated that mitochondrial organelle transplantation of isolated normal mitochondria into cancer cells decreased proliferation, lactate production and increased drug sensitivity of the cancer cells. Studies have shown that cellular uptake of exogenous mitochondria has restored functional recovery of defective recipient cells. Based on our experience with Mitochondrial Organelle Transfer (MOT) in cancer, we present this review commentary evidence that (MOT) might be a cell-based therapy for neurodegenerative diseases.

Keywords: Mitochondrial Organelle Transfer; Neurodegenerative Disease; Biotherapy

Introduction

Mitochondria are vital intracellular organelles involved in every important essential cellular function. They are involved in the synthesis of ATP for energy production, cell signaling, and production of reactive oxygen species, iron, and cellular metabolism, autophagy, and apoptosis. Paradoxically, they are involved in cell survival and cell death. Neurodegenerative diseases have long been associated with mitochondrial dysfunction, and many researchers are attempting to attack mitochondria as a therapeutic target for neurodegeneration.

In the 1930s, Warburg reported that cancer cells had altered cell metabolism due to defective mitochondrial respiration. He noticed a shift from oxidative phosphorylation to glycolysis with a marked increase in lactate production. There was a production of lactate in the presence of oxygen without an increase in oxidative phosphorylation. Thus, this aerobic glycolysis became known as the Warburg effect [1,2]. Our work on MOT has so far only involved cancer and has been encouraging. We have demonstrated that isolated normal mitochondria could easily enter into cancer cells when cocultured. The uptake of normal mitochondria into the cancer cells inhibited proliferation, increased drug sensitivity and decreased lactate production [3]. These results suggested that if MOT might be a cell-based therapy for cancer, it could also be a possible therapy for neurodegenerative diseases and other mitochondrial disorders. Mitochondria are involved in the longevity of the organism, and mitochondrial diseases are debilitating and occasionally fatal.

Disorders and Diseases Associated With Mitochondrial Dysfunction

Mitochondrial dysfunction is associated with many conditions besides cancer. Some of these conditions are aging, diabetes mellitus, Friedreich's Ataxia, Parkinson's disease, Amyotrophic Lateral Sclerosis (ALS), Alzheimer's and Huntington's diseases. This fact documents the important role mitochondria play in health and disease. This is discussed in detail in a very complex paper by Nunnari and Suomalainen entitled "Mitochondria in sickness and in health" [4]. This is a great article about all aspects of mitochondrial function in health and disease, and it is highly recommended. However, it is not for the novice. Now we know the result of MOT in cancer, it is appropriate to begin research to gain evidence that MOT could possible impact the treatment of these other diseases. If MOT could palliate and improve the condition of patients with ALS, it would be a tremendous hope for these fatal disease patients.

Intercellular Mitochondrial Migration: Evidence they Might be an Organelle for Cellular Biotherapy

Mitochondria are dynamic intracellular organelles that are active undergoing constant fission and fusion. They are intimately connected and networked with other cellular organelles. Their functions extend beyond cell membranes and controls organisms physiology by communications between cells, tissues and organs.

Parquier et al. [5] reported on the preferential transfer of mitochondria from endothelial cells to cancer cells through tunneling nanotubes. This transfer mediates cytoplasmic transfer and phenotype exchange between cancer and endothelial cells. This mitochondrial transfer resulted in an acquired chemoresistance. Another interesting paper supporting MOT as a possible treatment for neurodegenerative diseases was recently published by Kitani et al. [6]. The paper is entitled "Direct human mitochondrial transfer: a novel concept based on the endosymbiotic theory". They demonstrated isolated human mitochondria will internalize into isogeneic mesenchymal cells. They cited our paper [3] and their research support that transfer of exogenous mitochondria into diseased human cells could be a mechanism of cell-based therapy. This concept inspires new areas of research on tumorigenesis and allows for the development of new therapies for neurodegenerative disease and cancer.

Neuro Degeneration: Evidence of Mitochondrial Dysfunction

There is a strong evidence that mitochondrial dysfunction plays a significant role in neurodegenerative processes and diseases. We will attempt to present some of this strong evidence, and how MOT might play a role in therapy. Menzies, Cookson and Taylor et al. [7] have discussed in detail mitochondrial dysfunction in a cell culture model of Familial Amyotrophic Lateral Sclerosis (FALS). Evidence indicates mitochondrial abnormalities probably develop during motor neuron injury. They developed a cell culture model of FALS. The motor Neuron Cell Line (NSC-34) has been stably transfected to express normal or mutant Superoxide Dismutase (SOD1). They showed that presence of mutant (SOD1) results in the development of abnormally swollen and pale staining mitochondria. These morphological changes were associated with biochemical problems with a specific decrease in the function of complex II and IV of the mitochondrial Electron Transport Chain (ETC). Therefore, therapeutic measures protect mitochondrial respiration and may be helpful in SOD1 related familial and other forms of ALS.

Emerging evidence suggested motor neurons may die a programmed cell death pathway [8], and in relation to these factors, there are specific cell features which could cause motor neurons to be susceptible to injury [9]. There has been considerable interest in the possibility that mitochondrial damage may contribute to age-related motor neuronal injury [10,11]. Mitochondrial pathology of neurodegenerative disease includes generation of intracellular ATP, buffering of intracellular calcium, generation of intracellular free radical species and involvement of programmed cell death (Apoptosis). Mitochondrial proteins and DNA are shown to be susceptible to oxidative stress and free radicals that inhibit the function of specific mitochondrial enzymes [12]. Kong and Xu reported some early features of motor neuron injury occurring in transgenic mouse models of SOD1 related ALS. Before the animals developed clinical signs of motor dysfunction there was an appearance of morphologically abnormal mitochondria which were swollen and vacuolated [13]. All of this data clearly suggested that mitochondria may represent a subcellular organelle susceptible to damage in ALS. These mitochondrial alterations demonstrated in the cell culture model of SOD1 related FALS suggest that therapeutic measures targeted to the protection of mitochondrial respiratory chain function could be useful in that subgroup of ALS patients [7]. Why

not attempt normal isolated mitochondrial transfer?

A recent paper explains that how healthy mitochondria maintain proper neuronal function. This required a balance of biogenesis of new mitochondria and removal of damaged mitochondrial membranes, proteins and DNA (Mitophagy) [14,15]. Recent evidence suggests mitochondria are actively transported from the cell body and the axon, but a paper by Davis et al. [16] offers an interesting alternative. They state that axons expel mitochondria to the neighboring astrocytes for degeneration. This process represents a major route for mitochondrial disposal from these neurons. Davis et al. [16] have named this process of transcellular degeneration of mitochondria Trans mitophagy. Therefore, the health of the cell depends on critical mechanisms to remove damaged mitochondrial proteins and lipids. Mitochondrial quality control is essentially a complicated process, and there are many pathways of mitophagy.

Soubannier and Rippstein et al. [17] have reported the role of mitochondria derived vesicle formation in mitochondrial quality control and the molecular mechanisms that govern selective mitophagy. They describe a new pathway for the selective removal of proteins by Mitochondria Derived Vesicles (MDV) that are carriers of transit cargo to the lysosome. These vesicles carry a selective enrichment of oxidized cargo. They documented the ability of mitochondria to produce MDV that selectively transports their cargo to either the lysosomes or peroxisomes. Their conclusion states that the cargo into the MDV is very selective and can include one or both mitochondrial membranes and is enriched for oxidized protein. All in all MDV delivery of mitochondrial cargo to lysosomes is a significant new pathway for mitochondrial quality control especially in neurons. Any defect in this delicate mitophagy pathway could certainly contribute to mitochondrial dysfunction which might lead to neuro degeneration.

Mitochondrial physiology function is very delicate in the brain, spine, and their multiple neurons. It is easy to see how an imbalance in the mitochondrial function in certain neurons could lead to severe neuronal malfunction and various neurodegenerative diseases.

Mitochondria: Oxidative Stress a Therapeutic Target in Neurodegeneration

Federico, Cardaioli E, Da Pozzo et al. [18] have published a great paper on the role of mitochondrial oxidative stress and neurodegeneration. They did a great job explaining how the brain and central nervous system has high energy demands. This ATP energy is supplied by mitochondrial Oxidative Phosphorylation (OXPHOS). This OXPHOS requires many mitochondrial redox enzymes to prevent cellular damage from Reactive Oxygen Species (ROS). Any naturally occurring in efficiencies of mitochondrial OXPHOS can generate damaging ROS and cause defective mitochondrial respiration. Whatever the mechanism, a common feature of mitochondrial dysfunction is defective respiratory chain activity resulting in mitochondrial disorders and a wide range of clinical diseases. They also discuss in detail many of the individual mitochondrial disorders and diseases. The causes, mitochondrial defects and possible therapeutic measures are presented. This article is highly recommended.

George Perry has contributed tremendously to our knowledge of neurodegeneration and the role of defective mitochondria and mitochondrial dysfunction. He and his colleagues, Moreira, Zhu and Wang et al. have published a paper entitled "Mitochondria: a therapeutic target in neurodegeneration". They discuss the continuous energy requirement of the brain and neurons, and how even small periods of glucose and oxygen deprivation can result in neuronal cell death. They state that although the brain represents only 2% of body weight, it gets 15% of cardiac output and 20% of total body oxygen consumption. They mention how if oxidative stress occurs and excessive free radicals are produced, it can overwhelm the cell's antioxidant defenses to neutralize them. This is followed by mitochondrial dysfunction and neuronal damage.

The roles of mitochondrial dysfunction in neurodegeneration of the many neurodegenerative diseases are discussed in detail. Mitochondrial damage and the mechanism of the enzymes involved in these diseases are also presented. Multiple antioxidants are discussed as therapies targeted to the mitochondrial dysfunction in these conditions. The mechanism of an action in each of the enzyme defect in these individual neurodegenerative disorders is explained. They discuss improvements in some of these conditions and emphasize that probably some of these antioxidants should be used for prevention of some neurodegenerative disorders before the appearance of symptoms. The use of these agents for prevention possibly could delay the onset and alleviate symptoms of some of these dreaded diseases [19]. This article is a must read for those researchers interested in the pathogenesis and therapy of neurodegenerative diseases.

PGC1a: Role in Mitochondrial Metabolism and Neurodegenerative Disorders

PGC1a is an important transcriptional coactivator that is an inducer of mitochondrial biogenesis and metabolism. It is a promoter of oxidative metabolism and also a tremendous regulator of ROS control by increasing the expression of many ROS-detoxifying enzymes, thus playing a paradoxical role in mitochondrial ROS control. Therefore, PGC1a is involved in mitochondrial biogenesis and detoxification of ROS.

PGC1a is the peroxisome proliferator-activated receptor gamma coactivator 1 alpha and is a family member of transcriptional coactivators that are major regulators of metabolism. The importance of a balance in this role can be associated with an impaired mitochondrial function of ROS balance in ageing and neurodegenerative diseases.

Austin and St Pierre have done a tremendous job discussing the role of PGC1a in neurodegenerative disorders. Their paper is entitled "PGC1a and mitochondrial metabolism-emerging concepts and relevance in ageing and neurodegenerative disorders" [20]. This is a very complicated paper on the known roles of PGC1a family of transcriptional coactivators. Many roles are presented in detail, but this has also elicited unknown roles stimulating many questions that need to be answered.

They state the role of PGC1a is a regulation of oxidative metabolism and organelle biogenesis of mitochondria and peroxisomes. They showed that PGC1a increases the expression of Nuclear Respiratory Factors (NRES) which are transcriptional factors that regulate expression of many mitochondrial genes. This increase mediates a great induction of uncoupled respiration. This work identified PGC1a as a master inducer of mitochondrial biogenesis and respiration. They show how the peroxisome is a central organelle that supports mitochondrial function during oxidative metabolism. During oxidative metabolism mitochondria and peroxisomes cooperate in the metabolism of lipids which are important fuels from oxidative metabolism. The MDV we discussed earlier is important in the cooperative process.

ROS metabolism is discussed including the production of ROS and the many ROS-detoxifying enzymes in the compartments of mitochondria, peroxisomes, and the cytoplasm. PGC1a by controlling this dedicate balance has a global effect on mitochondrial functions. It confirms that PGC1a controls mitochondrial respiration in two ways,

(1)   Changing number of mitochondria in cells and

(2)   Changing the respiratory capacity of individual mitochondria. Therefore, PGC1a might provide a unified control of mitochondrial metabolism.

They explain because of the importance of PGC1a in energy homeostasis, how any imbalance could implicate PGC1a in many pathological conditions especially neurodegenerative disorders. They discuss pathological conditions of ageing, Huntington's disease, Parkinson's disease and Alzheimer's Disease (AD) and ALS. They postulate how PGC1a could play a potential role in many neurodegenerative conditions [20].

Wenz, Rossi, and Rotundo et al. [21] have stated that elevated mitochondrial PGC1a may have beneficial effects on ageing. An elevated expression of PGC1a in muscle tissue during life has been shown to delay some conditions, such as muscle loss. These improvements were thought to be less decrease in mitochondrial function and a reduction of oxidative damage. Some have shown a possible protective role of PGC1a in Parkinson's disease while others have implicated PGC1a in Alzheimer's disease, Duchenne muscle dystrophy and ALS [22,23]. These studies suggest that PGC1a has the potential as a therapeutic agent in various neurodegenerative diseases.

More Evidence for Mitochondrial Organelle Transfers as a Cell-Based Therapy

There is compelling evidence that MOT may be a viable technology to treat other organ dysfunctions. Masuzawa, Black and Parak et al. [24] have shown in a magnificent study that transplantation of autologously derived mitochondria protects that heart from ischemia-reperfusion injury. Mitochondrial dysfunction and damage occurs during cardiac ischemia and affects cardiac function during reperfusion. Mitochondrial transplantation reduced cardiac damage and markedly improved cardiac ventricular function. In vivo and in-vitro studies showed mitochondria in the interstitial spaces, but were internalized by the cardiomyocytes 2-8 hours after transplantation. These results confirmed that transplanted mitochondria are easily internalized into cardiomyocytes within a few hours of a transfer, and they maintain viability and cardiac function by increasing ATP levels.

Recently Kitani, Kami et al. [25] have presented more compelling evidence that functionally isolated mitochondria transfer might be a promising therapy for various diseases. Their paper was entitled "Internalization of isolated functional mitochondria: involvement of macropinocytosis." They demonstrated that isolated mitochondria could be transferred into xenogeneic and homogeneous cells by simple co-incubation and that the method of cellular uptake involved macropinocytosis. They proved the cellular uptake of the mitochondria by multiple sophisticated research techniques. The mitochondrial transfer improved the cellular viability in mitochondrial DNA-depleted cells and rescued mitochondrial respiratory function. These effects persisted for several days. Although problems exist and more needs to be done, these results are more evidence that MOT should be explored as a cellular-based therapy for serious diseases.

Conclusion

We have demonstrated in our research that isolated normal mitochondria can enter cancer cells and increase drug sensitivity and inhibit proliferation. This suggested that mitochondria could be powerful intracellular organelles for cell-based therapy. Further strong evidence has been presented in this commentary. This therapy has combined with other therapies and might improve or palliate patients with cancer and possibly neurodegenerative diseases. Organ transplantation has been done for years and now it is time to enter the era of cellular organelle transplantation. It has the potential to augment treatment of cancer, heart disease and possibly neurodegenerative disease.

We are beginning experiments with fibroblast mitochondria and neuronal cells. It is worth the effort to explore and perfect this technology. Especially if we can impact treatment of a disease like ALS. In order for MOT to be a viable technology, many obstacles must be overcome. We not only have to isolate normal mitochondria, but we also need to culture, expand and bank them for use when needed. We have cultured isolated normal mitochondria in our culture media (proprietary) for over 3 weeks; viability was documented by vital fluorescent stains. The cultured mitochondria also entered cancer cells and inhibited proliferation. The cultured mitochondria have been frozen for several weeks, thawed, and they still maintained viability. However, they stained less intense and the mitochondria population was less. We believe we are in the embryonic stage of MOT for cell-based therapy, but if we could enter the fetal stage and then birth, we possibly could impact the treatment of cancer and some neurodegenerative disorders. This is a difficult and complicated task, but if we reach our goal we might alleviate suffering and death from cancer and other dreaded diseases. The time for MOT is now!!

References

1. Warburg O, Wind F and Negleis E. On the Metabolism of Tumors in the Body. In: Warburg O. Ed, The Metabolism of Tumors, Constable, Princeton, 1930; 254-70.

2. Warburg O. On the Origin of Cancer Cells. Science. 1956; 123(3191): 309-14.

3. Elliott RL, Jiang XP, Head JF. Mitochondria organelle transplantation: introduction of normal mitochondria into human cancer cells inhibits proliferation and increases drug sensitivity. Breast Cancer Res Treat. 2012; 136(2): 347-54. Doi: 10.1007/s10549-012-2283-2.

4. Nunnari J, Suomalainen A. Mitochondria in sickness and in health. Cell. 2012 16; 148(6): 1145-59. Doi: 10.1016/ j.cell. 2012.02.035.

5. Pasquier J, Guerrouahen BS, Al Thawadi H, Ghiabi P, Maleki M, Abu-Kaoud N, et al. Preferential transfer of mitochondria from endothelial to cancer cells through tunneling nano tubes modulates chemo resistance. J Transl Med. 2013; 11: 94. Doi: 10.1186/1479-5876-11-94.

6. Kitani T, Kami D, Kawasaki T, Nakata M, Matoba S, Gojo S. Direct human mitochondrial transfer: a novel concept based on the endosymbiotic theory Transplant Proc. 2014; 46(4): 1233-6. Doi: 10.1016/j.transproceed.2013.11.133.

7. Menzies FM, Cookson MR, Taylor RW, Turnbull DM, Chrzanowska-Lightowlers ZM, Dong L, et al. Mitochondrial dysfunction in a cell culture model of familial amyotrophic lateral sclerosis. Brain. 2002; 125: 1522-33.

8. Sathasivam S, Ince PG, Shaw PJ. Apoptosis in amyotrophic lateral sclerosis: a review of the evidence. Neuropathol Appl Neurobiol. 2001; 27(4): 257-74.

9. Shaw PJ, Eggett CJ. Molecular factors underlying selective vulnerability of motor neurons to neurodegeneration in amyotrophic lateral sclerosis. J Neurol. 2000; 247 Suppl 1: I17-27.

10. Bead MF. Mitochondrial dysfunction in neurodegenerative diseases. J Pharmacol Exp Ther. 2012; 342(3): 619-30. Doi: 10.1124/jpct.112.192138.

11. Cortopassi GA, Wong A. Mitochondria in organism aging and degeneration. Biochim Biophys Acta. 1999; 1410(2): 183-93.

12. Zhang Y, Marcillat O, Giulivi C, Ernster L, Davies KJ. The oxidative inactivation of mitochondrial electron transport chain components and ATPase. J Biol Chem. 1990; 265(27): 16330-6.

13. Kong J, Xu Z. Massive mitochondrial degeneration in motor neurons triggers the onset of amyotrophic lateral sclerosis in mice expressing a mutant SOD1. J Neurosci. 1998; 18(9): 3241-50.

14. Youle RJ, van der Bliek AM. Mitochondrial fission fusion stress. Science. 2012; 337(6098): 1062-5. Doi: 10.1126/science.1219855.

15. Sheng ZH. Mitochondrial trafficking and anchoring in neurons: new insight and implication. J Cell Biol. 2014; 204(7): 1087-98. Doi: 10.1083/jcb.201312123.

16. Davis CH, Kim KY, Bushong EA, Mills EA, Boassa D, Shih T, et al. Transcellular degradation of axonal mitochondria. Proc Natl Acad Sci U S A. 2014; 111(26): 9633-8. Doi: 10.1073/pnas.1404651111.

17. Soubannier V, Rippstein P, Kaufman BA, Shoubridge EA, McBride HM. Reconstitution of mitochondria derived vesicle formation demonstrates selective enrichment of oxidized cargo. PLoS One. 2012; 7(12): e52830. Doi: 10.1371/journal.pone.0052830.

18. Federico A, Cardaioli E, Da Pozzo P, Formichi P, Gallus GN, Radi E. Mitochondria, oxidative stress and neurodegeneration. J Neurol Sci. 2012; 322(1-2): 254-62. Doi: 10.1016/j.jns.2012.05.030.

19. Moreira PI, Zhu X, Wang X, Lee HG, Nunomura A, Petersen RB, et al. Mitochondria: a therapeutic target in neurodegeneration. Biochim Biophys Acta. 2010; 1802(1): 212-20. Doi: 10.1016/j.bbadis.2009.10.007.

20. Austin S, St-Pierre J. PGC1a and mitochondrial metabolism-emerging concepts and relevance in aging and neurodegenerative diseases. J Cell Sci. 2012; 125(Pt 21): 4963-71. Doi: 10.1242/jcs.113662.

21. Wenz T, Rossi SG, Rotundo RL, Spiegelman BM, Moraes CT. Increased muscle PGC-1alpha expression protects from sarcopenia and metabolic disease during aging. Proc Natl Acad Sci U S A. 2009; 106(48): 20405-10. Doi: 10.1073/pnas.0911570106.

22. Gong B, Chen F, Pan Y, Arrieta-Cruz I, Yoshida Y, Haroutunian V, et al. SCFFbx2-E3-ligase-mediated degradation of BACE1 attenuates Alzheimer's disease amyloidosis and improves synaptic function. Aging Cell. 2010; 9(6): 1018-31. Doi: 10.1111/j.1474-9726.2010.00632.x.

23. Qin W, Haroutunian V, Katsel P, Cardozo CP, Ho L, Buxbaum JD, et al. PGC-1alpha expression decreases in the Alzheimer disease brain as a function of dementia. Arch Neurol. 2009; 66(3): 352-61. Doi: 10.1001/archneurol.2008.588.

24. Masuzawa A, Black KM, Pacak CA, Ericsson M, Barnett RJ, Drumm C, et al. Transplantation of autologously derived mitochondria protects the heart from ischemia-reperfusion injury. Am J Physiol Heart Circ Physiol. 2013; 304(7): H966-82. Doi: 10.1152/ajpheart.00883.2012.

25. Kitani T, Kami D, Matoba S, Gojo S. Internalization of isolated functional mitochondria: involvement of macropinocytosis. J Cell Mol Med. 2014; 18(8): 1694-703. Doi: 10.1111/jcmm.12316.

20

Saccharomyces cerevisiae Ghosts Using the Sponge-Like Re-Reduced Protocol

Amro Abd Al Fattah Amara*

Head of the Protein Research Department, Genetic Engineering and Biotechnology Research Institute, City for Scientific Research and Technological Applications, Egypt

Corresponding author: *Amro Abd Al Fattah Amara, Head of the Protein Research Department, Genetic Engineering and Biotechnology Research Institute, City for Scientific Research and Technological Applications, New Borg Al Arab, Alexandria, Egypt; E-mail: amroamara@web.de*

Abstract

Sponge-Like protocol for Ghosts preparation showing an increasing interest while it exceed the range of the *E* lysis gene protocol for Bacterial Ghosts preparation which is restricted only to gram-negative bacteria. The protocol can be used for nearly all bacterial strains. In this study, it succeeded for the first time to produce ghosts from Eukaryotic cell. In this study, *S. cerevisiae* yeast was turned to Ghost cells using Sponge-Like protocol for Ghosts preparation. However, centrifugation step was eliminated to avoid self-adhering or shrinking of the yeast empty cells. Instead, decantation was used. One unique properties of the yeast is that, it is able to decant. The protocol for ghost cells preparation was also reduced to use only the Minimum Inhibitory Concentration (MIC) of each of NaOH, SDS, NaHCO$_3$ and H$_2$O$_2$. NaHCO$_3$ was used instead of CaCO$_3$ due to the variation in the cell wall between the prokaryotic (Bacteria) and the eukaryotic (yeast) cells. Light and Scanning electron microscope were used to evaluate the quality of the *S. cerevisiae* Ghosts (*S.c.*Gs). Spectrophotometer was used to evaluate the amount of the realized DNA and Protein. The study show successful yeast ghost preparation with correct 3D structure.

Keywords: *Saccharomyces cerevisiae* Ghosts; Critical Chemical Concentration; Scanning Electron Microscope

Introduction

Microbial Ghosts are the microbes after their evacuation from their cytoplasmic content without damaging their 3D structures. Yeast is an old microbe used in traditional biotechnology [1]. Yeast has showing increasing interest in various biotechnological applications due to its different properties, which were extended by the genetic engineering and molecular biology tools [2-7]. Yeast is an eukaryotic microorganism which has some unique properties different from bacterial cells [2,8-10]. Being eukaryotic, yeasts enable the correct expression for different types of recombinant proteins [2,11,12]. For more information about the different human proteins produced and expressed in the yeasts refer to Service (2003) and the references within [13]. Regarding to its big size the yeast cell could be used as drug delivery system [14-19]. Yeast Ghosts preparation was reported using some conformational isomers of pancreatic [20]. This study

for the first time shows a new protocol for preparing *S. cerevisiae* Ghosts.

Material and Methods

Yeast Strain, Media and Growth Conditions

S. cerevisiae strain bought from local market in Egypt as dry granules (Germapan company-Casablanca-Morocco) was used in this study. The strain was purified using striking methods on Rose Bengal Chloramphenicol agar.

For regular cultivation, the following designed medium has been used: broth medium (one Liter); 5gm peptone; 3gm NaCl and 5gm glucose. For solid medium, 12gm/L agar was added. The cultivation temperature was 26°C.

Determination of the Minimum Inhibition Concentration (MIC)

Determination of the Minimum Inhibitory Concentration (MIC) for each of NaOH, SDS, NaHCO$_3$ and H$_2$O$_2$ was conducted using standard experiment for determining the MIC as described by Amara et al. [21].

NaOH, SDS, NaHCO$_3$ were prepared as 10% sterile stock solutions. SDS, NaHCO$_3$ were sterilized using the autoclave. NaOH did not autoclaved while it is sterile by itself at 10% concentration. H$_2$O$_2$ was purchased from local pharmacy as 30% sterile solution. 0.5ml of each (of the above-described compounds) was used to conduct the serial dilution experiment. For each compound 0.5ml was added to the first tube in which, the tube contains 4.5ml of the above described broth cultivation medium. 0.5ml was then transferred from the first test tube to the second one and so on until seven tubes for each compound. The tubes were then mixed gently. For the last tube, 0.5ml was discharged. Each tube was then inoculated with 100μl of the freshly prepared (10^8CFU/ml) *S. c.* culture. The tubes then mixed gently and left for 48h. The *S. c.* growth was observed by increasing the amount of sediment cells in the test tube bottom.

S.c.Gs Preparation Using Re-Reduced SL Protocol

One gram of the dried yeast cells was added to 10ml of the MIC of the NaHCO₃ and subjected to gentle shaking for one hr. Then the yeast cells left static to enable the cells to sediment. Sample from the supernatant was taken to determine the amount of the realized DNA and Protein. The supernatant then discharged and the cells washed by distilled water, left for decantation and the supernatant discharged. The same steps were repeated for each of NaOH, SDS, and H₂O₂. Finally, the cells washed by 60% ethanol and left suspended in ethanol at 4°C.

S.c.Gs Evaluation Using Light Microscope

Yeast smear for the treated cells was prepared using standard criteria followed by crystal violet stain. The cells were examined by the aid of the light microscope. The quality of the cells was determined based on the yeast 3D structure as either being correct or deformed.

Determination of the DNA Concentration

The concentration of the DNA was determined by measuring the absorption at 260nm. Quartz cuvette was used. An extinction 260 =1 corresponds to 50µg dsDNA mL^{-1} [22].

Determination of the Protein Concentration

Protein analysis of each experiment (the different supernatants) was determined using the spectrophotometer at 280nm. Quartz cuvette was used. The different protein concentrations were derived from Bovine Serum Albumin standard curve.

Sample Preparation for Electron Microscope Examination

For further study to the yeast, Ghosts quality electron microscope was used to scan the bacterial cells. Dry yeast smear was prepared and the smear surface then coated with approximately 15nm gold (SPI-Module Sputter Coater).

Scanning of the yeast Surface

The golden-coated sample was then scanned by analytical scanning electron microscope (Jeal JSM-6360LA) with secondary element at 10kv accelerating voltages at room temperature. The digital images then were adjusted and saved.

Result and Discussion

This study describes for first time the production of S. cerevisiae Ghosts. The main aim of this study is to prepare S. cerevisiae Ghosts. Recently, Amara et al. [21], have reported a new protocol for Ghosts preparations. The protocol is mainly based on using critical chemical concentration for some chemical compounds are able to introduce genteel pores in the E. coli cells. The original protocol including the use of Plackett Burman experimental design to map the best conditions for obtaining successful Ghosts cells. The protocol was validated in several published papers. The protocol show successful Ghosts preparation from E. coli BL21 and JM109, show sensitivity and prove different MIC between those both similar strains.

MIC concentration is a critical point where the used chemical compound kills the cells with the minimum expected damage. The protocol was reduced to be simpler and more practical where the best two experiments obtained from the Plackett-Burman experiment were used only and the MIC and MGC were used to specify the critical point for killing cells without causing any damage. The produced Ghosts from Salmonella typhimurium ATCC 14028 show the sensitivity of such protocol and correct surface antigens maintained during the preparation steps. Such protocol has extended the Bacterial Ghosts preparation, which was prepared using the bacteriophage E lysis gene. It extends its range from only gram-negative bacteria, which restricted due to the function activity of the E lysis gene to nearly all bacterial types. Additionally, Newcastle virus was prepared as Ghosts [23].

Moreover, this study adds the first Eukaryotic cell prepared using such protocol. The Ghosts preparation protocol was re-reduced in this study and only MIC concentration of the used chemical compounds were used. NaHCO₃ was used instead of CaCO₃ while it proves to be more powerful against Eukaryotic cells as proved by Amara and Steinbüchels 2014 [21]. Yeast cells are bigger than the bacterial cells and they have unique properties that they are able to sediment. For that, centrifugation was not used in this study and only decantation was used. This was planed also, because centrifugation cause Yeasts Ghosts to shrink and their empty inside space when subjected to the centrifugation force which cause the cells to deform or to come together as an empty balloon. As described in the original protocol the Yeast Ghosts DNA and protein losing were determined using the spectrophotometer at 260 and 280nm respectively (data not shown). The amount of the realized protein and DNA show correct evacuation of the yeast cells from their cytoplasmic content as described in the original protocol by Amara et al. [21]. The cells with different qualities were monitored using light microscope and electron microscope as in figures 1-3. Both of the light and the electron microscope show correct 3D structure for the prepared yeast Ghosts as in the Figures 1-3.

The study show correct preparation for yeast ghosts could be used in various applications.

Conclusion

This study show for the first time Yeast Ghosts prepared

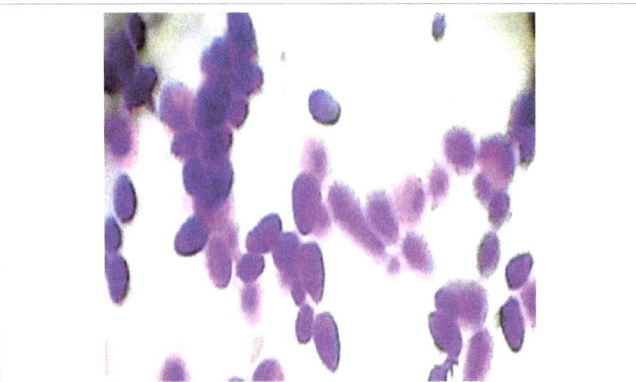

Figure 1: Viable S. cerevisiae cells (before treatment) stained by crystal violet using light microscope.

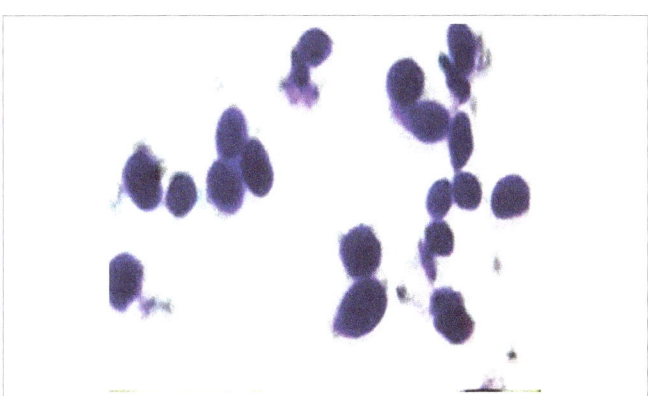

Figure 2: Ghost *S. cerevisiae* cells (after treatment) stained by crystal violet using light microscope .

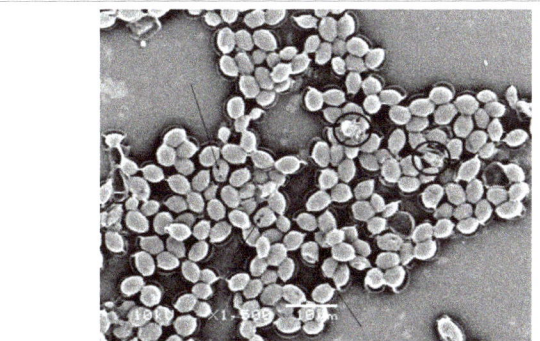

Figure 3: *S. cerevisiae* ghost cells, Image taken by scanning electron microscope (1500X). Arrows show pores in the Ghost cells.

using the critical chemical concentration of the NaOH, SDS, NaHCO$_3$ and H$_2$O$_2$. These compounds and by following the concept of the Sponge-Like protocol for preparing Bacterial Ghosts succeeded to prepare yeast Ghosts. Simply, MIC for each has been determined and has been used in steps to induce gentle pore(s) in the yeast cells as well as to evacuate their cytoplasmic contents. The light microscope as well as the electron microscope show correct 3D structure and correct pores in the yeast cells. The spectrophotometer proves correct release for both of the protein and the DNA, which are indications about the evacuation of the yeast cells from their cytoplasmic contents. Such a succeed in the yeast cells preparation open a new area for both drug delivery and vaccination using yeast cells either as carriers for surface foreign proteins or a package can be loaded with proteins as well as for a bigger candidates for loading different drugs and can be used successfully in drug delivery technology.

References

1. Maraz A. Impact of yeast genetics and molecular biology on traditional and new biotechnology. Acta Microbiol Immunol Hung. 1999; 46(2-3): 289-95.

2. Cereghino GP, Cregg JM. Applications of yeast in biotechnology: protein production and genetic analysis. Curr Opin Biotechnol. 1999; 10(5): 422-7.

3. Graf A, Dragosits M, Gasser B, Mattanovich D. Yeast systems biotechnology for the production of heterologous proteins. FEMS Yeast Res. 2009; 9(3): 335-48. doi: 10.1111/j.1567-1364.2009.00507.x.

4. Jacob Z. Yeast lipid biotechnology. Adv Appl Microbiol. 1993; 39: 185-212.

5. Maraz A. From yeast genetics to biotechnology. Acta Microbiol Immunol Hung. 2002; 49(4): 483-91.

6. Méndez B, Valenzuela P. Recombinant yeast as a production system in biotechnology. Bioprocess Technol. 1991; 13: 16-53.

7. Passoth V, Fredlund E, Druvefors UA, Schnürer J. Biotechnology, physiology and genetics of the yeast Pichia anomala. FEMS Yeast Res. 2006; 6(1): 3-13.

8. Hamasaki M, Noda T, Baba M, Ohsumi Y. Starvation triggers the delivery of the endoplasmic reticulum to the vacuole via autophagy in yeast. Traffic. 2005; 6(1): 56-65.

9. Harashima S, Kaneko Y. Application of the PHO5-gene-fusion technology to molecular genetics and biotechnology in yeast. J Biosci Bioeng. 2001; 91(4): 325-38.

10. Kajiwara K, Watanabe R, Pichler H, Ihara K, Murakami S, Riezman H, et al. Yeast ARV1 is required for efficient delivery of an early GPI intermediate to the first mannosyltransferase during GPI assembly and controls lipid flow from the endoplasmic reticulum. Mol Biol Cell. 2008; 19(5): 2069-82. doi: 10.1091/mbc.E07-08-0740.

11. Reyes-Becerril M, Salinas I, Cuesta A, Meseguer J, Tovar-Ramirez D, Ascencio-Valle F, et al. Oral delivery of live yeast Debaryomyces hansenii modulates the main innate immune parameters and the expression of immune-relevant genes in the gilthead seabream (Sparus aurata L.). Fish Shellfish Immunol. 2008; 25(6): 731-9. doi: 10.1016/j.fsi.2008.02.010.

12. Zhang M, Li S, Nyati MK, DeRemer S, Parsels J, Rehemtulla A, Regional delivery and selective expression of a high-activity yeast cytosine deaminase in an intrahepatic colon cancer model. Cancer Res. 2003; 63(3): 658-63.

13. Service RF. Biotechnology. Yeast engineered to produce sugared human proteins. Science. 2003; 301(5637): 1171.

14. Bianchi A, Negrini S, Shore D. Delivery of yeast telomerase to a DNA break depends on the recruitment functions of Cdc13 and Est1. Mol Cell. 2004; 16(1): 139-46.

15. Cowles CR, Snyder WB, Burd CG, Emr SD. Novel Golgi to vacuole delivery pathway in yeast: identification of a sorting determinant and required transport component. EMBO J. 1997; 16(10): 2769-82.

16. Henkel MK, Pott G, Henkel AW, Juliano L, Kam CM, Powers JC, Endocytic delivery of intramolecularly quenched substrates and inhibitors to the intracellular yeast Kex2 protease1. Biochem J. 1999; 341 (Pt 2): 445-52.

17. Kingsman AJ, Burns NR, Layton GT, Adams SE. Yeast retrotransposon particles as antigen delivery systems. Ann N Y Acad Sci. 1995; 754: 202-13.

18. Pedersen GT. Yeast flora in mother and child. A mycological-clinical study of women followed up during pregnancy, the puerperium and 5-12 months after delivery, and of their children on the 7th day of life and at the age of 5-12 months. Dan Med Bull. 1969; 16(7): 207-20.

19. Stepp JD, Huang K, Lemmon SK. The yeast adaptor protein complex, AP-3, is essential for the efficient delivery of alkaline phosphatase by the alternate pathway to the vacuole. J Cell Biol. 1997; 139(7): 1761-74.

21

Mammalian P5CR and P5CDH: Protein Structure and Disease Association

Chien-An A. Hu[1]* and Yongqing Hou[2]

[1]*Department of Biochemistry and Molecular Biology, University of New Mexico Health Sciences Center, Albuquerque, New Mexico 87131-0001, USA*

[2]*School of Animal Science and Nutritional Engineering, Wuhan Polytechnic University, Wuhan, Hubei, 430023, P. R. China*

*Corresponding author: Chien-An A. Hu, MSC08 4670, Department of Biochemistry and Molecular Biology, UNM HSC, Albuquerque, NM87131, USA, E-mail: AHu@salud.unm.edu

Abstract

The interconversions of L-proline (Pro) and L-glutamate (Glu) in mammalian cells involve an obligatory intermediate, Δ^1-pyrroline-5-carboxylate (P5C/PYC), and four uni-directional enzymes, proline oxidase/dehydrogenase (POX/PRODH), P5C dehydrogenase (P5CDH), P5C synthase (P5CS/PYCS), and P5C reductase (P5CR/PYCR). The catabolism of Pro by two dehydrogenation reactions catalyzed by POX and P5CDH is an important source of oxidizing signaling, whereas the anabolism of Pro through two reduction reactions catalyzed by P5CS and P5CR maintains redox homeostasis to promote cell growth. In addition, Pro is one of the "conditionally essential" amino acids in the neonatal intestine in both humans and animals. The homeostatic balance of Pro, Glu, and Arginine is critical for the growth, redox balance, immunomodulation, and development of mammals. Recent discoveries strongly suggest that Pro metabolic enzymes are tightly regulated by spatial-temporal gene expression, tissue and cell-specificity, substrate and/or inhibitor abundance, and subcellular compartmentalization. In terms of disease association, mutations in human P5CR1 gene have been identified in patients with autosomal recessive Cutis laxa type IIB and type IIIB/De Barsy syndrome, whereas mutations in P5CDH gene cause type II hyperprolinemia. Importantly, x-ray crystallographic studies have revealed the protein structures of human P5CR 1 and mammalian P5CDH. These new discoveries in structure function relationship may provide crucial guidelines in the treatment of the corresponding disorders.

Keywords: Cutis laxa Type IIB and Type IIIB; De Barsy syndrome; DJ-1; ORAOV1; P5CR/PYCR; P5CDH; Subcellular localization

Abbreviations

P5C/PYC: Δ^1-pyrroline-5-carboxylate; P5CDH: P5C Dehydrogenase; P5CR: P5C Reductase; P5CS: P5C Synthase; OAT: Ornithine Aminotransferase; GSA: Glutamic-γ-Semialdehyde

Introduction

In mammals, L-proline (Pro) and L-glutamate (Glu) are interconverted by four highly regulated, uni-directional enzymes, namely, Pro oxidase/dehydrogenase (POX/PRODH), Δ^1-pyrroline-5-carboxylate (P5C/PYC), P5C Dehydrogenase (P5CDH), P5C Synthase (P5CS/PYCS), and P5C Reductase (P5CR/PYCR), with P5C as the obligatory intermediate[1-4] (Table 1).

P5C is in tautomeric equilibrium with Glutamic-γ-Semialdehyde (GSA), which is reduced to Pro by the cytosolic and mitochondrial NAD(P)H-dependent P5CR isozymes. In addition, P5C/GSA is a substrate for two other enzymes, mitochondrial P5CDH, which converts P5C/GSA to Glu [5], and mitochondrial Ornithine (Orn) Aminotransferase (OAT), which catalyzes the interconversion of P5C and Orn [6]. Orn enters the urea cycle mainly for ammonia detoxification and Arginine (Arg) biosynthesis (Figure 1). It has been shown that Pro and Arg are two of the "conditionally essential" amino acids in the neonatal intestine of mammals. The interconversions of Pro, Orn, Glu, and Arg are critical for the growth and immunomodulation of mammals [1,4,7]. At the cellular level, the catabolism/degradation of Pro by POX and then P5CDH is an important source of redox signaling, whereas the anabolism/biosynthesis of Pro through P5CS and then P5CR maintains redox homeostasis to promote cell growth. In this review, we summarize the recent discoveries on human P5CR isozymes and P5CDH and their associated diseases/disorders.

Human P5CR: Three isozymes encoded by three different genes

Three isozymes of human P5CR/PYCR (EC1.5.1.2), have been identified, cloned, and characterized [1,2,8]. Human P5CR1, P5CR2, and P5CRL are encoded by three different genes, localized at three different chromosomal locations (Table 1). These three isozymes catalyze ATP and NAD(P)H-dependent reduction of P5C to Pro, which is important for the transfer of oxidizing potential across the cell [9,10]. The human P5CR1/PYCR1 structural gene [also known as Proliferation-Inducing protein 45 (PIG45)] is localized on chromosome 17q25.3, which encodes two protein isoforms, a 319-amino acid residue (aa) and a 316-aa polypeptide, respectively. The human P5CR2/PYCR2 structural gene is localized on chromosome 1q42.12, which encodes two protein isoforms, a 320-aa and a 246-aa polypeptide, respectively. Recently, De Ingeniis and colleagues [8] confirmed that there is a new isozyme of P5CR, P5CRL, in melanoma cells. The human P5CRL/PYCR3 structural gene is localized on chromosome 8q24.3, which encodes two isoforms, a 286-aa and a 266-aa polypeptide, respectively. In addition, it has

Table 1: Human P5CR Isozymes and P5CDH: from genes to protein isozymes to associated disorders.

Enzyme	Gene Name	Gene ID	Map Location	OMIM#	Isozyme	# Amino Acids
P5CR1	PYCR1/P5CR1	5831	17q25.3	179035	P5CR1.1	319
				612940	P5CR1.2	316
				614438		
P5CR2	PYCR2/P5CR2	29920	1q42.12	N/A	P5CR2.1	320
					P5CR2.2	246
P5CRL/P5CR3	PYCRL/P5CR3	65263	8q24.3	N/A	P5CRL.1	286
					P5CRL.2	266
P5CDH	ALDH4A1/P5CDH	8659	1p36	606811	P5CDH	563

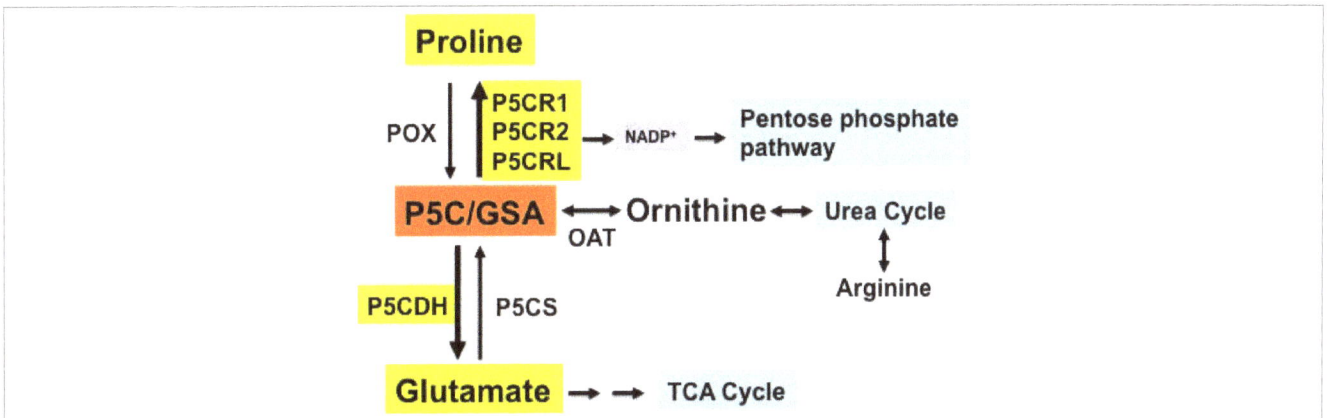

Figure 1: The interconversions of proline, glutamate, and ornithine. This tri-amino-acid-cycle also links with the urea cycle and the pentose phosphate pathway.

been demonstrated that P5CR1 and P5CR2 are localized in the mitochondria and are primarily involved in conversion of Glu to Pro, whereas P5CRL is localized in the cytosol and is exclusively linked to the conversion of Orn to Pro, and is not feedback inhibited by proline. Previously, Merrill and colleagues [11] showed that human P5CR in erythrocytes not only catalyzes the obligatory step in Pro biosynthesis, but also plays a physiological role in the generation of NADP+. The normal abundance of P5CR in the cell is maintained relatively low due to its high turnover [9]. A recent study by Krishnan and colleagues [12] showed over expression of P5CR1 resulted in 2-fold higher proline content, significantly lowered free radical levels, and increased cell survival. Another studies showed that increased P5CR1 activity was measurable in pulmonary and colorectal tumors [13,14]. In contrast, mammalian P5CR2 and P5CRL are relatively new and not well studied.

With regard to their interactomes, interestingly, PYCR1 has been identified as an interacting protein of two important regulatory proteins, DJ-1 [15] and Oral Cancer Over expressed 1 (ORAOV-1) [16]. DJ-1, encoded by the DJ-1/PARK7 gene, plays various functions involved in transcriptional regulation, anti-oxidative activity, and regulation of mitochondrial complex I. It has been shown that DJ-1 and PYCR1 interacts and colocalizes in mitochondria. DJ-1 increases the enzymatic activity of PYCR1 *in vitro* [15]. ORAOV1, encoded by ORAOV1 gene, is frequently amplified in esophageal squamous cell cancer, and has been shown

to regulate the cell cycle, apoptosis and angiogenesis. Cancer cells overexpressing ORAOV1 exhibited significantly increased tumorigenicity and larger tumors with poor differentiation. It has been shown that ORAOV1 also increases the enzymatic activity of PYCR1 and the production of Pro when it binds with P5CR1 [16]. However, whether P5CR1, DJ-1, and ORAOV1 interact with each other and function in the same complex in certain cell types is not known.

Human P5CR1 Protein structure and associated diseases

The crystal structure of human P5CR1 have been reported recently [13,14]. The 2.8 Angstroms (Å) resolution structure of the P5CR1 apo enzyme and its 3.1 Å resolution ternary complex with NAD(P)H and substrate-analog demonstrated that human P5CR1 possesses a decameric architecture with five homodimer subunits. It has been hypothesized that human P5CR1 possesses ten catalytic sites arranged around a peripheral circular groove. In terms of disease association, homozygous or compound heterozygous mutations in human P5CR1 have been identified in autosomal recessive Cutis laxa, Type IIB (ARCL2B or Cutis laxa with progeroid features; OMIM #612940; Table 1) [17-19]. The clinical phenotype of ARCL2B includes cutis laxa, abnormal growth and development, and associated skeletal abnormalities [18]. In addition, P5CR1 mutations have been linked to autosomal recessive cutis laxa type III (ARCL3, OMIM #614438), also known as De Barsy syndrome. ARCL3 is a rare autosomal recessive disorder, characterized by an aged appearance with distinctive facial features, sparse hair, ophthalmologic abnormalities, intrauterine growth retardation, and cutis laxa.

Human P5CDH: Protein structure and associated disease

Mammalian P5CDH (EC 1.5.1.12) is a mitochondrial matrix NAD$^+$-dependent dehydrogenase which converts P5C/GSA to Glu, and thus is a high Km/low affinity Aldehyde Dehydrogenase (ALDH) with GSA as a primary substrate. Mammalian P5CDH also exhibits activity with other aldehydes, and is dubbed as ALDH4A1 (ALDH, family 4, subfamily A, member 1) [1,2] (Table 1). Human P5CDH/ ALDH4A1 structural gene is localized on chromosome 1p36 and encodes a 563-aa polypeptide [1,5]. Importantly, it has been demonstrated in plants that Lack of P5CDH activity led to higher ROS production in the presence of Pro excess. Therefore, oxidation of P5C to Glu by P5CDH is critical to prevent P5C-Pro intensive cycling and avoid ROS production from electron run-off [20-22]. In additional, it has been shown that Drosophila deficient in P5CDH showed hyperprolinemia, swollen mitochondria, and early embryonic lethality [23]. Taken together, these observations suggest that P5CDH plays a protective role against ROS generation, mitochondria and cell damage, and apoptosis.

To understand the functions of P5CDH at the molecular level, how substrates and inhibitors interact with the enzyme, and how the substituted residues encoded by the mutant alleles of P5CDH can affect the enzymatic activity, the crystal structures of human and mouse P5CDH were determined recently [24,25]. Both wildtype P5CDH and mutant P5CDH proteins carrying S352A (2.4 Å) and S352L (2.85 Å) substitutions were resolved. In addition, 2.5-Å resolution Structures of the mouse P5CDH complexed with sulfate ion (1.3 Å resolution), glutamate (1.5 Å), and NAD$^+$ (1.5 Å) were determined in order to obtain high-resolution views of the active site. Together, the structures showed that single amino acid substitutions cause structural alterations and enzyme inactivation. Interestingly, the structure-activity relationship demonstrated that the semialdehyde carbon chain length and the position of the aldehyde group in relation to the cysteine nucleophile and oxanion hole of mouse P5CDH are critical. Efficient 4- and 5-carbon substrates share the common feature of being long enough to span the distance between the anchor loop at the bottom of the active site and the oxanion hole at the top of the active site. The inactive 2- and 3-carbon semialdehydes bind the anchor loop but are too short to reach the oxanion hole [25]. The K_i values are 0.27 mM for glyoxylate, 58 mM for succinate, 30 mM for glutarate, and 12 mM for L-glutamate. Interestingly, malonate is not an inhibitor [25]. With regard to its disease association, deficiency of P5CDH causes type II Hyperprolinemia (HPII), an autosomal recessive disorder characterized by accumulation of P5C and Pro [6,26,27]. Although HPII has been considered as a benign disorder, further research indicated that HPII may cause clinical manifestations, such as childhood febrile seizures.

Acknowledgements

This work was supported, in part, by the pilot projects (#030-2 and #0224 to CAAH) of UNM CTSC grant (8UL1TR000041).

References

1. Phang JM, Hu CA, Valle D. Disorders of proline and hydroxyproline metabolism. In: CR Scriver, AL Beaudet, WS Sly, D Valle, Metabolic and Molecular Basis of Inherited Disease. New York: McGraw Hill Press; 2001. 1821-1838.

2. Hu CA, Bart Williams D, Zhaorigetu S, Khalil S, Wan G, Valle D. Functional genomics and SNP analysis of human genes encoding proline metabolic enzymes. Amino Acids. 2008; 35(4): 655-64. doi: 10.1007/s00726-008-0107-9.

3. Hu CA, Khalil S, Zhaorigetu S, Liu Z, Tyler M, Wan G, Valle D. Human Delta1-pyrroline-5-carboxylate synthase: function and regulation. Amino Acids. 2008; 35(4): 665-72. doi: 10.1007/s00726-008-0075-0.

4. Wu G, Bazer FW, Burghardt RC, Johnson GA, Kim SW, Knabe DA, et al. Proline and hydroxyproline metabolism: implications for animal and human nutrition. Amino Acids. 2011; 40(4): 1053-1063. doi: 10.1007/s00726-010-0715-z.

5. Hu CA, Lin WW, Valle D. Cloning, characterization and expression of cDNAs encoding human delta 1-pyrroline-5-carboxylate dehydrogenase. J Biol Chem. 1996; 271(16): 9795-9800.

6. Valle D, Goodman SI, Harris SC, Phang JM. Genetic evidence for a common enzyme catalyzing the second step in the degradation of proline and hydroxyproline. J Clin Invest. 1979; 64(5):1365-1370.

7. Ren W, Zou L, Ruan Z, Li N, Wang Y, Peng Y, et al. Dietary L-proline supplementation confers immunostimulatory effects on inactivated Pasteurella multocida vaccine immunized mice. Amino Acids. 2013; 45(3): 555-61. doi: 10.1007/s00726-013-1490-4.

8. De Ingeniis J, Ratnikov B, Richardson AD, Scott DA, Aza-Blanc P, De SK, et al. Functional specialization in proline biosynthesis of melanoma. PLoS One. 2012; 7(9). e45190.

9. Phang JM. The regulatory functions of proline and pyrroline-5-carboxylic acid. Curr Top Cell Regul. 1985; 25: 91-132.

10. Dougherty KM, Brandriss MC, Valle D. Cloning human pyrroline-5-carboxylate reductase cDNA by complementation in Saccharomyces cerevisiae. J Biol Chem. 1992; 267(2): 871-875.

11. Merrill MJ, Yeh GC, Phang JM. Purified human erythrocyte pyrroline-5-carboxylate reductase. Preferential oxidation of NADPH. J Biol Chem. 1989; 264(16): 9352-9358.

12. Krishnan N, Dickman MB, Becker DF. Proline modulates the intracellular redox environment and protects mammalian cells against oxidative stress. Free Radic Biol Med. 2008; 44(4): 671-681.

13. Meng Z, Lou Z, Liu Z, Hui D, Bartlam M, Rao, Z. Purification, characterization, and crystallization of human pyrroline-5-carboxylate reductase. Protein Expr Purif. 2006; 49(1): 83-87.

14. Meng Z, Lou Z, Liu Z, Li M, Zhao X, Bartlam M, et al. Crystal Structure of Human Pyrroline-5-carboxylate Reductase. J Mol Biol. 2006; 359(5): 1364-1377.

15. Yasuda T, Kaji Y, Agatsuma T, Niki T, Arisawa M, Shuto S, et al. DJ-1 cooperates with PYCR1 in cell protection against oxidative stress. Biochem Biophys Res Commun. 2013; 436(2): 289-294. doi: 10.1016/j.bbrc.2013.05.095.

16. Togashi Y, Arao T, Kato H, Matsumoto K, Terashima M, Hayashi H, et al. Frequent amplification of ORAOV1 gene in esophageal squamous cell cancer promotes an aggressive phenotype via proline metabolism and ROS production. Oncotarget. 2014; 5(10): 2962-73.

17. Morava E, Guillard M, Lefeber DJ, Wevers RA. Autosomal recessive cutis laxa syndrome revisited. Eur J Hum Genet. 2009; 17(9), 1099-1110. doi: 10.1038/ejhg.2009.22.

18. Guernsey DL, Jiang H, Evans S C, Ferguson M, Matsuoka M, Nightingale M, et al. Mutation in pyrroline-5-carboxylate reductase 1 gene in families with cutis laxa type 2. Am J Hum Genet. 2009; 85(1): 120-129. doi: 10.1016/j.ajhg.2009.06.008.

19. Reversade B, Escande-Beillard N, Dimopoulou A, Fischer B, Chng SC, Li Y, et al. Mutations in PYCR1 cause cutis laxa with progeroid features. Nat Genet. 2009; 41(9): 1016-1021. doi: 10.1038/ng.413.

20. Deuschle K, Funck D, Hellmann H, Däschner K, Binder S, Frommer WB. A nuclear gene encoding mitochondrial Delta-pyrroline-5-carboxylate dehydrogenase and its potential role in protection from proline toxicity. Plant J. 2001; 27(4): 345-56.

21. Deuschle K, Funck D, Forlani G, Stransky H, Biehl A, Leister D, et al. The role of [Delta]1-pyrroline-5-carboxylate dehydrogenase in proline degradation. Plant Cell. 2004; 16(12): 3413-25.

22. Miller G, Honig A, Stein H, Suzuki N, Mittler R, Zilberstein A. Unraveling delta1-pyrroline-5-carboxylate-proline cycle in plants by uncoupled expression of proline oxidation enzymes. J Biol Chem. 2009; 284(39): 26482-92. doi: 10.1074/jbc.M109.009340.

23. He F, DiMario PJ. Drosophila delta-1-pyrroline-5-carboxylate dehydrogenase (P5CDh) is required for proline breakdown and mitochondrial integrity-Establishing a fly model for human type II hyperprolinemia. Mitochondrion. 2011; 11(3): 397-404. doi: 10.1016/j.mito.2010.12.001.

24. Srivastava D, Singh RK, Moxley MA, Henzl MT, Becker DF, Tanner JJ. The three-dimensional structural basis of type II hyperprolinemia. J Mol Biol. 2012; 420(3):176-89. doi: 10.1016/j.jmb.2012.04.010.

25. Pemberton TA, Tanner JJ. Structural basis of substrate selectivity of Δ(1)-pyrroline-5-carboxylate dehydrogenase (ALDH4A1): semialdehyde chain length. Arch Biochem Biophys. 2013; 538(1):34-40. doi: 10.1016/j.abb.2013.07.024.

26. Flynn MP, Martin MC, Moore PT, Stafford JA, Fleming GA, Phang JM. Type II hyperprolinaemia in a pedigree of Irish Travellers (nomads). Arch Dis Child. 1989; 64(12): 1699-1707.

27. Geraghty MT, Vaughn D, Nicholson AJ, Lin WW, Jimenez-Sanchez G, Obie C, et al. Mutations in the delta-1-pyrroline 5-carboxylase dehydrogenase gene cause type II hyperprolinemia. Hum Mol Genet. 1998; 7(9): 1411-1415.

22

Low-Temperature Trapping of Photointermediates of the Rhodopsin E181Q Mutant

**Megan N. Sandberg[1#], Jordan A. Greco[1#], Nicole L. Wagner[1], Tabitha L. Amora[1],
Lavoisier A. Ramos[1], Min-Hsuan Chen[2], Barry E. Knox[2*] and Robert R. Birge[1*]**

[1]*Departments of Chemistry and Molecular and Cell Biology, University of Connecticut, Storrs, CT 06269, USA*
[2]*Departments of Biochemistry and Molecular Biology and Ophthalmology, State University of New York Upstate Medical University, Syracuse, NY 13210, USA*
#Authors contributed equally to this article

Corresponding authors: *Robert R. Birge, Distinguished Chair of Chemistry, Departments of Chemistry and Molecular and Cell Biology, University of Connecticut, 55 North Eagleville Rd., Storrs, CT 06269, USA, E-mail: rbirge@uconn.edu*

Barry E. Knox, Professor, Department of Biochemistry & Molecular Biology, SUNY Upstate Medical University, 750 E. Adams St., Syracuse, NY 13210, USA, E-mail: knoxb@upstate.edu

Abstract

Three active-site components in rhodopsin play a key role in the stability and function of the protein: 1) the counter-ion residues which stabilize the protonated Schiff base, 2) water molecules, and 3) the hydrogen-bonding network. The ionizable residue Glu-181, which is involved in an extended hydrogen-bonding network with Ser-186, Tyr-268, Tyr-192, and key water molecules within the active site of rhodopsin, has been shown to be involved in a complex counter-ion switch mechanism with Glu-113 during the photobleaching sequence of the protein. Herein, we examine the photobleaching sequence of the E181Q rhodopsin mutant by using cryogenic UV-visible spectroscopy to further elucidate the role of Glu-181 during photoactivation of the protein. We find that lower temperatures are required to trap the early photostationary states of the E181Q mutant compared to native rhodopsin. Additionally, a Blue Shifted Intermediate (BSI, λ_{max} = 498 nm, 100 K) is observed after the formation of E181Q Bathorhodopsin (Batho, λ_{max} = 556 nm, 10 K) but prior to formation of E181Q Lumirhodopsin (Lumi, λ_{max} = 506 nm, 220 K). A potential energy diagram of the observed photointermediates suggests the E181Q Batho intermediate has an enthalpy value 7.99 KJ/mol higher than E181Q BSI, whereas in rhodopsin, the BSI is 10.02 KJ/mol higher in enthalpy than Batho. Thus, the Batho to BSI transition is enthalpically driven in E181Q and entropically driven in native rhodopsin. We conclude that the substitution of Glu-181 with Gln-181 results in a significant perturbation of the hydrogen-bonding network within the active site of rhodopsin. In addition, the removal of a key electrostatic interaction between the chromophore and the protein destabilizes the protein in both the dark state and Batho intermediate conformations while having a stabilizing effect on the BSI conformation. The observed destabilization upon this substitution further supports that Glu-181 is negatively charged in the early intermediates of the photobleaching sequence of rhodopsin.

Keywords: Rhodopsin; Photobleaching sequence; Photointermediates; Glu-181; E181Q; Blue-Shifted Intermediate (BSI); Absorption spectroscopy; Low-Temperature trapping

Introduction

Rhodopsin, a member of the G Protein-Coupled Receptor (GPCR) visual opsins, is located in the rod photoreceptor cells, which are responsible for scotopic (low-light) vision [1]. The visual pigment consists of a seven transmembrane helical apoprotein and an organic chromophore covalently bound to a conserved Lysine residue (Lys-296) in helix VII via a Protonated Schiff Base (PSB) linkage. Upon absorption of light, the chromophore, 11-*cis* retinal, isomerizes to an all-*trans* conformation, initiating a series of conformational changes within the protein, which are associated with the formation of a series of spectrally discrete photointermediates with known lifetimes [order of magnitude shown for 298 K]: Bathorhodopsin (Batho) [ns], Blue-Shifted Intermediate (BSI) [ns], Lumihodopsin (Lumi) [μs], Metahodopsin I (Meta I) [ms], and Metahodopsin II (Meta II). The Meta II intermediate is stable on the timescale of minutes and activates the heterotrimeric G protein, transducin [2-4].

The photointermediate Batho is the first intermediate stable at low temperatures and stores the energy needed (~32 Kcal/mol photon energy) to propagate the structural and conformational changes necessary to form the active state of the protein (Meta II), which in turn catalyzes the visual transduction process [5-7]. Although the exact mechanism of energy storage is unknown, it has been proposed that the energy storage in Batho involves conformational strain within the chromophore and electrostatic interactions between the chromophore and the protein [8]. At low temperatures, Batho is converted directly to Lumi by gradually warming rhodopsin in the dark. However, at room temperature, the formation of a BSI is observed after formation of Batho and prior to the formation of Lumi [9]. The first observation of a BSI during the Batho to Lumi transition occurred during nanosecond photolysis experiments on various artificial visual pigments, including 13-desmethyl rhodopsin and 5,6-dihydrorhodopsin

[10,11]. These rhodopsin analogs displayed a destabilization in the Batho intermediate as well as the accumulation of the BSI to sufficient concentrations as to be observed at low temperatures [8,10-13]. Time-resolved resonance Raman analysis suggests the Batho → BSI transition involves the relaxation of the chromophore in the C_{10}-C_{11} = C_{12}-C_{13} region [14], and it is important to note that Glu-181 is located directly above this region of the chromophore. The formation of the BSI in native rhodopsin has only been observed at room temperature, which is due to the equilibrium constant shifting towards the Batho intermediate resulting in too little formation of the BSI to be experimentally observed at low temperatures [8,9]. While changes in the steric and/or electrostatic interactions between the chromophore and the protein have been associated with the formation of BSI, the decay rate appears to be largely dependent on conformational changes within the protein, or protein relaxation [8].

Models of rhodopsin developed from recent crystal structure data provide evidence of an extended hydrogen-bonding network around the PSB involving residues Glu-113, Glu-181, Ser-186, Tyr-192, Tyr-268, and select water molecules [15,16], which has been shown to stabilize the dark state conformation of rhodopsin [17]. The spectral shifts observed during the early photointermediates (dark state, Batho, BSI, and Lumi) are due to modulations of the retinal PSB structure in and around this hydrogen-bonding network, in addition to the eletrostatic influence of the charged residues (Glu-113 and Glu-181), π-stacking, and van der Waals interactions with surrounding residues within the protein-binding pocket [18]. Titration experiments on native rhodopsin and site-directed mutants reveal that the primary counter-ion to the PSB in Meta I is Glu-181, implying there is a switch in the primary counter-ion from Glu-113 in the dark state to Glu-181 in the Meta I state via a rearrangement of the active site hydrogen-bonding network [19-21] and a rotation of helix VI to accommodate this shift [22]. During the transition from Meta I to Meta II, the PSB is deprotonated and the ionic interaction between the PSB and Glu-113, which serves to lock the protein in an inactive conformation, is broken [23]. This deprotonation of the PSB has been shown to be largely influenced by an enhanced torsional flexibility of the retinal polyene chain and a shift in the relative orientation of the β-ionone ring [24-27].

The ionizable residue Glu-181, which is located within the binding pocket along extracellular loop II connecting transmembrane helices IV and V, is directly involved in a hydrogen-bonding network with Tyr-268, Tyr-192, Ser-186, and Water molecule 14 (Wat-14) [28]. The protonation state of the carboxylic acid side chain of this residue during the early intermediates has been debated through numerous experimental and theoretical investigations [20,29-45]. Original models of the counter-ion switch predict that a proton is transferred from Glu-181 to Glu-113 during the Lumi to Meta I transition (Figures 1A–1D), however, our recent report on the excited state manifold of the rhodopsin mutant with a glutamine substitution at position 181 (E181Q) strongly supports a negatively charged Glu-181 during Batho [42]. Figures 1E–1H demonstrate that the counter-ion switch model is still applicable with a negatively charged Glu-181 during the dark state and Batho. Previous time-resolved

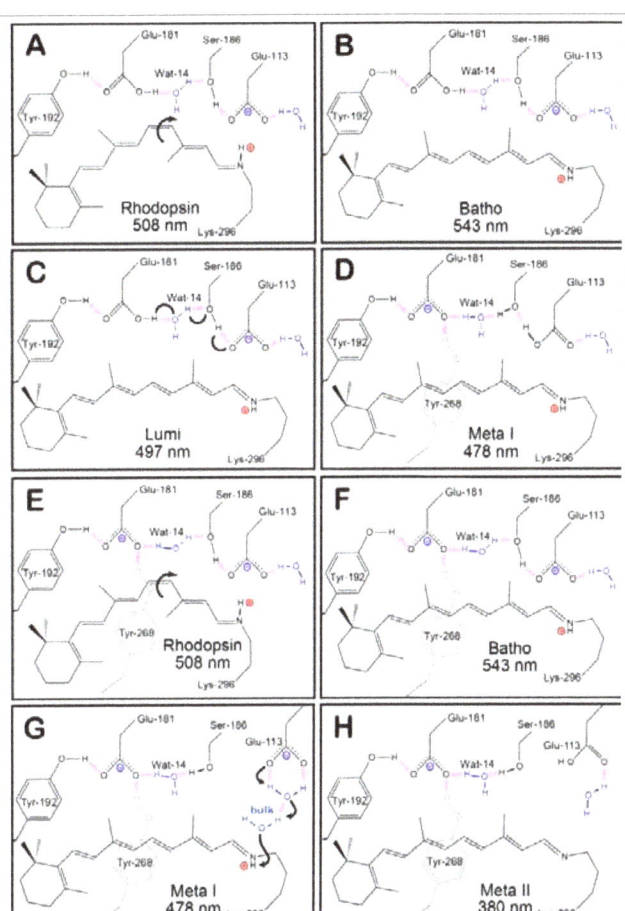

Figure 1: The hydrogen-bonding network of the protein-binding site of native rhodopsin at select intermediates during the photobleaching sequence. Panels A-D depict the original model of the counter-ion switch, in which Glu-181 is initially neutral. The primary photochemical event involves the isomerization of 11-*cis* retinal in rhodopsin (A) to all-*trans* retinal in Batho (B). During these early intermediates, Glu-113 serves as the primary counter-ion for the PSB. The transition from Lumi (C) to Meta I (D) is characterized by the transfer of a proton from Glu-181 to the hydrogen bonding network, which subsequently leads to the protonation of Glu-113. Panels E-H demonstrates our model, [42] which predicts that Glu-181 is also negatively charged during the early intermediates (E and F), and the hydrogen-bonding network rearranges to allow Glu-181 to serve as the primary counter-ion after the transition to Meta I (G). During the Meta II state (H), the PSB is deprotonated within the protein-binding site and the protein is activated in order to catalyze the visual transduction cascade. The purple dashed lines represent hydrogen bonding, and the positively and negatively charged species are indicated using red and blue labels, respectively.

studies on several Glu-181 mutants also suggest the residue plays a significant role in the early stages of the photobleaching sequence of rhodopsin [39]. At room temperature, the E181Q mutation results in the destabilization of the Batho intermediate and subsequently an accelerated decay from Batho to BSI (< 30 ns), resulting in the formation of a considerable amount of BSI [39].

Herein, the consequence of the mutational exchange of residue Glu-181 with Gln-181 on the structural stability and photoactivation mechanism of rhodopsin is explored using low-temperature trapping methods [46-48]. The mutant pigment E181Q was genetically engineered and spectroscopically characterized at cryogenic temperatures following purification. The formation and decay of each photointermediate was analyzed using UV-visible spectroscopy and notable differences were observed between the photobleaching sequences of E181Q and native rhodopsin. Plotting the potential energy surface of each intermediate of the photobleaching sequences provides insight into the observation of BSI in the rhodopsin mutant E181Q at low temperatures and possible mechanisms of BSI stability are discussed. The combination of experimental and theoretical data shown below lead us to conclude that a negatively charged carboxylic acid side chain at position 181 is crucial for the stability of the Batho and Lumi states in rhodopsin.

Materials and Methods

Visual pigment expression and purification

The pigment was constructed and isolated as previously reported [49]. The E181Q mutant was expressed in mammalian COS1 cells and purified by immunoaffinity chromatography techniques. The pigment was eluted in buffer Y1 [50 mM HEPES, 140 mM NaCl, 3 mM $MgCl_2$, pH 6.6] with 20% glycerol and 0.1% N-dodecyl-β-D-Maltoside (DM) and stored at 193 K until used.

Cryogenic experiments

The spectra of each photointermediate in the photobleaching sequence of rhodopsin and E181Q were measured using standard methods [42,47,48,50,51]. The samples were prepared in 67% glycerol, buffer Y1 [50 mM HEPES, 140 mM NaCl, 3 mM $MgCl_2$, pH 6.8], and 0.05% DM. Low-temperature experiments were conducted from 10 K to 220 K in a closed-cycle helium-refrigerated cryostat (APD Cryogenics) coupled to a Cary 50 UV-visible spectrophotometer (Varian, Inc.). For the mutant E181Q, a temperature below 20 K was required due to the instability of the E181Q Batho photoproduct. The 10 K temperature was chosen to prevent formation of any intermediates other than Batho. To generate each Photostationary State (PSS), samples were equilibrated to 10 K prior to illumination with a Photomax system equipped with a 200 W arc lamp and a monochromator (Oriel Instruments) tuned 20 nm to the blue of the absorption maximum (λ_{max}) of the dark state. Once photoconversion to the Batho photoproduct was complete, the temperature was then raised to 220 K in increments of 10 K. To avoid artifacts arising from temperature dependent baseline shifts, the temperature of the sample was cooled down to the starting temperature (10 K) before each spectrum was recorded. Semi-low temperature experiments were carried out from 233 K to 293 K in a Cary 5000 UV-visible spectrophotometer (Varian, Inc.) equipped with a temperature controlled sample holder (Quantum Northwest, Inc.). Samples were illuminated with 495 nm light until no further spectral shift was observed. The temperature of the sample was then raised to 293 K in increments of 5 K once photoconversion was completed.

Two methods were used in combination to determine the composition of each PSS. The first method involved warming the sample to ambient temperature to allow for the formation of Meta II, which has a λ_{max} for both rhodopsin and E181Q rhodopsin of approximately 380 nm. The spectrum of the resulting sample contains a mixture of the dark state and Meta II, separated in wavelength to a sufficient extent to allow reliable spectral deconvolution. The integral of the λ_{max} band, when compared to that observed for the pure dark state at the same temperature, permits accurate assignment of the amount of rhodopsin converted. Additionally, retinal oximes were extracted and analyzed using High Performance Liquid Chromatography (HPLC) following the methods and procedures reported previously [50] and described below. The isomeric compositions of each PSS are then used to deconvolute the measured spectra and determine the λ_{max} of the photointermediates of E181Q. The pure dark state is assumed to be comprised of 100% 11-cis retinal for spectral deconvolution, as previously described [52-54]. All absorption spectra presented are the average of three spectra normalized with respect to the protein aromatic residue band at 280 nm. Each difference spectrum was calculated by subtracting the selected PSS from the corresponding dark state or relevant PSS.

Chromophore extraction

The isomeric ratio of the chromophore was determined as follows. A mixture of 150 μL of ice-cold 1.0 M hydroxylamine (pH 7) solution, 1 mL of methanol, and 1 mL of dichloromethane was added to each glycerol/protein sample. The mixture was shaken vigorously for 1 min and put on ice. Hexane (1 mL) was then added and the sample was shaken and spun in a clinical centrifuge for 45 s. The hexane layer was removed, and another 5 mL hexane extraction was performed. The combined hexane layers were then dried (Na_2SO_4), filtered through a 0.2 μm filter, and evaporated in a clean tube under vacuum. The total volume of the hexane layer was brought to 100 μL.

A portion of the hexane layer (50 μL) was injected into an HPLC instrument (Waters Corporation), which was equipped with two HPLC columns (Waters Prep Nova-Pak HR silica columns, 3.9×300 mm, Waters catalog no. WAT038501) and a Waters 2487 dual wavelength absorbance detector monitoring at 360 nm. The mobile phase used to separate the retinal isomers was composed of 96% hexane, 3% tert-butyl methyl ether, 0.5% 1-octanol, and 0.5% 1,4-dioxane. All solvents used were HPLC grade (Fisher Scientific). The flow rate was fixed at 2.5 mL/min. Retinal oxime standards (all-trans, 11-cis and 9-cis retinal oxime) were used to assign the retention times of the peaks observed following the chromophore extractions. The syn oxime enantiomers were favored due to the low temperature and concentrations used for the extraction, and these peaks were used exclusively to determine the isomeric ratios of retinal within the chromatograms. The anti oxime enantiomers were observed, however, these species were found at longer retention times and were ignored for this analysis.

Computational methods

All molecular orbital calculations were carried out using Gaussian 09 [55]. The heavy atom coordinates for dark state rhodopsin, Batho, Lumi, and Meta II were taken from the 1U19

[16], 2G87 [56], 2HPY [57], and 2I37 [58] crystal structures of rhodopsin, respectively. Hydrogen atoms were added to all relevant atoms of residues within 5.6Å of the chromophore by using Anamol 5.6.4. The hydrogen atoms and these local residues, in addition to the chromophore, were optimized by using the Parameterized Model 3 (PM3) methods [59,60] in Gaussian 09, while holding all other heavy atoms at the crystal geometry coordinates. During these optimizations, the oxygen atoms of the nearby water molecules were locked while the hydrogen atoms were allowed to fully optimize. Subsequently, seven iterations of B3LYP/6-31G(d) procedures [61,62] in Gaussian 09 were used to generate the ground state structures of the photointermediates. The configuration of the glutamine residue was optimized by minimizing the two possible rotational geometries and selecting the geometry with the lowest energy. The electrostatic charge shifts of the protein-binding sites were generated in MathScriptor 3.5.0 (www.mathscriptor.org).

The rhodopsin photobleaching energy surface was taken from reference [63], and was generated based on photocalorimetry as the primary experimental method. The E181Q energy surface was generated by reference to the rhodopsin surface using the temperature ramping experiments to assign the barriers, and the theoretical calculations to estimate the energy minima. The energy barriers separating the various intermediates in E181Q, $E_{barrier}^{E181Q}$, were assigned by reference to the temperature of appearance, T_{obsvd}^{E181Q}, of the next intermediate:

$$E_{barrier}^{E181Q} \approx \frac{E_{barrier}^{RHO} \times T_{obsvd}^{E181Q}}{T_{obsvd}^{RHO}} \quad \text{Equation 1}$$

where $E_{barrier}^{RHO}$ and T_{obsvd}^{RHO} are the corresponding barrier and temperatures in the rhodopsin photobleaching sequence. Because the above method and the theoretical models are approximate, the E181Q surface should be viewed as qualitative.

Results

The raw spectra of the low-temperature and semi-low temperature spectroscopy studies of native rhodopsin and E181Q are described in detail below. At room temperature, E181Q displayed a significant red-shift in absorption maximum (λ_{max} = 508 nm) relative to native rhodopsin (λ_{max} = 499 nm) [42]. Lowering the temperature to 10 K resulted in a negligible shift for E181Q and a bathochromic shift of 4 nm for native rhodopsin (λ_{max} = 503 nm) at 70 K (Figure 2). Illumination of E181Q with 500 nm light initiated the formation of the first PSS, a mixture of resting state and Batho, which was red-shifted to a λ_{max} of 522 nm (PSS522). Figures 2A–2D demonstrate that the photoconversion to Batho for native rhodopsin and E181Q was complete in 1 hour at 70 K and 10 K, respectively, and involved the formation of a single species with the disappearance of the dark state. As the temperature of PSS522 was raised in 10 K increments, no spectral shift was observed from 10 K to 50 K (Figure 3). At 60 K, the spectrum begins to shift to form a second PSS at 498 nm (PSS498). The temperature was then gradually raised to 220 K, and the formation of a third intermediate was observed at 506 nm (PSS506). No further spectral change was measured after 200 K, indicating the formation of PSS506 was complete. HPLC analysis of PSS506 (Figure 4) revealed a retinal composition of

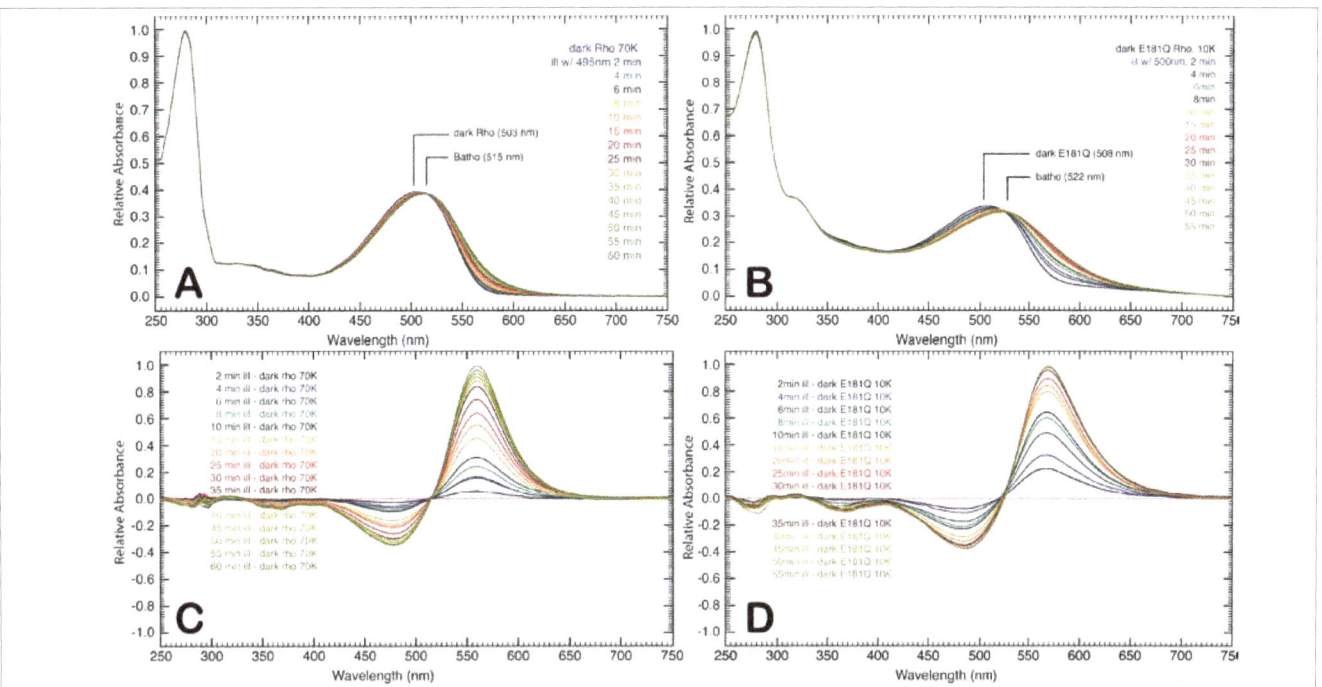

Figure 2: Time-resolved absorption spectra of rhodopsin at 70 K (A) and E181Q at 10 K (B) following illumination by using a 495 nm and 500 nm light source, respectively. The absorption spectra were collected at the given time points following illumination. The difference spectra of the absorption profiles are provided for rhodopsin (C) and E181Q (D) and were obtained by subtracting the respective spectra of the dark state from the corresponding time point throughout the experiment.

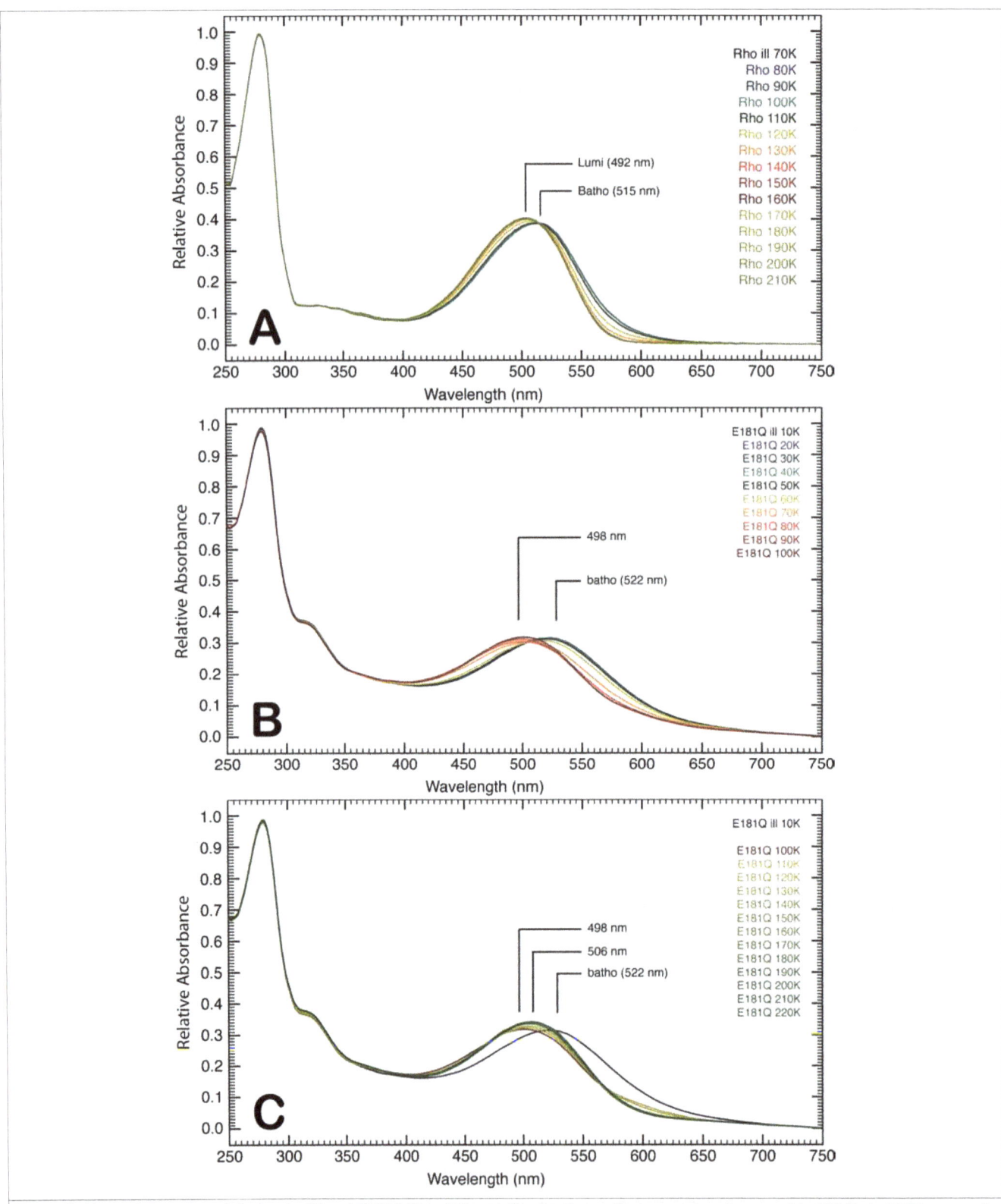

Figure 3: Absorption spectra of rhodopsin (A) and E181Q (B and C) post-illumination as the temperature is ramped to 220 K via 10 K increments. The initial formation of the Batho photointermediate for native rhodopsin was achieved at 70 K (A), and as the temperature was increased, the formation of a single Lumi photointermediate (492 nm) was observed. The absorption spectra for E181Q was first collected at 10 K, in which the PSS522 consisted of a mixture of the dark and Batho states of the mutant protein (B). As the temperature was increased, a second PSS formed (PSS498) at 60 K. The E181Q sample was then allowed to warm to 220 K (C), where a third PSS (PSS506) evolved. The PSS498 was found to be a mixture of Batho and BSI, whereas the PSS506 was a Lumi photointermediate of the mutant protein (see text).

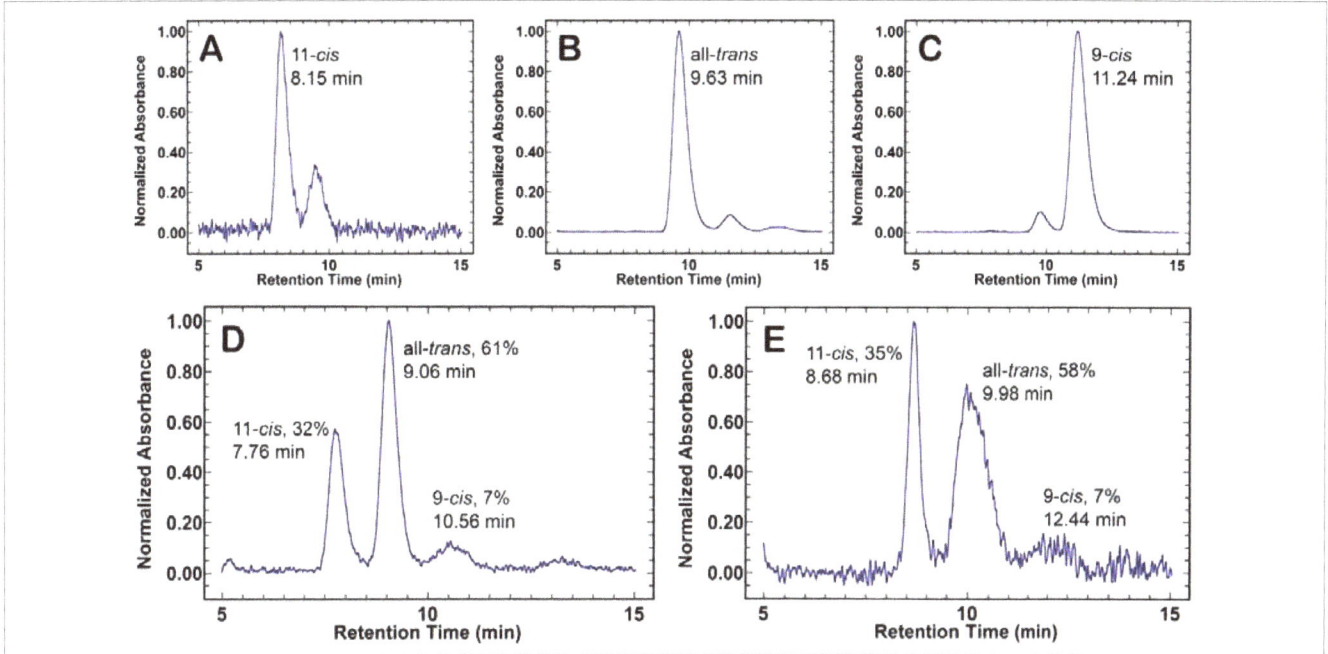

Figure 4: HPLC chromatograms of retinal oxime standards and retinal oxime extractions of PSS506 and PSS383. Panels A, B, and C show the chromatograms for the 11-*cis*, all-*trans*, and 9-*cis* retinal oxime standards, respectively. Each standard peak represents the retention time for the *syn* oxime enantiomer of the retinal isomers. The percent of retinal isomers formed for PSS506 (D) and PSS383 (E) were determined by integrating under each peak.

32% 11-*cis*, 61% all-*trans*, and 7% 9-*cis* retinal isomers.

The formation and decay of the late photointermediates (Lumi, Meta I, and Meta II) for rhodopsin and E181Q were studied using semi-low temperature spectroscopy, from 233 K to 293 K. Irradiation of E181Q with 495 nm light at 233 K promoted the formation of Lumi (λ_{max} = 495 nm) and no spectral shift was observed after 85 min of continuous illumination (Figure 5A). During the formation of Lumi, the formation and decay of a red-shifted difference spectrum species (λ_{max} = 575 nm) is seen along with the formation of a blue-shifted difference spectrum species (λ_{max} = 385 nm) (Figure 5B). Further temperature ramping experiments illustrated in Figure 6 indicate that the transition to the Meta I intermediate (λ_{max} = 480 nm) began at 238 K and was complete by 243 K. The Meta I spectrum is very broad and may be a mixture of Meta I and Lumi intermediate that has not decayed completely at 233 K. With increasing temperature (in 5 K increments), the spectrum continues to blue shift until the formation of Meta II (λ_{max} = 383 nm; PSS383) is complete. HPLC analysis of the retinal composition of PSS383 (Figure 4) shows the presence of 11-*cis* (35%) all-*trans* (58%), and 9-*cis* (7%) retinal isomers.

Discussion

Time-resolved UV-visible spectroscopy has been instrumental in determining the formation and decay of discrete photointermediates of rhodopsin, as well as elucidating the photobleaching pathway of the protein [9]. However, a key disadvantage of room-temperature time-resolved UV-visible spectroscopy for the rhodopsin mutant E181Q is the simultaneous formation and decay of multiple intermediates.

Low-temperature trapping experiments, which slow down the photobleaching process and ensure that each photointermediate is observed individually, were performed to avoid this problem [46-48]. In the photobleaching sequence of native rhodopsin, Batho is stable at 70 K and decays directly to Lumi upon gradual warming (Figure 3A). However, in the photobleaching sequence of E181Q, Batho is only stable at temperatures lower than 50 K and thermally equilibrates with BSI when the temperature is raised above 60 K. The final equilibrium mixture (PSS498), which is comprised mainly of BSI, is not established until 90 K. Spectral deconvolution was used to determine that pure BSI for E181Q has a λ_{max} of 479 nm. Raising the temperature to 220 K results in decay of the mixture to Lumi (PSS506), the identity of which is confirmed through the chromophore extraction of the PSS and HPLC analysis. The isomeric composition of PSS506 was found to be 32% 11-*cis* retinal, 61% all-*trans* retinal, and 7% 9-*cis* retinal. Pure Lumi is found to have a λ_{max} of 510 nm (at 220 K) by adding back 32% of the dark state spectrum of E181Q to the difference spectrum of PSS506 minus the dark state spectrum. Contributions from the 9-*cis* chromophore were ignored for this spectral deconvolution because so little is formed (7%). In comparison to the native rhodopsin spectral data, these observations suggest that the retinylidene binding pocket environment has been significantly perturbed during the early stages of the photoactivation mechanism.

In the semi-low temperature experiments, a single Lumi intermediate is seen for native rhodopsin (λ_{max} = 490 nm) at 233 K. Previous studies have shown that at low temperatures, the equilibrium constant between Lumi I and Lumi II for native rhodopsin is significantly shifted towards Lumi II and thus

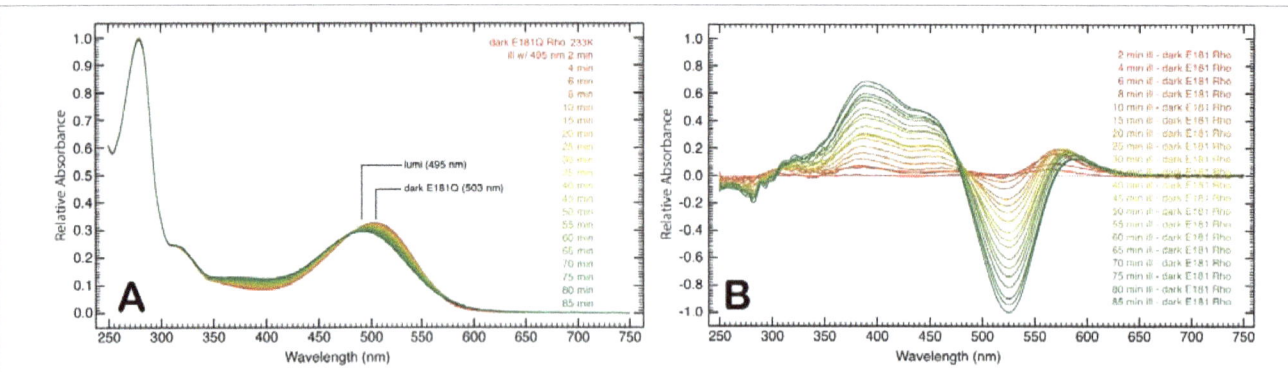

Figure 5: Time-resolved absorption spectra of E181Q at 233 K following illumination with 495 nm light (A). The difference spectra presented in panel (B) was obtained by subtracting the dark state spectrum from the spectra of each time point. While the blue-shifted species is accounted for by the evolution of Lumi (495 nm) over 85 min, the origin of the red-shifted species with a difference spectrum maximum at 575 nm is currently unknown.

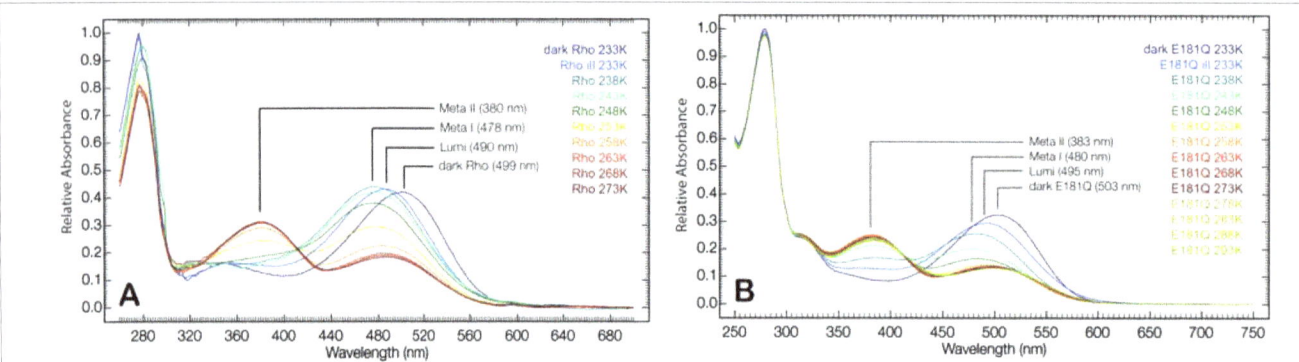

Figure 6: Absorption spectra of rhodopsin (A) and E181Q (B) at semi-low temperatures (233 K to 293 K) following illumination of the dark state to initiate the photobleaching sequence. The higher temperatures allowed for the formation and trapping of the late photointermediates (Meta I and Meta II), in addition to a single Lumi photointermediate.

only one Lumi intermediate is observed [64]. Similarly, we observe the formation of a single E181Q Lumi intermediate (λ_{max} = 495 nm). However, during the formation of E181Q Lumi, the formation and decay of a red-shifted difference spectrum species (λ_{max} = 575 nm) is observed (Figure 5). This species is not seen during Lumi formation in native rhodopsin and warrants further investigation. While the absorption spectra of the later intermediates of E181Q and native rhodopsin are similar (Figure 6), the Meta I and Meta II photointermediates of E181Q form more readily at lower temperatures compared to native rhodopsin. Recall that a negatively charged Glu-181 serves as the primary counter-ion during the late photointermediates [19-21], and thus the substitution with Gln-181 will serve to destabilize the retinal binding site during Meta I and Meta II. The isomeric composition of PSS383 (35% 11-*cis* retinal, 58% all-*trans* retinal, and 7% 9-*cis* retinal) is used in a similar fashion as for PSS506 in order to determine the λ_{max} of pure Meta II (λ_{max} = 384 nm at 273 K).

Lewis et al. [39], investigated the room temperature time-resolved spectra of Glu-181 mutants and observed a destabilization of the Batho intermediate, as well as a significant shift in equilibrium towards BSI in E181Q. During the late stages of the photoactivation mechanism, they also note the absence of a Lumi I to Lumi II transition for the E181Q mutant [39]. Because

the formation of these photointermediates are temperature dependent, only the formation of photointermediates that accumulate to appreciable amounts at room temperature are expected to form at low temperatures in observable concentrations. The notion that only a single Lumi intermediate was produced for this mutant at room temperature allows us to predict that we are observing a transition from BSI to Lumi at low temperatures (< 200 K) in E181Q and not a Lumi I to Lumi II transition. The proposed photobleaching sequence pathways of native rhodopsin and E181Q are summarized in Figure 7.

Recall that a destabilized Batho may result from a change in ionization or a disruption in the complex hydrogen-bonding network involving several binding pocket residues. Because Batho and BSI are in equilibrium, changes that destabilize Batho would result in the stabilization of BSI. While Lewis et al. [39] found that E181Q and other Glu-181 mutants lead to an accelerated decay of Batho, replacing Glu-181 with aspartic acid (E181D) resulted in a Batho lifetime similar to that of native rhodopsin. Aspartic acid is one carbon atom shorter than glutamic acid, however, the carboxyl group is maintained. Therefore, it is reasonable to predict that the stability of Batho relies on the presence of the carboxyl atoms. Although there is no clear agreement on the protonation state of Glu-181 [20,29-45], if we assume it is negatively charged in the dark state, then replacing Glu-181 with the neutral residue Gln-181 may provide insight into why the mutant E181Q leads to a faster decay of Batho and stabilizes the BSI. During the primary event, the distance between the C_{13}-methyl group and residue Glu-181 decreases from a distance of 5.7Å to 3.0Å [16,56]. The removal of a charged residue near the highly strained C_9-C_{13} portion of retinal may provide an increased flexibility in the polyene chain of the chromophore, which would allow the chromophore to more readily adopt a planar conformation and shift the equilibrium towards the BSI.

Moreover, substitution of Glu-181 with Gln-181 may also result in a destabilized Batho via perturbation of the active site hydrogen-bonding network. The notion that this mutational change would alter the hydrogen-bonding network is not surprising because the functional groups on these two residues contain different hydrogen-bonding character. Further support for a hydrogen-bonding rearrangement being responsible for the observed destabilized Batho comes from the fact that Wat-14, which is in a direct path of the C_{13}-methyl group of the chromphore during the primary event, is believed to contribute to the stability of Batho [65]. Thus, any perturbation of Wat-14 may also result in a destabilization of Batho. In addition to E181Q, a destabilized Batho intermediate has been observed in several artificial pigments, as well as select rhodopsin mutants and cone-type visual pigments [10,11,66-68]. Previous studies have shown the decay of Batho is dependent on the rotation barrier of the C_6-C_7 bond, which in turn depends on the steric interaction between the C_5-methyl and the C_8-hydrogen. In E181Q, a rearrangement in the hydrogen-bonding network during the photobleaching sequence causes a shift in the position of Tyr-268 towards Gln-181 and Tyr-192 (Figure 8). The repositioning

of Tyr-268 may lower the barrier to BSI by decreasing the steric interaction between the C_5-methyl and C_8-H groups. While the E181Q rhodopsin mutant is not the first to display a destabilized Batho [68], it is interesting to note that replacing Ser-186 with an alanine residue results in a normal Batho intermediate, despite the fact that Ser-186 is hydrogen-bonded to Glu-181 via Wat-14 in the dark state and early photointermediates of rhodopsin (Figures 1 and 8) [28]. Thus, in the case of E181Q, the observed destabilization of Batho may be caused in part by a rearrangement of the hydrogen-bonding network involving Tyr-268 and Tyr-192 rather than perturbations in the active site hydrogen-bonding network involving Wat-14 and Ser-186. Although the exact mechanism of Batho destabilization remains unclear, these results demonstrate that the carboxyl group at residue 181 is required for the stabilization of Batho in native rhodopsin. Furthermore, the dramatic influence of the glutamine substituion on the photobleaching kinetics suggests that this carboxyl group is likely negatively charged during the dark and Batho states.

Plotting the potential energy surface of the photobleaching pathway for both rhodopsin and E181Q provides further insight into the stabilization of BSI in E181Q (Figure 9). Our calculations, which utilize the temperature of appearance of the photointermediates and the calculated energies (Equation 1), are in agreement with the literature and predict the BSI in rhodopsin lies higher in energy than Batho [8,63]. When we overlay the energies of the E181Q photointermediates, shown in dashes, it is immediately apparent that the E181Q mutant is less stable than native rhodopsin throughout the photobleaching sequence. All

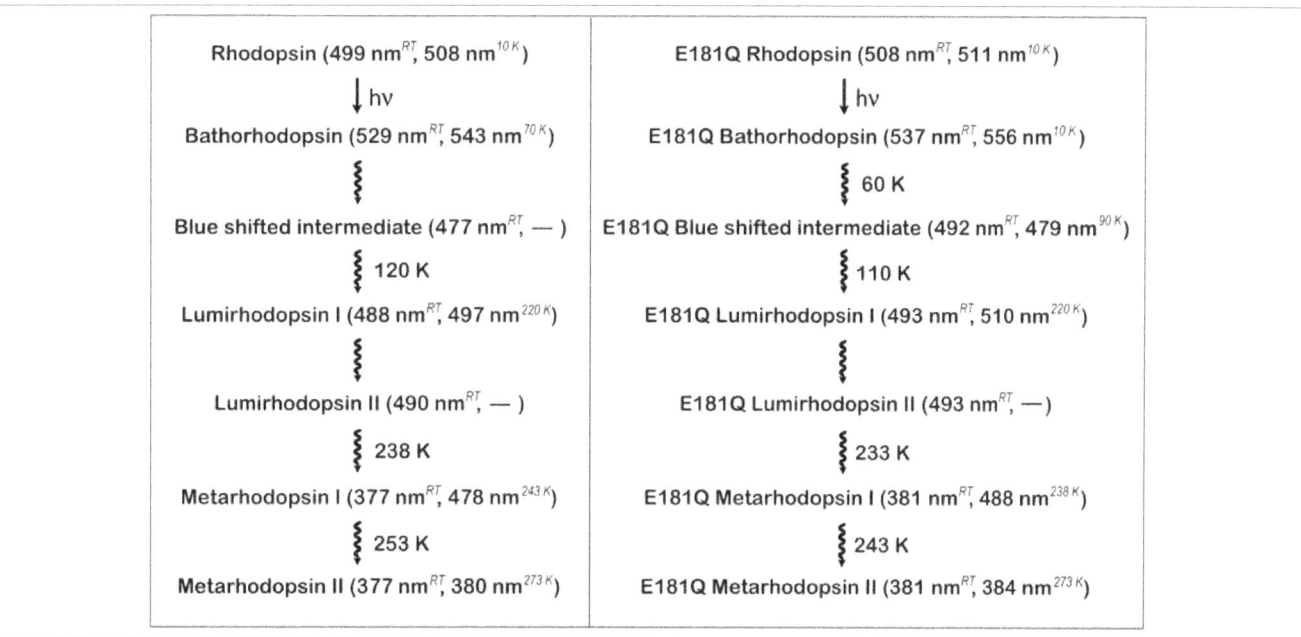

Figure 7: Photobleaching sequences of rhodopsin and E181Q rhodopsin. For each photointermediate, the λ_{max} is shown at both room temperature and low temperatures. The room temperatures are based on the values obtained by Lewis et al. [39] and the low-temperature data is based on the deconvolution of the spectra presented in this study. At low temperatures, no BSI was observed for rhodopsin, and no Lumi II was observed for either rhodopsin or E181Q.

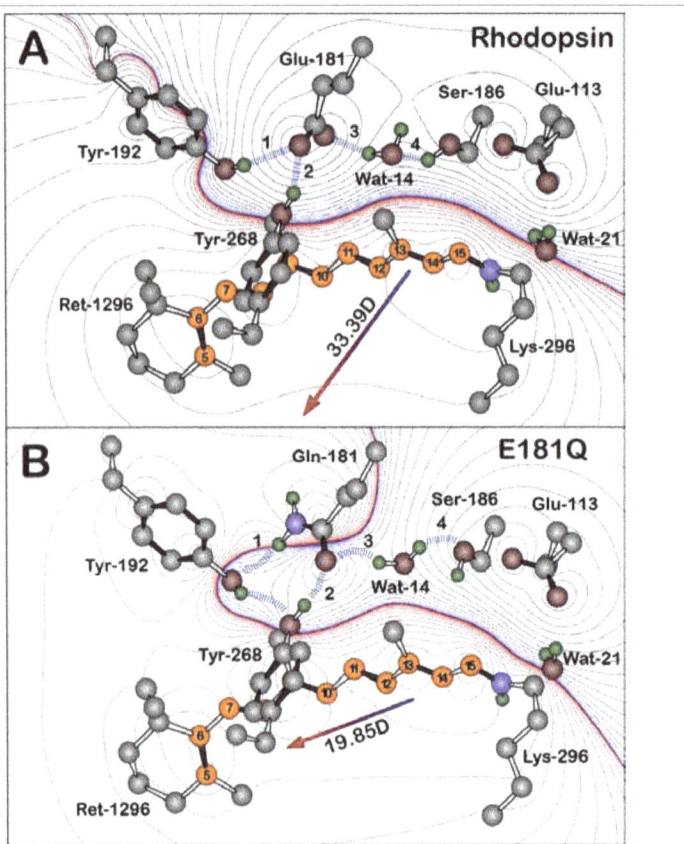

Figure 8: The hydrogen bonding network of the negatively charged Glu-181 residue of native rhodopsin (A) and of the Gln-181 residue of E181Q (B), both during the Lumi photointermediate. The blue dashed lines and the labels 1–4 highlight the key hydrogen-bonding network between residue 181, Ser-186, Tyr-192, Tyr-268, and Wat-14, which is perturbed upon the glutamine substitution. The Lumi structure (A) is based off of the 2HPY crystal structure [57] and a relaxed conformation of the Lumi photointermediate of E181Q (B) was obtained by minimizing the crystal structure with Gln-181. Polyene atoms of retinal (Ret-1296) are indicated in orange, and the numbering system shown here is used in the text. The water molecules are labeled using the Protein Data Bank (PDB) numbers minus 2000. All hydrogen atoms were included in the calculations and were optimized by using B3LYP/6-31G(d) methods, although only polar hydrogens are shown in the figure. Red and blue contours indicate regions of increased positive and negative charge, respectively. The contours are drawn by using the following first-order electrostatic energies: 0 (black), ± 0.282, ± 2.26, ± 7.63, ± 18, ± 35.3, ± 61, ± 96.9, ± 144, ± 206, ± 282, ± 376, ± 488, ± 621, ± 755 kJ/mol.

measured photointermediates of E181Q first appeared at lower temperatures than for the corresponding intermediates of native rhodopsin, which correlates with the predicted destabilization. This destabilization is further supported in the model of the Lumi photointermediate for both rhodopsin and E181Q provided in Figure 8, which indicates a significant modulation in the hydrogen-bonding network around the chromophore for the mutant. Most importantly, our model predicts the E181Q Batho intermediate has an enthalpy value 7.99 KJ/mol higher than E181Q BSI, whereas in rhodopsin, the BSI has an enthalpy that is 10.02 KJ/mol higher than Batho. From these results, we conclude the Batho to BSI transition is enthalpically driven in E181Q and entropically driven in native rhodopsin. Note, however, that these enthalpy differences and the potential energy surfaces depicted in Figure 9 are very approximated because they are based on Equation 1. A more rigorous theoretical study using a hybrid Quantum Mechanics/Molecular Mechanics (QM/MM) approach is currently being undertaken for further analysis of these models.

Conclusions

The photobleaching sequence of the rhodopsin mutant E181Q has been investigated by cryogenic studies in an attempt to further elucidate the role of Glu-181 during the photoactivation process of rhodopsin. We conclude that the photobleaching sequence of E181Q at low temperatures involves a two-step sequential decay from Batho to Lumi that includes an equilibrium between Batho and a subsequent BSI. The present study supports conclusions of other studies that have suggested Batho is destabilized in E181Q and that the Batho to BSI equilibrium lies toward the BSI [39]. The stabilization of the BSI in E181Q can be explained by the thermodynamics of the photobleaching process, which shows the Batho to the BSI transition is enthalpically driven in the rhodopsin mutant. Three key differences are noted for the photobleaching sequence of E181Q compared to rhodopsin collected at cryogenic temperatures: 1) lower temperatures are required to trap the primary photointermediate Batho in E181Q compared to native rhodopsin, 2) the decay of Batho occurs much more rapidly compared to native rhodopsin, and 3) the formation of the BSI is

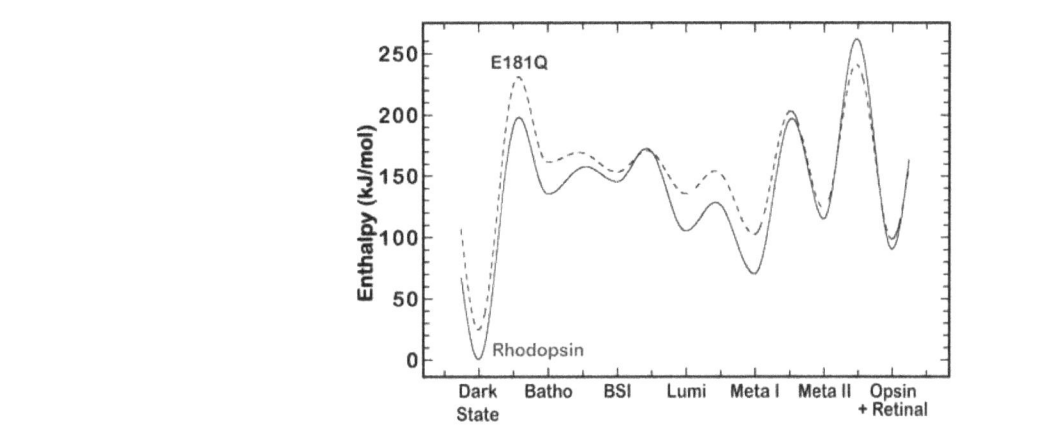

Figure 9: A potential energy surface of the photobleaching sequence of native rhodopsin (solid blue) and E181Q (dashed red). The rhodopsin surface is adopted from reference [63]. The E181Q surface is approximate and was generated using the methods described in the text.

observed during low temperature experiments. Additionally, the formation of the later intermediates (Meta I and Meta II) of E181Q all occur at lower temperatures compared to the formation of the later intermediates of native rhodopsin. We conclude that these differences in the photobleaching sequence of E181Q compared to native rhodopsin provide strong evidence that the negatively charged carboxylic acid side chain of residue Glu-181 plays a critical role in the early intermediates of the photobleaching sequence of native rhodopsin by maintaining the integrity of the active site hydrogen-bonding network which serves to stabilize the protein in the dark state and the primary photointermediate, Batho. Furthermore, by replacing the Glu-181 residue with Gln-181, a key electrostatic interaction is altered concurrent with a rearrangement of the hydrogen-bonding network within the binding pocket, thus allowing for the formation of the BSI to accumulate to concentrations observable at low temperatures.

Acknowledgements

This research was supported in part by grants from the National Institutes of Health to R.R.B. (GM-34548) and B.E.K. (EY-11256 and EY-12975), the National Science Foundation to R.R.B. (EMT-08517), the Harold S. Schwenk Sr. Distinguished Chair funds for support of specialized instrumentation at the University of Connecticut, Research to Prevent Blindness (Unrestricted Grant to SUNY UMU Department of Ophthalmology), and Lions of CNY (B.E.K.).

References

1. Baylor D. How photons start vision. Proc Natl Acad Sci USA. 1996; 93(2):560-565.

2. Fung BKK, Stryer L. Photolyzed rhodopsin catalyzes the exchange of GTP for bound GDP in retinal rod outer segments. Proc Natl Acad Sci USA. 1980; 77(5):2500-2504.

3. Hofmann KP, Jager S, Ernst OP. Structure and function of activated rhodopsin. Isr J Chem. 1995; 35:339-355. Doi: 10.1002/ijch.199500035.

4. Smith SO. Structure and activation of the visual pigment rhodopsin. Annu Rev Biophys. 2010; 39:309-28. doi: 10.1146/annurev-biophys-101209-104901.

5. Boucher F, Leblanc RM. Energy storage in the primary photoreaction of bovine rhodopsin. A photoacoustic study. Photochem Photobiol. 1985; 41:459-465.

6. Cooper A. Energy uptake in the first step of visual excitation. Nature. 1979; 282:531-533.

7. Schick GA, Cooper TM, Holloway RA, Murray LP, Birge RR. Energy storage in the primary photochemical events of rhodopsin and isorhodopsin. Biochemistry. 1987; 26(9):2556-2562.

8. Hug SJ, Lewis JW, Einterz CM, Thorgeirsson TE, Kliger DS. Nanosecond photolysis of rhodopsin: Evidence for a new blue-shifted intermediate. Biochemistry. 1990; 29(6):1475-1485.

9. Lewis JW, Kliger DS. Photointermediates of visual pigments. J Bioenerg Biomembr. 1992; 24(2):201-210.

10. Einterz CM, Hug SJ, Lewis JW, Kliger DS. Early photolysis intermediates of the artificial visual pigment 13-demethylrhodopsin. Biochemistry. 1990; 29(6):1485-1491.

11. Albeck A, Friedman N, Ottolenghi M, Sheves M, Einterz CM, Hug SJ, et al. Photolysis intermediates of the artificial visual pigment cis-5,6-dihydro-isorhodopsin. Biophys J. 1989; 55(2):233-241.

12. Shichida Y, Kropf A, Yoshizawa T. Photochemical reactions of 13-demethyl visual pigment analogs at low temperatures. Biochemistry. 1981; 20:1962-1968.

13. Randall CE, Lewis JW, Hug SJ, Bjorling SC, Eisner-Shanas I, Ottolenghi M, et al. A new photolysis intermediate in artificial and native visual pigments. J Am Chem Soc. 1991; 113:3473-3485.

14. Pan D, Ganim Z, Kim JE, Verhoeven MA, Lugtenburg J, Mathies RA. Time-resolved resonance raman analysis of chromophore structural changes in the formation and decay of rhodopsin's BSI intermediate. J Am Chem Soc. 2002; 124(17):4857-4864.

15. Palczewski K, Kumasaka T, Hori T, Behnke CA, Motoshima H, Fox BA, et al. Crystal structure of rhodopsin: A G protein-coupled receptor. Science. 2000; 289:739-745.

16. Okada T, Sugihara M, Bondar A, Elstner M, Entel P, Buss V. The retinal conformation and its environment in rhodopsin in light of a new 2.2 A crystal structure. J Mol Biol. 2004; 342(2):571-583.

17. Liu J, Liu MY, Nguyen JB, Bhagat A, Mooney V, Yan EC. Thermal decay of rhodopsin: Role of hydrogen bonds in thermal isomerization of 11-

cis retinal in the binding site and hydrolysis of protonated Schiff base. J Am Chem Soc. 2009; 131(25):8750-8751. doi: 10.1021/ja903154u.

18. Campomanes P, Neri M, Horta BA, Röhrig UF, Vanni S, Tavernelli I, et al. Origin of the spectral shifts among the early intermediates of the rhodopsin photocycle. J Am Chem Soc. 2014; 136(10): 3842-3851. doi: 10.1021/ja411303v.

19. Terakita A, Yamashita T, Shichida Y. Highly conserved glutamic acid in the extracellular IV-V loop in rhodopsins acts as the counterion in retinochrome, a member of the rhodopsin family. Proc Natl Acad Sci USA. 2000; 97(26): 14263-14267.

20. Yan ECY, Kazmi MA, Ganim Z, Hou JM, Pan D, Chang BS, et al. Retinal counterion switch in the photoactivation of the G protein-coupled receptor rhodopsin. Proc Natl Acad Sci USA 2003; 100(16):9262-9267.

21. Yan ECY, Kazmi MA, De S, Chang BS, Seibert C, Marin EP, et al. Function of extracellular loop 2 in rhodopsin: Glutamic acid 181 modulates stability and absorption wavelength of metarhodopsin II. Biochemistry. 2002; 41(11): 3620-3627.

22. Eilers M, Goncalves JA, Ahuja S, Kirkup C, Hirshfeld A, Simmerling C, et al. Structural transitions of transmembrane helix 6 in the formation of metarhodopsin I. J Phys Chem B. 2012; 116(35):10477-10489. doi: 10.1021/jp3019183.

23. Robinson P, Cohen G, Zhukovsky E, Oprian D. Constitutively active mutants of rhodopsin. Neuron 1992; 9(4): 719-725.

24. Zhu S, Brown MF, Feller SE. Retinal conformation governs pKa of protonated Schiff base in rhodopsin activation. J Am Chem Soc. 2013; 135(25):9391-9398. doi: 10.1021/ja4002986.

25. Leioatts N, Mertz B, Martínez-Mayorga K, Romo TD, Pitman MC, Feller SE, et al. Retinal ligand mobility explains internal hydration and reconciles active rhodopsin structures. Biochemistry. 2014; 53(2):376-385. doi: 10.1021/bi4013947.

26. Ahuja S, Eilers M, Hirshfeld A, Yan EC, Ziliox M, Sakmar TP, et al. 6-s-cis conformation and polar binding pocket of the retinal chromophore in the photoactivated state of rhodopsin. J Am Chem Soc 2009; 131(42):15160-15169.

27. Bartl FJ, Fritze O, Ritter E, Herrmann R, Kuksa V, Palczewski K, et al. Partial Agonism in a G Protein-coupled Receptor: Role of the Retinal Ring Structure in Rhodopsin Activation. J Biol Chem. 2005; 280(40):34259-34267.

28. Yan ECY, Epps J, Lewis JW, Szundi I, Bhagat A, Thomas P. Sakmar et al. Photointermediates of the rhodopsin S186A mutant as a probe of the hydrogen-bond network in the chromophore pocket and the mechanism of counterion switch. J Phys Chem C 2007; 111(25): 8843-8848. doi: 10.1021/jp067172o

29. Birge RR, Murray LP, Pierce BM, Balogh-Nair V, Findsen LA, Nakanishi K. Two-photon spectroscopy of locked-11-cis rhodopsin: Evidence for a protonated Schiff base in a neutral protein binding site. Proc Natl Acad Sci USA. 1985; 82(12):4117-4121.

30. Yan EC, Ganim Z, Kazmi MA, Chang BS, Sakmar TP, Mathies RA. Resonance raman analysis of the mechanism of energy storage and chromophore distortion in the primary visual photoproduct. Biochemistry. 2004; 43(34): 10867-10876.

31. Sekharan S, Buss V. Glutamic acid 181 is uncharged in dark-adapted visual rhodopsin. J Am Chem Soc. 2008; 130: 17220-17221. doi: 10.1021/ja805992d.

32. Mollevanger LC, Kentgens AP, Pardoen JA, Courtin JM, Veeman WS, Lugtenburg J, et al. High-resolution solid-state 13C-NMR study of carbons C-5 and C-12 of the chromophore of bovine rhodopsin. Evidence for a 6-s-cis conformation with negative-charge perturbation near C-12. Eur J Biochem. 1987; 163(1): 9-14.

33. Smith SO, Palings I, Miley ME, Courtin J, de Groot H, Lugtenburg J, et al. Solid state NMR studies of the mechanism of the opsin shift in the visual pigment rhodopsin. Biochemistry 1990; 29(35): 8158-8164.

34. Han M, Smith SO. NMR constraints on the location of the retinal chromophore in rhodopsin and bathorhodopsin. Biochemistry 1995; 34(4): 1425-1432.

35. Honig B, Dinur U, Nakanishi K, Valeria BN, Mary AG, Arnaboldi M, et al. An external point-charge for wavelength regulation in visual pigments. J Am Chem Soc 1979; 101(23): 7084-7086. doi: 10.1021/ja00517a060.

36. Nagata T, Terakita A, Kandori H, Shichida Y, Maeda A. The hydrogen-bonding network of water molecules and the peptide backbone in the region connecting Asp83, Gly120, and Glu113 in bovine rhodopsin. Biochemistry. 1998; 37(49): 17216-17222. doi: 10.1021/bi9810149

37. Ludeke S, Beck M, Yan EC, Sakmar TP, Siebert F, Vogel R. The role of Glu181 in the photoactivation of rhodopsin. J Mol Biol. 2005; 353(2): 345-356.

38. Rohrig UF, Guidoni L, Rothlisberger U. Early steps of the intramolecular signal transduction in rhodopsin explored by molecular dynamics simulations. Biochemistry. 2002; 41(35): 10799-10809. doi: 10.1021/bi026011h.

39. Lewis JW, Szundi I, Kazmi MA, Sakmar TP, Kliger DS. Time-resolved photointermediate changes in rhodopsin glutamic acid 181 mutants. Biochemistry. 2004; 43(39): 12614-12621.

40. Martinez-Mayorga K, Pitman MC, Grossfield A, Feller SE, Brown MF. Retinal counterion switch mechanism in vision evaluated by molecular simulations. J Am Chem Soc. 2006; 128(51): 16502-16503.

41. Frahmcke JS, Wanko M, Phatak P, Mroginski MA, Elstner M. The protonation state of Glu181 in rhodopsin revisited: Interpretation of experimental data on the basis of QM/MM calculations. J Phys Chem B 2010; 114(34):11338-11352. doi: 10.1021/jp104537w.

42. Sandberg MN, Amora TL, Ramos LS, Chen MH, Knox BE, Birge RR. Glutamic acid 181 is negatively charged in the bathorhodopsin photointermediate of visual rhodopsin. J Am Chem Soc. 2011; 133(9):2808-2811. doi: 10.1021/ja1094183.

43. Tomasello G, Olaso-Gonzalez G, Altoe P, Stenta M, Serrano-Andres L, Merchán M, et al. Electrostatic control of the photoisomerization efficiency and optical properties in visual pigments: On the role of counterion quenching. J Am Chem Soc. 2009; 131(14):5172-5186. doi: 10.1021/ja808424b.

44. Grossfield A, Pitman MC, Feller SE, Soubias O, Gawrisch K. Internal hydration increases during activation of the G-protein-coupled receptor rhodopsin. J Mol Biol. 2008; 381(2): 478-486. doi: 10.1016/j.jmb.2008.05.036.

45. Hall KF, Vreven T, Frisch MJ, Bearpark MJ. Three-layer ONIOM studies of the dark state of rhodopsin: the protonation state of Glu181. J Mol Biol. 2008; 383(1): 106-121. doi: 10.1016/j.jmb.2008.08.007.

46. Yoshizawa T, Shichida Y. Low-temperature spectroscopy of intermediates of rhodopsin. Methods Enzymol. 1982; 81: 333-54.

47. Kusnetzow A, Dukkipati A, Babu KR, Singh D, Vought BW, Knox BE, et al. The photobleaching sequence of a short-wavelength visual pigment. Biochemistry. 2001; 40(26): 7832-7844.

48. Ramos LS, Chen M-H, Knox BE, Birge RR. Regulation of Photoactivation in Vertebrate Short Wavelength Visual Pigments: Protonation of the retinylidene Schiff base and a counterion switch. Biochemistry. 2007; 46(18): 5330-5340.

49. Babu KR, Dukkipati A, Birge RR, Knox BE. Regulation of phototransduction in short wavelength cone visual pigments via the retinylidene Schiff base counterion. Biochemistry. 2001; 40: 13760-6.

50. Vought BW, Dukkipati A, Max M, Knox BE, Birge RR. Photochemistry of the primary event in short-wavelength visual opsins at low temperature. Biochemistry. 1999; 38(35): 11287-11297.

51. Kusnetzow AK, Dukkipati A, Babu KR, Ramos L, Knox BE, Birge RR. Vertebrate ultraviolet visual pigments: protonation of the retinylidene Schiff base and a counterion switch during photoactivation. Proc Natl Acad Sci USA. 2004; 101(4):941-946.

52. Birge RR, Einterz CM, Knapp HM, Murray LP. The nature of the primary photochemical events in rhodopsin and isorhodopsin. Biophys J. 1988; 53(3): 367-385.

53. Suzuki T, Callender RH. Primary photochemistry and photoisomerization of retinal at 77°K in cattle and squid rhodopsins. Biophys J. 1981; 34(2): 261-270.

54. Vought BW, Dukkipatti A, Max M, Knox BE, Birge RR. Photochemistry of the primary event in short-wavelength visual opsins at low temperature. Biochemistry. 1999; 38(35): 11287-11297.

55. Frisch MJ, Trucks GW, Schlegel HB et al. Gaussian 09, Revision A.02. Wallingford, CT: Gaussian, Inc. 2009.

56. Nakamichi H, Okada T. Crystallographic analysis of primary visual photochemistry. Angew Chem Int Ed Engl. 2006; 45(26):4270-4273.

57. Nakamichi H, Okada T. Local peptide movement in the photoreaction intermediate of rhodopsin. Proc Natl Acad Sci USA. 2006; 103(34):12729-12734.

58. Salom D, Lodowski DT, Stenkamp RE, Le Trong I, Golczak M, Jastrzebska B, et al. Crystal structure of a photoactivated deprotonated intermediates of rhodopsin. Proc Natl Acad Sci USA. 2006; 103(44): 16123-16128.

59. Stewart JJP. Optimization of parameters for semiempirical methods I. Method. J Comput Chem. 1989; 10(2):209-220.

60. Stewart JJP. Optimization of parameters for semiempirical methods II. Applications. J Comput Chem 1989; 10: 221-64.

61. Becke AD. Density-functional thermochemistry. III. The role of exact exchange. J Chem Phys. 1993; 98: 5648-52.

62. Lee C, Yang W, Parr RG. Development of the Colle-Salvetti correlation-energy formula into a functional of the electron density. Phys Rev B Condens Matter. 1988; 37(2): 785-789.

63. Birge RR, Vought BW. Energetics of rhodopsin photobleaching: photocalorimetric studies of energy storage in the early and later intermediates. Methods Enzymol. 2000; 315: 143-163.

64. Szundi I, Epps J, Lewis JW, Kliger DS. Temperature dependence of the lumirhodopsin I - lumirhodopsin II equilibrium. Biochemistry. 2010; 49(28): 5852-5858. doi: 10.1021/bi100566r.

65. Lewis JW, Fan GB, Sheves M, Szundi I, Kliger DS. Steric barrier to bathorhodopsin decay in 5-demethyl and mesityl analogues of rhodopsin. J Am Chem Soc. 2001; 123(41): 10024-10029.

66. Lewis JW, Liang J, Ebrey TG, Sheves M, Kliger DS. Chloride effect on the early photolysis intermediates of a gecko cone-type visual pigment. Biochemistry. 1995; 34(17): 5817-5823.

67. Shichida Y, Okada T, Kandori H, Fukada Y, Yoshizawa T. Nanosecond laser photolysis of iodopsin, a chicken red-sensitive cone visual pigment. Biochemistry. 1993; 32(40):10832-10838.

68. Jager S, Han M, Lewis JW, Szundi I, Sakmar TP, Kliger DS. Properties of early photoylsis intermediates of rhodopsin are affected by glycine 121 and phenylalanine 261. Biochemistry. 1997; 36(39): 11804-11810.

23

Decreased Proliferation and Abnormal Differentiation of Human Mesenchymal Stromal Cells in Steroid-Induced Osteonecrosis of Femoral Head

Changdong Wang[1], Wei Huang[2], Xuan Gong[3], Guoliang Ding[4], Xi Liang[2] and Ning Hu[2]*

[1]Department of Biochemistry and Molecular Biology, Molecular Medicine and Cancer Research Center, Chongqing Medical University, Chongqing 400016, China
[2]Department of Orthopaedic Surgery, the First Affiliated Hospital of Chongqing Medical University, Chongqing 400016, China
[3]Department of outpatient, Chongqing Zhongshan Hospital, Chongqing, 400013, China
[4]Department of Orthopaedic Surgery, the Second Affiliated Hospital of Baotou Medical College of Inner Mongolia University of Science and Technology, Baotou, 014030, China

*Corresponding author: Ning Hu, Department of Orthopedic Surgery, Chongqing Medical University, Chongqing, China,
E-mail: 1276321387@qq.com

Abstract

Objectives: To investigate if the pathogenesis of steroid-induced osteonecrosis of femoral head (ONFH) is associated with abnormal proliferation and differentiation of human mesenchymal stromal cells (MSCs) at the proximal femur.

Methods: Using isolated human MSCs and sections of whole femoral heads, we analyzed the proliferative capacity and osteogenic, angiogenic and adipogenic differentiation of human MSCs at the proximal femur.

Results: The proliferation of MSCs from patients with corticosteroid-induced ONFH is decreased. The down-regulated expression of BMP2, BMP7, BMP9 and Osteopontin provides supportive evidence corticosteroid-induced inhibition of osteogenesis. Down-regulation of HIF1α, VEGF and VWF by glucocorticoids is directly responsible for decreased angiogenesis. Over-expression of PPARγ2 and (442)aP2 suggests that the corticosteroids may induce MSCs to differentiate into adipocytes.

Conclusions: Our findings suggest steroids may reduce the proliferative activity of MSCs, down-regulate the expression of osteoblast differentiation factors such as BMP2, BMP7, BMP9 and OPN, decrease angiogenesis by suppressing HIF1α and VEGF, and up-regulate adipocyte transcription factor expression such as PPARγ2 and (442)aP2.

Keywords: Osteonecrosis; Mesenchymal Stromal Cells; Steroids; Bone Morphogenetic Proteins; Hypoxia-induciblefactor1α

Introduction

Osteonecrosis of femoral head (ONFH) is characterized by bone ischemia and microarchitectural deterioration, which lead a collapse of the femoral head during the late stage. ONFH represents a remarkable challenge to orthopedic surgeons and is a devastating disease for affected patients [1,2]. The use of corticosteroids such as dexamethasone and methylprednisolone for severe adult respiratory syndrome (SARS), rheumatoid diseases and organ transplantation has resulted in an increased risk of ONPH [3]. Intraosseous hypertension, thrombotic intravascular occlusion and extravascular compression by progressive accumulation of marrow fat stores are commonly accepted theories [4-6]. Although it is recognized that steroids administration and the development of steroids-induced ONFH are related, the precise pathogenesis remains largely unknown [7]. The pathobiological mechanism underlying the induction of adipogenesis and suppression of osteogenesis and angiogenesis by steroids has not been elucidated.

Human mesenchymal stromal cells (MSCs) are multipotent progenitors which undergo self-renewal and differentiate into osteogenic, angiogenic and adipogenic lineages [8]. It is suggested that ONFH is a disease of MSCs, due to abnormal proliferation or differentiation of MSCs [9]. Osteogenic differentiation is a sequential cascade that recapitulates most of the molecular events occurring during bone development and remodeling. Bone morphogenetic proteins (BMPs) play an important role during development and have been shown to regulate stem cell proliferation and osteogenic differentiation [10]. BMP2, 7 and 9 are the most potent BMPs among the 14 types of BMPs in inducing osteogenic differentiation of MSCs [11]. Normal and pathological bone physiology is inexorably tied to angiogenesis. The vasculature plays an important role for the mechanism of coupling resorption by osteoclasts and bone formation by osteoblasts [11]. Angiogenesis, the development of a microvascular network for blood supply, is critical for the development, remodeling and healing of bone [11-13]. Hypoxia can induce both apoptosis as well as necrosis of cells and is associated with vascular disease [12]. Hypoxia-inducible factor1α (HIF1α) is a well established regulator of angiogenic cascade, which usually coordinates with skeletal development [11,14]. Vascular endothelial growth factor (VEGF), transcriptionally targeted gene of HIF1α, is a key regulator of vasculogenesis in the embryo and angiogenesis in adult tissues [11,15,16]. Thus, it appeared promising to investigate the role of angiogenic factors such as HIF1α and VEGF. Small vessel occlusion by fatty emboli and the impedance of sinusoidal blood flow secondary to a rise in intraosseous pressure due to fatty infiltration following steroid therapy. In addition, steroid stimulates MSCs differentiation into adipocytes as well as the accumulation of fat in the marrow while suppressing cell differentiation into osteoblasts [17]. Peroxisome proliferator activated receptor γ2 (PPARγ2) gene expression destines cells for adipocyte differentiation [18,19].

Here, we investigate if the pathogenesis of steroid-induced osteonecrosis is associated with decreased proliferative capacity and abnormal differentiation of human MSCs at the proximal femur. The proliferation of cultured human MSCs in patients with corticosteroid-induced ONFH is depressed. We report the identification of abnormal differentiation in isolated MSCs and sections of whole femoral heads obtained during total hip replacement for glucocorticoid-induced osteonecrosis. The decreased expression of BMP2, BMP7, BMP9 and Osteopontin provide supportive evidence corticosteroid-induced inhibition of osteogenesis. Down-regulation of HIF1α, VEGF and VWF by glucocorticoids is directly responsible for disturbed angiogenesis. Over-expression of PPARγ2 and (442)aP2 suggests the corticosteroid-induced adipogenic differentiation. Taken together, our findings should not only expand our understanding of the molecular basis behind steroid-induced osteonecrosis, but also provide an opportunity to harness proliferative capacity and differentiation of MSCs in regenerative medicine.

Materials and Methods

Patients

The study was approved by the University Ethics Committee and informed consent was obtained from all patients. 12 ONFH patients (4 men and 8 women, aged from 38 to 65 years old) with history of corticosteroid usage for rheumatoid arthritis, systemic lupus erythematosus and uveitis were recruited. All Patients with stage III and stage IV ONFH underwent total hip replacement (THR). The exclusion criteria included metabolic bone diseases, such as Paget's disease, renal osteodystrophy, hyper- or hypoparathyroidism and malignant tumors. The necrotic bone in the center of split femoral head was harvested for study. Bone specimens from additional 12 patients who underwent THR for femoral neck fractures were harvested and enrolled as controls. No age-matched control was obtained because of the absence of surgical indication for THR in young patients with femoral neck fractures.

Harvest and identification of human mesenchymal stromal progenitor cells

Bone marrow was obtained at the femoral neck and adherent stromal cells were harvest and cultured in the conditions as described [20]. Cells were treated with 0.1% triton-X for 10 minutes and blocked with phosphate-buffered saline containing 1% bovine serum albumin and mouse serum for 30 minutes. Fluorescein isothiocyanate–labeled mouse anti-human CD29 antibody (1:100, Santa Cruz) and PE-labeled rat antihuman CD44 (1:100, Santa Cruz) were added simultaneously to 1 well. Mouse antihuman CD34, or CD14 antibodies (Santa Cruz)

were added; following incubation for 30 minutes, fluorescein isothiocyanate–labeled rabbit antimouse IgG antibody then was added. Identification of mesenchymal stem cells was performed as previously described [21].

Measurement of proliferation of human mesenchymal stromal cells using Flow Cytometry

When MSCs covered 90% of the flask, they were digested and resuspended at the density of 1×10^6/mL. Cells were fixed in 80% pre-cooled alcohol at 4 overnight. After washing with PBS and centrifuged for 5 minutes, the supernatant was discarded. 10 μL of RNase (100 μg/mL), 10 mL of propidium iodide (100 μg/mL) and 40 μL of PBS were added, and the cells were kept in darkness for 30 minutes. The cell cycles were measured with a flow cytometry instrument. The proliferation index (PI) was used to assess the levels of proliferation.

RNA isolation and semi-quantitative RT-PCR

Total RNA was isolated from subconfluent MSCs using TRIZOL Reagents (Invitrogen) and used to generate cDNA templates by RT reaction with hexamer and M-MuLV Reverse Transcriptase (New England Biolabs, Ipswich, MA). The first strand cDNA products were further diluted five to ten folds and used as PCR templates. Semiquantitative RT-PCR (sqPCR) was carried out as described [11]. PCR primers (supplementary material Table) were designed by using the Primer3 program to amplify the genes of interest (150–180 bp). A touchdown cycling program was as follows: 94°C for 2 minutes for 1 cycle; 92°C for 20 seconds, 68°C for 30 seconds, and 72°C for 12 cycles decreasing 1°C per cycle; and then at 92°C for 20 seconds, 57°C for 30 seconds, and 72°C for 20 seconds for 20–25 cycles, depending on the abundance of a given gene. PCR products were resolved on 1.5% agarose gels. All samples were normalized by the expression level of GAPDH.

Western blotting analysis

Western blotting was carried out as previously described [11]. Briefly, tissues were collected in Lysis Buffer. Cleared total cell lysate was denatured by boiling and resolved by 10% SDS-PAGE. After electrophoretic separation, proteins were transferred to an Immobilon-P membrane. Membrane was blocked with SuperBlock Blocking Buffer, and probed with anti-BMP2 (Proteintech), BMP7 (Proteintech), BMP9 (Proteintech), Osteopontin (ab8448), HIF1α (Proteintech), VEGF (Proteintech), PPARγ2 (Proteintech), vWF (Proteintech) and (442)aP2 (Proteintech) or anti-ß-actin (Santa Cruz), followed by incubation with a secondary antibody conjugated with horseradish peroxidase. The proteins of interest were detected by using SuperSignal West Pico Chemiluminescent Substrate kit.

Immunohistochemical and Immunofluorescent staining

Immunohistochemical staining on paraffin-embedded tissues was carried out with an anti-BMP2, BMP7, BMP9, HIF1α, VEGF, PPARγ2, vWF and (442)aP2 antibody. The presence of the expected protein was visualized by DAB staining and examined under a microscope as previously described [11]. The slides were then incubated with primary antibody diluted in PBS containing 1% BSA for 1h. The primary antibodies used were as follows:

anti-osteopontin, vWF and (442)aP2 antibody. After washing 3 times in PBS, AlexaFluor488, AlexaFluor555 (Invitrogen, Grand Island, NY) conjugated anti-rabbit or anti-mouse IgG was added in PBS with1% BSA for 1h. After the final wash, 6-diamidino-2-phenylindole (DAPI) (Sigma) was added and used as a counterstain for nuclei. Fluorescence images were acquired using an Olympus microscope with DP manager software.

Test concentration of triglycerid reagent

The tissues of the ONFH and control groups were lysed with NP40 buffer (1% NP-40, 0.15 M NaCl, 50 mM Tris, pH 8.0) containing protease inhibitors (Sigma). Protein quantitation was performed with BCA protein assay reagent (Pierce, Rockford, IL, USA). Equal amounts of protein from the different groups were added into the reaction of triglyceride reagent KIT for 5min at 37. The extracted dye was transferred to 96-well plates, and the optical density was measured at 600nm.

Statistical analysis

All quantitative experiments were performed in triplicate and/or repeated three times. Data were expressed as mean ± S.D. Statistical significances between vehicle treatments versus drug-treatment were determined by one-way analysis of variance and the Student's t-test. A value of $P < 0.05$ was considered statistically significant.

Results

Clinical data of patients

Representative radiograph and MRI showed femoral head erosion and collapse in the ONFH group and femoral neck fracture in the control group (Figure 1A and 1B). Bone marrow edema and femoral head necrosis were visible in the ONFH group (Figure 1C). This revealed significant accumulation of marrow fat stores of ONFH relative to the control of normal femoral head.

Inhibited proliferation of MSCs in corticosteroid-induced osteonecrosis of femoral head

We first identified the expression of typical MSCs markers. Immunofluorescent staining showed the cultured cells expressed typical MSCs markers such as CD29 and CD44 (Figure 2A and 2B), but not typical hematopoietic cell markers including CD34 and CD14 (data not shown). To investigate the potential role of steroid in affecting proliferative capability of MSCs, we examined growth period of MSCs between the ONFH and control groups using flow cytometry. The percentage of cells in the $G_2/M+S$ stages is taken as the proliferation index (PI) to indicate the proliferation of cells. Compared with the control group, the percentage of cells in the G_0/G_1 stages in the ONFH group was increased significantly, whereas the percentage in $G_2/M+S$ stages (PI) was decreased significantly ($P < 0.01$) (Figure 2C and 2D).

Corticosteroid down-regulates the expression of BMP2, 7, 9 and late osteogenic marker in the ONFH and inhibits osteogenesis of MSCs in vivo

Bone morphogenetic proteins (BMPs) play an important role in regulating stem cell proliferation and osteogenic differentiation.

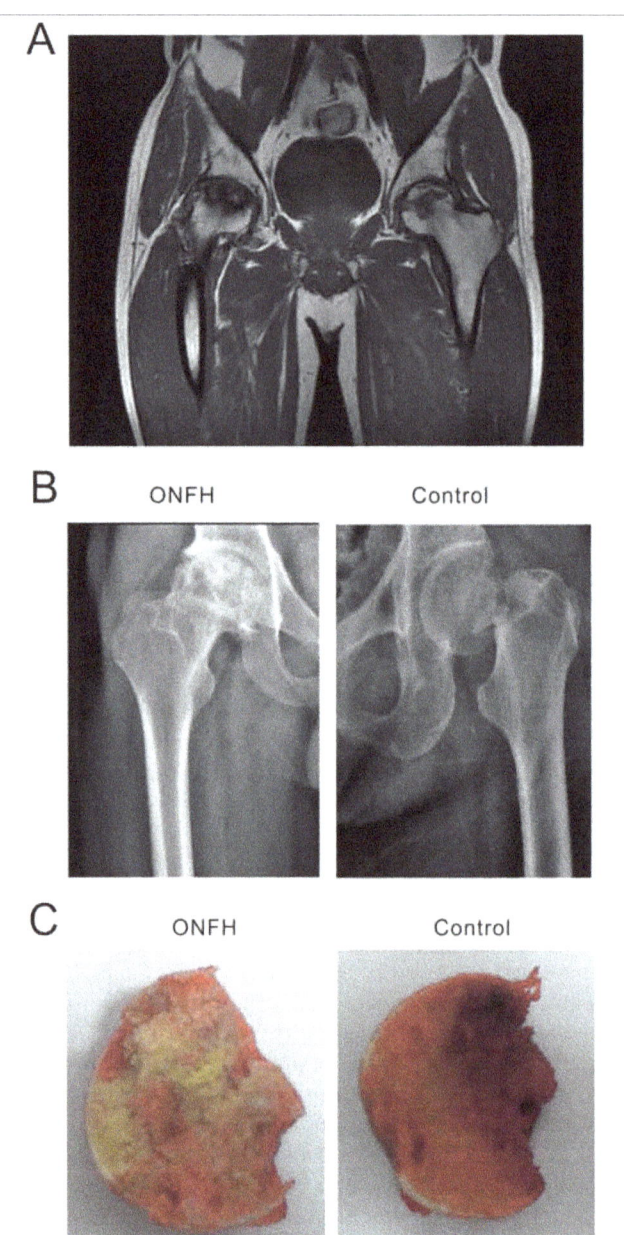

Figure 1: Clinical data of patients.
(A) MRI showed femoral head erosion and collapse in the ONFH group.
(B) Representative radiograph showed femoral head necrosis in the ONFH group and femoral neck fracture in the control group.
(C) The split femoral heads were harvested for study. In the ONFH group, Bone marrow edema and femoral head necrosis were visible. In the control group, bone specimens from the cases who underwent THR for femoral neck fractures were harvested.

BMP2, 7 and 9 are the most potent BMPs among the 14 types of BMPs in inducing osteogenic differentiation of MSCs. Compared with those in the control group, we found the transcriptions of BMP2, 7 and 9 were significantly reduced by 93.8%, 64.3% and 96.4% respectively in the ONFH group (Figure 3A). Similarly, the protein expression levels of BMP2, 7 and 9 were significantly reduced in the ONFH group (Figure 3B). Immunohistochemical staining also confirmed that corticosteroid inhibited expression

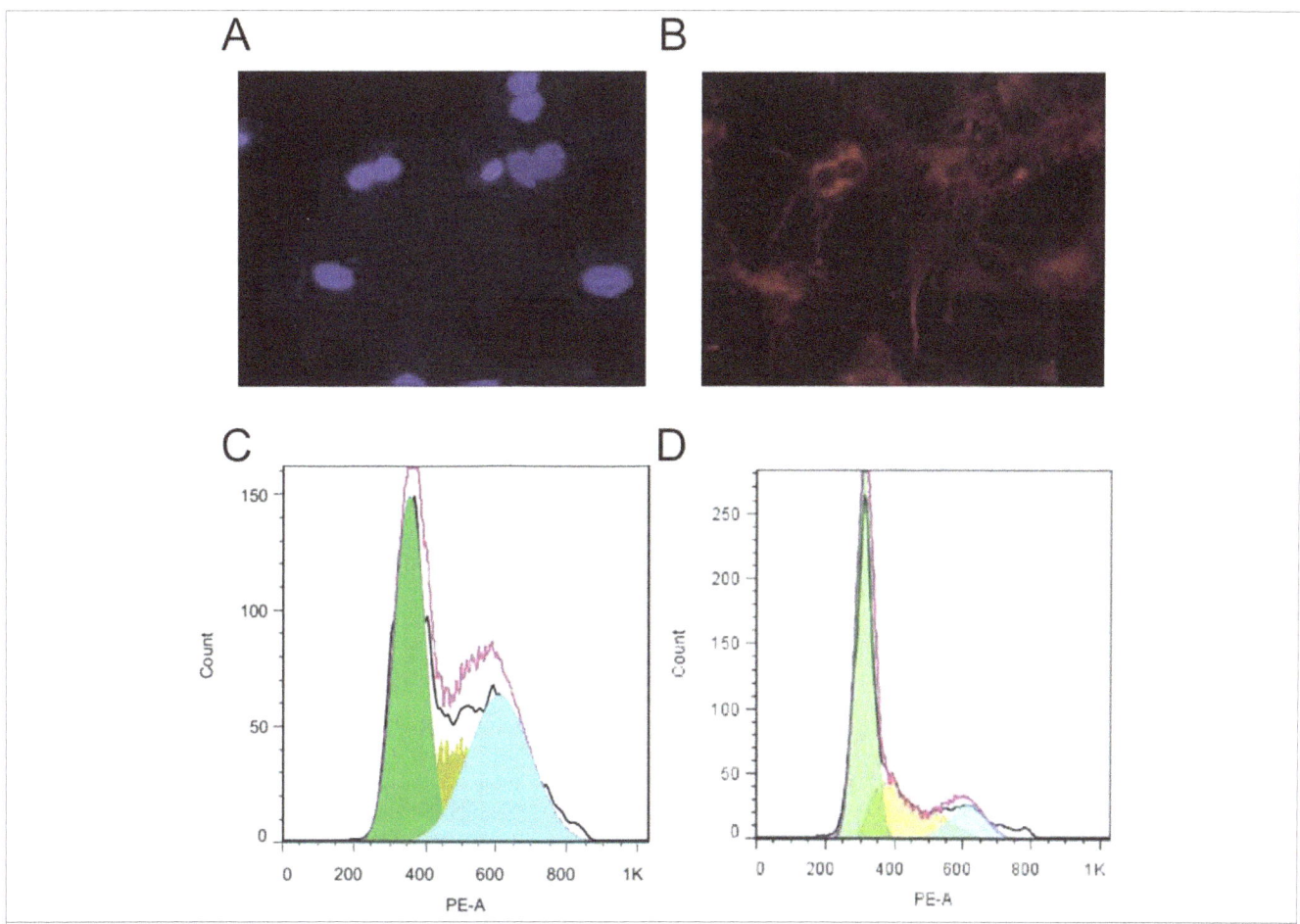

Figure 2: Decreased proliferation of MSCs in corticosteroid-induced ONFH.
The expression of the surface markers were determined by immunocytochemistry using the monoclonal antibodies CD29-FITC (A) and CD44-PE (B). Positive staining was observed in the cells (×200). Growth period of flow cytometry in the control group (C) and the ONFH group (D). The percentage of cells in G_2/M+S stages was 59.54% and the percentage in the G_0/G_1 stages was 40.46% in the control group (C). The percentage in G_2/M+S stages was 37.45% and the percentage in the G_0/G_1 stages was 62.55% in the ONFH group (D).

of BMP2, 7 and 9 at protein levels (Figure 3C). We also found that corticosteroid was able to suppress late osteogenic marker osteopontin (OPN) at mRNA (Figure 3A) and protein (Figure 3B) levels. Immunofluorescent staining results also confirmed that osteopontin protein was significantly decreased in ONFH group relative to the control (Figure 3D).

Corticosteroid down-regulates the expression of HIF1, VEFG and vWF in the ONFH and inhibits angiogenesis of MSCs in vivo

HIF1α is a well established regulator of angiogenic cascade. VEGF, transcriptionally targeted gene of HIF1α, is a key regulator of angiogenesis in adult tissues. And von Willebrand factor (vWF) is one of angiogenic markers. We found that mRNA of HIF1α, VEGF and vWF were inhibited by 98.2%, 62.3% and 69.93% respectively in the ONFH group, compared with those in the control group (Figure 4A). Western blot detected that protein expression levels of HIF1α, VEGF and vWF were inhibited significantly (Figure 4B). Immunohistochemical staining also confirmed that corticosteroid inhibited the expression of HIF1α,

VEGF and vWF at protein levels (Figure 4C). vWF protein expression level was also tested by Immunofluorescence and vWF was notably decreased in group of ONFH relative to the control(Figure 4D).

Corticosteroid upregulates the expression of PPARγ2, (442)aP2 and triglycerid in the ONFH and induced adipogenic differentiation of MSCs in vivo

The gene expression of PPARγ2 and (442)aP2 destine cells for adipocyte differentiation. We also found that mRNA level of PPARγ2 and (442)aP2 were augmented by 9.44 and 33.57 folds respectively in the ONFH group (Figure 5A). Protein expression of PPARγ2 and (442)aP2 were increased respectively (Figure 5B). Immunohistochemical staining also confirmed that corticosteroid augmented the expression of PPARγ2 and (442) aP2 at protein levels (Figure 5C). Immunofluorescent results displayed that (442)aP2 protein was significantly increased in ONFH group relative to the control(Figure 5D). In addition, triglycerid concentration was markedly increased in the ONFH (Figure 5E).

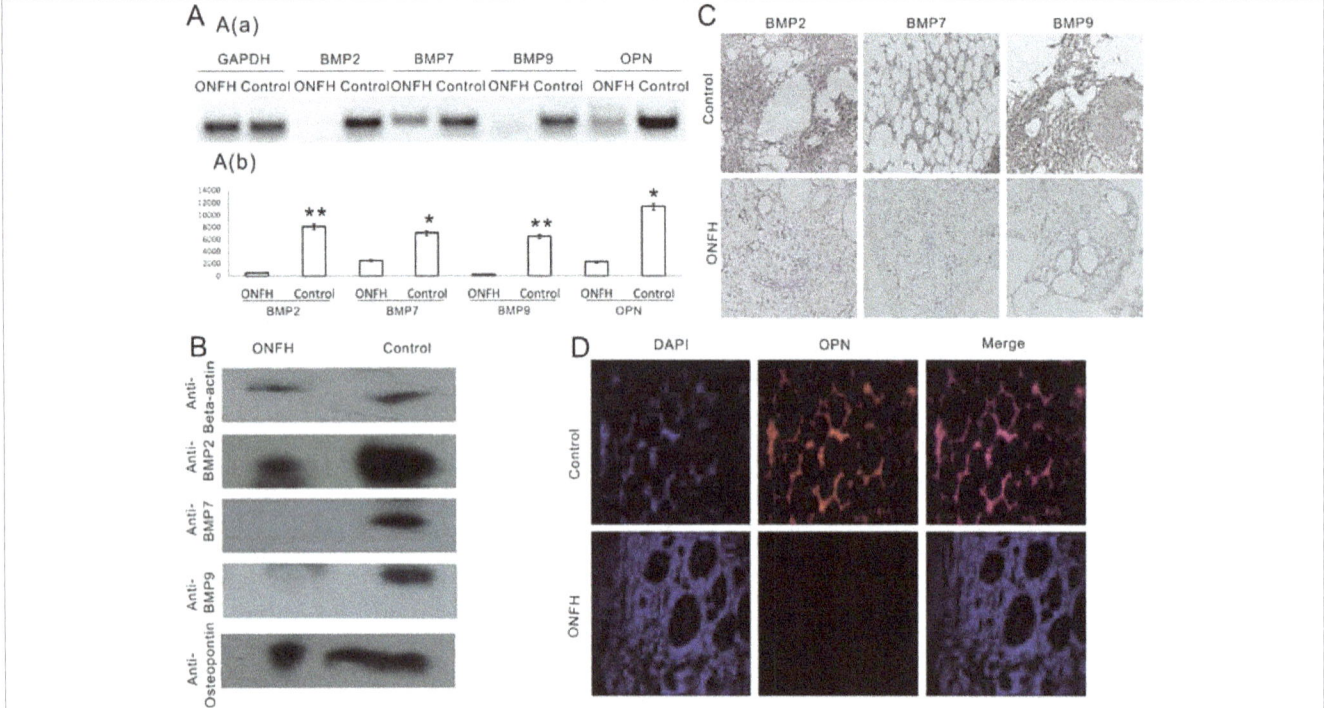

Figure 3: Corticosteroid down-regulates the expression of BMP2, 7, 9 and late osteogenic marker gene of osteopontin in the ONFH and inhibits osteogenesis of MSCs in vivo.
(A) The transcription expression of BMP2, BMP7, BMP9 and OPN in MSCs of the ONFH group and the control group. Total RNA was isolated from subconfluent cells and subjected to semi-quantitative PCR (sqPCR) analysis using primers specific for human BMP2, BMP7, BMP9 and OPN. Expected PCR products were resolved on agarose gels (a). The signal intensities of the expected products were quantitatively analyzed using the NIH ImageJ software (b). Each PCR condition was done in triplicate. Representative results are shown.
(B) The protein expression of BMP2, BMP7, BMP9 and OPN in MSCs of the ONFH group and the control group. Tissues were prepared in the same fashion as described in (a). Tissues were lyzed and subjected to SDS-PAGE and Western blotting with anti-BMP2, anti-BMP7, anti-BMP9 and anti-OPN antibodies. Expression levels of β-actin was used to assess equal loading of total lysate.
(C) Immunohistochemical staining of BMP2, BMP7 and BMP9 in the sections of the ONFH and control groups. Bone marrow was retrieved in a similar fashion as shown in Figure 1C. The retrieved tissues were subjected to BMP2, BMP7 or BMP9 antibody immunohistochemical staining. Isotype IgG was used as a negative control (not shown). Representative results are shown. Magnification, 40x.
(D) Immunofluorescent staining results also confirmed that osteopontin protein was significantly decreased in ONFH group relative to the control. Each assay condition was done in triplicate. * p < 0.05; ** p < 0.001.

Table 1: PCR Oligonucleotides.

human BMP2 Fwd:	ACTCGAAATTCCCCGTGACC
human BMP2 Rev:	CCACTTCCACCACGAATCCA
human BMP7 Fwd:	GACTTCAGCCTGGACAACGA
human BMP7 Rev:	TGTAGGGGTAGGAGAAGCCC
human BMP9 Fwd:	CTGTGGAGAGCCACAGGAAG
human BMP9 Rev:	CTCCTTTTCCGCCTGGCTAA
human OPN Fwd:	CATACAAGGCCATCCCCGTT
human OPN Rev:	TGGGTTTCAGCACTCTGGTC
human HIF1α Fwd:	TGCATCTCCATCTCCTACCC
human HIF1α Rev:	CGTTAGGGCTTCTTGGATGA
human VEGF Fwd:	CCCACTGAGGAGTCCAACAT
human VEGF Rev:	TTTCTTGCGCTTTCGTTTTT
human vWF Fwd:	CCACTTGCCACAACAACATC
human vWF Rev:	TGGACTCACAGGAGCAAGTG
human PPARγ2 Fwd:	ATCTTTCAGGGCTGCCAGT
human PPARγ2 Rev:	GGAGGCCAGCATTGTGTAA
human (442)aP2 Fwd:	GGTGCAGAAGTGGGATGG
human (442)aP2 Rev:	TGGCTCATGCCCTTTCAT
human GAPDH Fwd:	GCATGGCCTTCCGTGTCCCC
human GAPDH Rev:	GAGGGCAATGCCAGCCCCAG

Discussion

Corticosteroid has been the focus of studies on the pathogenesis of ONFH. Patients receiving steroid therapy have an approximately 20-fold increase in their likelihood of developing ONFH [22]. Though the dose effect of corticosteroid therapy on osteonecrosis remains largely unknown, recent studies suggest that corticosteroid doses above 25-40mg/day are significant risk factors for nontraumatic ON in renal transplant and SLE patients [22]. The number of MSCs at the proximal femoral was significantly decreased in the corticosteroid-induced osteonecrosis compared to other kinds of osteonecrosis [23]. Corticosteroid can promote apoptosis of osteoblasts and

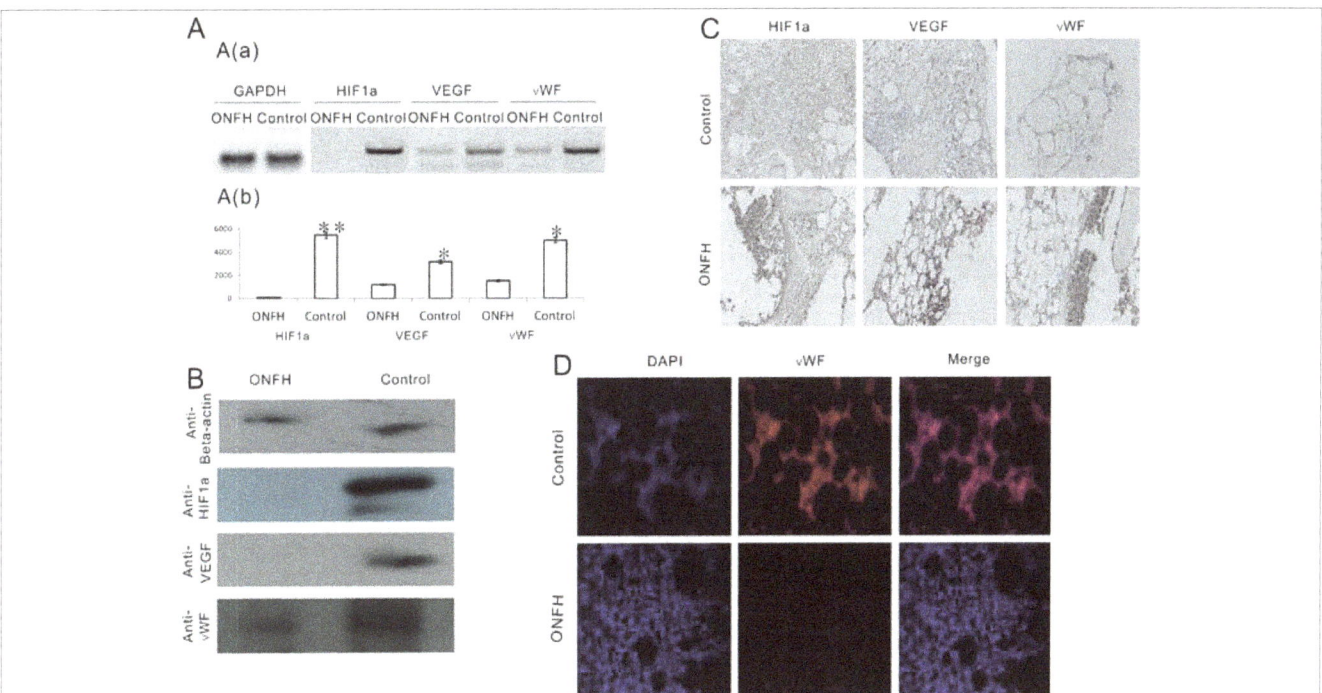

Figure 4: Corticosteroid down-regulates the expression of HIF1a, VEFG and vWF in the ONFH and inhibits angiogenesis of MSCs in vivo.
(A) The transcription expression of HIF1a, VEGF and vWF in MSCs of the ONFH group and the control group. Total RNA was isolated from subconfluent cells and subjected to semi-quantitative PCR (sqPCR) analysis using primers specific for human HIF1a, VEGF and vWF. Expected PCR products were resolved on agarose gels (a). The signal intensities of the expected products were quantitatively analyzed using the NIH ImageJ software (b). Each PCR condition was done in triplicate. Representative results are shown.
(B) The protein expression of HIF1a, VEGF and vWF in MSCs of the ONFH group and the control group. Tissues were prepared in the same fashion as described in (a). Tissues were lyzed and subjected to SDS-PAGE and Western blotting with anti-HIF1a, anti-VEGF and anti-vWF antibodies. Expression levels of β-actin was used to assess equal loading of total cell lysate.
(C) Immunohistochemical staining of HIF1a, VEGF and vWF in the sections of the ONFH and control groups. Bone marrow was retrieved in a similar fashion as shown in Figure 1. The retrieved tissues were subjected to HIF1a, VEGF or vWF antibody immunohistochemical staining. Isotype IgG was used as a negative control (not shown). Representative results are shown. Magnification, 40x.
(D) vWF protein expression level was also tested by Immunofluorescence and vWF was notably decreased in group of ONFH relative to the control. Each assay condition was done in triplicate. * $p < 0.05$; ** $p < 0.001$.

osteocytes and suppress the proliferative capacity of MSCs, leading to a decrease in osteocytes [24-26]. Our results also confirmed that corticosteroid induces osteonecrosis through lowering the proliferative activity or differential capacity as well as altering the differentiation direction of MSCs.

Osteogenic differentiation is a sequential cascade that recapitulates most of the molecular events occurring during embryonic skeletal development [10]. BMPs belong to the transforming growth factor β (TGFβ) superfamily and consist of at least 14 members in humans [8]. Genetic disruptions of BMPs have resulted in various skeletal and extraskeletal abnormalities during development [11]. BMP2, 7 and 9 are the most potent BMPs in inducing osteogenic differentiation of MSCs by regulating several important downstream targets during BMPs-induced osteoblast differentiation [8,11]. Our results confirmed that corticosteroid down-regulates the expression of BMP2, 7 and 9, thus reduces late osteogenic markers and inhibits osteogenesis of MSCs.

The process of bone development and repair depends on adequate formation of new capillaries from existing blood vessel

[11,12]. It has been reported that angiogenesis and osteogenesis are well coordinated processes during bone development [11]. We have demonstrated that BMP9 directly upregulates HIF1α expression in MSCs, which in turn induces both osteogenic factors and angiogenic factor VEGF [11]. Thus, potent osteogenic factors, such as BMP9, may induce a tightly-regulated convergence of osteogenic and angiogenic signaling in MSCs, and subsequently lead to efficient bone formation. HIF1 regulates target genes such as VEGF and vWF that mediate adaptive responses, such as angiogenesis, to reduced oxygen availability [11]. HIF1α polymorphisms are associated with idiopathic ONFH in men; and variations in HIF1α play a role in the pathogenesis and risk factor for ONFH [27]. Osteoblasts derived from femoral heads have been found to exhibit downregulation of VEGF within 24 hours of incubation with corticosteroid [28]. Down-regulation of HIF1α, VEGF and vWF by corticosteroid is directly responsible for disturbed angiogenesis resulting in the defects in capillary architecture, which eventually lead to osteonecrosis.

Corticosteroid can induce differentiation of MSCs into adipose cells. Since adipocytes and osteoblasts share a common progenitor pool, when exogenous stimulators shift the differentiation of

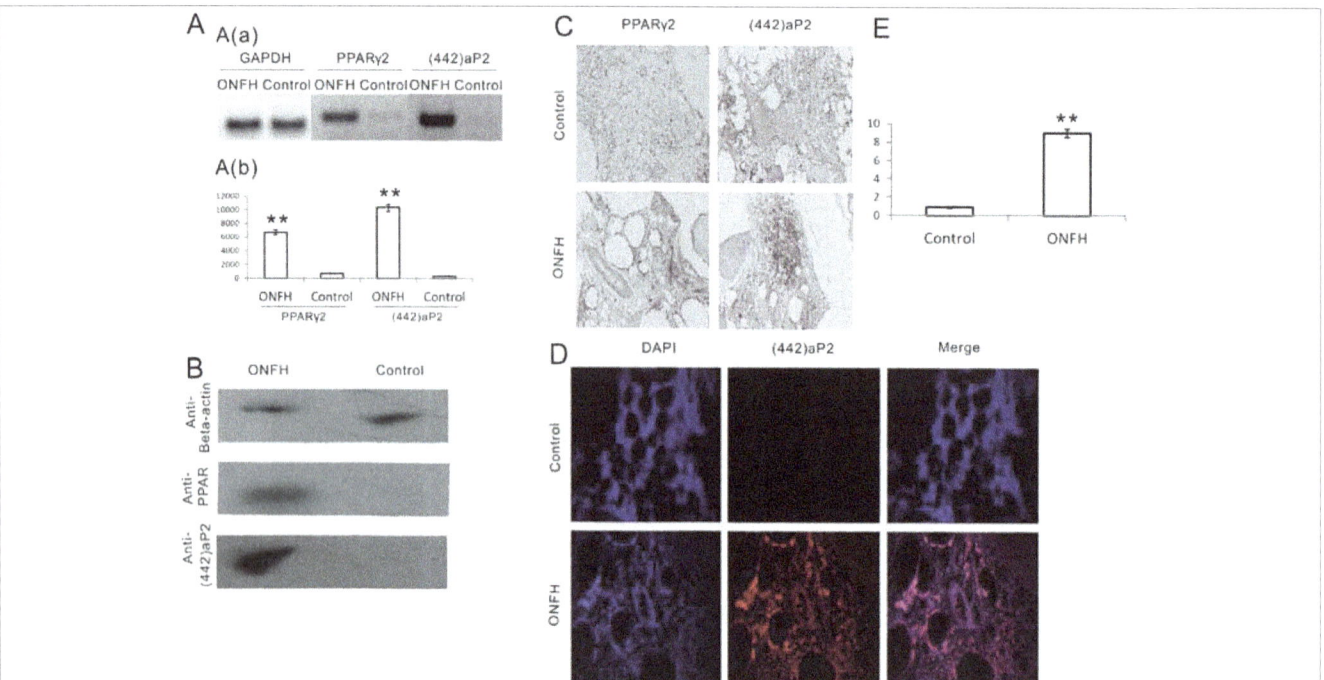

Figure 5: Corticosteroid upregulates the expression of PPARγ2, (442)aP2 and triglycerid in the ONFH and induced adipogenic differentiation of MSCs in vivo.
(A) The transcription expression of PPARγ2 and (442)aP2 in MSCs of the ONFH group and the control group. Total RNA was isolated from subconfluent cells and subjected to semi-quantitative PCR (sqPCR) analysis using primers specific for human PPARγ2 and (442)aP2. Expected PCR products were resolved on agarose gels (*a*). The signal intensities of the expected products were quantitatively analyzed using the NIH ImageJ software (*b*). Each PCR condition was done in triplicate. Representative results are shown.
(B) The protein expression of PPARγ2 and (442)aP2 in MSCs of the ONFH group and the control group. Tissues were prepared in the same fashion as described in (a). Tissues were lyzed and subjected to SDS-PAGE and Western blotting with anti- PPARγ2 and anti-(442)aP2 antibodies. Expression levels of β-actin was used to assess equal loading of total cell lysate.
(C) Immunohistochemical staining of PPARγ2 and (442)aP2 in the sections of the ONFH and control groups. Bone marrow was retrieved in a similar fashion as shown in Figure 1. The retrieved tissues were subjected to PPARγ2 or (442)aP2 antibody immunohistochemical staining. Isotype IgG was used as a negative control (not shown). Representative results are shown. Magnification, 40x.
(D) Immunofluorescent results displayed that (442)aP2 protein was significantly increased in ONFH group relative to the control.
(E) Detection of the concentration of triglyceride in bone tissues by TG reagent kit. The TG concentration of the control group was 0.87 ± 0.03mmol/l and the concentration of the ONFH group was 9.07 ± 0.28mmol/L. Each assay condition was done in triplicate. * *p < 0.05*; ** *p < 0.001*.

MSCs into the adipocyte lineage, the osteoprogenitor pool is not sufficient to provide enough osteoblasts in order to meet the need for bone remodeling and repair of ONFH [29]. Fat cell hypertrophy has been observed in histologic specimens of human femoral heads following treatment with dexamethasone for 5 days [30]. Dexamethasone affects osteoblasts by inhibit the expression of type-I collagen and osteocalcin, thereby inducing osteoblast and osteocyte apopotosis [31-33]. It also down-regulates the expression of Cbfa1/Runx2 and osteocalcin promoter activity while it increases the expression of PPARc2 [29, 34].

Conclusion

There may be several mechanisms that are involved in the pathogenesis of osteonecrosis. Corticosteroid reduces the proliferative activity of MSCs. It down-regulates osteoblast transcription factor gene expression such as BMPs, decreases angiogenesis by suppressing HIF1α and VEGF, and up-regulates adipocyte transcription factor expression. Consequently, the action impairs the differentiation of MSCs and decreases blood supply, leading to bone cell death.

Acknowledgement

The reported work was supported by research grants from Chongqing Science and Technology Commission foundation base and frontier research project (cstc2013jcyjA10033 to N. Hu, cstc2014jcyjA10024 to Wang) and the project of Chongqing Health Bureau (2013-2-005 to N. Hu), the doctoral funding of Ministry of Education of China (20125503120015 to C. Wang) and funding of Chongqing Medical University (0800280031 to C. Wang), and the Natural Sciences Foundation of China (#81391972 to W. Huang). The funders had no role in study design, data collection and analysis, decision to publish, or preparation of the manuscript.

References

1. Mont MA, Jones LC, Hungerford DS. Nontraumatic osteonecrosis of the femoral head: ten years later. J Bone Joint Surg Am. 2006; 88(5):1117-1132.

2. Mont MA, Marulanda GA, Jones LC, Saleh KJ, Gordon N, Hungerford DS, et al. Systematic analysis of classification systems for osteonecrosis of

the femoral head. J Bone Joint Surg Am. 2006; 88 Suppl 3:16-26.

3. Bradbury G, Benjamin J, Thompson J, Klees E, Copeland J. Avascular necrosis of bone after cardiac transplantation. Prevalence and relationship to administration and dosage of steroids. J Bone Joint Surg Am 1994; 76(9):1385–1388.

4. Hungerford DS, Lennox DW. The importance of increased intraosseous pressure in the development of osteonecrosis of the femoral head: implications for treatment. Orthop Clin North Am. 1985: 16(4):635–654.

5. Jones LC, Mont MA, Le TB, Petri M, Hungerford DS, Wang P, et al. Procoagulants and osteonecrosis. J Rheumatol. 2003; 30(4):783-791.

6. Wang GJ. The Frank Stinchfield Award paper. Improvement of femoral head blood flow in steroid-treated rabbits using lipid-clearing agent. Hip. 1987; 87–93.

7. Wang GJ, Cui Q, Balian G. The Nicolas Andry award. The pathogenesis and prevention of steroid-induced osteonecrosis. Clin Orthop Relat Res. 2000; 370:295–310.

8. Deng ZL, Sharff KA, Tang N, Song WX, Luo J, Luo X, et al. Regulation of osteogenic differentiation during skeletal development. Front Biosci. 2008; 13:2001-2021.

9. Gangji V, Hauzeur JP, Schoutens A, Hinsenkamp M, Appelboom T, Egrise D. Abnormalities in the replicative capacity of osteoblastic cells in the proximal femur of patients with osteonecrosis of the femoral head. J Rheumatol. 2003; 30(2):348-351.

10. Olsen BR, Reginato AM, Wang W. Bone development. Annu Rev Cell Dev Biol. 2000; 16:191-220.

11. Hu N, Jiang D, Huang E, Liu X, Li R, Liang X, et al. BMP9-regulated angiogenic signaling plays an important role in the osteogenic differentiation of mesenchymal progenitor cells. J Cell Sci. 2013; 126(Pt 2):532-41. doi: 10.1242/jcs.114231.

12. Streeten EA, Brandi ML. Biology of bone endothelial cells. Bone Miner. 1990; 10(2):85–94.

13. Colnot C, Thompson Z, Miclau T, Werb Z, Helms JA. Altered fracture repair in the absence of MMP9. Development. 2003; 130(17):4123–4133.

14. Majmundar AJ, Wong WJ, Simon MC. Hypoxia-inducible factors and the response to hypoxic stress. Mol Cell. 2010; 40(2):294-309.

15. Tatsuyama K, Maezawa Y, Baba H, Imamura Y, Fukuda M. Expression of various growth factors for cell proliferation and cytodifferentiation during fracture repair of bone. Eur J Histochem. 2000; 44(3):269–278.

16. Komatsu DE, Hadjiargyrou M. Activation of the transcription factor HIF-1 and its target genes, VEGF, HO-1, iNOS, during fracture repair. Bone. 2004; 34(4): 680-688.

17. Cui Q, Wang GJ, Balian G. Steroid-induced adipogenesis in a pluripotential cell line from bone marrow. J Bone Joint Surg Am. 1997; 79(7):1054–1063.

18. Mueller E, Drori S, Aiyer A, Yie J, Sarraf P, Chen H, et al. Genetic analysis of adipogenesis through peroxisome proliferator-activated receptor gamma isoforms. J Biol Chem. 2002; 277(44):41925–41930.

19. Shao D, Lazar MA. Peroxisome proliferator activated receptor gamma, CCAAT/enhancer-binding protein alpha, and cell cycle status regulate the commitment to adipocyte differentiation. J Biol Chem. 1997; 272(34):21473–21478.

20. Wang BL, Sun W, Shi ZC, Lou JN, Zhang NF, Shi SH, et al. Decreased proliferation of mesenchymal stem cells in corticosteroid-induced osteonecrosis of femoral head. Orthopedics. 2008; 31(5):444.

21. Suh KT, Kim SW, Roh HL, Youn MS, Jung JS. Decreased osteogenic differentiation of mesenchymal stem cells in alcohol-induced osteonecrosis. Clin Orthop Relat Res. 2005; (431):220-225.

22. Sakaguchi M, Tanaka T, Fukushima W, Kubo T, Hirota Y. Impact of oral corticosteroid use for idiopathic osteonecrosis of the femoral head: a nationwide multicenter case-control study in Japan. J Orthop Sci. 2010; 15(2):185-191. doi: 10.1007/s00776-009-1439-3.

23. Hernigou P, Beaujean F, Lambotte JC. Decrease in the mesenchymal stem-cell pool in the proximal femur in corticosteroid-induced osteonecrosis. J Bone Joint Surg Br. 1999; 81(2):349-355.

24. Weinstein RS, Nicholas RW, Manolagas SC. Apoptosis of osteocytes in glucocorticoid-induced osteonecrosis of the hip. J Clin Endocrinol Metab. 2000; 85(8):2907-2912.

25. Lee JS, Lee JS, Roh HL, Kim CH, Jung JS, Suh KT. Alterations in the differentiation ability of mesenchymal stem cells in patients with nontraumatic osteonecrosis of the femoral head: comparative analysis according to the risk factor. J Orthop Res. 2006; 24(4):604-609.

26. Yin L, Li YB, Wang YS. Dexamethasone-induced adipogenesis in primary marrow stromal cell cultures: mechanism of steroid-induced osteonecrosis. Chin Med J (Engl). 2006; 119(7):581-588.

27. Hong JM, Kim TH, Chae SC, Koo KH, Lee YJ, Park EK, et al. Association study of hypoxia inducible factor 1alpha (HIF1alpha) with osteonecrosis of femoral head in a Korean population. Osteoarthritis Cartilage. 2007; 15(6):688-694.

28. Varoga D, Drescher W, Pufe M, Groth G, Pufe T. Differential expression of vascular endothelial growth factor in glucocorticoid-related osteonecrosis of the femoral head. Clin Orthop Relat Res. 2009; 467(12):3273-3282. doi: 10.1007/s11999-009-1076-3.

29. Li X, Jin L, Cui Q, Wang GJ, Balian G. Steroid effects on osteogenesis through mesenchymal cell gene expression. Osteoporos Int. 2005;16(1):101-108.

30. Kitajima M, Shigematsu M, Ogawa K, Sugihara H, Hotokebuchi T. Effects of glucocorticoid on adipocyte size in human bone marrow. Med Mol Morphol. 2007; 40(3):150-156.

31. Canalis E, Delany AM. Mechanisms of glucocorticoid action in bone. Ann N Y Acad Sci. 2002; 966:73–81.

32. O'Brien CA, Jia D, Plotkin LI, Bellido T, Powers CC, Stewart SA, et al. Glucocorticoids act directly on osteoblasts and osteocytes to induce their apoptosis and reduce bone formation and strength. Endocrinology. 2004; 145(4):1835–1841.

33. Maes C, Carmeliet P, Moermans K, Stockmans I, Smets N, Collen D, et al. Impaired angiogenesis and endochondral bone formation in mice lacking the vascular endothelial growth factor isoforms VEGF164 and VEGF188. Mech Dev. 2002; 111(1-2):61–73.

34. Weinstein RS, Jia D, Powers CC, Stewart SA, Jilka RL, Parfitt AM, Manolagas SC. The skeletal effects of glucocorticoid excess override those of orchidectomy in mice. Endocrinology. 2004; 145(4):1980–1987.

Effects of Ethanolic Extract of *M. Oleifera* Seeds and Leaves on the Reproductive System of Female Albino Rats

Gogo Appolus Obediah[1*] and Gift Paago[2]

[1]Department of Biochemistry, Faculty of Science, Rivers State University, Port Harcourt, Nigeria

[2]Department of Human Pharmacology, Faculty of Basic Medical Sciences, College of Health Sciences, University of Port Harcourt, Nigeria

*Corresponding author: Gogo Appolus Obediah, Department of Biochemistry, Faculty of Science, Rivers State University, Port Harcourt, Nigeria, E-mail: gogoappolus@gmail.com

Abstract

Infertility has remained a major health problem among couples and sexually active individuals who seek to procreate, and the search for therapeutic solutions have remained endless. Therefore, this study investigated the effects of ethanolic extract of *M. oleifera* seeds and leaves on the reproductive system of female albino rats. Eighty-four (84) albino rats comparing of 56 females and 28 males, which acclimatised for two weeks and mated in ratio of 2 females to 1 male, the pregnant female rats were then divided into 7 Groups of 8; Group 1- Control (10 ml/kg body weight/day of vehicle [Tween 80] orally). Groups 2-4 (Seed extract at dose level of 100, 200 and 400 mg/kg body weight/day respectively), Groups 5-7 (Seed extract at dose level of 100, 200 and 400 mg/kg body weight/day respectively). Administration was carried out throughout the gestation period. The blood samples were collected for hormonal assay and under standard aesthetic conditions, the reproductive organs (uterus, ovaries and fallopian tube) excised for histological examination. From the results it was observed that the *M. oleifera* leave and seed extract caused resorption of the foetus with decrease in weight in a dose dependent manner; however, there was no disruption of the normal gestation. The levels of FSH and LH for animals treated with 400 mg/kg were significantly lower than those of 100 mg/kg, 200 mg/kg and the control group. From the histological slides, there was degeneration or atretic follicles in animals treated with extracts, which was intense at 400 mg/kg dose. In conclusion, the ethanol extract of both the leaf and seed of *M. oleifera* has shown abortifacient effect and therefore not advise for consumption during pregnancy.

Keywords: Moringa oleifera; abortifacient effect; albino rats; reproductive system

Introduction

For centuries and up until date, plants have remained an important and dependable source of medicine. World Health Organization (WHO) estimated that about 80% of the global population depends absolutely tradomedicine [8]. The therapeutic value of these plants is because of the variety of active phytochemicals and their essential composition. The role medicinal plants play in fighting and managing diseases have been attributed to presence of antioxidant in their constituents, often linked to numerous types of polyphenolic compounds [6]. Thus, the global interest in understanding the nature and dynamic of these natural antioxidants obtained from therapeutic plant materials for health care use has continued to grow.

Among the numerous medicinal plants which have shown great potentials is Moringa oleifera Lam (M. Oleifera); commonly known as Moringa [2]. Moringa is a versatile tropical tree popularly known for its culinary uses; however, it has wide range of application in the industry, medicine and agriculture, including animal feeding. For this purposes it has become increasingly popular in Asian, European and African continents, where its economical valuable is unprecedented [16, 19]. It has been dubbed the "Miracle tree" or "tree of life" by the media as every part of *M. oleifera* have been reported for one or more theurapetic uses as well as pharmaceutical and industrial byproducts [6,18].

Aside the culinary and other local uses various researchers have reviewed the numerous biochemical properties of various parts of *M. oleifera* in the past [6,10,15,16,18]. These parts contain both macro- and micro-nutrients, which are rich sources of natural antioxidants, which wide range of hormone modulation [9, 12, 14].

Across the globe, the importance of the reproductive system cannot be overlooked, as it is one of the most significant characteristics of humans and essential for the continuity of life, because of the continued exposure to life and attacks from environmental agents. The disease of the disease of the reproductive system is infertility and it has resulted in large cases of marital problems (WHO, 2013). This high burden of infertility has lead couples and individuals, who yearn but are unable to realize and sustain desired pregnancy to sort for assistance in tradomedicine especially in low resources countries.

The key hormones of the reproductive system are the Follicle-Stimulating Hormone (FSH) and Luteinizing Hormone (LH) are a gonadotropin (glycoprotein based) released by the anterior pituitary as a result of stimulation by Gonadotropin-Releasing Hormone (GnRH) and released by the hypothalamus. FSH and Luteinizing Hormone (LH) regulates the testis and ovary gonadal function by enhancing sex steroid production and gametogenesis. In women, follicle-stimulating hormone stimulates the growth of ovarian follicles in the ovary before the release of an egg at ovulation, and promotes oestra-diol production.

In the bid to manage infertility, various medicinal plants were often locally consumed and there have been reports of effective activities However, with respects to the reproductive system of females, there have been reported to biochemically and physiologically alter the reproductive cycle of female Wistar rats [3,5,19,20]. Hence the sole aim of this research work, which is to investigated the effects of ethanolic extract of *M. oleifera* seeds and leaves on the reproductive system of female albino rats. The objectives of this work are: to determine the level of influence the extract exerts on the pups' delivery and hormone, and to determine if any histological changes occur in the histology of the uterus, fallopian tube and ovaries of the female albino rat following administration.

Materials and Methods

Purchase, Identification and Extraction of Plant (*M. Oleifera*)

Moringa oleifera seed and leaf were purchased at Moringa House, 14, McAkini Road off Ada-George Road, Port Harcourt, Nigeria. The plant parts were harvested, identified and authenticated by a botanist. The seeds and leaves of *M. oleifera* were properly processed, grinded to powdered and subjected ethanol soxhlet extraction. The seed and leaf extracts were evaporated to near dryness on rotary evaporator (40°C), weighed and preserved at -4°C in a refrigerator until needed.

Experimental Design

Fifty-six (56) non-pregnant female albino rats (weighing between 180-200g) and twenty-eight (28) male albino rats (weighing between 200-220g) were obtained from the animal farm, University of Port Harcourt. The rats were housed four per cage and maintained under natural conditions. They were fed with laboratory feeds and clean tap water. They were allowed to acclimatize to laboratory environment for 14 days before commencement of research. All experimental protocols were in line with the approved guidelines of the University's Research Ethics Committee. After the 2weeks of acclimatization, the female animals were mated with males (2:1). Animals were checked for the presence of vaginal plug to confirm pregnancy. The pregnant rats were separated out for the main research. The pregnant female rats were divided into 7 groups of 7 animals each. The animals were treated with different dose of Moringa oleifera seed and leaf extract. Group 1 represented the control group, which was administered 10 ml/kg body weight/day of vehicle (Tween 80) orally. Groups 2-4 were administered suspension of ethanolic extract of Moringa oleifera seed in tween 80 orally at dose level of 100, 200 and 400 mg/kg body weight/day respectively. Groups 5-7 were administered suspension of ethanol extract of Moringa oleifera leaf in tween 80 at dose level of 100, 200 and 400 mg/kg body weight/day respectively. Extracts and vehicle were administered throughout the gestation period. The mothers were sacrificed after delivery under deep diethyl ether anaesthesia and the histology of the uterus and ovaries were conducted.

Determination of Lethal Dose

Doses used were based on the LD50 of the plant and previous studies done. The oral LD50 of the leaves of the plant in rats has been recorded as 6616.67 mg/kg while it is 5000mg/kg for the seeds [17,21]. The doses chosen are approximately 2, 4 and 8% of the LD50 respectively.

Histological Analysis of Reproductive Organs

The reproductive organs (uterus, ovaries and fallopian tube) of both the control and experimental groups were removed immediately after sacrifice and fixed in Bouin's solution for 24hrs and then dehydrated with ascending grade of alcohol (80% ethanol), cleared in xylene and embedded in paraffin wax. Thin sections of 7 microns thick were sectioned using a rotatory microtome. The sections were then de-paraffinized and stained using the routine haematoxylin and eosin (H & E). The sections were then examined under bright field light microscopy, and Photomicrographs of the results were obtained using digital research photographic microscope.

Hormonal Assay

The blood samples collected from the animals were subjected to hormonal analysis.

Data Analysis

All data generated were computed and analysed using Microsoft Office Excel 2013 and IBM SPSS version 23.0. Data was presented in tables of descriptive statistics as mean SEM (Standard Error of Mean). Analysis of Variance (ANOVA) was done to determine if significant difference exist between the groups, while Dunnett's multiple comparison test was done to determine the pair that differs (each group will be compared against the control; typical of Dunnettts). Comparison was carried out at three significant levels (95%, 99% and 99.9%). Hence $P < 0.05$, $P < 0.01$ and $P < 0.001$ respectively will be considered significant.

Discussions

Ethanol extract of Moringa oleifera was involved in the study, with their effects examined on the following: number of pups delivered, FSH, LH as well as the histology of the uterus, fallopian tube and ovaries of the female albino rats.

The ethanolic extract of the seed and leaf of Moringa oleifera was found to decrease the number of pups delivered in a dose dependent manner (Table 1). This may be related to the reported abortificient effect of the plant. On this regard several works have reported similar effect. Some of them include the report of and These works have also reported a dose dependent effect of the stem bark of the plant of the number of resorption of the pregnant animals, i.e. the higher the dose the greater the number of animals that experience resorption [1,22].

Table 1: Number of Pups delivered by female albino rats treated with extract of Moringa oleifera

Group	Mean SEM	
	Leaf Extract (µg/ml)	Seed Extract (µg/ml)
Control	7.14 0.34	7.14 0.34
100mg/kg	3.29 0.29*,**,***	4.14 0.51*,**,***
200mg/kg	2.14 0.26*,**,***	2.14 0.26*,**,***
400mg/kg	-	-
SEM =Standard Error of the Mean, * = $P < 0.05$, ** = $P < 0.01$, ***= $P < 0.001$		

Table 2 and 3 revealed that there was also a dose dependent decrease in the levels of FSH and LH in the tested animals; higher doses produce low hormonal level while the lower doses produce higher hormonal levels. The levels of FSH and LH for animals treated with 400 mg/kg were significantly lower than those of 100 mg/kg, 200 mg/kg and the control group. This is indicating that the ethanolic extract of both leaf and seed of moringa oleifera have the capacity of reducing the level of both hormones (FSH and LH) in biological systems.

Table 2: Level of FSH in albino rats treated with extract of *Moringa ole*

Group	Mean SEM	
	Leaf Extract (µg/ml)	Seed Extract (µg/ml)
Control	9.89 0.03	9.89 0.03
100mg/kg	6.11 0.17*,**,***	7.47 0.14*,**,***
200mg/kg	5.44 0.15*,**,***	6.83 0.13*,**,***
400mg/kg	3.13 0.04*,**,***	4.26 0.05*,**,***
SEM =Standard Error of the Mean, * = $P < 0.05$, ** = $P < 0.01$, *** = $P < 0.001$		

Table 3: Level of LH in albino rats treated with extract of Moringa oleifera

Group	Mean SEM	
	Leaf Extract (µg/ml)	Seed Extract (µg/ml)
Control	11.51 0.14	11.51 0.14
100mg/kg	9.76 0.07*,**,***	10.16 0.03*,**,***
200mg/kg	8.53 0.27*,**,***	8.62 0.09*,**,***
400mg/kg	5.34 0.10*,**,***	6.82 0.02*,**,***
SEM =Standard Error of the Mean, * = $P < 0.05$, ** = $P < 0.01$, *** = $P < 0.001$		

The result of the effect of the leaf and seed extract of Moringa oleifera on the ovaries showed different pathological changes when compared to the control group (Plates 1 - 7). There was no

Observed pathological change in the animals of the control group. Both the theca externa and the theca interna were intact. There was no degeneration of the ovarian follicles as well as distortion of the ovarian follicles. Animals treated with 100 mg/kg of the leaf and seed extract of the plant showed degeneration of the ovarian follicles or atretic follicles. The animals treated with 200 mg/kg of both the leaf and seed of the extract also showed the presence of follicular degeneration of atretic follicles. The presence of degeneration or atretic follicles in animals treated with ethanolic extract of Moringa oleifera indicates that the extract enhances the degeneration of pre-ovulatory follicles. This is in agreement with findings [13]. The animals in the group treated with 400 mg/kg of the extract also showed both distortion and degeneration of the ovarian follicle as well as atretic follicles thus indicating degeneration in the pre-ovulatory follicle. The degeneration of pre-ovulatory follicles occurs because of non-availability of steroidal hormones. This also agrees with result of the FSH and LH which are responsible for the production of both oestrogen and progesterone.

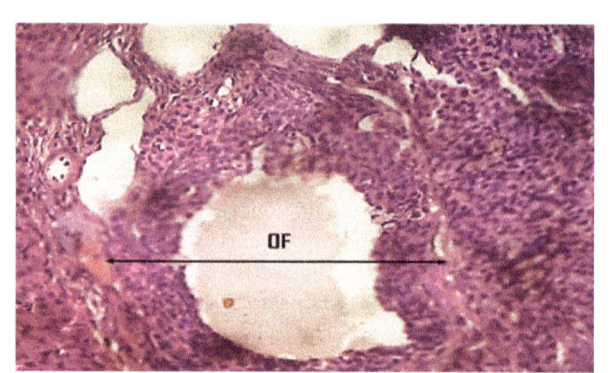

Plate 1: Photomicrograph normal histology of Ovary Showing Follicles (OF) in the cortical region. (Control)

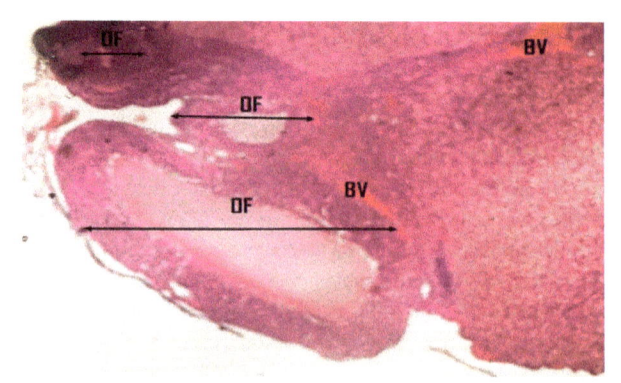

Plate 2: Photomicrograph of Ovary Showing Follicles (OF) in the cortical region and Blood vessels (BV), No ova are seen in the follicles may be due to degeneration (100 mg/kg of leaf extract)

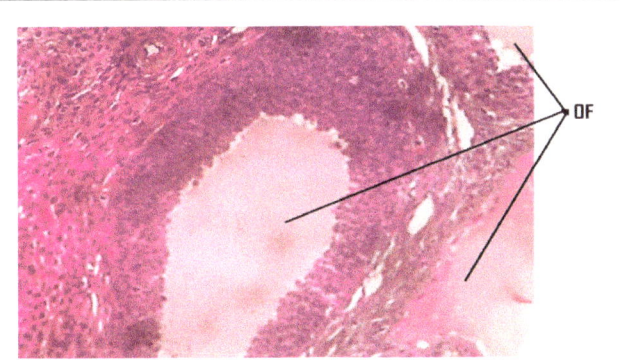

Plate 3: Photomicrograph ovary showing elongated ovarian follicles (OF) in the cortical area with dense granulosa cells. No ovum is seen in the follicle may be due to degeneration (200 mg/kg of leaf extract

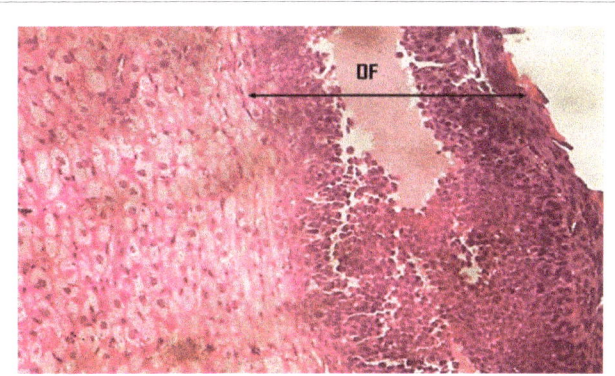

Plate 6: Photomicrograph of ovary showing elongated Ovarian Follicle (OF) with proliferation of the granulosa cells. The ovarian follicle is distorted. (200 mg/kg of extract)

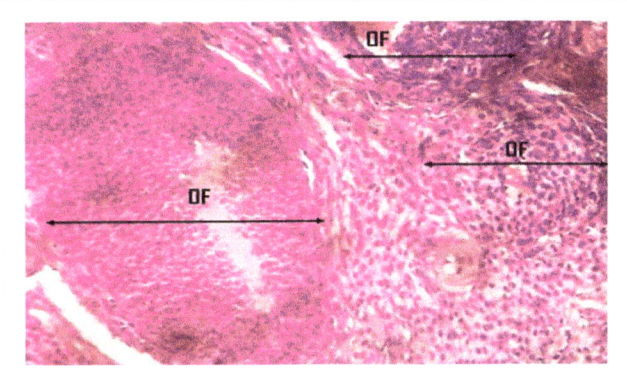

Plate 4: Photomicrograph of ovary showing Ovarian Follicles (OF) with proliferation of granulosa cells. The ovarian follicles are distorted. (400 mg/kg of leaf extract)

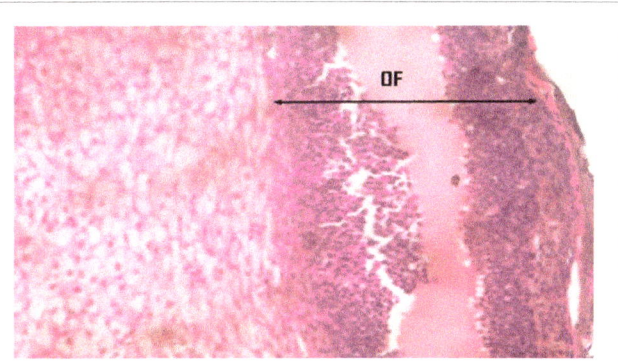

Plate 7: Photomicrograph of ovary showing elongated Ovarian Follicle (OF) with proliferation of granulosa cells. The ovarian follicle is distorted. No ovum is seen in the follicles may be due to degeneration. (400 mg/kg of seed extract)

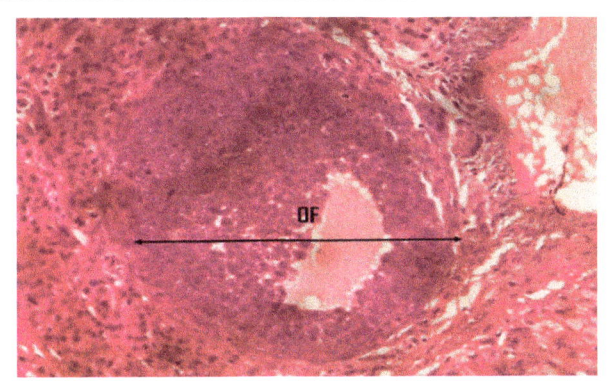

Plate 5: Photomicrograph of ovary showing Ovarian Follicles (OF) in the cortical region, No ovum is seen in the follicle. There is also proliferation of granulosa cells and slight distortion of the ovarian follicle (100 mg/kg of seed extract)

The result of histological examination of the uterus (Plates 8 - 14) showed no observable changes in the control group, it showed intact and normal simple cuboidal epithelial layer of the luminal border (C). the endomentrium (E) contain Blood Vessels (BV) and endomentrial glands (G) in the proliferative phase of development. Animals treated with 100mg/kg of leaf extract showed shrunk endomentrial glands (G), while those treated with 100 mg/kg of seed extract showed engorged endomentrial glands (G). Also animals treated with 200 mg/kg of leaf extract showed fewer endomentrial glands (G) that may be due to shrinkage of other glands, while those treated with 200 mg/kg of seed extract no observable gland which may have occurred due to shrinkage. The animals treated with 400 mg/kg of leaf showed lumen filled with mucus. Similarly, animals treated with 400 mg/kg of seed extract showed lumen mucus and engorged glands; these could be considered as polyps.

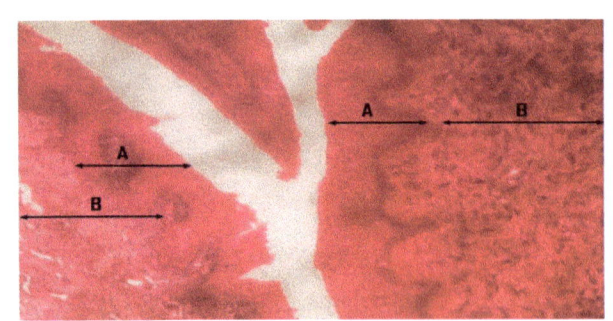

Plate 8: Photomicrograph of normal histology of fallopian tube showing simple columnar ciliated epithelium
(A) at the luminal border with layer of smooth muscles containing Blood Vessels (B) on the outside (Control)

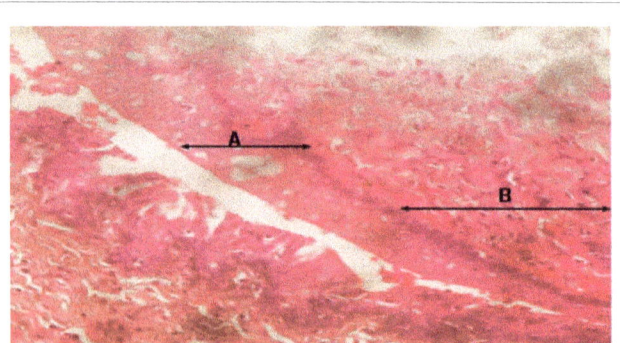

Plate 11: Photomicrograph of fallopian tube showing slight distortion of the tube of inner layer of simple columnar ciliated epithelium (A) and outer layer of smooth muscles (B) (400 mg/kg of leaf extract)

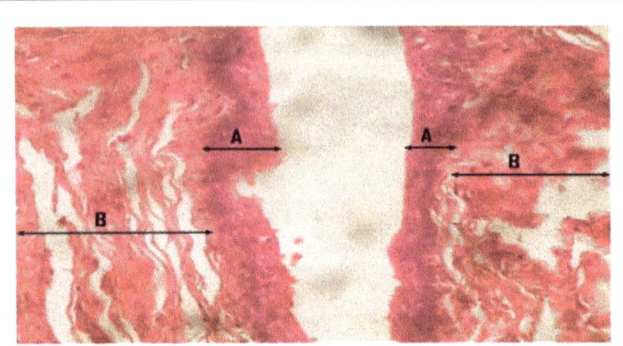

Plate 9: Photomicrograph of fallopian tube showing simple columnar ciliated epithelium (A) with an area (B) of slightly distorted smooth muscles. (100 mg/kg of leaf extract)

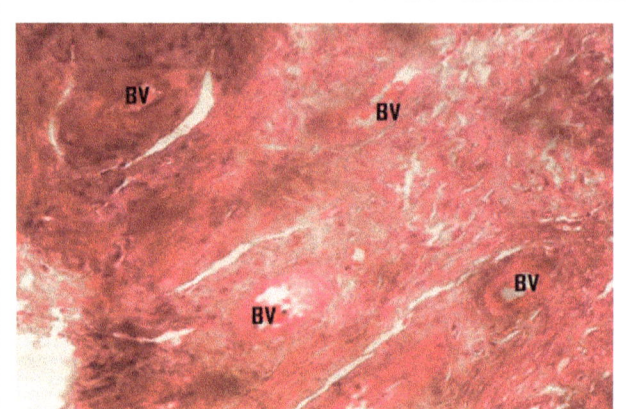

Plate 12: Photomicrograph of fallopian tube showing the outer/smooth layer of smooth muscles containing numerous Blood Vessels (BV), The tube section not clearly shown but with a slight distortion (100 mg/kg of seed extract)

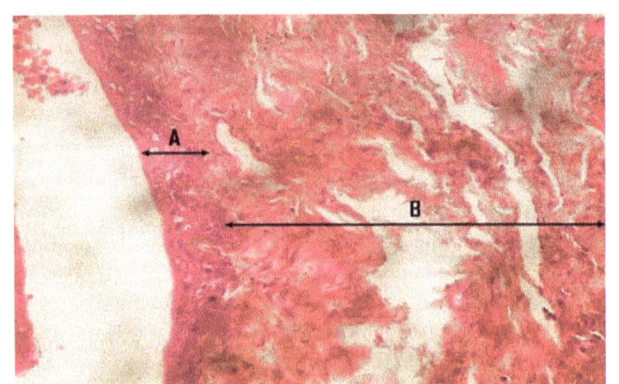

Plate 10: Photomicrograph of fallopian tube showing inner layer of simple columnar ciliated epithelium (A) and slightly distorted outer layer of smooth muscles (B) (200 mg/kg of leaf extract)

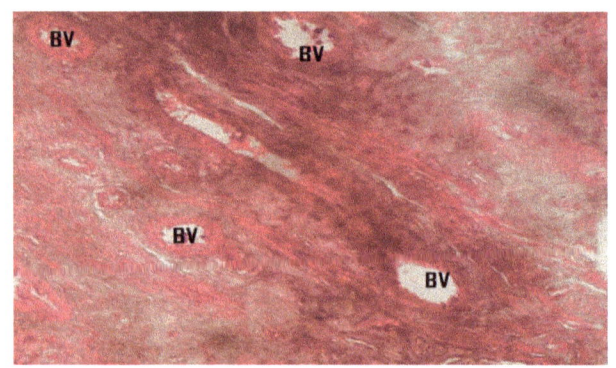

Plate 13: Photomicrograph of fallopian tube showing the outer/smooth layer of smooth muscles containing numerous Blood Vessels (BV), The tube section not clearly shown but with a slight distortion (200 mg/kg of seed extract)

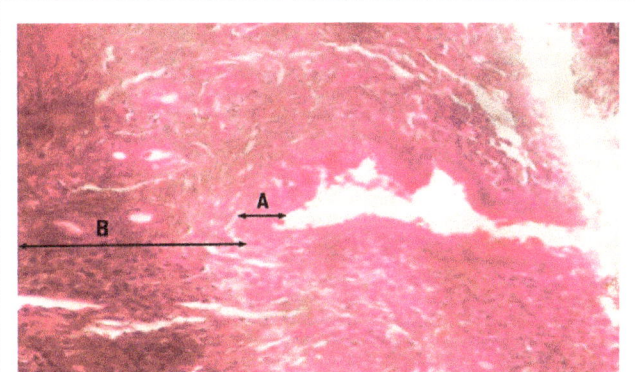

Plate 14: Photomicrograph of fallopian tube showing slight distortion of the tube of inner layer of simple columnar ciliated epithelium (A) and outer layer of smooth muscles (B) (400 mg/kg of leaf extract)

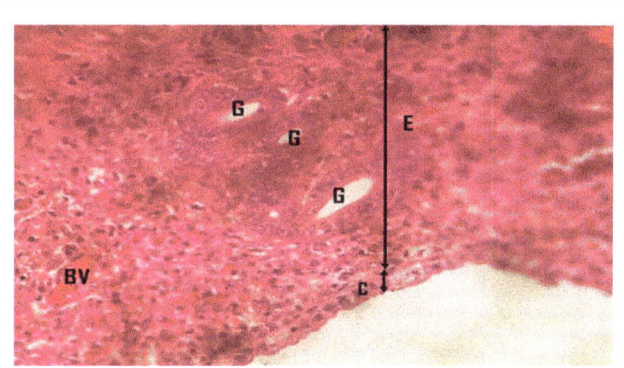

Plate 15: Photomicrograph of normal uterus showing simple cuboidal epithelial layer of the luminal border (C), The endomentrium (E) contain blood vessels (BV) and endomentrial glands (G) in the proliferative phase of development (control)

According to endometrial polyps are common spontaneous reproductive tract lesions that occur in aged rats, but because the animals used in this research are young, the presence of these lesions may be due to the effect of the extract. Another study from [7,4]. Revealed that certain agents such as quinacrine, can cause an increase in the incidence of endometrial hyperplasia and uterine stroma polyps. This also suggested that the phytochemicals in the ethanolic extract of both leaf and seed of Moringa oleifera are responsible for this effect.

In this study there was also an observed shrinkage of the uterine gland with increase in dose with the absence of extensive folding of luminal epithelium. In the higher dose of the extract the musculature was seen to be highly affected and stroma was compact with poor vascularity. The above changes in the uterine histology, after treatment with the extract may cause the endometrial milieu to become unfavourable for the implantation of the fertilized ovum and hence their antifertility effect. This agrees with other studies made by on Rumexsteudelii. These effects were seen to be dose dependent [11].

The histology of fallopian tube (Plates 15 - 21) in the control group show normal simple columnar ciliated epithelium at the luminal border with layer of smooth muscles containing blood vessels on the outside. On the contrary, the histology of fallopian tube in animals treated with 100 mg/kg of both leaf and seed extracts showed slight distortion of the smooth muscles. Also in animals treated with 200 mg/kg of both leaf and seed extracts, there was an observable slight distortion of the outer layer of smooth muscles. In animals treated with 400 mg/kg of leaf and seed extracts, there were observable distortion of the tube of inner layer of simple columnar ciliated epithelium and outer layer of smooth muscles. All these changes can cause obstruction of the smooth movement of a fertilized egg in the uterus.

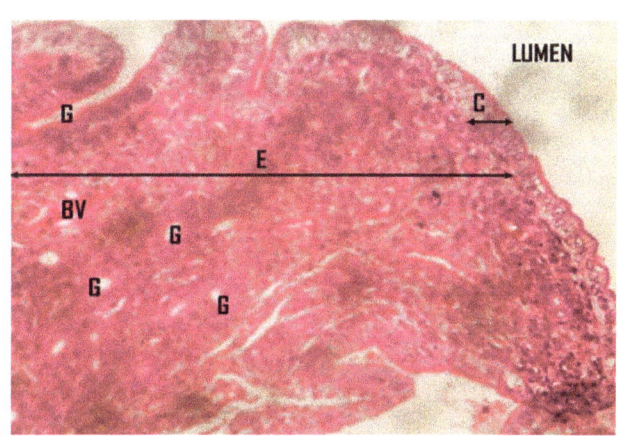

Plate 16: Photomicrograph of uterus showing simple cuboidal epithelial layer of the luminal border (C), The endomentrium (E) contain Blood Vessels (BV) and shrunk endomentrial glands (G) in the proliferative phase of development (100 mg/kg of leaf extract)

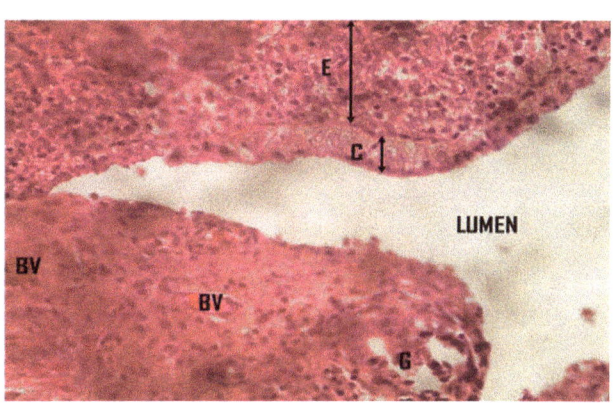

Plate 17: Photomicrograph of uterus showing simple cuboidal epithelial layer of the luminal border (C), The endomentrium (E) contain Blood Vessels (BV) and fewer endomentrial glands (G) in the proliferative phase of development. This may be due to shrinkage (200 mg/kg of leaf extract)

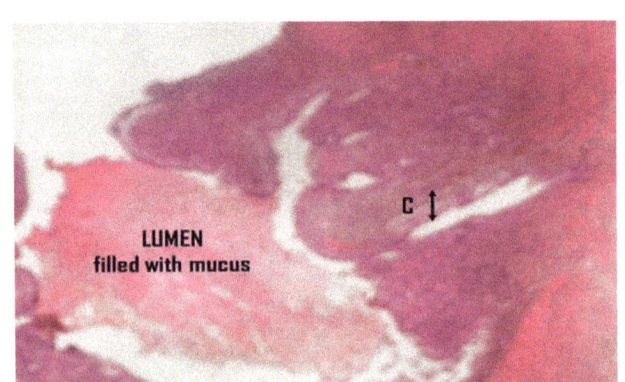

Plate 18: Photomicrograph of uterus showing simple cuboidal epithelial layer of the luminal border (C) with lumen filled with mucus in the secretory phase of development. No gland was found may be due to shrinkage. The presence of mucus could be an evidence of fluid resorption. (400 mg/kg of leaf extract)

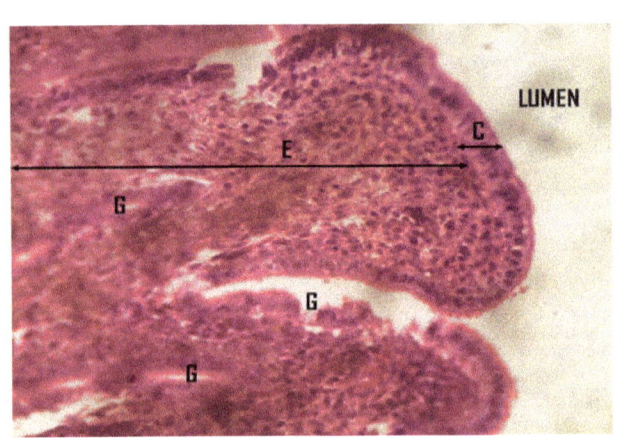

Plate 19: Photomicrograph of normal uterus showing simple cuboidal epithelial layer of the luminal border (C),The endomentrium (E) contain blood vessels (BV) and engorged endomentrial glands (G) in the proliferative phase of development. (100 mg/kg of seed extract)

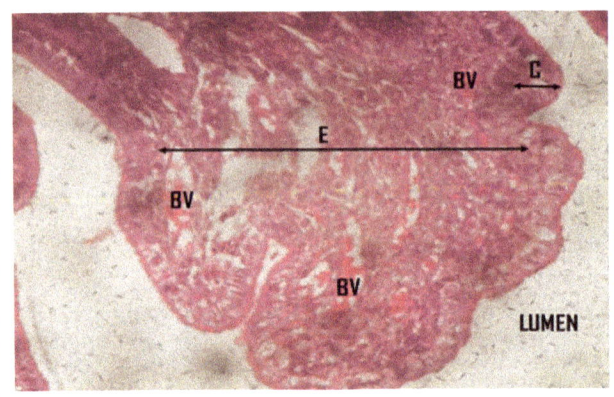

Plate 20: Photomicrograph of normal uterus showing simple cuboidal epithelial layer of the luminal border(C), the endomentrium (E) contain blood vessels (BV) and with no observable glands (G) in the proliferative phase of development .This may be due to shrinkage(200 mg/kg of seed extract)

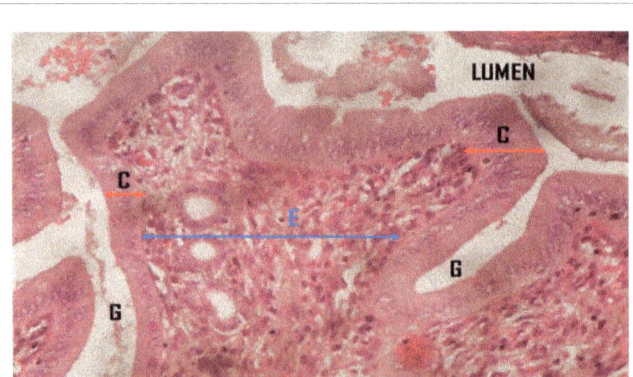

Plate 21: Photomicrograph of uterus showing simple cuboidal epithelial layer of the luminal border (C) with lumen containing mucus in the proliferative phase of development. The presence of the mucus could be an evidence of fluid resorption. The glands are seen to be enlarged or engorge (400 mg/kg of leaf extract)

Summary of Findings

Dose dependent decrease was observed in the levels of FSH and LH in the treated animals; the higher the dose the lower the hormonal levels and vice versa.

Different pathological changes (degeneration of the ovarian follicles or atretic follicles) were observed on the ovaries of the rats treated with the extracts (leaf and seed) as compared to the control group.

There were no observable histological changes in the uterus of the control group, while changes ranges from shrunken endomentrial to engorged as well as fewer endomentrial glands as the dose increases (from 100 mg/kg to 200 mg/kg as well as 400 mg/kg).

Conclusions

The ethanol extract of both the leaf and seed of moringa oleifera has shown abortificient effect in that they caused a decrease in the number of litters from animals treated with the extract. The ethanol extract also causes atretic follicle and tissue engorgement follicle in a dose dependent manner in the ovaries as a sign of high levels of degenerating pre-ovulatory follicle and an absence of the steroid hormones. The abortificient effect was also observed in the uterus where it causes endometrial polyps and shrinkage of the uterine gland as well as making the endometrial milieu to become unfavourable for the implantation of the fertilized ovum. While in the fallopian tube, the both extract caused distortion of the tube of inner layer of simple columnar ciliated epithelium and outer layer of smooth muscles.

Recommendations

Pregnant mothers no matter the anticipated beneficial effect should not use the plant. More research should be conducted to know the effects on other organs and hormones not covered in this work.

References

1. Awe S. O, Makinde J. M, Olajide O. A. Cathartic effect of the leaf extract of Vernonia amygdalina. Fitoterapia.1999; 70: 161- 165.

2. Bosch CH. Moringa oleifera Lam. In: Grubben GJH, Denton OA. (Eds). PROTA (Plant Resources of Tropical Africa / Ressourcesvégétales de l'Afriquetropicale), Wageningen, Netherlands. 2004.

3. Bose Ck. Possible Role of Moringa oleifera Lam. Root in Epithelial Ovarian Cancer. Medscape General Medicine, 2007; 9(1): 26-32.

4. Cancel AM, Smith T, Rehkemper U, Dillberger JE, Sokal D and McClain RM. One Year neonatal mouse carcinogenesis study of quinaccrine dihydrachloride. Int. J. Toxicol. 2006; 25(2): 109-118.

5. Choi JH, Choi KC, Auersperg N, Leung PC. (2004). Overexpression of follicle-stimulating hormone receptor activates oncogenic pathways in preneoplastic ovarian surface epithelial cells. Journal of Clinical Endocrinology and Metabolism, 2004; 89(11): 5508-5516.

6. Demiray S, Pintado ME, and Castro PML. Evaluation of phenolic profiles and antioxidant activities of Turkish medicinal plants: Tilia argentea, Crataegi folium leaves and Polygonum bistorta roots. World Acad. Sci. Eng. Technol., 2009; 54: 312-317.

7. Dinse GE, Peddada SD, Harris SE, Elmore SA. Comparison of NTP historical control tumor incidencerates in female Harlan Sprague Dawley and Fisher344/N rats. Toxicol. Pathol. 2010; 38(5): 765-775. doi: 10.1177/0192623310373777.

8. Ekor M. The growing use of herbal medicines: issues relating to adverse reactions and challenges in monitoring safety," Frontiers in Pharmacology, 2014;10;4:177.doi: 10.3389/fphar.2013.00177.

9. El-Missiry MA. Enhanced testicular antioxidant system by ascorbic acid in alloxan diabetic rats. Comparative Biochemistry and Physiology, 1999; 124: 233-237.

10. Fahey JW Moringa oleifera :A review of the medical evidence for its nutritional, therapeutic, and prophylactic properties, (2005).

11. Gebrie E, Makonnen, Debella A, Zerihum L .Phytochemical screening and pharmacological evaluations for the antifertility effect of the methanolic root extract of Rumexsteudelii. Journal of Ethnopharmacol.2005; 96(1-2): 139-43.

12. Ghosh D, Das UB, and Misro M. Protective role of alphatochopherol-succinate in cyclophosphamide induced testicular gametogenic steroidogenic disorders: a correlative approach to oxidative stress. Free Radical Research. 2002; 36:1199-1208.

13. Koneri R, Balaraman R, Saraswati C. D. Anti-ovulatory and abortifacient potential of the ethanolic extract of root of Mormordicacymbalaria Fenzl in rats. Indian J Pharmacol.2006;38(2): 111- 114.

14. Kojo S. Vitamin C: basic metabolism and its function as an index of oxidative stress .Current Medicinal Chemistry. 2004; 11(8):1041-1064.

15. Nwamarah JU, Otitoju O, and Otitoju GT. Effects of Moringa oleifera Lam. aqueous leaf extracts on follicle stimulating hormone and serum cholesterol in Wistar rats. African Journal of Biotechnology, 2015; 14 (3): 181-186.

16. Orwa C, Mutua A, Kindt R, Jamnadass R, Anthony S. Agroforestree Database: a tree reference and selection guide version 4.0. World Agroforestry Centre, Kenya.2009.

17. Osman A, Alsomait H, Seshadri S, El-Toukhy T, Khalaf Y. The effect of sperm DNA fragmentation on live birth rate after IVF or ICSI: a systematic review and meta-analysis.2015; 30(2): 120-127. Doi: 10.1016/j.rbmo.2014.10.018

18. Radovich T. Farm and forestry production and marketing profile for Moringa. In: Elevitch, C.R. (Ed.) Specialty Crops for Pacific Island Agroforestry. Permanent Agriculture Resources (PAR), Holualoa, Hawai .2009.

19. Shukla S, Mathur R, Prakash AO. Antifertility profile of the aqueous extract of Moringa oleifera roots. Journal of Ethnopharmacology. 1988;22(1):51-62.

20. Shukla S, Mathur R, Prakash AO. Biochemical and physiological alterations in female reproductive organs of cyclic rats treated with aqueous extract of Moringa oleifera Lam. Acta Eur Fertil.1988;19(4): 225-232.

21. Songpol C, Pornchai S, Anudep R, Suphan P, Vanida C, Praw Suppajariyawat . Safety Evaluations of Ethanolic Extract of Moringa oleifera Lam. Seed in Experimental Animals. 2012;42(3):343-352.

22. Zade V, Dabhadkar D. Abortifacient efficacy of Moringa oleifera stem bark on female albino rats. World Journal of Pharmaceutical Research. 2014;3(3):4666-4679.

25

Chemical Reactivity Properties of Standard Aromatic Amino Acids Studied by Means of Conceptual Density Functional Theory

Norma Flores-Holguin[1], Juan Frau[2] and Daniel Glossman-Mitnik[*1, 2]

[1]NANOCOSMOS Virtual Lab, Department of Environment and Energy, Advanced Materials Research Center, Miguel de Cervantes 120, Complejo Industrial Chihuahua, Chihuahua Chih 31136, Mexico.

[2]Departament of Chemistry, University of the Balearic Islands, Palma de Mallorca 07122, Spain.

*Corresponding author: Dr. Daniel Glossman-Mitnik, NANOCOSMOS Virtual Lab, Department of Environment and Energy, Advanced Materials Research Center, Miguel de Cervantes 120, Complejo Industrial Chihuahua, Chihuahua Chih 31136, Mexico
E-mail: daniel.glossman@cimav.edu.mx or dglossman@gmail.com

Abstract

This study assessed eight density functionals that include CAM-B3LYP, LC- ωPBE, M11, MN12SX, N12SX, ωB97, ωB97X and ωB97XD related to the Def2TZVP basis sets together with the SMD solvation model in the calculation of the molecular properties and structures of the four standard aromatic amino acids: Histidine, Phenylalanine, Tryptophan and Tyrosine.. The global chemical reactivity descriptors for the systems are calculated via the Conceptual Density Functional Theory (CDFT). The prediction of the maximum absorption wavelength directly from the HOMO-LUMO tends to be considerably accurate relative to the experimental values for the MN12SX density functional. Additionally, the ability of the studied molecules in acting as efficient inhibitors of the formation of Advanced Glycation Endproducts (AGEs) (perhaps as neutraceuticals), which constitutes a useful knowledge for the development of drugs for fighting Diabetes, Alzheimer and Parkinson diseases. Finally, the bioactivity scores for the four standard aromatic amino acids are predicted through different methodologies.

Keywords: Conceptual DFT; Chemical Reactivity Theory; Histidine; Phenylalanine; Tryptophan; Tyrosine; Aromatic Amino Acids.

Introduction

The sea is an inexhaustible source of natural resources that give rise to molecules that can serve as a guide for the development of new medicines. For this reason, numerous investigations have been carried out in recent years dedicated to the search for new natural products that can be obtained from the knowledge of marine species [1].

Among the chemical species that can be obtained from natural products of marine origin stand out the peptides that are molecules of intermediate size between amino acids and proteins. The therapeutic application of these peptides, called for this reason therapeutic peptides, is currently one of the most active fields of research due to the great possibilities they represent as aids for the treatment of numerous diseases [2].

For the consideration of therapeutic peptides from the point of view of medicine it is necessary to know their molecular properties and their bioactivity. It is our belief that the bioactivity of these peptides is intimately related to their chemical reactivity from a molecular perspective [3, 4]. For this reason, we consider it essential to study the chemical reactivity of natural products that have the potential to become medicines through the tools provided by Computational Chemistry and Molecular Modeling. Probably the most powerful tool currently available to study the chemical reactivity of molecular systems from the point of view of Computational Chemistry and Molecular Modeling is the Conceptual DFT (CDFT), also called Chemical Reactivity Theory, which using a series of global descriptors allow to predict the interactions between molecules and understand the way in that chemical reactions proceed [5-7, 9].

The objective of this work is to study the chemical reactivity of the four standard aromatic amino acids: Histidine, Phenylalanine, Tryptophan and Tyrosine using the techniques of CDFT, determining its global properties, that is, of the molecule as a whole. Likewise, the potential ability of these amino acids to act as inhibitors of the formation of Advanced Glycation Endproducts (AGEs) will be established according to our previous ideas, and the descriptors of bioavailability and bioactivity (Bioactivity Scores) will be calculated through different procedures described in the literature [10-12].

Theoretical Background

As this work is a part of an ongoing study related to our project on Computational Medicinal Nanochemistry, the theoretical background will be similar to that presented in previous works and will be shown here again for completeness reasons [13-20]. As in those previous works, we will be using the Kohn-Sham theory which involves the calculation of the molecular density, energy of the system, and the orbital energies particularly associated with the frontier orbitals including the Highest Occupied Molecular Orbital (HOMO) and Lowest Unoccupied Molecular Orbital (LUMO) [22-24]. This theory is necessary for establishing the quantitative values of the various

CDFT descriptors. Recently, there has been an increased interest in using range-separated (RS) exchange correlation functionals in Kohn-Sham DFT [25-28]. These functionals tend to partition the r_{12}^{-1} operator and exchange the parts into long- and short-ranged parts, whose range separation parameter, ω, controls the rate of attaining the long-range behavior. It is possible to fix the value of ω or "tune" it by a system-by-system mechanism that minimizes a tuning norm. The basis of the optimal tuning approach is the fact that the energy of the HOMO, $\epsilon_H(N)$, in case of the exact Kohn-Sham (KS) theory as well as generalized KS theory for an N electron system should be -IP(N). Here, IP represents the vertical ionization potential, which is calculated as the energy difference, $E(N-1) - E(N)$, by considering a particular functional. If approximate functionals are used, it would possibly lead to considerable differences between $E(N)$ and -IP(N). Optimal tuning involves determining the system-specific range-separation parameter, ω, non-empirically with an RSE functional.

Alternatively, it also implies that several other parameters including $\epsilon_H(N) = $ -IP(N) are optimally satisfied [29-36]. Even though there is no equivalent form to match this prescription for deriving the electron affinity (EA) together with the LUMO in case of neutral species, it is possible to say that $\epsilon_H(N+1) = $ -EA(N), which facilitates obtaining the optimized value of ω, which is then optimized to establish both properties. This would make it easy to predict the CDFT descriptors. In the past, a simultaneous prescription referred to as the "KID procedure", owing to its correspondences with the Koopmans' theorem, was proposed by the authors [13-20]. As it has been explained in the last referenced works, KID stands for "Koopmans in DFT" and is a procedure to check the verification of the $\epsilon_H(N) = $ -IP(N) satisfaction and at the same time a comparison between the $\epsilon_L(N)$ of the neutral species (the LUMO) and the $\epsilon_H(N-1)$ for the anionic system (the SOMO). The descriptor related to this comparison is called ΔSL [13-20].

Settings and Computational Methods

This study obtained the molecular structure of the four standard aromatic amino acids: Histidine, Phenylalanine, Tryptophan and Tyrosine from Pub-Chem (https://pubchem.ncbi.nlm.nih.gov), a website that serves as the public repository for information pertaining chemical substances along with their associated biological activities. The pre-optimization of the resultant system involved selecting the most stable conformers. The selection was done using random sampling that involved molecular mechanics techniques and inclusion of the various torsional angles via the general MMFF94 force field involving the Marvin View 17.15 program, which constitutes as an advanced chemical viewer suited to multiple and single chemical queries, structures, and reactions (https://www.chemaxon.com) [37-41]. After that, the chemistry of the structures was checked and the 3D structures of the stereoisomers were generated using the same Marvin View 17.15 program. The chirality at the stereogenic centers was verified in accordance to the Cahn-Ingold-Prelog

priority rules. The resulting geometries were further refined as it was explained before and the lowest energy conformation for each molecule was chosen to calculate the electronic energy and the HOMO and LUMO orbitals at the DFT functional level as mentioned in the next paragraph.

Consistent with our previous work, the computational studies were performed with the Gaussian 09 series of programs that implement density functional methods [13-20, 42]. The basis set Def2SVP was used in this work for geometry optimization and frequency determination, while the Def2TZVP basis set was used for calculating electronic properties [43, 44]. All calculations were performed in the presence of water as solvent under the Solvation Model Density (SMD) parameterization of the Integral Equation Formalism- Polarized Continuum Model (IEF-PCM) [45].

To calculate the molecular structure and properties of the studied systems, we have chosen eight density functionals which is known to consistently provide satisfactory results for several structural and thermodynamic properties: CAM-B3LYP [27], LC-ωPBE [46], M11 [47], MN12SX [48], N12SX [48], ωB97, ωB97X and ωB97XD [26].

The SMILES notations of the studied compounds were fed in the online Molinspiration software from Molinspiration Cheminformatics (www.molinspiration.com) for the calculation of the molecular properties (Log P, Total polar surface area, number of hydrogen bond donors and acceptors, molecular weight, number of atoms, number of rotatable bonds, etc.) and for the prediction of the bioactivity score for different drug targets (GPCR ligands, Kinase inhibitors, Ion channel modulators, Enzymes and Nuclear receptors). The bioactivity scores were compared with those obtained through the use of other software like MolSoft from Molsoft L.L.C. (http://molsoft.com/mprop/) and Chem Doodle Version 9.02 from iChem Labs L.L.C. (www.chemdoodle.com).

Results and Discussion

Molecular Structure Optimization and Verification of the KID Procedure

The molecular structures of the optimized conformers of the four standard aromatic amino acids obtained as mentioned in the Settings and Computational Methods section, and whose graphical sketches are shown in Figure 1, were reoptimized in the gas phase by considering the DFTBA model available in Gaussian 09 and then optimized again using the eight density functionals mentioned in the previous section together with the Def2SVP basis set and the SMD solvent model using water as the solvent.

After verifying that each of the structures corresponded to the minimum energy conformations through a frequency calculation analysis, the electronic properties were determined by using the same model chemistry but with the Def2TZVP basis set instead of that used for the geometry optimization.

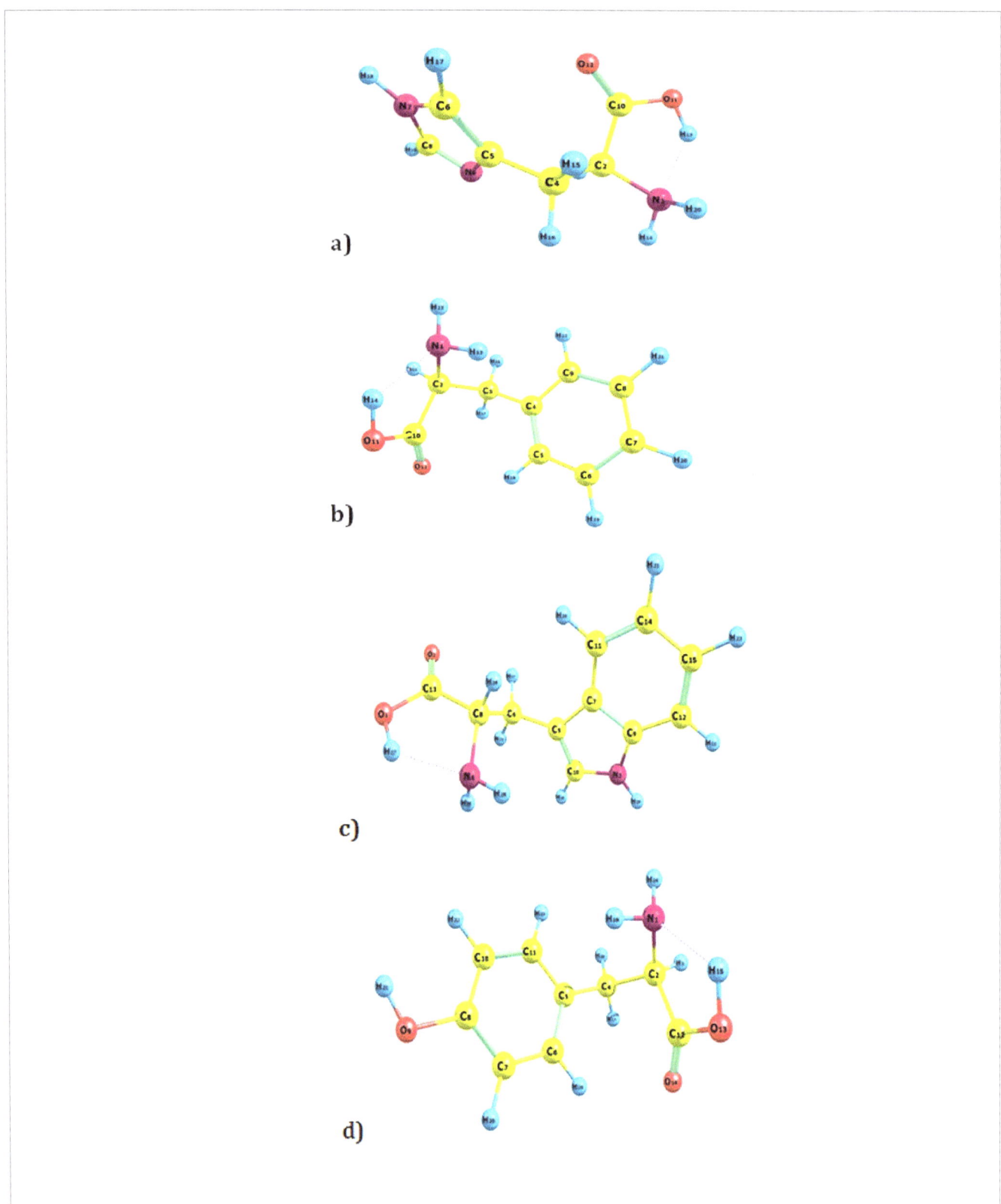

Figure 1: Graphical sketches of the optimized molecular structures of a) Histidine, b) Phenylalanine, c) Tryptophan and d) Tyrosine showing the numbering of the atoms.

The analysis of the results obtained in the study aimed at verifying that the KID procedure was fulfilled. On doing it previously, several descriptors associated with the results that the HOMO and LUMO calculations obtained are related with results obtained using the vertical I and A following the ΔSCF procedure. A link exists between the three main descriptors and the simplest conformity to the Koopmans' theorem by linking $\in_H$ with -I, $\in_L$ with -A, and their behavior in describing the HOMO-LUMO gap as $J_I = |\in_H + Egs(N-1) - Egs(N)|$, $J_I = |\in_L + Egs(N) - Egs(N+1)|$ and $J_{HL} \cdot \sqrt{J_I^2 + J_A^2}$ 2. Notably, the J_A descriptor consists of an approximation that remains valid only when the HOMO that a radical anion has (the SOMO) shares similarity with the LUMO of the neutral system. Consequently, we decided to design another descriptor ΔSL, to guide in verifying the accuracy of the approximation [13-20]. The results of this analysis are presented in Table 1 to 4 for Histidine, Phenylalanine, Tryptophan and Tyrosine, respectively.

Table 1: Electronic energies of the neutral, positive and negative molecular systems (in au) of Histidine, the HOMO, LUMO and SOMO orbital energies (also in au), J_I, J_A, J_{HL} and ΔSL descriptors calculated with the eight density functionals and the Def2TZVP basis set using water as solvent simulated with the SMD parameterization of the IEF-PCM model.

	Eo	E+	E-	HOMO	LUMO	SOMO	J(I)	J(A)	J(HL)	ΔSL
CAM-B3LYP	-548.81	-548.59	-548.84	-0.277	0.033	-0.094	0.059	0.063	0.086	0.126
LC-ωPBE	-548.70	-548.48	-548.74	-0.326	0.078	-0.143	0.105	0.110	0.152	0.222
M11	-548.83	-548.61	-548.84	-0.319	0.054	-0.116	0.095	0.058	0.111	0.170
MN12SX	-548.63	-548.42	-548.66	-0.219	-0.043	-0.035	0.002	0.014	0.014	0.008
N12SX	-548.81	-548.60	-548.84	-0.215	-0.023	-0.032	0.002	0.005	0.005	0.008
ωB97	-548.96	-548.75	-548.99	-0.321	0.083	-0.136	0.105	0.108	0.151	0.219
ωB97X	-548.91	-548.70	-548.28	-0.313	0.074	-0.103	0.096	0.563	0.572	0.178
ωB97XD	-548.87	-548.66	-548.90	-0.297	0.059	-0.110	0.080	0.085	0.117	0.170

Table 2: Electronic energies of the neutral, positive and negative molecular systems (in au) of Phenylalanine, the HOMO, LUMO and SOMO orbital energies (also in au), J_I, J_A, J_{HL} and ΔSL descriptors calculated with the eight density functionals and the Def2TZVP basis set using water as solvent simulated with the SMD parameterization of the IEF-PCM model.

	Eo	E+	E-	HOMO	LUMO	SOMO	J(I)	J(A)	J(HL)	ΔSL
CAM-B3LYP	-554.8	-554.55	-554.82	-0.304	0.024	-0.078	0.058	0.05	0.077	0.102
LC-ωPBE	-554.69	-554.44	-554.73	-0.353	0.062	-0.135	0.102	0.097	0.141	0.196
M11	-554.81	-554.56	-554.84	-0.345	0.05	-0.105	0.093	0.074	0.119	0.155
MN12SX	-554.63	-554.39	-554.67	-0.249	-0.042	-0.041	0.002	0.005	0.005	0.001
N12SX	-554.83	-554.59	-554.86	-0.243	-0.03	-0.037	0.003	0.004	0.005	0.007
ωB97	-554.96	-554.99	-554.99	-0.347	0.066	-0.127	0.376	0.095	0.388	0.193
ωB97X	-554.92	-554.67	-554.3	-0.34	0.06	-0.113	0.094	0.561	0.569	0.173
ωB97XD	-554.88	-554.91	-554.91	-0.324	0.048	-0.109	0.353	0.077	0.362	0.157

Table 3: Electronic energies of the neutral, positive and negative molecular systems (in au) of Tryptophan, the HOMO, LUMO and SOMO orbital energies (also in au), J_I, J_A, J_{HL} and ΔSL descriptors calculated with the eight density functionals and the Def2TZVP basis set using water as solvent simulated with the SMD parameterization of the IEF-PCM model.

	Eo	E+	E-	HOMO	LUMO	SOMO	J(I)	J(A)	J(HL)	ΔSL
CAM-B3LYP	-686.37	-686.17	-686.41	-0.259	0.018	-0.092	0.054	0.055	0.077	0.11
LC-ωPBE	-686.24	-686.03	-686.28	-0.305	0.052	-0.136	0.095	0.093	0.133	0.188
M11	-686.39	-686.18	-686.43	-0.299	0.043	-0.125	0.086	0.083	0.119	0.168
MN12SX	-686.18	-685.97	-686.21	-0.206	-0.044	-0.042	0.002	0.005	0.006	0.002
N12SX	-686.42	-686.22	-686.45	-0.201	-0.032	-0.04	0.002	0.004	0.004	0.008
ωB97	-686.58	-686.37	-686.61	-0.3	0.057	-0.128	0.094	0.092	0.131	0.185
ωB97X	-686.52	-686.31	-685.76	-0.293	0.051	-0.115	0.087	0.709	0.714	0.167
ωB97XD	-686.48	-686.27	-686.51	-0.279	0.041	-0.112	0.075	0.076	0.107	0.153

Table 4: Electronic energies of the neutral, positive and negative molecular systems (in au) of Tyrosine, the HOMO, LUMO and SOMO orbital energies (also in au), J_I, J_A, J_{HL} and ΔSL descriptors calculated with the eight density functionals and the Def2TZVP basis set using water as solvent simulated with the SMD parameterization of the IEF-PCM model.

	Eo	E+	E-	HOMO	LUMO	SOMO	J(I)	J(A)	J(HL)	ΔSL
CAM-B3LYP	-630.05	-629.82	-630.08	-0.28	0.021	-0.093	0.057	0.056	0.08	0.114
LC-ωPBE	-629.92	-629.69	-629.96	-0.327	0.059	-0.138	0.1	0.097	0.139	0.197
M11	-630.07	-629.84	-630.1	-0.321	0.048	-0.127	0.091	0.085	0.124	0.174
MN12SX	-629.85	-629.63	-629.89	-0.226	-0.042	-0.043	0.002	0.003	0.003	0.001
N12SX	-630.07	-629.85	-630.1	-0.22	-0.032	-0.039	0.002	0.004	0.004	0.008
ωB97	-630.22	-630	-630.26	-0.321	0.063	-0.13	0.099	0.095	0.137	0.193
ωB97X	-630.17	-629.95	-629.46	-0.315	0.057	-0.117	0.092	0.656	0.663	0.174
ωB97XD	-630.13	-629.91	-630.16	-0.3	0.046	-0.112	0.078	0.078	0.11	0.158

As Tables 1 to 4 provide, the KID procedure applies accurately for the MN12SX and N12SX density functionals that are range-separated hybrid meta-NGA as well as range-separated hybrid NGA density functionals respectively. In fact, the values of J_I, J_A and J_{HL} are actually not zero. Nevertheless, the results tend to be impressive especially for the MN12SX density functional. As well, the ΔSL descriptor reaches the minimum values when MN12SX and N12SX density functionals are used in the calculations. This implies that there are sufficient justifications to assume that the LUMO of the neutral approximates the electron affinity.

Calculation of the Maximum Absorption Wavelength

Being aromatic amino acids, the molecules considered here will absorb energy in the UV region of the electromagnetic spectrum and they would be best studied using the Time-Dependent Density Functional Theory (TDDFT). In the past, various TDDFT studies of molecules of different size have used optimally-tuned RSH density functionals [29, 30, 32, 33, 35, 49-63]. The considerable success of the approach is however undermined by the issue of tuning being system dependent. Therefore, focus should be on establishing the effectiveness of the behaviors of the fixed RSH density functionals in describing the excitation characteristics. In his works, Becke has recently mentioned that the adiabatic connection and the ideas of Hohenberg, Kohn, and Sham apply only to electronic ground states is a common misconception [64]. Furthermore, consistent with Baerends et al., KS model is not appreciated for being superior because of its lowest excitation energy in molecules. Physically, it amounts to an excitation of the KS system rather than electron addition as would be the case in Hartree-Fock. Thus, it can be effectively be used as a measure of the optical gap and is an effective approximation to the gap (in molecules) [65]. In their conclusion, van Meer et al. advanced that the HOMO-LUMO gap associated with the KS model tends to be an approximation of the lowest excitation energy, a desirable characteristic with no concerns regarding it [66] (Figure 2).

Therefore, the calculation of the maximum wavelength absorption of these amino acids involved conducting ground state calculations with the aforementioned eight density functionals at the same level of model chemistry and theory and determining the HOMO-LUMO gap. Figure 1 provides an illustration that

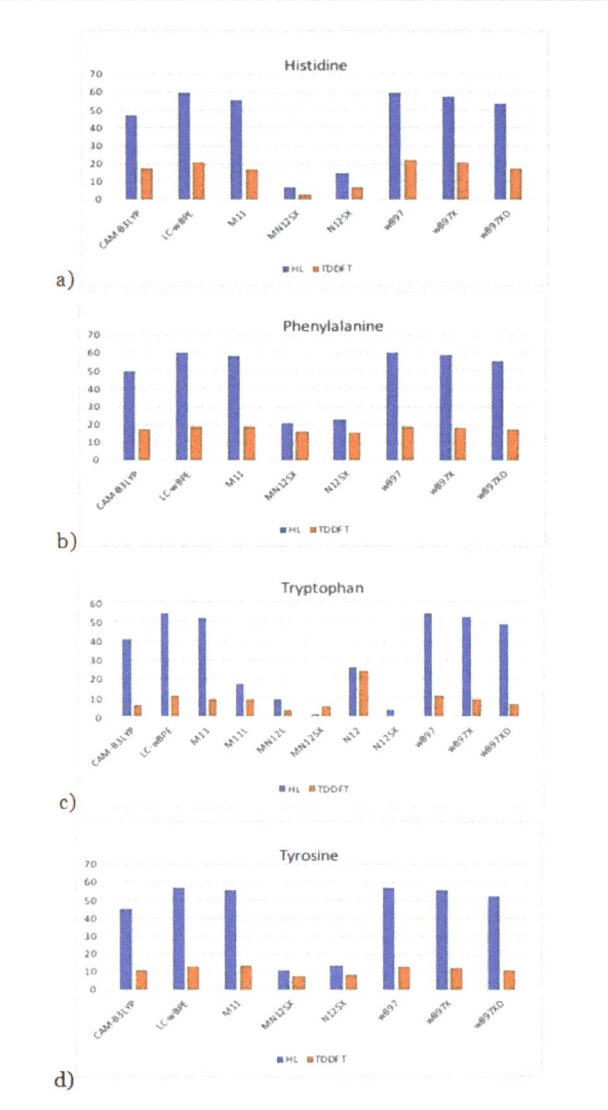

Figure 2: Graphical sketches of the differences between the calculated maximum absorption wavelengths and experimental values (in nm) of a) Histidine, b) Phenylalanine, c) Tryptophan and d) Tyrosine by direct comparison with the value of the HOMO-LUMO gap (HL) or through a TDDFT calculation (TDDFT).

compares graphically the results involved in the ground-state approximation derived from the HOMO-LUMO gap together with the TDDFT results with the known experimental values for the molecules (Histidine = 211 nm, Phenylalanine = 257.5 nm, Tryptophan = 278 nm and Tyrosine = 274.25 nm) [67].

Notably, the presented results suggest that the differences with the experimental value for λ_{max} tend to have the same order in the various functionals that the current study considers. If the λ_{max} values that the HOMO-LUMO gap generates were the ones considered, MN12SX and N12SX would appear to be specially accurate in predicting this value. Such finding does not apply in the rest of the density functionals that this study considered.

Global Descriptors Calculation

As can be seen from Tables 1 to 4, the results for the descriptors show values that are consistent with our previous findings for the case of the melanoidins and peptides of marine origin, that is, the MN12SX density functional is capable of giving HOMO and LUMO energies that allow to verify the agreement with the approximate Koopmans' theorem [13-20]. This is not only true because the J_{HL} values are almost, zero, but due to the fact that the ΔSL descriptor, which relates to the difference between the LUMO of the neutral and the HOMO of the anion, is also close to zero. Indeed, these values cannot be exactly equal to zero, but the small differences mean that errors in the prediction of the global reactivity descriptors will be negligible. Moreover, it can be seen from Tables 1 to 4 that the MN12SX density functional predict negative values for the LUMO energies which will represent positive values of the electron affinity A.

The KID procedure has its foundations on the behavior of the four descriptors J_I, J_A, J_{HL}, and ΔSL : the closer they are to zero the better agreement of a density functional in giving accurate CDFT descriptors calculated only from the HOMO and LUMO.

This allows to avoid the calculation of the energies of the cation and anion species which being open systems are more difficult to converge than the parent neutral molecule, that is inconvenient when studying large systems like those considered in this study.

By taking into account the KID procedure presented in our previous works together with the finite difference approximation, the global reactivity descriptors can be expressed as:

Electronegativity $\quad \chi = -\frac{1}{2}(I + A) \approx \frac{1}{2}(\in_L + \in_H)$ [5,6]

Global Hardness $\quad \eta = (I - A) \approx (\in_L - \in_H)$ [5, 6]

Electrophilicity $\quad \omega = \frac{\mu^2}{2\eta} = \frac{(I+A)^2}{4(I-A)} \approx \frac{(\in_L + \in_H)^2}{4(\in_L - \in_H)}$ [68]

Electrodonating Power $\quad \omega^- = \frac{(3I+A)^2}{16(I-A)} \approx \frac{(3\in_H + \in_L)^2}{16\eta}$ [69]

Electroaccepting Power $\quad \omega^+ = \frac{(I+3A)^2}{16(I-A)} \approx \frac{(\in_H + 3\in_L)^2}{16\eta}$ [69]

Net Electrophilicity $\quad \Delta\omega^\pm = \omega^+ - (-\omega^-) = \omega^+ + \omega^-$ [70]

where $\in_H$ and $\in_L$ are the energies of the HOMO and LUMO, respectively. According to our previous discussion, the results for the global reactivity descriptors based on the values of the HOMO and LUMO energies calculated with the MN12SX density functional are presented in Table 5 which illustrates the results obtained after calculating for the electronegativity χ, chemical hardness η, global electrophilicity ω, electroaccepting ($\omega+$) and electrodonating ($\omega-$) powers as well as net electrophilicity with the MN12SX density. The Def2TZVP basis set is used with water acting as a solvent in line with the SMD solvation model.

Table 5: Global reactivity descriptors for the four standard aromatic amino acids calculated with the MN12SX density functional.

	Electronegativity (χ)	Chemical Hardness (η)	Electrophilicity (ω)
Histidine	3.5551	4.7941	1.3182
Phenylalanine	3.9521	5.6202	1.3896
Tryptophan	3.398	4.4069	1.31
Tyrosine	3.6407	5.0186	1.3206
	Electrodonating Power (ω−)	Electroaccepting Power (ω+)	Net Electrophilicity (Δω±)
Histidine	2.7716	1.374	4.1457
Phenylalanine	3.0027	1.4276	4.4303
Tryptophan	2.7016	1.3787	4.0803
Tyrosine	2.8079	1.3686	4.1764

Quantification of the AGEs Inhibitor Ability

The Maillard reaction between a reducing carbonyl and the amino group of a peptide or protein leads to the formation of a Schiff base which through a series of steps renders different molecules known as Advanced Glycation Endproducts or AGEs.

It is believed that the presence of these AGEs is one of the main reasons for the developing of some diseases like Diabetes, Alzheimer and Parkinson [71].

Among several strategies that have been considered for the prevention of the formation of AGEs, it is worth to mention the use of compounds presenting amino groups in their structure capable of interacting with the reducing carbonyl of carbohydrates and being competitive with the amino acids, pep- tides and proteins present in our body. Many compounds have been devised as drugs to achieve this goal and to name a few; we can include Pyridoxamine, Aminoguanidine, Carnosine, Metformin, Pioglitazone and Tenilsetam [72, 73].

It can be proposed that peptides having amino and amido groups could be thought as potential therapeutic drugs for preventing the formation of AGEs because they could in the Maillard reaction with reducing carbohydrates before than the peptides and proteins of our body. Although this a merely speculative proposal, we believe that it is worth to explore this possibility by following a methodology earlier presented by us. In a previous work, we have studied the ability of a group of proposed molecules to act as inhibitors of the formation of AGEs by quantifying their behavior in terms of CDFT reactivity descriptors [10]. It was concluded that the key factor in the study of the chemical reactivity of the potential AGEs inhibitors was on their nucleophilic character and although there are several definitions of nucleophilicity, our results suggested that the inverse of the net electrophilicity could be a good definition for the nucleophilicity N [74]. On the basis of the mentioned analysis, we were able to find some qualitative trends for the studied molecular systems.

In this work, we will extend this correlation to the four standard aromatic amino acids in order to see if they can be considered as precursors of therapeutic drugs as nutraceuticals for the inhibition of the formation of AGEs. As the model

chemistry employed in both works is the same, the comparison is straightforward:

Aminoguanidine > Metformin > Tryptophan > Histidine > Tyrosine > Phenylalanine > Carnosine > Tenilsetam > Pyridoxamine > Pioglitazone

This qualitative trend is representative of the known pharmacological properties of the studied AGEs inhibitors [72, 73] and it can be seen that the studied amino acids possess AGEs Inhibitor abilities similar to that of Metformin or Carnosine if we rely only in the mentioned criteria. However, the AGEs inhibition ability could also depend on many other factors. Thus, additional information will be needed to fully support this conclusion.

Bioactivity Scores

When considering a given molecular system as a potential therapeutic drug, it is customary to check if the considered species follows the Lipinsky Rule of Five which is used to predict whether a compound has or not has a drug-like character [75]. The molecular properties related to the drug-like character were calculated with the aid of the MolSoft and Molinspiration soft- ware and are presented in Table 6 where miLogP represents the octanol/water partition coefficient, TPSA is the molecular polar surface area, n atoms is the number of atom of the molecule, nON and nOHNH are the number of hydrogen bond acceptors and hydrogen bond donors respectively, nviol is the number of violations of the Lipinsky Rule of Five, nrotb is the number of rotatable bonds, volume is the molecular volume, and MW is the molecular weight of the studied system.

Table 6: Molecular properties of the four standard aromatic amino acids calculated to verify the Lipinsky Rule of Five.

Molecule	miLogP	TPSA	nAtoms	nON	nOHNH	nviol	nrotb	volume	MW
Histidine	-3.00	92.00	11	5	4	0	3	136.79	155.16
Phenylalanine	-1.23	63.32	12	3	3	0	3	155.96	165.19
Tryptophan	-1.08	79.11	15	4	4	0	3	184.94	204.23
Tyrosine	-1.71	83.55	13	4	4	0	3	163.98	181.19

However, what the Lipinsky Rule of Five really measures is the oral bioavailability of a potential drug because this is desired property for a molecule having drug-like character. Indeed, this criteria cannot be applied to peptides, even when they are small, as we can see from Table 1-6, due to the inherent molecular weight and number of hydrogen bonds.

In a more recent work, Martin have developed what she called "A Bioavailability Score" (ABS) for avoiding these problems [76]. The rule for the ABS established that the Bioavailability Score for neutral organic molecules must be 0.55 if they pass the Lipinsky Rule of Five and 0.170 if they fail. The ABS value for all the amino acids considered in this work have been calculated by using the ChemDoodle software and the results were equal to 0.170 for all of them.

Then, a different approach was followed by considering similarity searches in the chemical space of compounds with structures that can be compared to those that are being studied and with known pharmacological properties.

As has been mentioned in the Settings and Computational Methods section, this task can be accomplished using the online Molinspiration software for the prediction of the bioactivity score for different drug targets (GPCR ligands, kinase inhibitors, ion channel modulators, enzymes and nuclear receptors). The results are named Bioactivity Scores and the values for the standard aromatic amino acids are presented in Table 7.

Table 7: Bioactivity scores of the standard aromatic amino acids calculated on the basis of GPCR Ligand, Ion Channel Modulator, Nuclear Receptor Ligand, Kinase Inhibitor, Protease Inhibitor and Enzyme Inhibitor interactions.

Molecule	GPCR Ligand	Ion Channel Modulator	Kinase Inhibitor	Nuclear Receptor Ligand	Protease Inhibitor	Enzyme Inhibitor
Histidine	0.36	0.77	−0.45	-1.84	0.33	0.84
Phenylalanine	−0.22	0.34	-0.89	-0.53	-0.09	0.16
Tryptophan	0.33	0.54	-0.13	-0.22	0.15	0.44
Tyrosine	-0.08	0.41	-0.68	-0.20	-0.04	0.27

These bioactivity scores for organic molecules can be interpreted as active (when the bioactivity score > 0), moderately active (when the bioactivity score lies between -5.0 and 0.0) and inactive (when the bioactivity score < -5.0). All the Mirabamides A-H were found to be moderately bioactive towards all the enzymes considered for the study.

Conclusions

This study assessed eight density functionals that include CAM-B3LYP, LC-ωPBE, M11, MN12SX, N12SX, ωB97, ωB97X and ωB97XD related to the Def2TZVP basis sets together with the SMD solvation model in the calculation of the molecular properties and structures of the four standard aromatic amino acids: Histidine, Phenylalanine, Tryptophan and Tyrosine.

The global chemical reactivity descriptors for the systems were calculated via the Conceptual Density Functional Theory (CDFT). The prediction of the maximum absorption wavelength directly from the HOMO-LUMO tends to be considerably accurate relative to the experimental values for the MN12SX density functional.

Otherwise, the ability of the studied molecules in acting as efficient inhibitors of the formation of Advanced Glycation Endproducts (AGEs) (perhaps as nutraceuticals), which constitutes a useful knowledge for the development of drugs for fighting Diabetes, Alzheimer and Parkinson diseases.

Finally, the bioactivity scores for the four standard aromatic amino acids are predicted through different methodologies.

Acknowledgements

This work has been partially supported by CIMAV, SC and Consejo Nacional de Ciencia y Tecnologia (CONACYT, Mexico) through Grant 219566- 2014 for Basic Science Research. Daniel Glossman-Mitnik conducted this work while a Visiting Lecturer at the University of the Balearic Islands from which support is gratefully acknowledged. Norma Flores-Holguin and Daniel Glossman-Mitnik are researchers of CIMAV and CONACYT. This work was cofunded by the Ministerio de Economia y Competitividad (MINECO) and the European Fund for Regional Development (FEDER) (CTQ2014-55835-R

References

1. La Barre S and Kornprobst JM. Outstanding Marine Molecules. Wiley-Blackwell, Weinheim, 2014.

2. Kim S K. Marine Proteins and Peptides - Biological Activities and Applications. Wiley-Blackwell, Chichester, UK, 2013.

3. Rekka E and Kourounakis P. Chemistry and Molecular Aspects of Drug Design and Action. CRC Press, Boca Raton, 2008.

4. N'aray-Szab'o G and Warshel A. Computational Approaches to Biochemical Reactivity. Kluwer Academic Publishers, New York, 2002.

5. Parr R, Yang W. Density-Functional Theory of Atoms and Molecules. Oxford University Press, New York, 1989.

6. Geerlings P, De Proft F, Langenaeker W. Conceptual Density Functional Theory. Chem Rev. 2003;103(5):1793-1873.

7. Ayers P, Parr R. The Variational Principles for Describing Chemical Reactions: The Fukui Function and Chemical Hardness Revisited. J Am Chem Soc. 2000;122(9):2010-2018.

8. Poater A, Saliner AG, Carbó-Dorca R, Poater J, Solà M, Cavallo L, et al. Modeling the Structure-Property Relationships of Nanoneedles: A Journey Toward Nanomedicine. J Comput Chem. 2009;30(2):275-284. doi: 10.1002/jcc.21041

9. Poater A, Gallegos Saliner A, Solà M, Cavallo L, Worth AP. Computational methods to predict the reactivity of nanoparticles through structure-property relationships. Expert Opin Drug Deliv. 2010;7(3):295-305. doi: 10.1517/17425240903508756

10. Frau J, Glossman-Mitnik D. Chemical Reactivity Theory Study of Advanced Glycation Endproduct Inhibitors. Molecules. 2017;22(1):226.

11. Gupta GK and Kumar V. Chemical Drug Design. Walter de Gruyter GmbH, Berlin, 2016.

12. Gore M and Jagtap UB, Computational Drug Discovery and Design. Springer Science+Business Media, LLC, New York, 2018.

13. Frau J, Glossman-Mitnik D. Molecular Reactivity and Absorption Properties of Melanoidin Blue-G1 through Conceptual DFT. Molecules. 2018;23(3):E559. doi: 10.3390/molecules23030559

14. Frau J, Glossman-Mitnik D. Conceptual DFT Study of the Local Chemical Reactivity of the Dilysyldipyrrolones A and B Intermediate Melanoidins. Theoretical Chemistry Accounts. 2018;137(5):67.

15. Frau J, Glossman-Mitnik D. Conceptual DFT Study of the Local Chemical Reactivity of the Colored BISARG Melanoidin and Its Protonated Derivative. Front Chem. 2018;6(136):1-9. doi: 10.3389/fchem.2018.00136

16. Frau J, Glossman-Mitnik D. Molecular Reactivity of some Maillard Reaction Products Studied through Conceptual DFT. Contemporary Chemistry. 2018;1(1):1-14.

17. Frau J, Glossman-Mitnik D. Computational Study of the Chemical Reactivity of the Blue-M1 Intermediate Melanoidin. Computational and Theoretical Chemistry. 2018;1134:22-29.

18. Frau J, Glossman-Mitnik D. Chemical Reactivity Theory Applied to the Calculation of the Local Reactivity Descriptors of a Colored Maillard Reaction Product. Chemical Science International Journal. 2018;22(4):1-14.

19. Frau J, Glossman-Mitnik D. Blue M2: An Intermediate Melanoidin Studied via Conceptual DFT. J Mol Model. 2018;24(6):138. doi: 10.1007/s00894-018-3673-0

20. Frau J, Flores-Holguin N, Glossman-Mitnik D. Chemical Reactivity Properties, pKa Values, AGEs Inhibitor Abilities and Bioactivity Scores of the Mirabamides A–H Peptides of Marine Origin Studied by Means of Conceptual DFT. Mar Drugs. 2018;16(9):302-319. doi: 10.3390/md16090302

21. Lewars E. Computational Chemistry - Introduction to the Theory and Applications of Molecular and Quantum Mechanics. Kluwer Academic Publishers, Dordrecht, 2003.

22. Young D. Computational Chemistry - A Practical Guide for Applying Techniques to Real-World Problems. John Wiley & Sons, New York, 2001.

23. Jensen F. Introduction to Computational Chemistry. 2nd Edition, John Wiley & Sons, Chichester, England, 2007.

24. Cramer. Essentials of Computational Chemistry - Theories and Models, 2nd Edition, John Wiley & Sons, Chichester, England, 2004.

25. Iikura H, Tsuneda T, Yanai T, Hirao K. A Long-Range Correction Scheme for Generalized-Gradient-Approximation Exchange Functionals. The Journal of Chemical Physics. 2001;115(8):3540-3544.

26. Chai JD, Head-Gordon M. Long-Range Corrected Hybrid Density Functionals with Damped Atom-Atom Dispersion Corrections. Physical Chemistry Chemical Physics. 2008;10:6615-6620.

27. Yanai T, Tew DP, Handy NC. A New Hybrid Exchange-Correlation Functional Using the Coulomb-Attenuating Method (CAM-B3LYP). Chemical Physics Letters. 2004;393(1-3):51-57.

28. Heyd J, Scuseria GE. Efficient Hybrid Density Functional Calculations in Solids: Assessment of the Heyd-Scuseria-Ernzerhof Screened Coulomb Hybrid Functional. J Chem Phys. 2004;121(3):1187-1192. doi: 10.1063/1.1760074

29. Stein T, Kronik L, Baer R. Reliable Prediction of Charge Transfer Excitations in Molecular Complexes Using Time-Dependent Density Functional Theory. J Am Chem Soc. 2009;131(8):2818-2820. doi: 10.1021/ja8087482

30. Stein T, Kronik L, Baer R. Prediction of Charge-Transfer Excitations in Coumarin-Based Dyes Using a Range-Separated Functional Tuned From First Principles. J Chem Phys. 2009;131(24):244119. doi: 10.1063/1.3269029

31. Stein T, Eisenberg H, Kronik L, Baer R. Fundamental Gaps in Finite Systems from Eigenvalues of a Generalized Kohn-Sham Method. Phys Rev Lett. 2010;105(26):266802-266804. doi: 10.1103/PhysRevLett.105.266802

32. Karolewski A, Stein T, Baer R, Kummel S. Communication: Tailoring the Optical Gap in Light-Harvesting Molecules. J Chem Phys. 2011;134(15):151101-151105. doi: 10.1063/1.3581788

33. Kuritz N, Stein T, Baer R, Kronik L. Charge-Transfer-Like $\pi \rightarrow \pi^*$ Excitations in Time-Dependent Density Functional Theory: A Conundrum and Its Solution. J Chem Theory Comput. 2011;7(8):2408-2415. doi: 10.1021/ct2002804

34. Ansbacher T, Srivastava HK, Stein T, Baer R, Merkx M, Shurki A. Calculation of Transition Dipole Moment in Fluorescent Proteins-Towards Efficient Energy Transfer. Phys Chem Chem Phys. 2012;14(12):4109-4117. doi: 10.1039/c2cp23351g

35. Kronik L, Stein T, Refaely-Abramson S, Baer R. Excitation Gaps of Finite-Sized Systems from Optimally Tuned Range-Separated Hybrid Functionals. J Chem Theory Comput. 2012;8(5):1515-1531. doi: 10.1021/ct2009363

36. Stein T, Autschbach J, Govind N, Kronik L, Baer R. Curvature and Frontier Orbital Energies in Density Functional Theory. Journal of Physical Chemistry Letters. 2012;3(24):3740-3744. doi: 10.1021/jz3015937

37. Halgren TA. Merck Molecular Force Field. I. Basis, Form, Scope, Parameterization, and Performance of MMFF94. Journal of Computational Chemistry. 1996;17(5-6):490-519.

38. Halgren TA. Merck Molecular Force Field. II. MMFF94 van der Waals and Electrostatic Parameters for Intermolecular Interactions. Journal of Computational Chemistry. 1996;17(5-6):520-552.

39. Halgren TA. MMFF VI. MMFF94s Option for Energy Minimization Studies. Journal of Computational Chemistry. 1999;20(7):720-729.

40. Halgren TA, Nachbar RB. Merck Molecular Force Field. IV. Conformational Energies and Geometries for MMFF94. Journal of Computational Chemistry. 1996;17(5-6):587-615.

41. Halgren TA. Merck Molecular Force field. V. Extension of MMFF94 Using Experimental Data, Additional Computational Data, and Empirical Rules. Journal of Computational Chemistry. 1996;17(5-6):616-641.

42. Frisch MJ, Trucks GW, Schlegel HB, Scuseria GE, Robb MA and Cheeseman R et al. Gaussian 09 Revision E.01, Gaussian Inc., Wallingford CT, 2016.

43. Weigend F, Ahlrichs R. Balanced Basis Sets of Split Valence, Triple Zeta Valence and Quadruple Zeta Valence Quality for H to Rn: Design and Assessment of Accuracy. Phys Chem Chem Phys. 2005;7(18):3297-3305. doi: 10.1039/b508541a

44. Weigend F. Accurate Coulomb-fitting Basis Sets for H to Rn. Phys Chem Chem Phys. 2006;8(9):1057-1065. doi: 10.1039/b515623h

45. Marenich AV, Cramer CJ, Truhlar DG. Universal Solvation Model Based on Solute Electron Density and a Continuum Model of the Solvent Defined by the Bulk Dielectric Constant and Atomic Surface Tensions. J Phys Chem B.1 2009;113(18):6378-6396. doi: 10.1021/jp810292n

46. Henderson TM, Izmaylov AF, Scalmani G, Scuseria GE. Can Short-Range Hybrids Describe Long-Range-Dependent Properties? J Chem Phys. 2009;131(4):044108. doi: 10.1063/1.3185673

47. Peverati R, Truhlar DG. Improving the Accuracy of Hybrid Meta-GGA Density Functionals by Range Separation. The Journal of Physical Chemistry Letters. 2011;2(21):2810-2817.

48. Peverati R, Truhlar DG. Screened-Exchange Density Functionals with Broad Accuracy for Chemistry and Solid-State Physics. Physical Chemistry Chemical Physics. 2012;14(47):16187-16191.

49. Egger DA, Weissman S, Refaely-Abramson S, Sharifzadeh S, Dauth M, Baer R, et al. Outer-Valence Electron Spectra of Prototypical Aromatic Heterocycles From an Optimally Tuned Range-Separated Hybrid Functional. J Chem Theory Comput. 2014;10(5):1934-1952. doi: 10.1021/ct400956h

50. Foster ME, Wong BM. Nonempirically Tuned Range-Separated DFT Accurately Predicts Both Fundamental and Excitation Gaps in DNA and RNA Nucleobases. Journal of Chemical Theory and Computation. 2012;8(8):2682-2687.

51. Foster ME, Azoulay JD, Wong BM, Allendorf MD. Novel Metal–Organic Framework Linkers for Light Harvesting Applications. Chemical Science. 2014;5(5):2081-2090.

52. Jacquemin D, Moore B, Planchat A, Adamo C, Autschbach J. Performance of an Optimally Tuned Range-Separated Hybrid Functional for 0-0 Electronic Excitation Energies. J Chem Theory Comput. 2014;10(4):1677-1685. doi: 10.1021/ct5000617

53. Karolewski A, Kronik L, Kummel S. Using Optimally Tuned Range Separated Hybrid Functionals in Ground-State Calculations: Consequences and Caveats. J Chem Phys. 2013;138(20):204115. doi: 10.1063/1.4807325

54. Koppen JV, Hapka M, Szczeniak MM, Chalasinski G. Optical Absorption Spectra of Gold Clusters Au(n) (n = 4, 6, 8,12, 20) From Long-Range Corrected Functionals with Optimal Tuning. J Chem Phys. 2012;137(11):114302.

55. Lima IT, Prado Ada S, Martins JB, de Oliveira Neto PH, Ceschin AM, da Cunha WF, et al. Improving the Description of the Optical Properties of Carotenoids by Tuning the Long-Range Corrected Functionals. J Phys Chem A. 2016;120(27):4944-4950. doi: 10.1021/acs.jpca.5b12570

56. Manna AK, Lee MH, McMahon KL, Dunietz BD. Calculating High Energy Charge Transfer States Using Optimally Tuned Range- Separated Hybrid Functionals. J Chem Theory Comput. 2015;11(3):1110-1117. doi: 10.1021/ct501018n

57. Li BM, Autschbach J. Longest-Wavelength Electronic Excitations of Linear Cyanines: The Role of Electron Delocalization and of Approximations in Time-Dependent Density Functional Theory. J Chem Theory Comput. 2013;9(11):4991-5003. doi: 10.1021/ct400649r

58. Niskanen M, Hukka TI. Modeling of Photoactive Conjugated Donor-Acceptor Copolymers: the Effect of the Exact HF Exchange in DFT Functionals on Geometries and Gap Energies of Oligomer and Periodic Models. Phys Chem Chem Phys. 2014;16(26):13294-13305.

59. Pereira TL, Leal LA, da Cunha WF, Timoteo de Sousa Junior R, Ribeiro Junior LA, Antonio da Silva Filho D. Optimally Tuned Functionals Improving the Description of Optical and Electronic Properties of the Phthalocyanine Molecule. J Mol Model. 2017;23(3):71. doi: 10.1007/s00894-017-3246-7

60. Phillips H, Zheng S, Hyla A, Laine R, Geva E, Dunietz BD, et al. Ab Initio Calculation of the Electronic Absorption of Functionalized Octahedral Silsesquioxanes via Time-Dependent Density Functional Theory with Range-Separated Hybrid Functionals. J Phys Chem A. 2012;116(4):1137-1145. doi: 10.1021/jp208316t

61. Phillips H, Geva E, Dunietz BD. Calculating Off-Site Excitations in Symmetric Donor-Acceptor Systems via Time-Dependent Density Functional Theory with Range-Separated Density Functionals. J Chem Theory Comput. 2012;8(8):2661-2668. doi: 10.1021/ct300318g

62. Refaely-Abramson S, Baer R, Kronik L. Fundamental and Excitation Gaps in Molecules of Relevance for Organic Photovoltaics from an Optimally Tuned Range-Separated Hybrid Functional. Physical Review B. 2011;84(7):075144-075148.

63. Sun H, Autschbach J. Electronic Energy Gaps for π-Conjugated Oligomers and Polymers Calculated with Density Functional Theory. J Chem Theory Comput. 2014;10(3):1035-1047. doi: 10.1021/ct4009975

64. Becke AD. Vertical Excitation Energies from the Adiabatic Connection. J Chem Phys. 2016;145(19):194107. doi: 10.1063/1.4967813

65. Baerends EJ, Gritsenko OV, van Meer R. The Kohn-Sham Gap, the Fundamental Gap and the Optical Gap: The Physical Meaning of Occupied and Virtual Kohn-Sham Orbital Energies. Phys Chem Chem Phys. 2013;15(39):16408-16425. doi: 10.1039/c3cp52547c

66. Van Meer R, Gritsenko OV, Baerends EJ. Physical Meaning of Virtual Kohn-Sham Orbital's and Orbital Energies: An Ideal Basis for the Description of Molecular Excitations. J Chem Theory Comput. 2014;10(10):4432–4441. doi: 10.1021/ct500727c

67. Taniguchi M, Lindsey JS. Database of Absorption and Fluorescence Spectra of >300 Common Compounds for use in PhotochemCAD. Photochem Photobiol. 2018;94(2):290-327. doi: 10.1111/php.12860

68. Parr RG, Szentpaly LV, Liu S. Electrophilicity Index. Journal of the American Chemical Society. 1999;121(9):1922-1924.

69. Gazquez JL, Cedillo A, Vela A. Electrodonating and Electroaccepting Powers. J Phys Chem A. 2007;111(10):1966-1970. doi: 10.1021/jp065459f

70. Chattaraj PK, Chakraborty A, Giri S. Net Electrophilicity. J Phys Chem A. 2009;113(37):10068-10074. doi: 10.1021/jp904674x

71. Ahmed N. Advanced Glycation Endproducts - Role in Pathology of diabetic Complications. Diabetes Res Clin Pract. 2005;67(1):3-21. doi: 10.1016/j.diabres.2004.09.004

72. Rahbar S, Figarola JL. Novel Inhibitors of Advanced Glycation Endproducts. Arch Biochem Biophys. 2003;419(1):63-79.

73. Peyroux J, Sternberg M. Advanced glycation end products (AGEs): Pharmacological Inhibition in diabetes. Pathol Biol (Paris). 2006;54(7):405-419. doi: 10.1016/j.patbio.2006.07.006

74. Domingo LR, Perez P. The Nucleophilicity N index in Organic Chemistry. Org Biomol Chem. 2011;9(20):7168-7175. doi: 10.1039/c1ob05856h

75. Leeson P. Drug Discovery: Chemical Beauty Contest. Nature. 2012;481(7382):455-456.

76. Martin YC. A Bioavailability Score. Journal of Medicinal Chemistry. 2005;48(9):3164-3170.

26

Effects of Alkaloids of *Cocos Nucifera* Husk fiber on Some Selected Enzymes in the Albino Mice

Oluwayemi Joshua Bamikole[1], Godwin Okwori Adikwu[2]

[1]*Institute of Child Health, University of Ibadan. Nigeria*

[2]*Department of Biochemistry, University of Ilorin. Nigeria*

Corresponding author: *Oluwayemi Joshua Bamikole, Institute of Child Health, University of Ibadan, Nigeria , E-mail: bamikoleyemi@gmail.com*

Abstract

Cocos nucifera is a medicinal plant used in Nigeria for treatment of malaria. In this study, the effects of administration of alkaloids of *C. nucifera* husk fiber on the activities of some selected albino mice enzymes were investigated. The mice were administered the alkaloids orally at varying concentrations of 31.25, 62.50, 125, 250 and 500 mg/kg body weights while the controls were administered with distilled water. At the end of 7 days, the animals were sacrificed, and the assays for the activities of Alkaline Phosphatase (ALP), Gamma-glutamyl transferase (GGT), Aspartate Amino transferase (AST), Alanine Amino transferase (ALT), Glutamate Dehydrogenase (GLDH) in serum and kidney were carried out. From the results, there was significant decrease ($p < 0.5$) in the activities of ALP and GGT in the kidney at doses of 125, 250 and 500 mg/kg body weights while there was significant increase ($p < 0.05$) in their activities at 31.25, 62.50 and 125 mg/kg body weights in the serum. Generally, there was no significant change ($p > 0.05$) in the activities of AST, ALP and GLDH in the serum and kidney. The findings in the study suggest that the alkaloids have adverse effect on the integrity of plasma membranes of kidney cells at higher doses which can cause a significant adverse effect on the heart.

Introduction

Alkaloids generally exert pharmacological activities particularly in mammals such as humans. Even today many of our most commonly used drugs are alkaloids from natural sources and new alkaloidal drugs are still being developed for clinical use. The decoctions of root stem barks, leaves and fruits of plants are resorted to and are used extensively as anti malarial remedies basically because of its cheaper costs without attention being paid to their possible toxicological consequence [2]. In this study, the effects of the alkaloids of *Cocos nucifera* husk fiber ethanolic extract (West African tall variety) on selected enzymes were evaluated in relationship to toxicity and body function (Kidney, Heart).

Cocos Nucifera

The coconut palm (*Cocos nucifera*) is in the tropics. It is particularly important in the low islands of the Pacific where, in the absence of land-based natural resources, it provides almost all the necessities of life. It has been nicknamed the 'tree of heaven' and 'tree of life'. Today, it remains an important economic and subsistence crop in many small Pacific island states [4].

Coconut is believed to have its origins in the Indo-Malayan region, from whence it spread throughout the tropics. Its natural habitat is the narrow sandy coast, but it is now found on soils ranging from pure sand to clays and from moderately acidic to alkaline. It thrives under warm and humid conditions but tolerates short periods of temperatures below 21°C (70°F) [4].

Common Uses of Coconut Husk

Coconut husks are used as mulch. They decompose slowly but are a good source of potassium. In low-rainfall areas, husks are buried in trenches to serve as water reservoirs during drought. Decomposed husk is placed in holes when planting coconut seedlings on sandy soils. Placed on the ground convex surface up, husks are commonly used as mulch around coconut seedlings and other plants to control weeds [4].

Materials and Methods

The extract was prepared according to the method of *Adebayo et, al* [1]. The samples were shade dried at room temperature and pulverized into powder. 500 grams (500 g) of the powder was percolated in 2.1 L of n-hexane for 72 hours in a tightly stoppered glass container. This was shaken at intervals. The resulting mixture was filtered with Whatmann filter paper (110mm). The filtrate was then concentrated under pressure using rotary evaporator at 40 0C, thereby generating a semisolid extract fraction of approximately 162ml, used for subsequent extraction.

The alkaloid was extracted using the method of Manske [8]. To 162ml of the semisolid extract was 22ml of 1M HCl added, after which it was basified by the addition of 6.5 ml of 5M NaOH yielding a white precipitate alongside with the solidification of the extract. The basified solution together with 18ml of 0.9% NaCl and 125ml of chloroform were separated with a separating funnel thrice, at each interval producing an upper (aqueous) layer and a lower (organic) layer. A total of three organic layers containing the alkaloids were concentrated on a water bath at a temperature of 37^0C. The percentage yield was 0.335% alkaloid.

Animal Grouping and Administration of Alkaloids

The forty-eight mice were randomly assigned into five groups, of eight Mice each. Daily administration of 200µl of the alkaloid of ethanolic extract fraction of Cocos nucifera was done orally for seven days as follows:

Group A (Control): Administered appropriate volume of distilled water solution.

Group B: Administered 31.25 mg/Kg body weight of extract fraction.

Group C: Administered 62.5 mg/Kg body weight of extract fraction.

Group D: Administered 125 mg/Kg body weight of extract fraction.

Group E: Administered 250 mg/Kg body weight of extract fraction.

Group F: Administered 500 mg/Kg body weight of extract fraction.

The extract fraction was dissolved in warm distilled water solution to form a suspension before administration.

Toxicological Studies

At the end of the 7-day experimental period, the Mice were sacrificed by slight diethyl ether anaesthesia, the neck area was quickly cleared of fur and the jugular veins exposed, from which blood was collected into EDTA bottle to prevent clotting. The EDTA blood sample was centrifuged at 3000 rpm for 10 minutes and the serum pipetted out. This was stored frozen until needed for analysis.

Also, the mice were quickly dissected and the kidneys were isolated, cleaned of blood stains, weighed and suspended in an ice-cold 0.25M sucrose solution (an isotonic solution) to maintain the organs' integrity. Each organ was homogenized separately, using mortar and pestle, in ice-cold 0.25 M sucrose solution (1:6, w/v). The homogenates were frozen overnight to allow cell lysis. This ensures maximum release of enzymes.

The Kidney-body ratio, Total Protein Concentration, Total Protein concentration, Alkaline Phosphatase activity Aspartate Amino transferase activity, Alanine Amino transferase activity, AST activity, Gamma-glutamyl Transferase Activity, Glutamate Dehydrogenase Activity in serum and tissue homogenate, ALT activity in tissue supernatant and serum were determined.

Statistical Analysis

Experimental data are presented as Mean ± Standard Error of Mean (SEM). Statistical analysis was implemented using computer software SPSS 20.0 version statistical package program (SPSS, Chicago, IL). One-way analysis of variance was used to compare variables among the different groups. Level of significance (Post hoc comparisons) among the various treatments was determined by Duncan's Multiple Range Test. The values were considered statistically significant at p<0.05.

Results

Effect of alkaloids of Cocos nucifera Husk Fiber on kidney to Body Weight Ratio

The administration of alkaloid of Cocos nucifera husk fibre (WAT) at all doses investigated in this study did not cause any significant change at (p > 0.05) in the kidney-body weight ratio of mice compared to control Table 1.

Table 1: Effect of alkaloids of Cocos nucifera husk fibre on Kidney to Body Weight Ratio (%) of mice

Treatment (mg/Kg body weight)	Kidney/Body Weight (%)
CONTROL(distilled water)	1.096±0.0329[ab]
31.25	1.191±0.0244[a]
62.50	1.245±0.0379[ab]
125	1.061±0.0705[b]
250	1.204±0.0374[ab]
500	1.307±0.125[ab]

Values are expressed as Mean±SEM (n=8). Values in each column with different superscript are significantly different (p < 0.05).

Alkaline Phosphatase

The alkaloid fraction significantly increased (p <0.05) kidney ALP activity at the dose of 31.25 mg/Kg body weight, decreased it at doses of 125 mg/kg, 250mg/kg and 500 mg/kg body weights while it did not significantly alter (p>0.05) it at 62.50 mg/kg body weight dose compared to control. Serum ALP activity was significantly increased (P<0.05) at doses of 62.5, 125 mg/kg body weights, deceased at 500 mg/kg body weight of the alkaloids but was not significantly altered (P>0.05) at other doses compared to the control Table 2.

Gamma-glutamyl transferase

The activity of gamma-glutamyl transferase in albino mice kidney was significantly decreased (p<0.05) at doses of 62.50mg/kg and 125mg/kg body weights of alkaloids administered while it was not significantly changed (p>0.05) at other doses compared to the control. Serum GGT activity increased significantly (p<0.05) at doses of 31.25mg/kg and 62.50mg/kg body weights and decreased significantly (p<0.05) at other doses compared to the control Table 3.

Aspartate Amino Transferase

The extract fraction significantly increased (P<0.05) kidney and serum aspartate amino transferase activity at the dose of 31.25mg/kg and 62.50 body weights of alkaloids respectively while it was not significantly altered (P>0.05) at other doses of the alkaloids compared to control Table 4.

Table 2: Alkaline Phosphatase activities in the kidney and serum of mice administered alkaloids of Cocos nucifera husk fibre

GROUP	Specific Activity (nmol/min/mg protein)	
	Kidney	Serum
CONTROL(distilled water)	254.58±13.13[b]	368.7943±8.5[b]
31.25mg/kg b.wt	548.37±59.74[c]	346.6312±27.8[b]
62.50mg/kg b.wt	275.45±34.28[b]	734.2199±29.2[d]
125mg/kg b.wt	164.86±8.911[a]	518.2624±38.1[c]
250mg/kg b.wt	145.41±12.65[a]	331.9149±27.1[b]
500mg/kg b.wt	90.72±11.57[a]	217.4941±24.4[a]

Values are expressed as Mean±SEM (n=8). Values in each column with different superscript are significantly different ($p < 0.05$).

Table 3: Gamma-glutamyl transferase activities in kidney and serum of mice administered alkaloids of Cocos nucifera husk fibre

GROUP	Specific Activity (U/L)	
	Kidney	Serum
CONTROL(distilled water)	5.211±0.58[c]	2817.80±568.60[c]
31.25mg/kg b.wt	4.053±0.58[bc]	10036.00±4034.90[e]
62.50mg/kg b.wt	2.316±1.16[ab]	6098.80±4888.50[de]
125mg/kg b.wt	1.158±0.00[a]	617.60±77.20[a]
250mg/kg b.wt	4.632±0.00[c]	1273.80±176.90[b]
500mg/kg b.wt	4.632±0.00[c]	231.60±66.80[a]

Values are expressed as Mean±SEM (n=8). Values in each column with different superscript are significantly different ($p < 0.05$).

Table 4: Aspartate Aminotransferase activities in kidney and serum of mice administered alkaloids of Cocos nucifera husk fibre

GROUP	Specific Activity (nmol/ml/mg protein)	
	Kidney	Serum
CONTROL(distilled water)	0.29±0.05[a]	0.007±0.002[a]
31.25mg/kg b.wt	0.43±0.05[b]	0.011±.002[ab]
62.50mg/kg b.wt	0.29±0.05[a]	0.017±0.002[b]
125mg/kg b.wt	0.22±0.02[a]	0.013±0.002[ab]
250mg/kg b.wt	0.22±0.03[a]	0.013±0.003[ab]
500mg/kg b.wt	0.27±0.03[a]	0.012±0.002[ab]

Values are expressed as Mean±SEM (n=8). Values in each column with different superscript are significantly different ($p < 0.05$).

Alanine Amino transferase

TThe alkaloids of *Cocos nucifera* husk fibre at all doses administered did not cause any significant change in alanine amino transferase activities in the kidney and serum   ($p > 0.05$) at all doses when compared with control Table 5.

Glutamate Dehydrogenase

The alkaloids at a dose of 250mg/kg body weight caused a significant increase ($p < 0.05$) in kidney's glutamate dehydrogenase activity but did not alter the activity at other doses compared to the control Table 6.

Table 5: Alanine Aminotransferase activities in the kidney and serum of mice administered alkaloids of Cocos nucifera husk fibre

GROUP	Specific Activity (nmol/ml/mg protein)	
	Kidney	Serum
CONTROL(distilled water)	0.27±0.05[ab]	0.03±0.006[a]
31.25mg/kg b.wt	0.15±0.01[a]	0.04±.008[a]
62.50mg/kg b.wt	0.19±0.04[ab]	0.05±.005[a]
125mg/kg b.wt	0.30±0.02[b]	0.04±0.002[a]
250mg/kg b.wt	0.27±0.06[ab]	0.04±.006[a]
500mg/kg b.wt	0.18±0.04[ab]	0.04±.007[a]

Values are expressed as Mean±SEM (n=8). Values in each column with different superscript are significantly different ($p > 0.05$).

Table 6: Glutamate Dehydrogenase activity in kidney of mice administered alkaloids of Cocos nucifera husk fibre

Group	Specific Activity (µmol/min/mg protein)
	Kidney
CONTROL(distilled water)	0.145±0.020[a]
31.25mg/kg b.wt	0.171±0.029[a]
62.50mg/kg b.wt	0.330±0.050[a]
125mg/kg b.wt	0.375±0.034[a]
250mg/kg b.wt	0.726±0.176[b]
500mg/kg b.wt	0.304±0.037[a]

Values are expressed as Mean±SEM (n=8). Values in each column with different superscript are significantly different ($p < 0.05$).

Histological Observation of the Kidney

In the photomicrographs, the presence of few renal corpuscles confirms that the area of the kidney sectioned and stained is the renal cortex. The renal cortex is easily identified even at low magnification by the presence of renal corpuscles, which are absent in the renal medulla. However, the bowman capsule shown in B reveals the obliteration of the bowman space by the alkaloids at 31.25 mg/kg body weight, possibly due to inflammation.

Discussion

Organ-Body weight ratio

Organ-body weight ratios are normally investigated to determine whether the size of the organ has changed in relation to the weight of the whole animal [1]. Changes in the organ body weight ratio may be an indication of cell constriction or inflammation since the cells are the unit components of the organs. The constriction in the organ may occur as a result of loss of fluid from the organ due to damage, while increase in organ-body weight ratio may suggest inflammation (Moore and Dalley,

1999). Thus, since there was no significant change in the kidney-body weight ratio as a result of administration of alkaloids of *Cocos nucifera* husk fibre in this study, it may imply that the alkaloids do not cause inflammation or constriction in the kidney.

Enzyme Studies

The measurement of the activities of the marker enzymes in tissues and body fluid plays a significant and important role in diagnosis, disease investigation and in the assessment of drug or plant extract for safety/toxicity risk [7]. Tissue enzyme assay can also indicate tissue cellular damage long before structural damage can be picked by conventional histological techniques. Such measurement can also give an insight to the site of cellular tissue damage as a result of assault by the plant extract [2].

Membrane Bound Enzymes

Alkaline phosphatase (ALP) has been reported to be a marker enzyme for plasma membrane and endoplasmic reticulum (Wright and Plummer, 1974), it is often used to access the integrity of the plasma membrane such that any alteration in the activity of the enzyme in the tissue and serum would indicate likely damage to the plasma membrane [14]. Any perturbation in the membrane caused by interaction with xenobiotics could lead to alteration in ALP activity [9]. Decrease in ALP activity caused by stressors probably indicates an altered transport of phosphate and an inhibitory effect on the cell growth and proliferation (Goldfischer, 1964) [5]. Inhibition of ALP activity reflects alteration in protein synthesis and uncoupling of oxidative phosphorylation [13].

Gamma-glutamyl transferase (GGT) is predominantly used as a diagonistic marker for liver disease. Elevated serum GGT serum activity can be found in diseases of the liver, biliary track and pancreas [3].

Alkaline Phosphatase

The decrease in activity of ALP in the kidney at higher doses of alkaloids administered when compared with the control might be due to the *in situ* inactivation of the enzyme molecule by the alkaloid or inhibition of the enzyme activity at the cellular level either at the transcriptional or translational levels or damage to the cell membrane resulting to leakage. The increased serum activity may be due to leakage from kidney tissue and the decrease observed at a higher dose may be as a result of increased clearance from the blood.

Gamma-Glutamyl Transferase

Decreased activity of serum and kidney's gamma-glutamyl transferase observed may be as a result of decreased GGT synthesis by inhibition at either transcriptional or translational levels or allosteric inhibition of the enzyme in situ. It could also be as a result of leakage from membrane damage. This can be observed in the increased serum activity at lower doses, though it became reduced at higher doses possibly due to increased clearance.

Amino Acid-Metabolizing Enzymes

It is well known that amino transferases play crucial role in amino acid metabolism and in providing intermediates for gluconeogenesis [7]. Amino transferases are active both in the cytoplasm and mitochondria of tissue cells where they form an important bridge between protein and carbohydrate metabolism [10]. Amino transferases respond to any stress or altered physiological condition [6]. Stress generally is known to elevate amino transferase activity [12]. Under stress conditions, animals need more energy resulting in higher demand for carbohydrate and their precursors to keep the glycolytic pathway and TCA cycle at sustained levels [11].

Aspartate Amino Transferase and Alanine Amino Transferase

The kidney is known as the secondary site of amino acid catabolism. Generally, AST and ALT activities were not significantly changed in the kidney and serum. This implies that the energy needs of the kidney remain fairly unchanged and the alkaloids did not adversely affect the inter conversion of the amino acids to their respective intermediates in carbohydrate metabolism.

Glutamate Dehydrogenase

Generally, GLDH activity was not affected by the alkaloid. The increased activity observed at 250 mg/kg body weight of alkaloids in the kidney may be as a result of increased synthesis of the enzyme or activation of the enzyme in situ. This may lead to increased catabolism of amino acids at this dose.

Conclusion

The present study evaluated the toxicity potentials of alkaloid fraction of Cocos nucifera husk fiber ethanolic extract. The results of the study suggest the following:

i.   The alkaloids of Cocos nucifera husk fiber may adversely affect the plasma membrane of kidney at higher doses which might cause an adverse effect on the heart.

ii.   They may have little or no adverse effect on amino acid metabolism in the kidney at all doses.

Thus, the consumption of the ethanolic extract fraction of Cocos nucifera husk fiber at higher doses may cause kidney dysfunction. Efforts should be directed at the isolation of the specific active principles (alkaloid) in order to reduce the risk of their toxicity.

Acknowledgement

We acknowledge Dr. Adebayo, Adewumi Muyiwa, and the Lab technicians of the Department of Biochemistry, University of Ilorin.Nigeria

Funding

None

References

1. Adebayo JO, Balogun EA, Malomo SO, Soladoye AO, Olatunji LA, Kolawole OM, et al. Antimalarial Activity of Cocos nucifera Husk Fibre: Further Studies. Evidence-Based Complementary and Alternative Medicine. 2013;2013:1-9.

2. Adebayo JO & Krettli AU. Potential anti malarials from Nigerian plants: a review. J Ethnopharmacol. 2011;133(2):289-302. doi:10.1016/j.jep.2010.11.024.

3. Betro MG, Oon RC, Edward JB. "Gamma-glutamyl transpeptidase in diseases of the liver and bone". Am. J. Clin. Pathol. 1973; 60 (5): 672-678. doi:10.1093/ajcp/60.5.672.

4. Chan E, and Elevitch C R. Cocos nucifera(coconut). Species Profiles for Pacific Island Agroforestry. Permanent Agriculture Resources (PAR). Hōlualoa, Hawai; 2006; 2(1): 1-27.

5. Engstrom L. Studies on bovine liver alkaline phosphatase, phosphate incorporation. Biochimicaet Biophysica Acta. 1964; 92: 71-74.

6. Knox WE. and Greengard O. An Introduction to enzyme physiology. In: Advanced Enzyme Regulation. Pergamon Press,New York, London. 1965; 257-248.

7. Malomo S O. Toxicological implication of ceftriaxone administration in rats. Nig. Journal Biochem Molecular Biol. Is 2000;15(1): 33-38. doi:10.1155/2013/803835.

8. Manske R H. The Alkaloids. Chemistry and physiology, New York Academic press. P.675. 1965; 8.

9. Molina R, Morenol, Pichardo S. et.al. Acid and alkaline phosphatase activities and pathological changes induced in Tilapia fish (Oreochromis sp.) exposed subchronically to microcystins from toxic cyanobacterial blooms under laboratory conditions. Toxicon; 2005; 46(7):725–735. doi: 10.1016/j.toxicon.2005.07.012.

10. Rafelson JM E, Hayashi JA. and Beckoro-vainy A. Basic Biochemistry. Macmillan Publishing Co. Inc. New York 1980;121-122. doi:10.1016/0307-4412(80)90131-4.

11. Tiwari S. and Singh A. Piscicidal activity of alcoholic extract of Nerium indicum leaf and their biochemical stress response on fish metabolism. African Journal of Traditional Complementary and Alternative Medicines; 2004; 1: 15-29.doi: 10.4314/ajtcam.v1i1.31092

12. Velisek J, Inlason T, Gomulka P, Svoboda Z, et al. Effects of cypermethin on rainbow trout (Onchorhynchus mykiss). Veterinani Medicina; 2006; 51(10): 469-476.

13. Verma, S., Saxena, M. and Tonk, I. The influence of Ider 20 on the biochemical and enzymes in the liver of Clarias batrachus. Environmental Pollution; 1984; 33: 245–255.

14. Yakubu, M.T. Aphrodisiac and toxicological potentials of aqueous extratct of Fadogia agretis stem in male albino rats. Ph. D. Thesis, University of Ilorin, Nigeria. 2006; 7(4): 399-404.doi: 10.1111/j.1745-7262.2005.00052.x.

27

How Flagella Expression May be Regulated by the Carbon and Energy Source?

Anibal R. Lodeiro*

Laboratory of Interactions between Rhizobia and Soybean (LIRyS). IBBM-Faculty of Sciences. National University of La Plata and CONICET CCT-La Plata, Argentina

Corresponding author: *Anibal R. Lodeiro, IBBM-Faculty of Exact Sciences, UNLP-CONICET. Calles 47 y 115 (1900) La Plata, Argentina. E-mail: lodeiro@biol.unlp.edu.ar*

Abstract

Bradyrhizobium diazoefficiens has two flagellar systems: a subpolar, constitutive system and a lateral, inducible system. Contrary to other bacterial species, the lateral system is induced in liquid medium in response to the carbon and energy source. Since both flagella are moved by the proton-motive force, a relationship between the energy status of the cell and the signal that triggers lateral flagella expression might exist. Here I discuss how this relationship may control the induction of the lateral flagellar system, and its implicancies for improvement of *Bradyrhizobium*-based inoculants for soybean plants.

Keywords: Flagella; *Bradyrhizobium diazoefficiens*; Soyabean; Symbiosis

Introduction

Bradyrhizobium is a soil bacterial genus that includes several species of importance due to their use as biofertilizers for soybean crops worldwide. Among these species, *B. diazoefficiens* and *B. japonicum* stand out as being the most widely employed.

These bacteria fix atmospheric N_2 in symbiosis with soybean plants, by reducing N_2 to NH_4^+ in a reaction catalyzed by bradyrhizobial nitrogenase. The NH_4^+ thus produced is supplied as N-source to the plant, in such a rate that all its N-needs may be satisfied. To this end, the *Bradyrhizobium* bacteria are inoculated to soybean seeds before sowing with the aim that these bacteria infect the roots and develop the N_2-fixing symbiosis inside them. Since the symbiosis only occurs in specialized organs in the roots, seed-inoculated bacteria need moving from the site of inoculation to the sites of infection, and this movement has to occur in the soil, a porous and tortuous medium, which not always contain water enough for bacterial swimming. Therefore, the study of bradyrhizobial motility is of prime importance to improve this ecologically sustainable technology for soybean fertilization.

B. diazoefficiens USDA 110 is the type strain of this species [1], and its genome was completely sequenced in 2002 [2]. In addition, this strain is the most studied biochemically, genetically, and physiologically, as well as in the relevant aspects of its symbiosis with soybean plants. The motility of this strain was characterized in two kinds of bacterial movement: swimming [3-5] and swarming [6]. Both movements are propelled by flagella, while other movements such as e.g. twitching or gliding were not reported. Although there exists evidence of the presence of pili, which constitute the device required for twitching, these appendages were studied only for their role in cell adhesion [7].

Remarkably, *B. diazoefficiens* USDA 110 possesses two entirely different flagellar systems: a subpolar system and a lateral system [4-6]. These systems are encoded in different gene clusters, and it seems that each one possesses its own regulatory system for the control if its expression. Indeed, the expression of the subpolar system seems constitutive in planktonic cells, while the lateral system is inducible [5,6]. Induction of the lateral system was observed as obeying to the carbon and energy source of the growth medium: when the sole carbon and energy source is arabinose, the lateral system is expressed, but it is inhibited when the only carbon and energy source is mannitol [6]. Although several other bacterial species are known to possess inducible lateral flagellar systems, in general the inducer is the medium viscosity or the proximity of a surface, which are perceived by the polar/subpolar flagellar system that under these circumstances behaves as a mechanosensor [8]. However, *B. diazoefficiens* is the only example known where the lateral flagellar system is induced by the carbon and energy source, and therefore the identity of the signal transducer is a complete enigma.

Structure and Rotation of Flagella

The flagellum consists in three main structures: the flagellar filament, the hook, and the basal body, which contains the motor [9]. Although the flagellar motor was not studied in *B. diazoefficiens*, there exists a great deal of knowledge in other species, in particular *Escherichia coli* [10]. The flagellar motor is embedded in the inner cell membrane and has two main rings: a stator formed mainly by the proteins MotA and MotB, and a rotor to which the rest of the flagellum is attached. The rotor is composed mainly by the proteins FliF, FliG, FliM and FliN, which play a central role in flagellar rotation. The Protonmotive Force (PMF) that is generated during cell respiration is the energy source for flagellar motor rotation. Protons pass from the periplasmic space towards the cytoplasm through a channel formed between FliG and MotA/MotB. Recent studies indicate that the rotor contains a ring of 26 FliG subunits faced against an external ring of MotA/

MotB subunits in such a way that an array of negatively charged amino acids in MotA, MotB and FliG interact with the protons that traverse the channel, producing changes in the conformations of the ring proteins, allowing movement of the rotor. These amino acids are disposed in such a way that the passage of protons from the periplasm to the cytoplasm moves the rotor against the stator thus producing a torque sufficient to impulse the cell body into the liquid environment [10-12].

In general, bacterial species that possess two flagellar systems use the PMF to move one of the flagella, while the other is moved by a Na^+ gradient that is also formed between the periplasm and the cytoplasm. However, in *B. diazoefficiens* both flagellar systems seem to be moved by the PMF [4], thus sharing this energy source with the synthesis of ATP. Ultimately, the PMF comes from the oxidation of organic carbon and energy sources, and therefore the availability of energy in the cell might be in connection with the regulation of the lateral flagellum expression by the carbon and energy source.

Catabolism of Arabinose and Mannitol

L-arabinose is catabolized through a pathway that resembles the Entner-Doudoroff (ED) pathway for catabolism of hexoses [13,14] (Figure 1). The first step is oxidation of L-arabinose to

L-arabonate with formation of one mole of NADH per mole of L-arabinose. Then, L-2-keto 3-deoxy arabonate is formed, which splits in pyruvate and glycolaldehyde in a reaction catalyzed by an aldolase. The pyruvate continues through TCA cycle, while glycolaldehyde is oxidized in two sequential steps of NADH-producing reactions to glyoxylate, which is finally converted to formate and oxidized to CO_2 with production of another NADH [13-15]. Hence, a total of eight NADH moles plus one $FADH_2$ mole are produced per mole of arabinose completely oxidized to CO_2.

D-mannitol is oxidized to fructose with production of one NADH mole per mole of mannitol in a reaction catalyzed by mannitol dehydrogenase [16]. Then, the fructose produced may be catabolized by the ED or the Emden-Meyerhof-Parnas (EMP) pathways [17], or the Pentose-Phosphate (PP) pathway [18] with production of 10 additional NADH moles plus two $FADH_2$ moles (ED, EMP pathways), or 11 additional NADH moles plus two $FADH_2$ moles (PP pathway) in the complete oxidation of one mole of fructose to CO_2. In Figure 1, the yields of NADH plus $FADH_2$ with arabinose are compared with those with mannitol catabolized by the ED pathway as an example.

Thus, assuming that 10 H^+ moles are passed to the periplasm per mole of NADH oxidized and 6 H^+ moles are passed per mole of

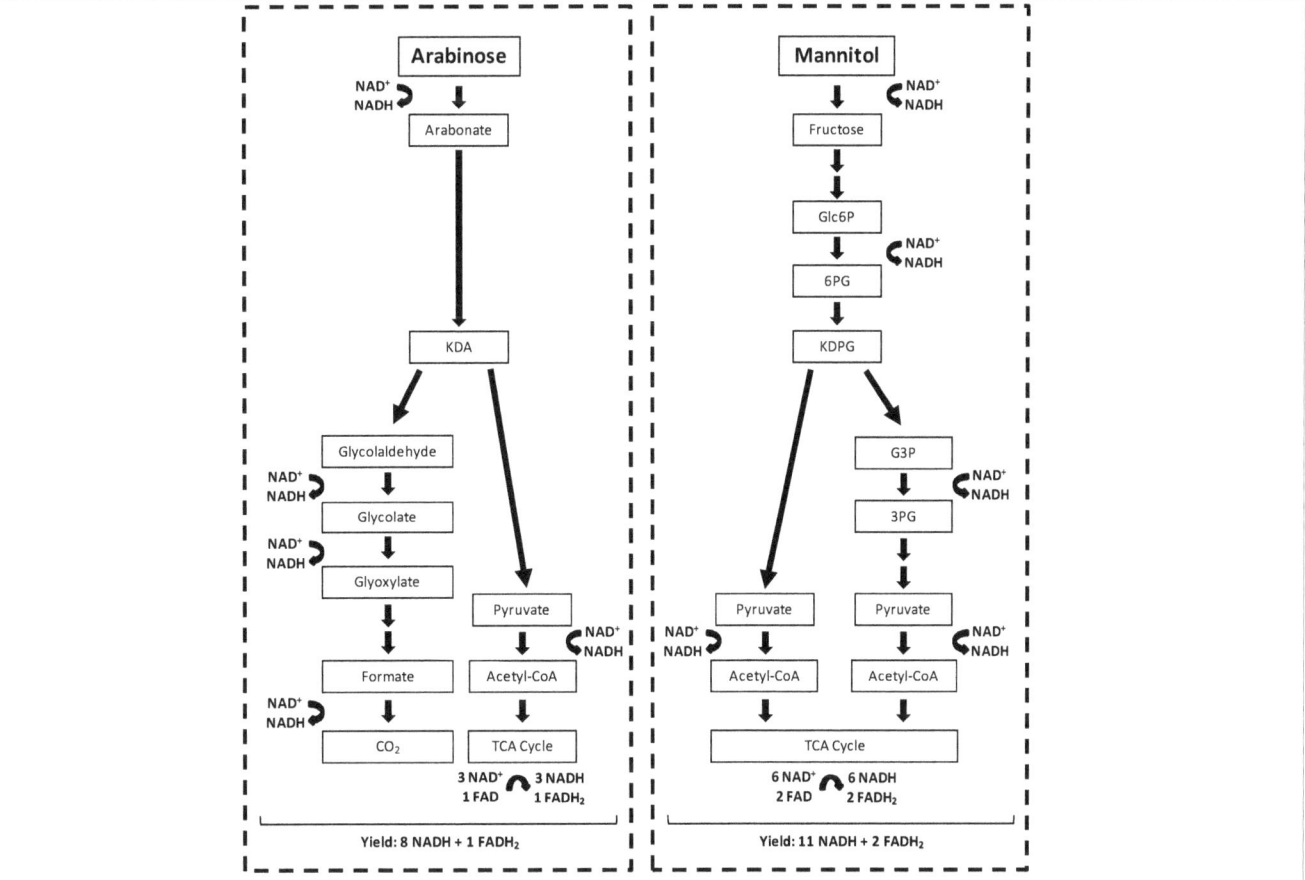

Figure 1: Comparison of the catabolism of arabinose (left) and mannitol (through the ED pathway, right) in *Bradyrhizobium diazoefficiens* with emphasis in the reactions where reducing power is generated. KDA: 2-keto-3-deoxyarabonate; Glc6P: glucose-6-phosphate; 6PG: 6- phosphogluconate; KDPG: 2-keto-3-deoxyphosphogluconate; G3P: glyceraldehyde-3-phosphate; 3PG: 3-phosphoglycerate.

$FADH_2$ oxidized, it results that 86 H^+ moles are passed per mole of arabinose or 122-132 H^+ moles are passed per mole of mannitol completely oxidized to CO_2. Despite this higher yield of mannitol with respect to arabinose, O_2 consumption rate was reported many years ago as roughly twice higher with arabinose than with mannitol [19], but these observations should be repeated with *B. diazoefficiens* USDA 110, and modern culture conditions and analytic technologies. Anyway, these data indicate that although the molar yield of H^+ is lower with arabinose than with mannitol, the net rate of H^+ passage from cytoplasm to periplasm per mole of carbon source oxidized still might be around 50% higher with arabinose. Since growth rates in minimal medium with arabinose or mannitol are similar [6], energy consumption rates for growth should also be similar, and according to the above estimates, a higher PMF may remain available for maintenance functions when arabinose is the carbon and energy source.

Perspectives

We could envisage that the cell senses the conditions in which PMF is sufficient for ATP synthesis and motion of both flagella systems at the same time and only if these conditions are met, lateral flagella expression is allowed. The conditions need not necessarily involve high viscosity of the medium because the induction of the lateral flagellar system by arabinose was observed in liquid medium. If arabinose is present in the root exudates near the infection sites [3], the expression of lateral flagella in response to this carbohydrate might be useful for the bacteria to stabilize their swimming direction towards such sites [20]. To respond to the cell energy status, the regulator(s) of lateral flagella expression should perform some measure of the PMF. There exist some ways of measuring PMF in connection with motility. For instance, a group of chemoreceptors specialized in sensing the energy status of the cell is known. These chemoreceptors bind FAD and are able to sense the redox state of the electron transport chain to elicit energy taxis, i.e. the orientation of the bacterial cell swimming towards an energy-rich environment [21]. Another candidate is the Phosphotransferase System (PTS), which also participates in chemotaxis [22]. Despite these systems being known as sensors of energy status in relation with motility, they do not display a clear relationship with the control of transcription or translation. Whether these systems, or a yet unknown signal transduction system, play a role in the control of lateral flagellar expression in response to the carbon and energy source is a research issue that might provide new knowledge about regulation of energy use in bacteria.

This issue is of special importance in the *Bradyrhizobium*-soybean symbiosis. For instance, in Argentina more than 20 million hectares are cultivated with soybean, and 94% of producers use *Bradyrhizobium*-based inoculants to achieve N-nutrition through biological N_2 fixation in their crops [23]. Motility of *Bradyrhizobium* bacteria in the soil is essential to achieve a successful symbiotic interaction [5] and therefore, understanding the control of motility and its stimulation by root-exuded compounds is one key for the development of improved inoculants for agriculture.

Acknowledgements

The author is member of the Scientific Career of CONICET, Argentina, and his work is financed by ANPCyT, CONICET, and UNLP, all from Argentina.

References

1. Delamuta JR, Ribeiro RA, Ormeño-Orrillo E, Melo IS, Martínez-Romero E and Hungria M. Polyphasic evidence supporting the reclassification of *Bradyrhizobium japonicum* group Ia strains as *Bradyrhizobium diazoefficiens* sp. nov. Int J Syst Evol Microbiol. 2013; 63: 3342-51. doi: 10.1099/ijs.0.049130-0.

2. Kaneko T, Nakamura Y, Sato S, Minamisawa K, Uchiumi T, Sasamoto S, et al. Complete genomic sequence of nitrogen-fixing symbiotic bacterium *Bradyrhizobium japonicum* USDA110. DNA Res. 2002; 9(6): 189-97.

3. Barbour WM, Hattermann DR and Stacey G. Chemotaxis of *Bradyrhizobium japonicum* to soybean exudates. Appl Environ Microbiol. 1991; 57(9): 2635-39.

4. Kanbe M, Yagasaki J, Zehner S, Göttfert M and Aizawa S. Characterization of two sets of subpolar flagella in *Bradyrhizobium japonicum*. J Bacteriol. 2007; 189(3): 1083-89.

5. Althabegoiti MJ, Covelli JM, Perez-Gimenez J, Quelas JI, Mongiardini EJ, Lopez MF, et al. Analysis of the role of the two flagella of *Bradyrhizobium japonicum* in competition for nodulation of soybean. FEMS Microbiol Lett. 2011; 319(2): 133-39. doi: 10.1111/j.1574-6968.2011.02280.x.

6. Covelli JM, Althabegoiti MJ, Lopez MF and Lodeiro AR. Swarming motility in *Bradyrhizobium japonicum*. Res Microbiol. 2013; 164(2): 136-44. doi: 10.1016/j.resmic.2012.10.014.

7. Vesper SJ and Bauer WD. Role of pili (Fimbriae) in attachment of *bradyrhizobium japonicum* to soybean roots. Appl Environ Microbiol. 1986; 52(1): 134-41.

8. Merino S and Tomás JM. Lateral Flagella Systems. In: Jarrell KF, editor. Pili and Flagella. Current Research and Future Trends. Kingston, UK: Caister Academic Press; 2009. p 173-90.

9. McCarter LL. Regulation of flagella. Curr Opin Microbiol. 2006; 9(2): 180-86.

10. Berg HC. The rotary motor of bacterial flagella. Annu Rev Biochem. 2003; 72: 19-54.

11. Stock D, Namba K and Lee LK. Nanorotors and self-assembling macromolecular machines: the torque ring of the bacterial flagellar motor. Current Opin Biotechnol. 2012; 23(4): 545-54. doi: 10.1016/j.copbio.2012.01.008.

12. Attmannspacher U, Scharf B and Schmitt R. Control of speed modulation (chemokinesis) in the unidirectional rotary motor of *Sinorhizobium meliloti*. Mol Microbiol. 2005; 56(3): 708-18.

13. Pedrosa FO and Zancan GT. L-Arabinose metabolism in *Rhizobium japonicum*. J Bacteriol. 1974; 119(1): 336-38.

14. Watanabe S, Shimada N, Tajima K, Kodaki T and Makino K. Identification and characterization of L-arabonate dehydratase, L-2-keto-3-deoxyarabonate dehydratase, and L-arabinolactonase involved in an alternative pathway of L-arabinose metabolism. Novel evolutionary insight into sugar metabolism. J Biol Chem. 2006; 281(44): 33521-36.

15. Koch M, Delmotte N, Ahrens CH, Omasits U, Schneider K, Danza F, et al. A link between arabinose utilization and oxalotrophy in *Bradyrhizobium japonicum*. Appl Environ Microbiol. 2014; 80(7): 2094-101.

16. Kuykendall LD and Elkan GH. Some features of mannitol metabolism in *Rhizobium japonicum*. J Gen Microbiol. 1977; 98(1): 291-95.

17. Mulongoy K and Elkan GH. Glucose catabolism in two derivatives of a *Rhizobium japonicum* strain differing in nitrogen-fixing efficiency. J Bacteriol. 1977; 131(1): 179-87.

18. Sosa-Saavedra F, León-Barrios M and Pérez-Galdona R. Pentose phosphate pathway as the main route for hexose catabolism in *Bradyrhizobium* sp. lacking Entner-Doudoroff pathway. A role for NAD$^+$-dependent 6-phosphogluconate dehydrogenase (decarboxylating). Soil Biol Biochem. 2001; 33(3): 339-43. doi: 10.1016/S0038-0717(00)00146-2

19. Thorne DW and Burris RH. Respiratory enzyme systems in symbiotic nitrogen fixation: III. The respiration of *Rhizobium* from legume nodules and laboratory cultures. J Bacteriol. 1940; 39(2): 187-96.

20. Bubendorfer S, Koltai M, Rossmann F, Sourjik V and Thormann KM. Secondary bacterial flagellar system improves bacterial spreading by increasing the directional persistence of swimming. Proc Natl Acad Sci USA. 2014; 111(31):11485-90. doi: 10.1073/pnas.1405820111.

21. Alexandre G. Coupling metabolism and chemotaxis-dependent behaviours by energy taxis receptors. Microbiology. 2010; 156: 2283-93. doi: 10.1099/mic.0.039214-0.

22. Neumann S, Grosse K and Sourjik V. Chemotactic signaling via carbohydrate phosphotransferase systems in *Escherichia coli*. Proc Natl Acad Sci USA. 2012; 109(30): 12159-64. doi: 10.1073/pnas.1205307109.

23. Piccinetti C, Arias N, Ventimiglia L et al. Efectos Positivos de la Inoculación de Soja sobre la Nodulación, la FBN y en los Parámetros de Producción del Cultivo. In Albanesi AS (ed). *Microbiología Agrícola. Un aporte de la Investigación en Argentina, 2da ed*. Tucumán, Argentina. Magna Publicaciones, 2013;283-97.

28

Naringin, a Grape Fruit Bioflavonoid Protects Mice Bone Marrow Cells against the Doxorubicin-Induced Oxidative Stress

Ganesh Chandra Jagetia[*1] and C. Lalrinengi[1]

[1]Department of Zoology, Mizoram University, Aizawl-796 004, India

*Corresponding author: Dr. Ganesh Chandra Jagetia, Professor, Department of Zoology, Mizoram University, Tanhril, Aizawl-796 004, Mizoram, India. Email: gc.jagetia@gmail.com

Abstract

Doxorubicin an anthracylcine group of antibiotics is under frequent clinical use to treat numerous neoplastic disorders and it kills cancer cells by inducing different reactive oxygen species that increase oxidative stress in the neoplastic cells. Since there is no preferential accumulation of doxorubicin it is also taken up by normal cells which also suffer from the oxidative stress induced by it leading to their killing or development of second malignances. The doxorubicin is known to kill bone marrow stem cells leading to its suppression. Therefore, present study was designed to study the effect of naringin treatment on doxorubicin-induced oxidative stress. The mice were administered with 0, 1, 5 or 10 mg/kg body weight of doxorubicin or the animals were given 10 mg/kg body weight naringin either before one hour or after one hour of doxorubicin treatment. The bone marrow was extracted at 0.5, 1, 2 and 4 h post-drug treatment and was processed for the estimation of glutathione (GSH), glutathione-s-transferase (GST), catalase and superoxide dismutase (SOD) and lipid peroxidation (LOO). Treatment of mice with doxorubicin alone led to a significant decline in the GSH concentration, and activities of GST, catalase and SOD accompanied by an elevation in the lipid peroxidation in the bone marrow cells of mice. Treatment of mice with naringin before and after doxorubicin administration resulted in a significant elevation in all antioxidants at all times, except lipid peroxidation that did not reveal any significant alteration. The GSH showed a maximum decline at half hour whereas GST and SOD continued to decline until 4 h post-treatment. However, a maximum decline in catalase activity was recorded at 1 h post treatment. In contrast LOO increased at 1 h post treatment and remained almost unaltered up to 4 h post-treatment in all the three groups. The pretreatment of naringin was more effective than the post-treatment. Our study demonstrates that naringin alleviated the doxorubicin-induced oxidative stress by raising the antioxidant status and marginally reducing lipid peroxidation.

Keywords: Mice, Bone marrow, Doxorubicin, Naringin, Glutathione, Glutathione-S-Transferase, Catalase, Lipid peroxidation.

Introduction

Doxorubicin (Adriamycin; also known as hydroxydaunorubicin) is a photosensitive antibiotic isolated from *Streptomyces peucetius*, a soil bacteria in the 1960s [1]. Doxorubicin belongs to an anthracycline group of antibiotics, and is closely related to the natural product daunomycin. Doxorubicin is one of the highly effective anthracyline groups of cancer chemotherapeutic drugs, which acts on the specific phase of the cell cycle, especially the S phase of cell division [2]. The doxorubicin has been introduced in the clinics to treat various malignant diseases in the early 1970s and remains an integral part of various modern chemotherapeutic regimens where it is used to treat different malignant neoplasia including Hodgkin's and non-Hodgkin's lymphomas, myeloblastic leukemias, breast cancer, small cell lung cancer, ovarian cancer, childhood solid tumors, hepatocarcinomas, soft tissue sarcomas, Kaposi's sarcoma, and bone tumors [3-7].

Though the mechanism of action of doxorubicin remains a bit controversial, it is well established that the doxorubicin kills neoplastic cells by binding to nucleic acids and by inhibiting topoisomerase enzymes, presumably by specific intercalation of the planar anthracycline nucleus with the DNA double helix [8,9]. The recent reports indicate that doxorubicin induces oxidative stress by free radical production and mitochondrial dysfunction [10]. The presence of iron in mitochondria helps to increase free radical production leading to cardiotoxicity, because heart tissue has only a limited defense system and cannot cope up with the free radical induced toxicity of doxorubicin [10-14]. Another adverse effect of DNA intercalating agents is production of second malignancies in the long-term survivors [15]. Apart from cardiotoxicity the use of doxorubicin is also associated with bone marrow and hematological toxicities. Since doxorubicin is highly effective anticancer agent its full potential can only be realized if its adverse effects are considerably reduced. Therefore, it is necessary to screen natural products that can alleviate the doxorubicin induced adverse side effects.

The grapefruit (*Citrus paradisii*), arose in Barbados as an accidental cross between an orange (*C. sinensis*) and a shaddock or pomelo (*C. maxima*), both of which were introduced from Asia in the seventeenth century, and was first introduced into Florida in the 1820s. The three major types of grapefruits that exist today are white, pink/red and ruby/rio red varieties. Grapefruit juice combines the sweet and tangy flavour of the orange and shaddock and also provides up to 69% of the recommended dietary allowance (RDA) for vitamin C. The grapefruit juice contains flavonoids in the form of glycosides. Naringin is the most abundant flavonoid present in the juice, flower, and rind of grapefruit and constitutes up to 10% of the dry weight. It is relatively soluble in water and grapefruit juice contains up to 100 to 867 mg/L of naringin [16]. Upon ingestion, the naringin is converted into aglycone form known as naringenin and sugars by the action of intestinal flora [17]. Naringin has a wide range of biological actions including cholesterol-lowering, antiatherogenic and anti-inflammatory [18,19]. It has been reported to act as a cardioprotective, radioprotective, neuroprotective and antimutagenic agent [20-27]. Naringin has been found to reduce belomycin and doxorubicin induced genotoxicity in mice bone marrow, inhibit chemical carcinogenesis in mice, protect against iron-induced oxidative stress, lung fibrosis, osteoporosis, lipodystrophy and dyslipidemia in rats [28-34]. The administration of 16 g/kg b. wt. naringin to rats in acute toxicity studies has been reported to be nontoxic and its daily administration for 13 weeks at a dose of 1250 mg/kg did not induce any adverse side effects in rats indicating its safety [35]. Apart from cardiotoxicity doxorubicin also causes bone marrow depression that hampers utilization of its full potential in cancer treatment regimens. Therefore, it was desired to study the protective effect of naringin on the doxorubicin induced oxidative stress in the bone marrow of mice [36].

Materials and Methods

Chemicals and reagents

Naringin ((2S)-7-[(2S,3R,4S,5S,6R)-4,5-dihydroxy-6-(hydroxymethyl)-3-[(2S,3R,4R,5R,6S)-3,4,5-trihydroxy-6-methyloxan-2-yl]oxyoxan-2-yl]oxy-5-hydroxy-2-(4-hydroxyphenyl)-2,3-dihydro-chromen-4-one), was procured from Acros Organics Ltd., Geel (Belgium), whereas the doxorubicin was supplied by Getwell Pharmaceuticals, Gurgaon, India. The glutathione, 2-thiobarbituric acid (TBA), 5,5-dithiobis(2-nitrobenzoic acid) (DTNB), phenazine methosulphate, ethylenediaminetertaacetic acid (EDTA), diethylenetriaminopenta-acetic acid (DETAPAC) nitroblue tetrazolium (NBT), 5-thio-(2-nitrobenzoic acid)] (TNB), 1-chloro-2,4-dinitrobenzene (CDNB), nitroblue tetrazolium, tetraethoxypropane, nicotinamide adenine dinucleotide (NADH), and nicotinamide adenine dinucleotide phosphate (NADPH) were purchased from Sigma Chemicals Co. St. Louis, USA. Other routine chemicals were procured from Merck India, Mumbai.

Animals care and Handling

The animal care and handling were done according to the guidelines issued by the World Health Organization, Geneva, Switzerland and the INSA (Indian National Science Academy, New Delhi, India). Eight to ten weeks old male Swiss albino mice weighing 25-30g were selected from an inbred colony maintained under the controlled conditions of temperature (23±2°C), humidity (50±5%) and light (12 of light and dark, respectively). The animals had ready access to sterile food and water throughout the study. The animals were kept in a polypropylene cage containing sterile paddy husk (procured locally) as bedding throughout the experiment. The study was cleared by the Animal Ethics Committee, of the Mizoram University, Aizawl, India.

Preparation of the drug and mode of administration

Naringin and doxorubicin was dissolved in sterile double distilled water (DDW) freshly each time immediately before use. The naringin (NIN) or doxorubicin (DOX) was administered intraperitoneally

Experimental

The animals were divided into the following groups:

1. DOX group:- The animals of this group were intraperitoneally administered with 0, 1, 5 or 10 mg/kg b. wt. of doxorubicin alone.

2. NIN+DOX:- This group of animals was intraperitoneally administered with 10mg/kg b. wt. Naringin one hour before administration of 0, 1, 5 or 10 mg/kg b. wt. doxorubicin. The 10 mg/kg naringin was selected based on our earlier study where 10 mg/kg b. wt. of naringin was found to provide maximum protection against the DOX-induced DNA damage, when compared to other doses [30].

3. DOX+NIN:- The animals of this group were given 10 mg/kg b. wt. naringin intraperitoneally one hour after the administration of 0, 1, 5 or 10 mg/kg b. wt. doxorubicin.

The animals of all groups were killed by cervical dislocation at 0.5, 1, 2 and 4 h posttreatment for determination of glutathione (GSH), glutathione-s-transferase (GST), catalase, superoxide dismutase (SOD) and lipid peroxidation (LOO) in the bone marrow of mice. Usually five animals were used for each dose of DOX in each group at each time point and a total of 240 animals were used for the whole experiment. The bone marrow cells were extracted by removing the femora of each animal at the above specified posttreatment times and placed onto a wet Whatman filter paper. The femora were freed from muscles and other tissues, cleaned and the bone marrow was flushed using PBS into individual tubes.

Preparation of homogenate

The tubes were centrifuged resuspended in PBS and one million cells were homogenized for the estimation of GSH, GST, catalase, SOD and LOO.

Total Proteins

The protein contents were determined according to the modified method of Lowry's [37].

Glutathione

Glutathione contents were measured by the method of Moron et al., with minor modifications. Briefly, the proteins in 1 ml supernatant were precipitated by 0.5 ml ice cold 10% 5-sulfosalicylic acid [38]. The tubes were kept on ice for 10 min, centrifuged (Sorvall Instruments RC5C, DuPont, Minnesota, USA) at 15,000 rpm at 4°C for 15 min and the supernatant free of proteins was collected. The entire supernatant was immediately mixed with 0.5 ml of NADPH (4 mg of reduced form was dissolved in 100 ml of 0.5% NaHCO3), 0.5 ml of glutathione reductase (6 units/ml in 0.1 M phosphate buffer, pH 7.0) and 1 ml of 0.6 M DTNB [prepared in 0.2 M phosphate buffer (pH 8)]. The formation of TNB was read against blank at 412 nm in a UV-Visible double beam spectrophotometer (Shimadzu Corporation, Tokyo, Japan). A sample blank lacking GSH was used to determine the background rate and the resulting background rate of product formation was subtracted from the sample values prior to GSH quantification. The GSH concentration has been expressed as μmol/million cells. Standard curve was prepared from a stock solution of 10 mM GSH (30.7 mg GSH/10 ml) in 5% 5-sulfosalicylic acid diluted to 1-10 nM GSH/ml.

Glutathione-S-Transferase

Glutathione-S-transferase was estimated by the method of Habig and Pabst [39]. Briefly, the bone marrow cell homogenate was mixed with 0.1 M potassium phosphate buffer, CDNB and 10 mM GSH, and incubated for 10 min at 37°C. The absorbance was read against the blank at 340 nm using a double beam UV-VIS spectrophotometer. The absorbance of the samples was read at 1min intervals. GST activity has been expressed as nmol/mg of protein.

GST activity = Absorbance of sample – Absorbance of blank× 1000/9.6 × Vol of sample

Catalase

The catalase activity was estimated by the catalytic reduction of hydrogen peroxide [40]. Briefly, hydrogen peroxide was added to the tissue homogenate and incubated at 37°C. The decomposition of hydrogen peroxide was monitored at specific time intervals by recording the absorbance against the blank at 240 nm using a double beam UV-VIS spectrophotometer at the intervals of 0.5 and 10s intervals up to 30s. The average difference in absorbance in 30s was calculated.

Superoxide Dismutase

The SOD was estimated by the method as described earlier [41]. Briefly, the cell homogenate was mixed with reagent consisting of phenazine methosulphate, nitroblue tetrazolium and NADH and incubated for 90 sec. at 30°C and the reaction was stopped by adding acetic acid and n-butanol. Blank was prepared by adding the reagent without the sample and incubated for 90 sec. at 30°C and the reaction was stopped by adding acetic acid and n-butanol. The absorbance of sample was measured against the blank at 560 nm in a UV-VIS spectrophotometer.

Lipid Peroxidation

The level of various thiobarbituric acid reactive substances (TBARS) including malondialdehyde, lipid hydroperoxides and aldehydes in the cell homogenates was measured by the method of Ohkawa et al., [42]. The method is based on the spectroscopic estimation of malondialdehyde and thiobarbituric acid 1:2 adduct formation in the reaction mixture. Briefly, 1 ml supernatant was heated with thiobarbituric acid (0.8%), sodium dodecyl sulphate (0.1%) and acetic acid (20%) in a boiling water bath for 30 min (lipoproteins get precipitated). The resultant mixture was cooled, extracted with n-butanol-pyridine, and the absorbance of the butanol layer was recorded at 532 nm using UV-Visible double beam spectrophotometer. The resulting concentration of TBA reactive substances is expressed as nmol/million cells obtained from a standard curve of tetraethoxypropane.

Statistical analyses

The statistical significance among all the groups was determined by Students "t" test and analysis of variance (ANOVA). The results are the average of three individual experiments. The data of each experiment did not differ significantly from one another and hence, all the data are combined and means calculated. A p value of < 0.05 was considered statistically significant.

Results

The results are expressed as mean ± SEM (standard error of the mean) in Table 1-5.

Table 1: The effect of 10 mg/kg body weight naringin on the doxorubicin-induced depletion in the glutathione concentration (μmol/million cells) in the bone marrow of mice at various post treatment times

Post treatment time (h)	Doxorubicin(mg/kg body weight)									
	0	1			5			10		
	Control	DOX	NIN+DOX	DOX+NIN	DOX	NIN+DOX	DOX+NIN	DOX	NIN+DOX	DOX+NIN
0.5	1.80±0.03	0.36±0.01‡	0.53±0.013*	0.43±0.013	0.18±0.018‡	0.29±0.018	0.23±0.019	0.12±0.003‡	0.20±0.006	0.18±0.009
1	1.80±0.03	0.42±0.02‡	0.57±0.016*	0.50±0.022	0.23±0.035‡	0.35±0.037	0.29±0.037	0.13±0.022‡	0.18±0.073	0.16±0.015
2	1.80±0.03	0.48±0.05‡	0.65±0.05*	0.55±0.05	0.30±0.024‡†	0.42±0.022†	0.36±0.023	0.14±0.011‡	0.19±0.011	0.16±0.011
4	1.80±0.09	0.51±0.01‡†	0.68±0.018*	0.58±0.015†	0.31±0.017‡†	0.43±0.018†	0.30±0.075	0.11±0.011‡	0.17±0.011	0.13±0.016

$P < 0.05$, No symbol=Non-significant

‡- statistical significance was calculated against control with the positive control i.e DOX treatment alone.

*- statistical significance was calculated against DOX alone treatment in their respective treatment hours.

#- statistical significance was calculated against DOX+NIN combination treatment in their respective treatment hours.

†- statistical significance was calculated against their corresponding 0.5h treatment combinations.

Table 2: The effect of 10 mg/kg body weight naringin on the doxorubicin-induced alleviation in the glutathione–s–transferase activity (nmol/million cells) in the bone marrow of mice at different post treatment times

Post treatment times (h)	Doxorubicin(mg/kg body weight)									
	0	1			5			10		
	Control	DOX	NIN+DOX	DOX+NIN	DOX	NIN+DOX	DOX+NIN	DOX	NIN+DOX	DOX+NIN
0.5	4.25±0.01	3.99±0.19	4.70±0.10*	4.61±0.057	3.42±0.26‡	4.11±0.06*	3.99±0.11	2.76±0.02‡	3.28±0.18†	3.23±0.01
1	4.25±0.01	4.07±0.13	4.52±0.18	4.30±0.33	3.36±0.14‡	3.82±0.23	3.75±0.12	2.35±0.22‡	3.01±0.19*	3.07±0.2*
2	4.25±0.01	3.67±0.01	4.29±0.05*	3.97±0.12*	3.12±0.15‡	3.56±0.06	3.55±0.10	2.31±0.04‡	2.81±0.02	2.81±0.02
4	4.25±0.01	3.56±0.14	4.20±0.11†	3.85±0.02	2.98±0.12‡†	3.53±0.02	3.46±0.13	2.33±0.10‡	2.71±0.12	2.65±0.11

$P < 0.05$, No symbol=Non-significant

‡- statistical significance was calculated against control with the positive control i.e DOX treatment alone.

*- statistical significance was calculated against DOX alone treatment in their respective treatment hours.

#- statistical significance was calculated against DOX+NIN combination treatment in their respective treatment hours.

†- statistical significance was calculated against their corresponding 0.5h treatment combinations.

Table 3: The effect of 10 mg/kg body weight naringin on the doxorubicin-induced depletion in the catalase activity (nmol/million cells)in the bone marrow of mice at various post treatment times

Post-treatment times (h)	Doxorubicin (mg/kg body weight)									
	0	1			5			10		
	Control	DOX	NIN+DOX	DOX+NIN	DOX	NIN+DOX	DOX+NIN	DOX	NIN+DOX	DOX+NIN
0.5	29.17±0.03	11.63±1.21‡	19.49±0.70*#	18.49±1 16*	10.63±0.60‡	14.63±1.91*#	12.73±0.69*	8.71±1.30‡	13.86±0.31*	11.85±0.51*
1	29.17±0.03	8.54±0.62‡†	15.79±0.66*†	15.14±0.15*†	7.87±0.23‡†	10.95±1.92*†	9.91±0.42*†	7.54±0.27‡†	8.97±0.37*†	8.23±0.40*†
2	29.17±0.03	10.07±0.57‡†	16.28±1.02*#†	14.52±1.14*†	8.67±1.39‡	10.54±0.71*†	10.03±1.40†	7.38±0.85‡†	8.91±1.27*	8.22±1.36*†
4	29.17±0.03	9.83±0.53‡†	16.29±0.41*#†	15.18±0.87*†	8.99±0.43‡†	10.50±0.45*#†	9.29±0.47*†	5.52±0.49‡	9.94±0.82*†	8.85±0.56*†

$P < 0.05$, No symbol=Non-significant

‡- statistical significance was calculated against control with the positive control i.e DOX treatment alone.

*- statistical significance was calculated against DOX alone treatment in their respective treatment hours.

#- statistical significance was calculated against DOX+NIN combination treatment in their respective treatment hours.

†- statistical significance was calculated against their corresponding 0.5h treatment combinations.

Table 4: The effect of 10 mg/kg body weight naringin on the doxorubicin-induced decline in the superoxide dismutase activity(Units/million cells)in the bone marrow of mice at different post treatment times.

Post treatment times (h)	Doxorubicin(mg/kg body weight)									
	0	1			5			10		
	Control	DOX	NIN+DOX	DOX+NIN	DOX	NIN+DOX	DOX+NIN	DOX	NIN+DOX	DOX+NIN
0.5	5.12±0.03	3.03±0.07‡	4.55±0.08*	4.43±0.13*	2.78±0.09‡	3.95±0.09*	3.62±0.10*	1.83±0.11‡	3.12±0.10*	2.90±0.12*
1	5.12±0.03	2.61±0.06‡†	4.13±0.07*	4.01±0.12*	2.36±0.08‡	3.53±0.08*	3.20±0.09*	1.41±0.10‡	2.70±0.09*	2.48±0.11*
2	5.12±0.03	2.30±0.08‡†	3.81±0.09*	3.69±0.14	2.04±0.10‡	3.21±0.10*	2.88±0.11	1.09±0.12‡†	2.38±0.10*	2.16±0.13*

| 4 | 5.12±0.03 | 2.21±0.07‡† | 4.02±0.08* | 3.78±0.13* | 1.96±0.09‡† | 3.42±0.09* | 2.97±0.10* | 1.01±0.11‡† | 2.59±0.10* | 2.25±0.12* |

$P < 0.05$, No symbol=Non-significant

‡- statistical significance was calculated against control with the positive control i.e DOX treatment alone.

*- statistical significance was calculated against DOX alone treatment in their respective treatment hours.

" statistical significance was calculated against DOX+NIN combination treatment in their respective treatment hours.

†- statistical significance was calculated against their corresponding 0.5h treatment combinations.

Table 5: The effect of 10 mg/kg body weight naringin on the doxorubicin-induced increase inlipid peroxidation(TBARs (nmol/million cells)in the bone marrow of mice at various post treatment times.

Post treatment times (h)	Doxorubicin (mg/kg body weight)									
	0	1			5			10		
	Control	DOX	NIN+DOX	DOX+NIN	DOX	NIN+DOX	DOX+NIN	DOX	NIN+DOX	DOX+NIN
0.5	0.09±0.002	0.12±0.002‡	0.12±0.008	0.10±0.01	0.17±0.005‡	0.14±0.001	0.13±0.007	0.20±0.005‡	0.195±0.01	0.194±0.01
1	0.09±0.002	0.13±0.003‡	0.13±0.001	0.12±0.003	0.16±0.007‡	0.162±0.01	0.161±0.004	0.24±0.02‡	0.23±0.02	0.24±0.004
2	0.09±0.002	0.14±0.01‡	0.14±0.006	0.13±0.004	0.19±0.015‡	0.18±0.00	0.18±0.01	0.24±0.02‡	0.23±0.02	0.22±0.02
4	0.09±0.002	0.13±0.006‡	0.12±0.01	0.11±0.008	0.18±0.003‡	0.18±0.01	0.17±0.01	0.23±0.005‡	0.22±0.007	0.22±0.01

$P < 0.05$, No symbol=Non-significant

‡- statistical significance was calculated against control with the positive control i.e DOX treatment alone.

*- statistical significance was calculated against DOX alone treatment in their respective treatment hours.

" statistical significance was calculated against DOX+NIN combination treatment in their respective treatment hours.

†- statistical significance was calculated against their corresponding 0.5h treatment combinations.

Glutathione

The spontaneous glutathione concentration in the non-drug treated control Swiss albino bone marrow cells is 1.80 ± 0.03 nmol/million cells (Table 1). Administration of naringin did not alter the GSH concentration significantly in mice bone marrow cells as compared to control (data not shown). When the mice were given different concentrations of doxorubicin it caused a dose dependent decline in the GSH concentration and a maximum depletion in GSH concentration was observed for 10 mg/kg b. wt. DOX (Table 1). GSH concentration showed highest decrement at 1/2 h post-treatment that continued to elevate until 4 h posttreatment without restoration to normal level (Table 1). Administration of naringin 1 h before DOX treatment resulted in a significant rise in the GSH concentration as early as 0.5 h post treatment, which continued to rise up to 4 h posttreatment where the GSH concentration was greatest. The trend in GSH rise was almost similar in the bone marrow cells of mice that received naringin 1 h after DOX treatment however; the amount was lesser than the pretreatment group (Table 1). The increasing dose of DOX tapered this effect and when 10 mg/kg DOX was given the elevation in GSH concentration was not significantly greater in both the NIN+DOX and DOX+ NIN groups (Table 1).

Glutathione-S-transferase

The glutathione-s-transferase activity in the bone marrow cells of non-drug treated control mice is 4.25±0.01 nmol/million cells and naringin administration alone did not alter the glutathione-s-transferase activity significantly as compared to control (data not shown) (Table 2). Treatment of mice with different doses of DOX reduced GST activity significantly at 5 and 10 mg/kg DOX at all the posttreatment times (Table2). The activity of GST declined continuously with time and the lowest GST activity was observed at 4 hours posttreatment (Table 2). Treatment of naringin before and after DOX administration resulted in an elevation in the GST activity that depended on the assay time and DOX dose (Table 2). A maximum elevation in GST activity was recorded for NIN+DOX group at 0.5 h posttreatment in NIN+DOX as well as DOX+NIN group The pattern of decline in the GST activity was similar to that of DOX treatment (Table 2).

Catalase

The catalase activity in the bone marrow of normal untreated mice is 29.17nmol/million cells, which remained unchanged after naringin administration (data not shown) (Table 3). The treatment of mice with different doses of DOX reduced the catalase activity in a dose dependent manner and the greatest reduction in the catalase activity was observed for 10 mg/kg DOX (Table 3). The DOX treatment also caused a time dependent attrition in the catalase activity and maximum reduction was observed at 1 h post treatment with an attempt of recovery thereafter (Table 3). Despite the signs of recovery the catalase activity was far from normal (Table 3). The catalase activity continued to decline in the bone marrow of mice treated with 10 mg/kg DOX until 4 h posttreatment where it was lowest in these animals (Table 3). This reduction in catalase activity by DOX was 3, 3.25 and 5.3 fold at 4 h posttreatment for 1, 5 and 10 mg/kg DOX, respectively (Table 3). The administration of naringin before and after DOX treatment significantly elevated the catalase activity and maximum elevation was observed at 0.5 h post treatment (Table 3). Thereafter catalase activity declined at 2 h post-treatment and increased thereafter (Table 3).

Superoxide dismutase

The base line SOD activity in the mice bone marrow cells is 5.12 ± 0.03 Units/million cells and naringin treatment did alter the spontaneous SOD activity significantly (data not shown) (Table 4). The administration of different doses of doxorubicin in mice resulted in a dose dependent alleviation in the SOD activity at all post treatment assay times as compared to control (Table 4). The lowest SOD activity was observed for 10 mg/kg DOX, which was almost five fold lower at 4 h post treatment (Table 4). The SOD activity showed a drastic decline at 0.5 h and continued to decline until 4 h post-treatment where a nadir was reached (Table 4). Naringin treatment before DOX administration did not alter the pattern of decline in the SOD activity in NIN+DOX group, except that the SOD activity was significantly enhanced (Table 4). The SOD activity was 2 fold greater at 0.5 h, whereas this elevation was approximately 3.5 fold at 1 and 2 h post-DOX-treatment, respectively (Table 1). Treatment of naringin before or after DOX administration caused a significant elevation in the SOD activity at all post treatment times (Table 4). At lower DOX doses the elevation was greater when compared to the higher dose of DOX (Table 4). Despite this rise in the SOD activity by naringin it was far from normal (Table 4).

Lipid peroxidation

The rate of lipid peroxidation in the bone marrow cells of Swiss albino mice is 0.09±0.002 nmol/million cells (Table 5). Naringin treatment did not significantly alter the spontaneous lipid peroxidation in mice bone marrow (data not shown). The doxorubicin administration resulted in a dose dependent but significant rise in the lipid peroxidation when compared to the control, and a highest lipid peroxidation was observed in the bone marrow of mice receiving 10 mg/kg DOX (Table 5). The analysis of lipid peroxidation with time did not show any significant alteration as the differences were not statistically significant (Table 5). Administration of naringin before and after DOX treatment did not significantly reduce lipid peroxidation although it was marginally lesser than the DOX alone treated groups (Table 5).

Discussion

Doxorubicin is widely used anthracycline group of antibiotics in clinical condition to treat different type of neoplastic disorders. It is used alone or in combination with other chemotherapeutic drugs to treat difficult neoplasia. However, its use is associated with severe bone marrow depression and life threatening cardiomyopathy [10, 36]. One of the important mechanisms by which DOX induces oxidative stress is production of free radicals therefore it is desirable to reduce the oxidative stress by alleviating free radical production. The interaction of DOX free radicals in the presence of iron further aggravates the oxidative stress and naringin has been reported to suppress free radicals and chelate iron [24,31]. The first line of antioxidants in cellular milieu is glutathione reductase, catalase, superoxide dismutase and trace elements like selenium, copper, zinc etc. The second line of defence includes cellular antioxidants like

glutathione and third defence include a set of complex enzymes that carry out the repair of damages of macromolecules vital for cell survival [43]. Therefore present study was undertaken to study the modulatory effect of naringin given before and after DOX treatment on the glutathione, glutathione-s-transferase, catalase, superoxide dismutase and lipid peroxidation in mice bone marrow.

The GSH (γ-glutamylcysteinylglycine), a low molecular, water soluble non-protein thiol is synthesized in all living cells from three amino acids including glutamic acid, cysteine and glycine, which play a crucial role in combating oxidative stress in the cells by donating one electron and donation of two electrons by GSH results in the formation of GSSG, which is converted back to GSH by glutathione reductase [44-47]. The doxorubicin treatment has increased the oxidative stress in the bone marrow cells of mice by alleviating the GSH contents in a dose dependent manner. A similar effect has been observed earlier in the liver of mouse and rats treated with DOX [25,48-50]. Treatment of mice with naringin before and after DOX administration raised the GSH contents and the effect was better in the naringin pretreated group than the naringin post-treatment. Earlier investigations conform to our observations where a similar effect has been observed in mice heart and rat livers pretreated with naringin however, reports on post-treatment studies are lacking [25,26,50,51]. The other agents like *Agele marmelos* and glutamine, an amino acid have been reported to increase the GSH concentration in mice and rats treated with DOX [48,49]. The reduction in oxidative stress by GSH is the outcome of its reaction with free radicals directly or its participation through other enzymatic reaction [52]. It also conjugates with NO radical to form S-nitrosoglutathione adduct. This product is cleaved by thioredoxin system to generate GSH and NO back [47]. The other functions of GSH include its conjugation with electrophiles and physiological metabolites, which are essential in cell physiology [53].

The observation that DOX inhibited the GST activity in mouse bone marrow is in accordance with earlier findings where DOX has been reported to reduce GST activity in mouse heart and mice and rat liver [25, 51]. The treatment of naringin before and after DOX treatment elevated the GST activity and the effect of both pre and post naringin treatment was almost similar. The earlier studies have shown that naringin pretreatment elevated the GST activity in mouse heart and liver and rat liver [25, 50]. DOX-induced GST alleviation may have played a significant role in inducing oxidative stress as GSTs are essential in protecting against the toxic insult and DNA damage [54]. GSTs catalyze conjugation of a variety of substances with GSH causing detoxification and it also modulates cell proliferation and death [55]. The reduction of GSTs by DOX may be crucial in suppressing cell death and enhanced cell killing. Their overexpression in cancer cells have been considered as a cause of chemotherapy resistance [56].

Catalase or oxidoreducatase is a tetrameric intra cellular enzyme present in animals, plants and microorganisms. The accumulation of hydrogen peroxide in cellular milieu is detrimental to the health of cells and catalase mainly cleaves hydrogen peroxide into non-toxic products like water and molecular hydrogen, neutralizing the toxic effect of hydrogen peroxide [57]. Treatment of mice with different doses of DOX resulted in a dose dependent reduction in the catalase activity in their bone marrow cells, which could be one of the reasons of myelosuppression in the patients. The earlier studies have reported a reduction in the catalase activity after DOX treatment in the mice and rat heart and liver [25,26,49,50]. The administration of mice with naringin before and after DOX treatment elevated the catalase activity, where both the pre- and post-treatment were equally effective especially at higher DOX doses. Likewise, early studies have indicated an increase in the catalase activity in the mouse and rat heart and liver [25,26,50,51]. The extract of *Aegle marmelos* has been found to reduce the DOX induced catalase activity in mouse heart earlier [49]. This increase in catalase activity by naringin may have played an important role in reducing the oxidative stress triggered by DOX in the present study.

The superoxide ions are generated during respiration, which are converted into the less harmful product hydrogen peroxide, which is neutralized by glutathione peroxidase and catalase [58-60]. However, DOX enhances the generation of these radicals and presence of iron converts the superoxide radicals into the more reactive, toxic and damaging species of OH radicals [10]. The DOX treatment has been found to deplete SOD activity in a dose dependent manner in the bone marrow of mice in the present study. A similar effect has been reported earlier in the mice heart and liver and rat liver [25,49,50]. The naringin treatment before and after DOX administration has been reported to retard DOX-induced attrition in the SOD activity. Earlier naringin treatment has been reported to protect against the DOX-induced depletion in the SOD activity in mice and rats [25,26,50,51]. The elevated level of SOD may have protected mice bone marrow cells from DOX-induced oxidative stress.

Lipid peroxidation is produced as a result of membrane damage as the lipids of the cell membrane interact with free radicals produced and undergo lipid peroxidation [61]. The production of lipid peroxidation has been considered as one of the hallmarks of oxidative stress. The DOX increased lipid peroxidation in the mouse bone marrow which is in conformation to earlier studies where DOX has been found to increase lipid peroxidation in the mouse and rat heart and liver [25,26,48-51,62]. Earlier studies have shown that naringin pretreatment resulted in the reduction of DOX induced lipid peroxidation in vitro and in mouse and rat heart and liver [23,26, 29,31,50,51]. Similarly, naringin alleviated the bleomycin-induced lipid peroxidation in rat lung earlier [33]. However, naringin treatment was not very effective in arresting the lipid peroxidation in bone marrow cells in the present study.

The exact mechanism of reduction of oxidative stress by naringin in the bone marrow cells of mice is not clearly understood. The DOX generates free radicals by the activation of NADPH oxidase system and naringin has been reported to inhibit the activation of NADPH oxidase leading to the attrition

of free radical formation [63,64]. Therefore suppression of DOX-induced free radical generation by naringin may be one of the important mechanisms that may have helped to keep the activity of GST, catalase and SOD higher accompanied by more availability of glutathione. The presence of naringin may have neutralized free radicals immediately after their production since it has been reported to scavenge free radicals earlier [24]. It is well established that DOX generates higher amount of free radicals in the presence of iron and chelation of iron by naringin may have arrested the DOX induced free radicals reducing oxidative stress. Naringin has been reported to chelate iron [31]. DOX has been reported to transcriptionally activate NF-κB and COX-II in cardiomyocytes which is responsible for increased oxidative stress and inhibition of NF-κB and COX-II by naringin may have reduced DOX-induced oxidative stress [65-67]. Naringin has been reported to arrest transcriptional activation of of NF-κB and COX-II in vitro [64,68]. Nrf2 is essential in the expression of various antioxidant genes and its inhibition by DOX may have contributed to raise the oxidative stress in mice bone marrow cells leading to decline in all these antioxidants. The activation of Nrf2 elements by naringin may have been responsible for increased activities of GST, catalase and SOD along with raised GSH. The naringin has been reported to activate Nrf2 signalling pathway earlier [69].

Conclusions

The present study demonstrates that DOX has been able to increase the oxidative stress in mice bone marrow cells by increasing lipid peroxidation and reducing the activities of GST, catalase and SOD accompanied by a decreased concentration of GSH, whereas naringin pre- and post-DOX-treatment increased the activities of GST, catalase and SOD and also elevated GSH contents in the bone marrow of mice reducing the oxidative stress. The inhibition of the activation of NADPH oxidase, NF-κB, and COX-II may have contributed to the protective effect exerted by naringin along with activation of Nrf2 signaling at molecular level.

Acknowledgements

This work was supported by a grant No.F4-10/2010(BSR) UGC from the University Grants Commission, New Delhi, India

References

1. Arcamone F, Cassinelli G, Fantini G, Grein A, Orezzi P, Pol C, et al. Adriamycin, 14-hydroxydaunomycin, a new antitumor antibiotic from S. peucetius var. caesius. Biotechnol Bioeng. 1969;11(6):1101-1110.

2. Kusyk CJ, Hsu TC. Adriamycin-induced chromosome damage: elevated frequencies of isochromatid aberrations in G2 and S phases. Experientia. 1976;32(12):1513-1514.

3. Minotti G, Menna P, Salvatorelli E, Cairo G, Gianni L. Anthracyclines: molecular advances and pharmacologic developments in antitumor activity and cardiotoxicity. Pharmacol Rev. 2004;56(2):185-229.

4. Quiles JL, Ochoa JJ, Huertas JR, Lopes-Frias M, Mataix J. Olive oil and mitochondrial oxidative stress: studies on adriamycin toxicity, physical

exercise and ageing. In Quiles JL, CABI Publishing, Oxford. 2006;119-151.

5. Tam K. The Roles of Doxorubicin in Hepatocellular Carcinoma. ADMET & DMPK. 2013;1(3):29-44.

6. Volkova M, Russell R. Anthracycline cardiotoxicity: prevalence, pathogenesis and treatment. Curr Cardiol Rev. 2011; 7(4):214–220.

7. Ewer MS, Ewer SM. Cardiotoxicity of anticancer treatments. Nat Rev Cardiol. 2015;12(9):547-558.

8. Pommier Y, Leo E, Zhang H-L, Marchand C. DNA Topoisomerases and their poisoning by anticancer and antibacterial drugs. Chem Biol. 2010;17(5):421-433.

9. Agudelo D, Bourassa P, Bérubé G, Tajmir-Riahi HA. Intercalation of antitumor drug doxorubicin and its analogue by DNA duplex: Structural features and biological implications. Int J Biol Macromol. 2014;66:144-150.

10. Ichikawa Y, Ghanefar M, Wu R, Khechaduri A, Prasad SVN, Mutharasan RK, et al. Cardiotoxicity of doxorubicin is mediated through mitochondrial iron accumulation. J Clin Invest. 2014;124(2):617-630.

11. DeAtley SM, Aksenov MY, Aksenova MV, Jordan B, Carney JM, Butterfield DA. Adriamycin-induced changes of creatine kinase activity in vivo and in cardiomyocyte culture. Toxicology. 1999;134(1):51-62.

12. Gewirtz DA. A critical evaluation of the mechanisms of action proposed for the antitumor effects of the anthracycline antibiotics adriamycin and daunorubicin. Biochem Pharmacol. 1999;57(7):727-741.

13. Gammella E, Maccarinelli F, Buratti P, Recalcati S, Cairo G. The role of iron in anthracycline cardiotoxicity. Front Pharmacol. 2014;5:25.

14. Goszcz K, Deakin SJ, Duthie GG, Stewart D, Leslie SJ, Megson IL. Antioxidants in cardiovascular therapy: panacea or false hope? Front Cardiovasc Med. 2015;2:29.

15. Pendleton M, Lindsey RH Jr, Felix CA, Grimwade D,Osheroff N. Topoisomerase II and leukemia. Ann N Y Acad Sci. 2014;1310:98-110.

16. Zhang J. Flavonoid in grape fruit and commercially grape fruit juices: concentration, distribution, and potential health benefits. Proc. Fla State HortSoc. 2007;120:288-294.

17. Kanaze FI, Bounartzi MI, Georgarakis M Niopas I. Pharmacokinetics of the citrus flavanone aglycones hesperetin and naringenin after single oral administration in human subjects. Eur J Clin Nutr. 2007;61(4):472-477.

18. Lee CH, Jeong TS, Choi MS, Hyun BH, Oh GT, Kim JR, et al. Anti-atherogenic effect of citrus flavonoids, naringin and naringenin, associated with hepatic CAT and aortic VCAM-1 and MCP-1 in high cholesterol-fed rabbits. Biochem Biophys Res Commun. 2001;284(3):681-688.

19. Nie YC, Wu H, Li PB, Luo YL, Long K, Xie LM, et al. Anti-inflammatory effects of naringin in chronic pulmonary neutrophilic inflammation in cigarette smoke-exposed rats. J Med Food. 2012;15(10):894-900.

20. Calomme M, Pieters L, Vlietinck A, VandenBerghe D. Inhibition of bacterial mutagenesis by Citrus flavonoids. Planta Med. 1996;62(3):222-226.

21. Jeon SM, Bok SH, Jang MK, Kim YH, Nam KT, Jeong TS, et al. Comparison of antioxidant effects of naringin and probucol in cholesterol-fed rabbits. Clin Chim Acta. 2002;317(1-2):181-190.

22. Jagetia GC, Reddy TK. The grapefruit flavanone naringin protects against the radiation-induced genomic instability in the mice bone marrow: a micronucleus study. Mutat Res. 2002;519(1-2):37-48.

23. Jagetia GC, Reddy TK. Modulation of radiation-induced alteration in the antioxidant status of mice by naringin. Life Sci. 2005;77(7):780-794.

24. Jagetia GC, Venkatesha VA, Reddy TK. Naringin, a citrus flavonone, protects against radiation-induced chromosome damage in mouse bone marrow. Mutagenesis. 2003;18(4):337-343.

25. Jagetia GC, Reddy TK. The grape fruit flavonone naringin protects mice against doxorubicin-induced cardiotoxicity. J Mol Biochem. 2014;3:34-49.

26. Kwatra M, Kumar V, Jangra A, Mishra M, Ahmed S, Ghosh P, et al. Ameliorative effect of naringin against doxorubicin-induced acute cardiac toxicity in rats. Pharm Biol. 2016;54(4):637-647.

27. Kim HD, Jeong KH, Jung UJ, Kim SR. Naringin treatment induces neuroprotective effects in a mouse model of Parkinson's disease in vivo, but not enough to restore the lesioned dopaminergic system. J Nutr Biochem. 2016;28:140-146.

28. Jagetia A, Jagetia GC, Jha S. Naringin, a grape fruit flavanone, protects V79 cells against the bleomycin-induced genotoxicity and decline in the survival. J Appl Toxicol. 2007;27(2):122-132.

29. Jagetia GC, Reddy TK. Chemopreventive effect of naringin on the benzo(a)pyrene-induced forestomach carcinoma in mice. Int J Cancer Preven. 2004;1(6);429-444.

30. Jagetia GC, Reddy TK. The grape fruit bioflavonoid naringin protects against the doxorubicin-induced micronuclei formation in mouse bone marrow. Int J MolBiol. 2016;1(1):00006.

31. Jagetia GC, Reddy TK. Alleviation of iron induced oxidative stress by the grape fruit flavanone naringin in vitro. Chem Biol Interact. 2011;190(2-3):121-128.

32. Adebiyi OO, Adebiyi OA, Owira PM. Naringin reverses hepatocyte apoptosis and oxidative stress associated with hiv-1 nucleotide reverse transcriptase inhibitors-induced metabolic complications. Nutrients. 2015;7(12):10352-10368.

33. Turgut NH, Kara H, Elagoz S, Deveci K, Gungor H, Arslanbas E. The protective effect of naringin against bleomycin-induced pulmonary fibrosis in Wistar rats. Pulm Med. 2016;2016:7601393.

34. Ma X, Lv J, Sun S, Ma J, Xing G, Wang Y, et al. Naringin ameliorates bone loss induced by sciatic neurectomy and increases Semaphorin 3A expression in denervated bone. Sci Rep. 2016;6:24562.

35. Li D, Li J, An Y, Yang Y, Zhang SQ. Doxorubicin-induced apoptosis in H9c2 cardiomyocytes by NF-κB dependent PUMA upregulation. Eur Rev Med Pharmacol Sci. 2013;17(17):2323-2329.

36. Bhinge KN, Gupta V, Hosain SB, Satyanarayanajois SD, Meyer SA, Blaylock B, et al. The opposite effects of Doxorubicin on bone marrow

stem cells versus breast cancer stem cells depend on glucosylceramide synthase. Int J Biochem Cell Biol. 2012;44(11):1770-1778.

37. Sandermann H, Stromiger JL. Purification and properties of C 55-isoprenoidalcohol phosphor kinase from *Staphylococcus aureus*. J Biol Chem. 1972;247(16):5123-5131.

38. Moron MS, Depierre JW, Mannervik B. Levels of glutathione, glutathione reductase and glutathione S-transferase activities in rat lung and liver. Biochim Biophys Acta. 1979;582(1):67-78.

39. Habig WH, Pabst MJ, Jakoby WB. Glutathione S-transferases. The first enzymatic step in mercapturic acid formation. J Biol Chem. 1974;249(22):7130-7139.

40. Abei H. Catalase in vitro. Methods Enzymol. 1984;105:121-126.

41. Marklund S, Marklund G. Involvement of the superoxide anion radical in the autooxidation of pyrogallol and a convenient assay for superoxide dismutase. Eur J Biochem. 1974;47(3):469-474.

42. Ohkawa H, Ohishi N, Yagi K. Assay for lipid peroxides in animal tissues by thiobarbituric acid reaction. Anal Biochem. 1979;95(2):351-358.

43. Valko M, Leibfritz D, Moncol J, Cronin MT, Mazur M, Telser J. Free radicals and antioxidants in normal physiological functions and human disease. Int J Biochem Cell Biol. 2007;39(1):44-84.

44. Lushchak VI. Glutathione Homeostasis and Functions: Potential Targets for Medical Interventions. J Amino Acids. 2012;Article ID736837:26 pages.

45. Lu SC. Glutathione synthesis. Biochim Biophys Acta. 2013;1830(5): 3143–3153.

46. Schumacker PT. Reactive oxygen species in cancer: A dance with the devil. Cancer Cell. 2015;27(2):156-157.

47. Benhar M, Shytaj IL, Stamler JS, Savarino A. Dual targeting of the thioredoxin and glutathione systems in cancer and HIV. J Clin Invest. 2016;126(5):1630-1639.

48. Todorova VK, Kaufmann Y, Hennings L, Klimberg VS. Oral glutamine protects against acute doxorubicin-induced cardiotoxicity of tumor-bearing rats. J Nutr. 2010;140(1):44-48.

49. Jagetia GC, Venkatesh P. An Indigenous plant Bael (*Aegle marmelos* (L.) Correa) Extract Protects Against the Doxorubicin-Induced Cardiotoxicity in Mice. BiochemPhysiol. 4:163.

50. Jagetia GC, Lalnuntluangi V. The Citrus flavanone naringin enhances antioxidant status in the albino rat liver treated with doxorubicin. BiochemMolBiol J. 2016;2(2):1-9.

51. Papasani VMR, Hanumantharayappa B, Annapurna A (2014) Cardioprotective effect of naringin against doxorubicin induced cardiomyopathy in rats. Indo Am J Pharmaceut Res; 4(5): 2593-2598.

52. Fiser B, Jójárt B, Csizmadia IG, Viskolcz B. Glutathione – Hydroxyl radical interaction: A theoretical study on radical recognition process. PLoS One. 2013;8(9):e73652.

53. Wu G, Fang Y-Z, Yang S, Lupton JR, Turner ND. Glutathione metabolism and its implications for health. J Nutr. 2004;134(3):489-492.

54. Schnekenburger M, Karius T, Diederich M. Regulation of epigenetic traits of the glutathione-S-transferase P1 gene: from detoxification toward cancer prevention and diagnosis. Front Pharmacol. 2014;5:170.

55. Laborde E. Glutathione transferases as mediators of signalling pathways involved in cell proliferation and cell death. Cell Death Differ. 2010;17(9):1373-1380.

56. McIlwain CC, Townsend DM, Tew KD. Glutathione-S-transferase polymorphisms: Cancer incidence and therapy. Oncogene. 2006;25(11):1639-1648.

57. Kodydková J, Vávrová L, Kocík M, Žák A. Human catalase, its polymorphisms, regulation and changes of its activity in different diseases. Folia Biol (Praha). 2014;60(4):153-167.

58. Fattman CL, Schaefer LM, Oury TD. Extracellular superoxide dismutase in biology and medicine. Free Radic Biol Med. 2003;35(3):236-256.

59. Miller AF. Superoxide dismutases: Ancient enzymes and new insights. FEBS Lett. 2012;586(5):585-595.

60. Carillon J, Rouanet J-M, Cristol J-P, BrionRchard. Superoxide dismutase administration, a potential therapy against oxidative stress related diseases: several routes of supplementation and proposal of an original mechanism of action. Pharm Res. 2013;30(11):2718-2728.

61. Yin H, Xu L, Porter NA. Free radical lipid peroxidation: Mechanisms and analysis. Chem Rev. 2011;111(10):5944-5972.

62. Pieniążek A, Czepas J, Piasecka-Zelga J, Gwoździński K, Koceva-Chyła A. Oxidative stress induced in rat liver by anticancer drugs doxorubicin, paclitaxel and docetaxel. Adv Med Sci. 2013;58(1):104-111.

63. Gilleron M, Marechal X, Montaigne D, Franczak J, Neviere R, Lancel S. NADPH oxidases participate to doxorubicin-induced cardiac myocyte apoptosis. Biochem Biophys Res Commun. 2009;388(4):727-731.

64. Li W, Wang C, Peng J, Liang J, Jin Y, Liu Q, et al. Naringin inhibits TNF-α induced oxidative stress and inflammatory response in HUVECs via Nox4/NF-κ B and PI3K/Akt pathways. Curr Pharm Biotechnol. 2014;15(12):1173-1182.

65. Octavia Y, Tocchetti CG, Gabrielson KL, Janssens S, Crijns HJ, Moens AL. Doxorubicin-induced cardiomyopathy: from molecular mechanisms to therapeutic strategies. J Mol Cell Cardiol. 2012;52(6):1213-1225.

66. Li D, Li J, An Y, Yang Y, Zhang SQ. Doxorubicin-induced apoptosis in H9c2 cardiomyocytes by NF-κB dependent PUMA upregulation. Eur Rev Med Pharmacol Sci. 2013;17(17):2323-2329.

67. Zhang DX, Ma DY, Yao ZQ, Fu CY, Shi YX, Wang QL, et al. ERK1/2/ p53 and NF-κB dependent-PUMA activation involves in doxorubicin-induced cardiomyocyte apoptosis. Eur Rev Med Pharmacol Sci. 2016;20(11):2435-2442.

68. Liang J, Wang C, Peng J, Li W, Jin Y, Liu Q, et al. Naringin regulates cholesterol homeostasis and inhibits inflammation via modulating NF-κB and ERK signaling pathways in vitro. Pharmazie. 2016;71(2):101-108.

69. Kulasekaran G, Ganapasam S. Neuroprotective efficacy of naringin on 3-nitropropionic acid-induced mitochondrial dysfunction through the modulation of Nrf2 signaling pathway in PC12 cells. Mol Cell Biochem. 2015;409(1-2):199-211.

PERMISSIONS

LIST OF CONTRIBUTORS

Robert S. Matson
QuantiScientifics LLC, 1920 E. Katella Ave. Suite S, Orange, CA 92867, USA

Amro Abd Al Fattah Amara
The head of the Protein Research Department, Genetic Engineering and Biotechnology Research Institute, City for Scientific Research and Technological Applications, Universities and Research Centre District, New Borg El-Arab, Egypt

Tarik Ainane
Superior School of Technology - Khenifra (EST-Khenifra), University of Moulay Ismail, PB 170, 54000 Khenifra, Morocco

Said Gharby
Laboratory of Chemistry of Plants, Organic Synthesis and Bioorganic, Faculty of Science, University Mohammed V-Agdal, Rabat, Morocco

Mohammed Talbi and Mohamed Elkouali
Laboratory of Analytical Chemistry and Physical Chemistry of Materials, Faculty of Sciences Ben Msik, University Hassan II, BP 7955 Casablanca 20660, Morocco

Abdelmjid Abourriche and Ahmed Bennamara
Biomolecules and organic synthesis laboratory, Faculty of Sciences Ben Msik, University Hassan II, BP 7955 Casablanca 20660, Morocco

Naoual Oukkache
Laboratory of Venoms and Toxins, Pasteur Institute of Morocco, 1 Place Louis Pasteur, Casablanca 20360, Morocco

Hassan Lamdini
Department of Infectious Diseases, IbnRochd Hospital University Center, Casablanca 20270Morocco

Saumya Dhup
Centre for Bioresources and Biotechnology, TERI University, 10, Institutional Area, Vasant Kunj, New Delhi 110070, India

Dheeban C. Kannan and Vibha Dhawan
Biotechnology and Management of Bioresources Division, the Energy and Resources Institute (TERI), India Habitat Centre, Lodhi Road, New Delhi 110003, India

Hanbai Liang, Shota Kurimoto and Yoshio Minabe
Department of Restorative Neurosurgery and Psychiatry

Tetsumori Yamashima
Department of Restorative Neurosurgery and Psychiatry
Department of Psychiatry and Neurobiology, Kanazawa University Graduate School of Medical Science. 13-1 Takara-machi, 920-8640, Kanazawa, Ishikawa, Japan

Kosuke R Shima, Hiroki Shimizu and Tsuguhito Ota
Department of Cell Metabolism and Nutrition, Brain/Liver Interface Medicine Research Center

Amro Abd Al Fattah Amara
Head of the Protein Research Department and the office of the Scientific Publishing, Genetic Engineering and Biotechnology Research Institute, City for Scientific Research and Technological Applications, Universities and Research Center district, New Borg El-Arab, Egypt

Gagné F, André C, Turcotte P and Gagnon C
Aquatic Contaminants Research Division, Environment and Climate Change Canada, 105 McGill Street, Montreal, Quebec, Canada

Auclair J and Gagne F
Aquatic Contaminants Research Division, Environment and Climate Change Canada, 105 McGill, Montréal, QC, Canada

Morel E and Wilkinson KJ
Department of Chemistry, University of Montreal, C.P. 6128, Succ, Centre-Ville, Montreal, Canada H3C 3J7

Amrita Kumari, Tetsuo Koyama, Ken Hatanoand and Koji Matsuoka
Division of Material Science, Graduate School of Science and Engineering, Saitama University, Sakura, Saitama 338-8570, Japan

Mao Mao
Key Laboratory for Special Functional Materials, Henan University, Kaifeng 475004, P. R. China

Feng Wu
Key Laboratory for Special Functional Materials, Henan University, Kaifeng 475004, P. R. China

Division of Life Science and Health, Graduate School at Shenzhen, Tsinghua University, Shenzhen, 518055, P. R. China

Qian Liu, Yu Cen and Lan Ma
Division of Life Science and Health, Graduate School at Shenzhen, Tsinghua University, Shenzhen, 518055, P. R. China

Lei Shi and Zhifeng Qin
Shenzhen Entry-Exit Inspection and Quarantine Bureau of the People's Republic of China (SZCIQ), Shenzhen, 518045, P.R. China

Houda Hanana, Patrice Turcotte, Martin Pilote, Joëlle Auclair, Christian Gagnon and François Gagné
Aquatic Contaminant Research Division, Environment and Climate Change Canada, 105 McGill, Montreal, Quebec, Canada H2Y 2E7

Ebrahim Ghiamati and Samieh Oliaei
Chemistry Department, University of Birjand, Birjand, South Khorasan, Iran

Amro Abd Al Fattah Amara
Protein Research Department, Genetic Engineering and Biotechnology Research Institute, City for Scientific Research and Technological Applications, New Borg Al Arab, Alexandria, Egypt

Suarez N, Ferrara F, Rial M and Chabalgoity A
Department of Biotechnology Development, Institute of Hygiene, Faculty of Medicine, Udelar, Uruguay

Pirez M
Chair of Immunology, Faculty of Chemistry, Udelar, Uruguay

Yashpalsinh N Girase
Research Scholar, Pacific Academy of higher Education and Research University, Udaipur, India

Srinivasrao V
Department of Research and Development, Pacific University, Udaipur, India

DiptiSoni
Department of Chemistry, Pacific Academy of higher Education and Research University, Udaipur, India

Nawal Abd El-Baky
Protective and Therapeutic Protein Laboratory, Protein Research Department, Genetic Engineering and Biotechnology Research Institute, City for Scientific Research and Technology Applications, New Borg El Arab, Alexandria 21934, Egypt

Nawal Abd El-Baky, Mona M. Sharaf, Eman Amer, Hoda Reda Kholef and Amro A. Amara
Protein Research Department, Genetic Engineering and Biotechnology Research Institute, City for Scientific Research and Technological Applications, Universities and Research Center District, New Borg El-Arab, Alexandria, Egypt

Mohamed Zakaria Hussain
Department of Microbiology and Immunology, Faculty of Medicine, Tanta University, Tanta

Zhu Hua
Center for Integrative Conservation, Xishuangbanna Tropical Botanical Garden, Chinese Academy of Sciences, Mengla, Yunnan 666303, P. R. China

RL Elliott, XP Jiang and JF Head
Elliott-Baucom-Head Breast Cancer Research and Treatment Center, Baton Rouge, LA 70806, USA

Amro Abd Al Fattah Amara
Head of the Protein Research Department, Genetic Engineering and Biotechnology Research Institute, City for Scientific Research and Technological Applications, Egypt

Chien-An A. Hu
Department of Biochemistry and Molecular Biology, University of New Mexico Health Sciences Center, Albuquerque, New Mexico 87131-0001, USA

Yongqing Hou
School of Animal Science and Nutritional Engineering, Wuhan Polytechnic University, Wuhan, Hubei, 430023, P. R. China

Megan N. Sandberg, Jordan A. Greco, Nicole L. Wagner, Tabitha L. Amora, Lavoisier A. Ramos and Robert R. Birge
Departments of Chemistry and Molecular and Cell Biology, University of Connecticut, Storrs, CT 06269, USA

Min-Hsuan Chen and Barry E. Knox
Departments of Biochemistry and Molecular Biology and Ophthalmology, State University of New York Upstate Medical University, Syracuse, NY 13210, USA

Changdong Wang
Department of Biochemistry and Molecular Biology, Molecular Medicine and Cancer Research Center, Chongqing Medical University, Chongqing 400016, China

Wei Huang, Xi Liang and Ning Hu
Department of Orthopaedic Surgery, the First Affiliated Hospital of Chongqing Medical University, Chongqing 400016, China

Xuan Gong
Department of outpatient, Chongqing Zhongshan Hospital, Chongqing, 400013, China

Guoliang Ding
Department of Orthopaedic Surgery, the Second Affiliated Hospital of Baotou Medical College of Inner Mongolia University of Science and Technology, Baotou, 014030, China

Gogo Appolus Obediah
Department of Biochemistry, Faculty of Science, Rivers State University, Port Harcourt, Nigeria

Gift Paago
Department of Human Pharmacology, Faculty of Basic Medical Sciences, College of Health Sciences, University of Port Harcourt, Nigeria

Daniel Glossman-Mitnik
NANOCOSMOS Virtual Lab, Department of Environment and Energy, Advanced Materials ResearchCenter, Miguel de Cervantes 120, Complejo Industrial Chihuahua, Chihuahua Chih 31136, Mexico
Departament of Chemistry, University of the Balearic Islands, Palma de Mallorca 07122, Spain

Norma Flores-Holguin
NANOCOSMOS Virtual Lab, Department of Environment and Energy, Advanced Materials ResearchCenter, Miguel de Cervantes 120, Complejo Industrial Chihuahua, Chihuahua Chih 31136, Mexico

Juan Frau
Departament of Chemistry, University of the Balearic Islands, Palma de Mallorca 07122, Spain

Oluwayemi Joshua Bamikole
Institute of Child Health, University of Ibadan. Nigeria

Godwin Okwori Adikwu
Department of Biochemistry, University of Ilorin. Nigeria

Anibal R. Lodeiro
Laboratory of Interactions between Rhizobia and Soybean (LIRyS). IBBM-Faculty of Sciences. National University of La Plata and CONICET CCT-La Plata, Argentina

Ganesh Chandra Jagetia and C. Lalrinengi
Department of Zoology, Mizoram University, Aizawl-796 004, India

Index